D1000759

APPLIED COMBUSTION

MECHANICAL ENGINEERING

A Series of Textbooks and Reference Books

Editor

L. L. Faulkner

Columbus Division, Battelle Memorial Institute
and Department of Mechanical Engineering
The Ohio State University
Columbus, Ohio

1. *Spring Designer's Handbook,* Harold Carlson
2. *Computer-Aided Graphics and Design,* Daniel L. Ryan
3. *Lubrication Fundamentals,* J. George Wills
4. *Solar Engineering for Domestic Buildings,* William A. Himmelman
5. *Applied Engineering Mechanics: Statics and Dynamics,* G. Boothroyd and C. Poli
6. *Centrifugal Pump Clinic,* Igor J. Karassik
7. *Computer-Aided Kinetics for Machine Design,* Daniel L. Ryan
8. *Plastics Products Design Handbook, Part A: Materials and Components; Part B: Processes and Design for Processes,* edited by Edward Miller
9. *Turbomachinery: Basic Theory and Applications,* Earl Logan, Jr.
10. *Vibrations of Shells and Plates,* Werner Soedel
11. *Flat and Corrugated Diaphragm Design Handbook,* Mario Di Giovanni
12. *Practical Stress Analysis in Engineering Design,* Alexander Blake
13. *An Introduction to the Design and Behavior of Bolted Joints,* John H. Bickford
14. *Optimal Engineering Design: Principles and Applications,* James N. Siddall
15. *Spring Manufacturing Handbook,* Harold Carlson
16. *Industrial Noise Control: Fundamentals and Applications,* edited by Lewis H. Bell
17. *Gears and Their Vibration: A Basic Approach to Understanding Gear Noise,* J. Derek Smith
18. *Chains for Power Transmission and Material Handling: Design and Applications Handbook,* American Chain Association
19. *Corrosion and Corrosion Protection Handbook,* edited by Philip A. Schweitzer
20. *Gear Drive Systems: Design and Application,* Peter Lynwander

21. *Controlling In-Plant Airborne Contaminants: Systems Design and Calculations,* John D. Constance
22. *CAD/CAM Systems Planning and Implementation,* Charles S. Knox
23. *Probabilistic Engineering Design: Principles and Applications,* James N. Siddall
24. *Traction Drives: Selection and Application,* Frederick W. Heilich III and Eugene E. Shube
25. *Finite Element Methods: An Introduction,* Ronald L. Huston and Chris E. Passerello
26. *Mechanical Fastening of Plastics: An Engineering Handbook,* Brayton Lincoln, Kenneth J. Gomes, and James F. Braden
27. *Lubrication in Practice: Second Edition,* edited by W. S. Robertson
28. *Principles of Automated Drafting,* Daniel L. Ryan
29. *Practical Seal Design,* edited by Leonard J. Martini
30. *Engineering Documentation for CAD/CAM Applications,* Charles S. Knox
31. *Design Dimensioning with Computer Graphics Applications,* Jerome C. Lange
32. *Mechanism Analysis: Simplified Graphical and Analytical Techniques,* Lyndon O. Barton
33. *CAD/CAM Systems: Justification, Implementation, Productivity Measurement,* Edward J. Preston, George W. Crawford, and Mark E. Coticchia
34. *Steam Plant Calculations Manual,* V. Ganapathy
35. *Design Assurance for Engineers and Managers,* John A. Burgess
36. *Heat Transfer Fluids and Systems for Process and Energy Applications,* Jasbir Singh
37. *Potential Flows: Computer Graphic Solutions,* Robert H. Kirchhoff
38. *Computer-Aided Graphics and Design: Second Edition,* Daniel L. Ryan
39. *Electronically Controlled Proportional Valves: Selection and Application,* Michael J. Tonyan, edited by Tobi Goldoftas
40. *Pressure Gauge Handbook,* AMETEK, U.S. Gauge Division, edited by Philip W. Harland
41. *Fabric Filtration for Combustion Sources: Fundamentals and Basic Technology,* R. P. Donovan
42. *Design of Mechanical Joints,* Alexander Blake
43. *CAD/CAM Dictionary,* Edward J. Preston, George W. Crawford, and Mark E. Coticchia
44. *Machinery Adhesives for Locking, Retaining, and Sealing,* Girard S. Haviland
45. *Couplings and Joints: Design, Selection, and Application,* Jon R. Mancuso
46. *Shaft Alignment Handbook,* John Piotrowski

47. *BASIC Programs for Steam Plant Engineers: Boilers, Combustion, Fluid Flow, and Heat Transfer,* V. Ganapathy
48. *Solving Mechanical Design Problems with Computer Graphics,* Jerome C. Lange
49. *Plastics Gearing: Selection and Application,* Clifford E. Adams
50. *Clutches and Brakes: Design and Selection,* William C. Orthwein
51. *Transducers in Mechanical and Electronic Design,* Harry L. Trietley
52. *Metallurgical Applications of Shock-Wave and High-Strain-Rate Phenomena,* edited by Lawrence E. Murr, Karl P. Staudhammer, and Marc A. Meyers
53. *Magnesium Products Design,* Robert S. Busk
54. *How to Integrate CAD/CAM Systems: Management and Technology,* William D. Engelke
55. *Cam Design and Manufacture: Second Edition; with cam design software for the IBM PC and compatibles, disk included,* Preben W. Jensen
56. *Solid-State AC Motor Controls: Selection and Application*, Sylvester Campbell
57. *Fundamentals of Robotics,* David D. Ardayfio
58. *Belt Selection and Application for Engineers,* edited by Wallace D. Erickson
59. *Developing Three-Dimensional CAD Software with the IBM PC,* C. Stan Wei
60. *Organizing Data for CIM Applications,* Charles S. Knox, with contributions by Thomas C. Boos, Ross S. Culverhouse, and Paul F. Muchnicki
61. *Computer-Aided Simulation in Railway Dynamics,* by Rao V. Dukkipati and Joseph R. Amyot
62. *Fiber-Reinforced Composites: Materials, Manufacturing, and Design,* P. K. Mallick
63. *Photoelectric Sensors and Controls: Selection and Application,* Scott M. Juds
64. *Finite Element Analysis with Personal Computers,* Edward R. Champion, Jr., and J. Michael Ensminger
65. *Ultrasonics: Fundamentals, Technology, Applications: Second Edition, Revised and Expanded,* Dale Ensminger
66. *Applied Finite Element Modeling: Practical Problem Solving for Engineers,* Jeffrey M. Steele
67. *Measurement and Instrumentation in Engineering: Principles and Basic Laboratory Experiments,* Francis S. Tse and Ivan E. Morse
68. *Centrifugal Pump Clinic: Second Edition, Revised and Expanded,* Igor J. Karassik
69. *Practical Stress Analysis in Engineering Design: Second Edition, Revised and Expanded,* Alexander Blake

70. *An Introduction to the Design and Behavior of Bolted Joints: Second Edition, Revised and Expanded,* John H. Bickford
71. *High Vacuum Technology: A Practical Guide,* Marsbed H. Hablanian
72. *Pressure Sensors: Selection and Application,* Duane Tandeske
73. *Zinc Handbook: Properties, Processing, and Use in Design,* Frank Porter
74. *Thermal Fatigue of Metals,* Andrzej Weronski and Tadeuz Hejwowski
75. *Classical and Modern Mechanisms for Engineers and Inventors,* Preben W. Jensen
76. *Handbook of Electronic Package Design,* edited by Michael Pecht
77. *Shock-Wave and High-Strain-Rate Phenomena in Materials*, edited by Marc A. Meyers, Lawrence E. Murr, and Karl P. Staudhammer
78. *Industrial Refrigeration: Principles, Design and Applications*, P. C. Koelet
79. *Applied Combustion*, Eugene L. Keating
80. *Engine Oils and Automotive Lubrication*, edited by Wilfried J. Bartz
81. *Mechanism Analysis: Simplified and Graphical Techniques, Second Edition, Revised and Expanded*, Lyndon O. Barton
82. *Fundamental Fluid Mechanics for the Practicing Engineer*, James W. Murdock

Additional Volumes in Preparation

Fiber-Reinforced Composites: Materials, Manufacturing, and Design, Second Edition, Revised and Expanded, P. K. Mallick

Introduction to Engineering Materials: Behavior, Properties, and Selection, G. T. Murray

Numerical Methods for Engineers, Edward R. Champion, Jr.

Vibrations of Shells and Plates: Second Edition, Revised and Expanded, Werner Soedel

Mechanical Properties of Polymers and Composites: Second Edition, Revised and Expanded, Lawrence E. Nielsen and Robert F. Landel

Turbomachinery: Second Edition, Revised and Expanded, Earl Logan

Mechanical Engineering Software

Spring Design with an IBM PC, Al Dietrich

Mechanical Design Failure Analysis: With Failure Analysis System Software for the IBM PC, David G. Ullman

APPLIED COMBUSTION

EUGENE L. KEATING

U.S. Naval Academy
Annapolis, Maryland

Marcel Dekker, Inc. New York • Basel • Hong Kong

Library of Congress Cataloging-in-Publication Data

Keating, Eugene L.,
Applied combustion / Eugene L. Keating.
p. cm. — (Mechanical engineering ; 79)
Includes bibliographical references and index.
ISBN 0-8247-8127-9 (acid-free)
1. Combustion engineering. I. Title. II. Series: Mechanical engineering (Marcel Dekker, Inc.) ; 79.
TJ254.5K43 1993
621.402'3—dc20 92-17715
CIP

This book is printed on acid-free paper.

MARCEL DEKKER, INC.
270 Madison Avenue, New York, New York 10016

Current printing (last digit):
10 9 8 7 6 5 4 3 2 1

PRINTED IN THE UNITED STATES OF AMERICA

Preface

Today's world requires that undergraduate engineering education produce graduates having a broad foundation in technical as well as scientific principles of energy conservation and conversion. In addition, engineers and scientists are now needed who can utilize concepts of physics and chemistry to design, develop, operate, and/or maintain current or future energy-consuming machinery. Many texts and references have been written in specific areas of energy, such as thermodynamics, internal combustion engines, gas turbines, chemical kinetics, and flame theory. Other recent publications have attempted to address, in a single volume, the entire field of current energy conversion engineering, covering topics such as solar, nuclear, geothermal, and wind energy systems. *Applied Combustion* provides its readers with a broad engineering introduction to principles of chemical energy conversion, or combustion, as well as practical applications of those laws to a variety of chemical heat engines.

In the early chapters, fundamentals of combustion are formulated from general principles of chemistry, physics, and thermodynamics. Characteristics of chemical energy resources, or fuels, such as coal, distillates derived from crude oil, alcohols, syncrudes, bio-gas, natural gas, and hydrogen, are treated in detail in the following chapters. The final chapters apply these fundamentals to several combustion engines. Basic precepts found in earlier chapters are arranged so that they are progressively developed and utilized in later chapters.

A variety of subject areas are covered in this book, including the ideal oxidation-reaction equation, fuel heat release rates, chemical equilibrium, incomplete combustion, chemical kinetics, theory of detonation and thermal explosions, as well as basic flame theory. Energy characteristics of equipment utilizing chemical fuels, including boilers, gas turbine combustors, and compression- or spark-ignition internal combustion engines, are reviewed. Example problems illustrate a proper engineering analysis and solution technique as well as important principles relevant to understanding many combustion processes and devices. Emphasis is placed on describing general performance in terms of the concept of a fuel–engine compatibility, i.e., fuel consumption rates, pollution characteristics, and various energy conversion efficiency interactions of heat and power machinery.

Currently, only a few publications are available that provide as wide an engineering science introduction to the general subject of combustion as does this work. The use of dual SI and Engineers' dimensions and units, numerous key examples and exercises at the end of each chapter, as well as numerous appendixes, make this book suitable for use as a text in many engineering programs. In addition, the unique bridge between combustion science and combustion technology, thermochemical engineering data, and design formulation of basic performance relationships found in this work should make it a valuable technical reference for many personal and professional engineering libraries.

I hope that this book will stimulate its readers to pursue further this important field of study. Furthermore, I hope that this material will provide a basic foundation for those whose careers involve efforts in current and related areas of engineering activity through graduate education, research and development, and professional engineering societies. Future requirements and advances in applied combustion may have an even greater impact on the power and propulsion fields of aerospace, chemical, marine, and mechanical engineering than have many breakthroughs of the recent past.

I am indebted to numerous individuals and institutions that have contributed over many years to my career. Special acknowledgment is due to my fellow faculty members at the U.S. Merchant Marine and U.S. Naval Academies, as well as members of the Society of Automotive Engineers. It was while working with my colleagues in these institutions that much of the philosophy and specific material for this manuscript evolved.

Eugene L. Keating, Ph.D., P.E.

Contents

Nomenclature

a	acceleration, m/sec^2, ft/sec^2
a	specific Hemholtz function, J/kg, Btu/lbm
A	generalized chemical compound, i.e., methane, water, etc.,
A	area, m^2, ft^2
A	Arrhenius rate constant prefactor
A	total Hemholtz function, J, Btu
AF	air–fuel ratio
$\bar{a}_f^0$	Hemholtz function of formation, J/kg mole, Btu/lb mole
ΔA_c	Hemholtz function of combustion kJ/kg mole, Btu/lb mole
ΔA_f	Hemholtz function of formation, kJ/kg mole, Btu/lb mole
ΔA_r	Hemholtz function of reaction, kJ/kg mole, Btu/lb mole
$BMEP$	brake mean effective pressure
c	sonic velocity, m/sec, ft/sec
C_p	constant-pressure specific heat, J/kg K, Btu/lbm °R
C_v	constant-volume specific heat, J/kg K, Btu/lbm °R
d	diameter m, ft
D	total differential operator
F	force, N, lbf
FA	fuel–air ratio
g	gravitational acceleration, m/sec^2, ft/sec^2
g_0	dimensional constant, 1.0 kg m/N sec^2, 32.1724 lbm/lbf sec^2

g	Gibbs function, J/kg, Btu/lbm
G	total Gibbs function, J, Btu
G	mass flow per unit area, kg/m^2 sec, lbm/ft^2 sec
ΔG_c	Gibbs function of combustion, kJ/kg mole, Btu/lb mole
ΔG_f	Gibbs function of formation, kJ/kg mole, Btu/lb mole
ΔG_r	Gibbs function of reaction, kJ/kg mole, Btu/lb mole
h	specific enthalpy, J/kg, Btu/lbm
H	total enthalpy, J, Btu
h_{fg}	latent heat of vaporization, J/kg, Btu/lbm
$\bar{h}_f^0$	molar enthalpy of formation, J/kg mole, Btu/lb mole
HHV	fuel higher heating value, J/kg, Btu/lbm
ΔH_c	heat of combustion, kJ/kg mole, Btu/lb mole
ΔH_f	heat of formation, kJ/kg mole, Btu/lb mole
ΔH_r	heat of reaction, kJ/kg mole, Btu/lb mole
$IMEP$	indicated mean effective pressure, kPa, psi
k	Arrhenius elemental rate constant
$\not{k}$	Boltzmann's constant
K_c	equilibrium constant based on species concentrations
K_p	equilibrium constant based on partial pressures
L	length, m, ft
LHV	fuel lower heating value, J, Btu
M	total mass, kg, lbm
m_i	mass of species *i*, kg, lbm
mf_i	mass fraction
MW	molecular weight, kg/kg mole, lbm/lb mole
MEP	mean effective pressure, kPa, lbf
N	number of moles
N	rotational speed, rev/sec
N_m	Mach number
P	pressure, N/m^2 or Pa, lbf/in.^{2}
P_r	relative pressure
q	heat transfer per unit mass, J/kg, Btu/lbm
Q	total heat transfer, J, Btu
R	elementary reaction rate
R_i	reactant species *i*
R	specific gas constant, J/kg K, Btu/lbm °R
r_e	expansion ratio
r_c	cutoff ratio
r_p	pressure ratio
r_v	compression ratio
s	specific entropy, J/kg K, Btu/lbm °R
S	total entropy, J, Btu

$\bar{s}_f^0$	molar entropy of formation, J/kg mole, Btu/lb mole
ΔS_c	entropy of combustion, kJ/kg mole, Btu/lb mole
ΔS_f	entropy of formation, kJ/kg mole, Btu/lb mole
ΔS_r	entropy of reaction, kJ/kg mole, Btu/lb mole
$S.G.$	specific gravity
t	time, sec
T	temperature, °C and K, °F and °R
u	specific internal energy, J/kg K, Btu/lbm °R
U	total internal energy, J, Btu
$\bar{u}_f^0$	molar internal energy of formation, J/kg mole, Btu/lb mole
ΔU_c	internal energy of combustion, kJ/kg mole, Btu/lb mole
ΔU_f	internal energy of formation, kJ/kg mole, Btu/lb mole
ΔU_r	internal energy of reaction, kJ/kg mole, Btu/lb mole
v	specific volume, m^3/kg, ft^3/lbm
V	total volume, m^3, ft^3
V	velocity, m/sec, ft/sec
V_C	clearance volume, m^3/stroke, ft^3/stroke
V_D	displacement volume, m^3/stroke, ft^3/stroke
w	specific work *m*, J/kg, ft lbf/lbm
W	total work, J, ft lbf
W	total weight, N, lbf
$\bar{x}_i$	mole fraction of species 1
Z	elevation, m, ft
Z_{AB}	collision frequency, number of collisions/sec
δ	differential operator
Δ	finite difference in parameters
γ	specific heat ratio
γ	specific weight, N/m^3, lbf/ft^3
λ	mean free path, cm/collision
η	efficiency (subscript will designate the particular value)
ρ	density, kg/m^3, lbm/ft^3
σ	molecular diameter, m^3, ft^3
Σ	summation operator
τ	torque N m, ft lbf
Φ	equivalence ratio
υ	moles of species

Notes on Nomenclature Rules A dot over a symbol indicates a derivate with respect to time;

$\dot{Q}$ heat flux, W, Btu/hr
$\dot{W}$ power W, hp

A bar over a symbol indicates that the parameter is on a mole basis;

$\bar{h}$ molar specific enthalpy, J/kg mole, Btu/lb mole
$\bar{R}$ universal gas constant, J/kg mole K, Btu/lb mole °R

An angle bracket $<\quad>$ is an operator that implies that the quantity in question is a function of the parameters contained within;

$C_p<T>$ constant specific heat is a function of temperature
$V<t>$ velocity if a function of time

A square-bracket [ ] is an operator that implies that the quantity in question is the concentration of the parameter contained within.

Subscript *i* or *j* references the quantity of a species *i* or *j*:

MW_i
mf_j

An arrow through a term in an equation means that it vanishes in that equation:

~~dE~~

APPLIED COMBUSTION

1

Introduction to Applied Combustion

1.1 INTRODUCTION

The birth, growth, decay, and even ultimate death of many civilizations can be described in terms of their specific cultural understanding and application of the principles of combustion. For example, in both ancient and primitive cultures, fire was primarily a mystery to be feared and therefore worshiped. In such circumstances, the control of fire provided certain groups with the power either to improve or to destroy human life. This belief concerning the mystical nature of fire, or more precisely the combustion process that causes it, was forever changed when Empedocles (ca. 490–430 B.C.) postulated that fire was but one of the four constituent elements of all matter: earth, air, fire, and water.

This idea was commonly held until the Renaissance and the Age of Enlightenment, when men such as Carnot (1796–1832) began to study the key nature of matter, energy, and combustion. In his "Reflections on the Motive Power of Fire," Carnot postulated his now-famous thermodynamic cycle, which would convert a fraction of energy transfer from a source, such as fire, into work, with the remaining energy, of necessity, being rejected into a sink. This publication provided a theoretical basis for an absolute temperature scale, contributed to the formulation of classical thermodynamics, and spurred development of the modern heat engine, which was a major factor in the Industrial Revolution.

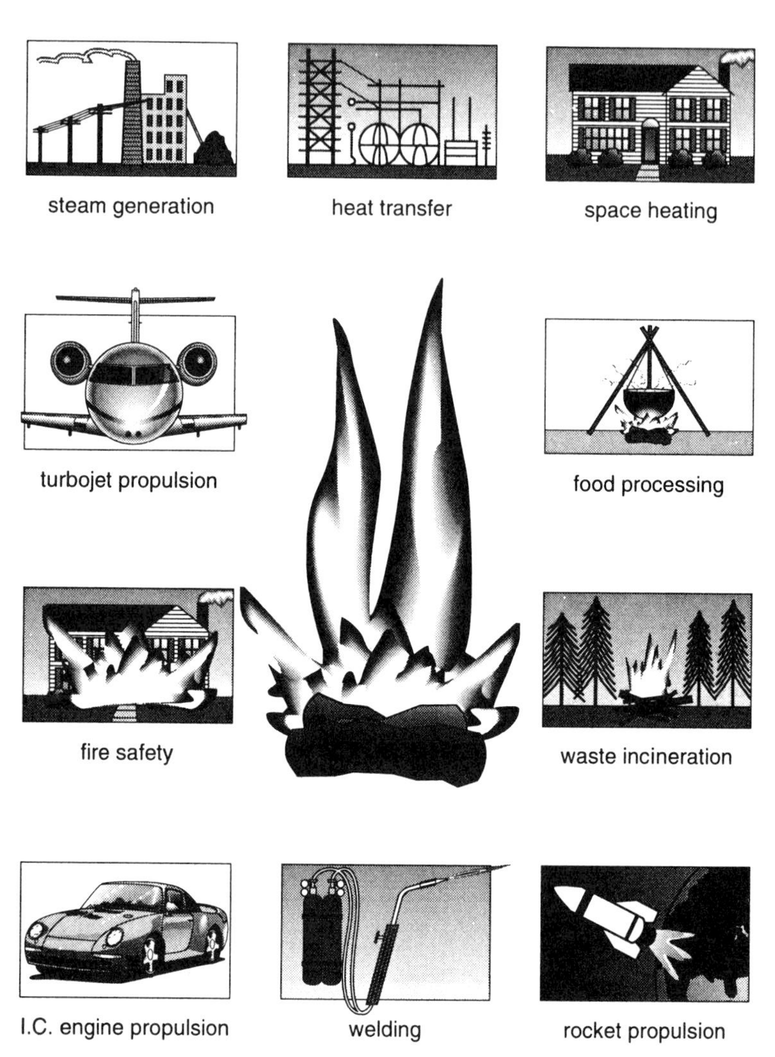

Figure 1.1 Typical applications of combustion science.

Giant steps have been made during this century to push frontiers of combustion science to their outer bounds, enabling technology to move mankind rapidly and efficiently over the surface of the earth, through the air, as well as on and under the sea. Mobility platforms have also taken men into orbit around the earth and out to the surface of the moon. Well-known nonpower applications of combustion principles can also be cited which, together with the development of power systems, have made the material quality of life in America today far better than that of any culture in the recorded history of civilization (see Figure 1.1).

EXAMPLE 1.1 Carnot postulated an ideal thermodynamic cycle having the maximum theoretical thermal efficiency, i.e., desired work output to required heat input. The thermal efficiency η for the cycle shown below can be expressed as

$$\eta = \frac{\text{Desired energy output}}{\text{Required energy input}} = \frac{\text{Net work (power)}}{\text{Heat (flux) added}} = 1 - \frac{T_L}{T_H}$$

where

T_L = lowest cycle absolute temperature

T_H = highest cycle absolute temperature

For a T_L of 70°F calculate the required T_H for a cycle thermal efficiency of (a) 20%, (b) 40%, (c) 60%, (d) 80%, and (e) 100%.

Solution

1. Thermal efficiency:

$$\eta = 1 - \frac{T_L}{T_H} \qquad \text{where} \qquad T_H = \frac{T_L}{(1 - \eta)}$$

$$T_L = 70 + 460 = 530°\text{R}$$

η	T_H	
0.20	662.5	368.1
0.40	883.3	490.7
0.60	1325.0	736.1
0.80	2650.0	1472.2
1.00	—	—
	°R	K

Comments A high Carnot cycle efficiency implies high cycle temperatures, which is consistent with the combustion process.

1.2 ENERGY AND COMBUSTION

The material covered in this text will focus principally on an engineering study of the energy conversion aspects of combustion. *Combustion* in this dissertation will refer to any relatively fast gas-phase chemical reaction that liberates substantial energy as heat. The chemistry and physics of combustion, i.e., destruction and rearrangement of certain molecules, which rapidly release energy, require temperatures of 1620–2200°C (3000–4000°F), take place within a few millionths of a second, and occur within a characteristic length of a few angstroms. Until analytical and physical tools such as the modern high-speed computer and laser became available, the microscopic nature of the very combustion process itself remained uncertain and impeded major progress in both scientific investigations and engineering applications of this most basic and universal phenomona. Currently, combustion is a mature discipline and an integral element of diverse research and development programs ranging from fundamental studies of the physics of flames and high-temperature molecular chemistry to applied engineering projects involved with developments such as advanced coal-burning machinery and improved combustion engine emission controls.

Recall the classical thermodynamic concept of an energy *source* defined as a nondepleting thermal energy reservoir and a *sink* as an unfillable thermal energy repository. Prior to the present age of jet and space travel, environmental consciousness, and critical shortfalls in certain crucial fuel resources, energy demands were relatively low and energy was cheap. Until recent times, very little motivation, or even need, existed to develop thoroughly all engineering aspects of combustion. Engineering and technology now recognize that the earth is a finite ecosphere having limited terrestrial resources and a delicately balanced environment. To appreciate the significance of this issue, consider the conflicting requirements today for an energy resource such as a chemical fuel and its combustion. It must be:

Nondepleting
Nonpolluting
Readily available
Economically viable
Politically neutral
Technically accessible
Legally valid

Socially acceptable
Aesthetically pleasing

A *fuel* can be considered as a finite resource of chemical potential energy in which energy stored in the molecular structure of particular compounds is released via complex chemical reactions. Chemical fuels can be classified in a variety of ways, including by phase, availability, or even application (see Table 1.1). Some of the basic requirements of a fuel include:

High energy density (content)
High heat of combustion (release)
Good thermal stability (storage)
Low vapor pressure (volatility)
Nontoxicity (environmental impact)

Pollution from internal and external combustion systems, such as internal combustion engines, stationary power plants, and the incineration of waste, results from an inability to completely rearrange fuel molecules into complete and stable products. Combustion engineering, therefore, is a key element needed to understand and deal with environmental issues of pollution.

Improved combustion, i.e., an ability to oxidize certain fuels more completely (such as hydrocarbon compounds like coal, gasoline, and natural gas), will improve power machinery performance characteristics, reduce fuel consumption, and clean emissions from stationary and mobility propulsion systems. This, in turn, could save billions of dollars in foreign oil purchases. An engineering assessment of fuel utilization now requires more than a general knowledge of the physical and chemical nature of specific natural and synthetic fuel resources and must introduce additional engineering and economic issues such as:

Source energy conversion penalties (drilling, mining, or growing)
Fuel preparation penalties (distillation, gasification, or milling)
Transportation penalties (transportation from source to application)
Fuel storage penalties (liquefaction, compression, cooling, or required environmental controls)
Energy generation/conversion penalties (heat-transfer and power losses)
Environmental pollution control penalties (emissions reduction)

Table 1.1 Classification of Some Common Chemical Fuels

By phase		By application	
Naturally available	Synthetically produced	Heat transfer	Power
Solid		Space heating	Stationary power plants
Coal	Coke	Process heat transfer	Mobile propulsion systems
Wood	Charcoal		
Vegetation	Inorganic solid waste		
Organic solid waste			
Liquid			
Crude oil	Syncrudes		
Biological oils	Petroleum distillates		
Fuel plants	Alcohols		
	Colloidal fuels		
	Benzene		
Gas			
Natural gas	Natural gas		
Marsh gas	Hydrogen		
Biogas	Methane		
	Propane		
	Coal gasification		

1.3 THE FUEL-ENGINE INTERFACE

A need exists today to improve in a practical way the state of the art in engineering of current power and propulsion systems in use by increasing their combustion and thermal efficiencies and to develop new fuels and new engines to operate thereon. As Professor Oppenheim at the University of California has so succinctly stated, "Further advances in engine technology should (1) minimize pollution emissions, (2) maximize engine efficiency, and (3) optimize tolerance to a wider variety of fuels." Engineering design, development, operation, and maintenance of heat engines now requires more than a comprehension of both fuels and engines. Many crucial aspects of combustion-driven heat engines are due in fact to their *fuel-engine interface;* see Figure 1.2, including: compatibility of fuels with specific engines, pollutants generated by burning particular fuels in certain engines, general energy input/output performance characteristics of engines operating on a given fuel, and various interactions and efficiencies associated with those interfaces.

Heat engines can be categorized in several different ways such as: internal versus external combustion engines, continuous versus intermittent combustion engines, or spark versus compression-ignition combustion engines; see Figure 1.3. The performance and emissions characteristics differ between these various categories. Many outstanding reference texts have been written dealing principally with traditional topics covered in fundamentals of fuels and combustion. Other writers have produced excellent dissertations treating in detail the particulars of certain heat engines such as boilers and furnaces, spark-ignition engines, compression-ignition engines, and the gas turbine. Literature and technology in these fields are expanding at an exponential rate. The professional journals are now presenting areas of interest not covered in many texts: items such as potential alternate power fuels (coal-derived, alcohols, and hydrogen); stratified charge engines, fast lean-burn engines, computer modeling of these advanced engine systems; and topics of availability, maintainability, and reliability pertinent to extended operation of these machines.

George Huebner, chairman of the Environmental Research Institute of Michigan, in an SAE paper entitled "Future Automotive Power Plants," stated: "Technical choices are being made that affect our nation's major engineering decisions for the future by those unskilled in the engineering profession and science of combustion." He further stated in the paper that "too much discussion of and consideration [is] given to alternative power plants without consideration of the fuel type."

A new approach is needed in engineering education to introduce important aspects of combustion in such a way as to present a proper overview of

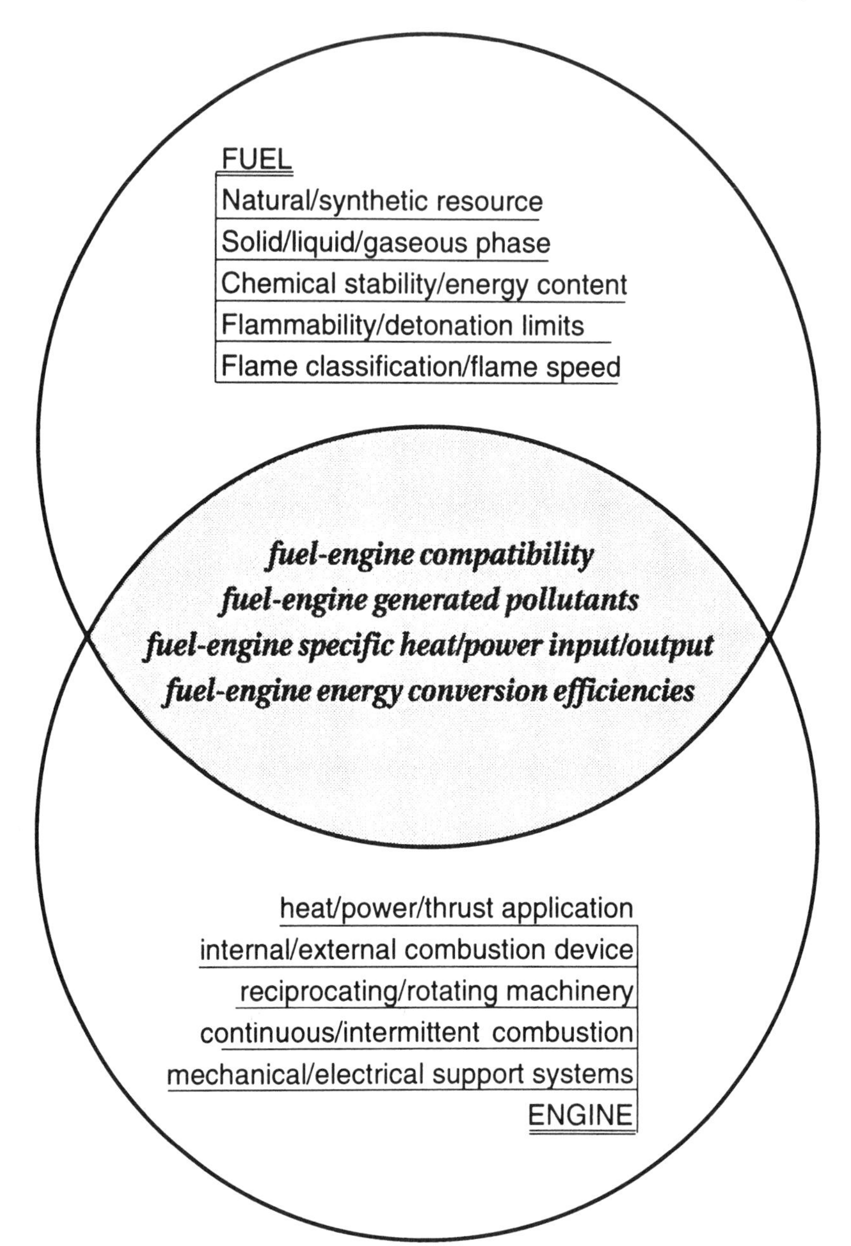

Figure 1.2 The critical fuel-engine interface.

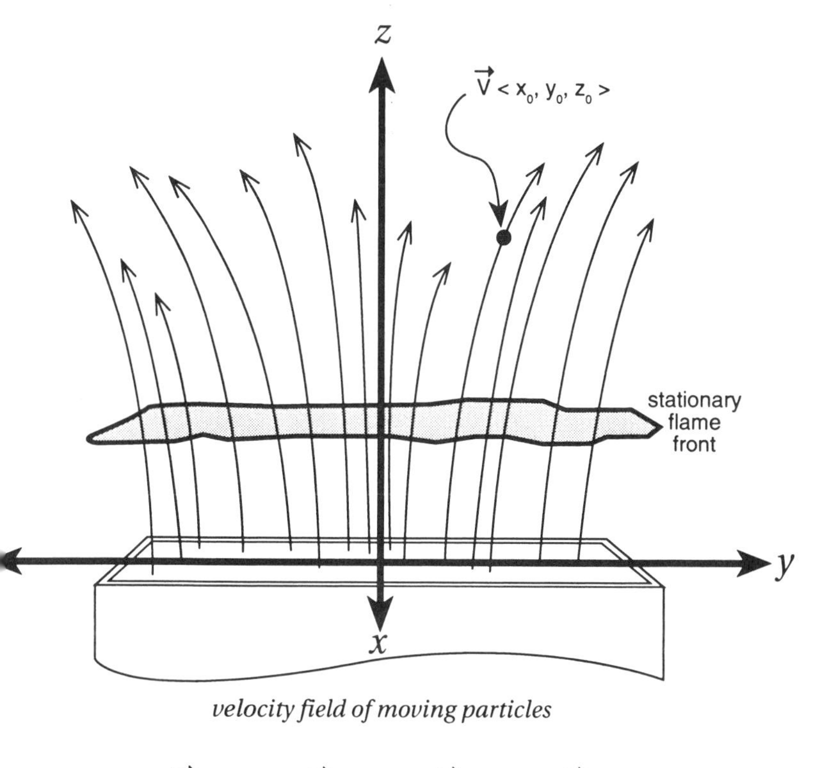

$$\vec{V} = V_x\langle x,y,z,t\rangle \vec{i} + V_y\langle x,y,z,t\rangle \vec{j} + V_z\langle x,y,z,t\rangle \vec{k}$$

igure 1.5 Eulerian (field) representation of combustion.

mple physical phenomena may require very complex analytical tools; hereas, in other instances, one can use a very simple analysis to provide sight into a very complex phenomenon.

Various physical parameters, termed *dimensions,* represented with brack-:s, [ ], are selected to express the physics of combustion quantitatively. A elected group of defined principal or primary dimensions such as length nd time are used to express other secondary dimensions such as velocity, e., length divided by time. Arbitrary and useful magnitudes, or *units,* for ese physical parameters are established and maintained under the super-sion of the General Conference on Weights and Measures. Two systems f dimensions and units will be used in this book, Système International I) and English Engineers. Table 1.3 lists some basic and secondary di-ensions and units of combustion. Often it is necessary to change units in

major areas in the discipline, to create motivation for further study, and to provide basic knowledge that will enable practicing engineers to stay current in the field.

1.4 ENGINEERING SCIENCE AND COMBUSTION

The physical requirements for reactive mixtures and the thermochemical path that fuel and oxidant, or *reactants,* should follow to form *products of combustion* while releasing energy will be developed in this text; see Table 1.2. Application of these concepts will be made in later chapters. In order to formulate basic theories of combustion correctly, analysis must be correct both qualitatively (physics) and quantitatively (mathematics). Classical physics and the conversion laws of mass, momentum, and energy are used to describe a fixed mass of material (closed system or control mass). In this instance, specific material is identified and followed using particle or *Lagrangian* analysis. Consider, for example, the motion of a solid exhaust gas particulate as shown in Figure 1.4. Position and velocity of the particle are functions of space and time, and velocity $\vec{V}$ is a function (represented by $< >$) of space and time $V<x,y,z,t>$. Because of the nondescript nature

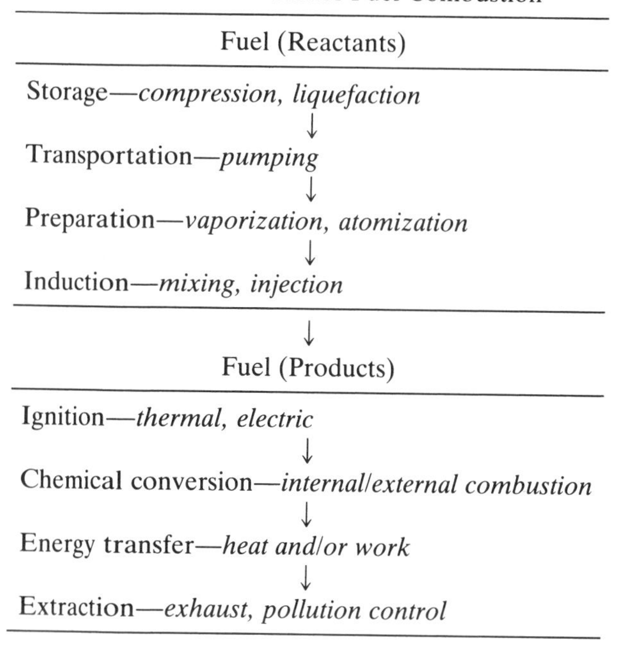

Table 1.2 Chemical Path for Fuel Combustion

Fuel (Reactants)
Storage—*compression, liquefaction*
↓
Transportation—*pumping*
↓
Preparation—*vaporization, atomization*
↓
Induction—*mixing, injection*
↓
Fuel (Products)
Ignition—*thermal, electric*
↓
Chemical conversion—*internal/external combustion*
↓
Energy transfer—*heat and/or work*
↓
Extraction—*exhaust, pollution control*

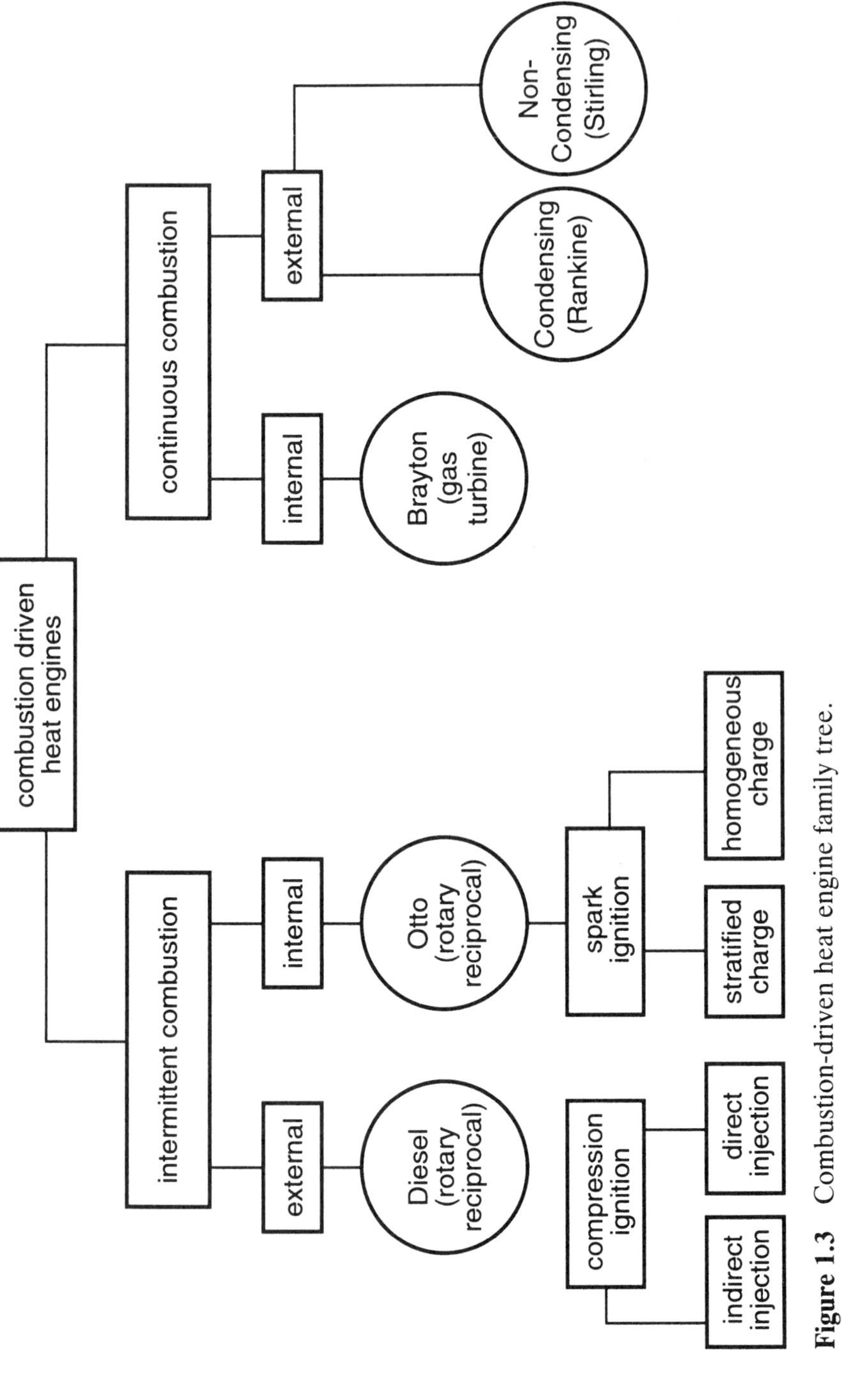

Figure 1.3 Combustion-driven heat engine family tree.

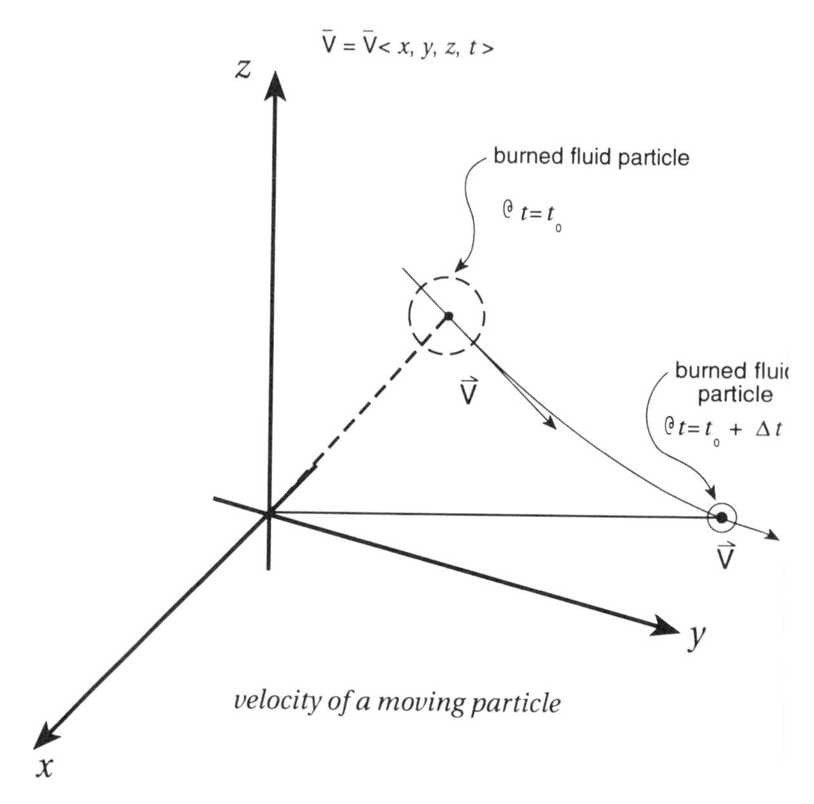

Figure 1.4 Lagrangian (particle) representation of combustion.

of fluid elements in combustion flowfields and the complexity of cl
activity therein, the unsteady three-dimensional nature of combu
written in *Eulerian* or field theory. In this instance, a particular p
space is identified and observed. Figure 1.5 represents the flow of r
gases through a stationary flame front in which the field velocity ve
in this instance, describes a velocity vector at a point (x_0, y_0, z_0) i
Note also that basic conservation laws were developed using a fixed
(open system or control volume) system terminology. Mathematica
mulation of the conservation laws, such as mass conservation, will be
ent depending on type of analysis used, i.e., Lagrangian versus Eule
addition, many problems are readily defined in terms of an infinite
small system using differential calculus; whereas others are more
stated in finite geometries using integral calculus. From this, it sho
apparent that, in some cases, engineering treatment of very bas

Table 1.3 Basic Dimensions and Units

Dimensions		Units			
		Système International (SI)		English Engineers	
Primary					
Force	[*F*]			Pound force	(lbf)
Mass	[*M*]	Kilogram	(kg)	Pound mass	(lbm)
Length	[*L*]	Meter	(m)	Feet	(ft)
Time	[*t*]	Second	(sec)	Second	(sec)
Temperature	[*T*]	Kelvin	(K)	Rankine	(°R)
Secondary					
Force	[*F*]	newton	(kg m/sec^2)		
Pressure	[*F*]/[*L*]^2	pascal	(N/m^2)	psf	(lbf/ft^2)
Energy	[*F*][*L*]	joule	(N m)	Btu	(ft lbf)
Power	[*F*][*L*]/[*t*]	watt	(J sec)	horsepower or hp	(ft lbf/sec)
Viscosity	[*M*]/[*L*][*t*]	poises	(kg/m sec)		(lbm/ft sec)
Kinematic viscosity	[*L*]^2/[*t*]	stokes	(m^2/sec)		(ft^2/sec)
Some principal constants					
Standard pressure		101 kPa (1 atm)		14.7 lbf/in^2 (1 atm)	
Absolute temperature		K = °C + 273		°R = °F + 460	
Standard acceleration of gravity		g = 9.8066 m/sec^2		g = 32.1740 ft/sec^2	
g_0		1.0 kg m/N sec^2		32.1740 ft lbm/lbf sec^2	

engineering calculations by use of identities termed *conversion factors*. Values for several conversions are found in Appendix A.1.

To illustrate a proper use of dimensions and units, consider Newton's second law, which describes the physical relationship between force $[F]$, mass $[M]$, and acceleration $[a]$ for a fixed mass.

$$\vec{F} \sim M \cdot \vec{a} \tag{1.1}$$

where dimensionally

$$[F] \sim [M][L]/[t]^2$$

Since $[F]$, $[M]$, $[L]$, and $[t]$ are distinct dimensions, Newton's second law is more appropriately written quantitatively as

$$\vec{F} = M \cdot \vec{a}/g_0 \tag{1.1a}$$

where

$$g_0 = M \cdot \vec{a}/\vec{F} \tag{1.1b}$$

Dimensionally, the constant g_0 becomes

$$g_0 = [M][L]/[F][t]^2$$

which gives Newton's second law dimensional equality since

$$[F] = \frac{[M][L]/[t]^2}{[M][L]/[F][t]^2} = [F]$$

In SI, the primary dimensions and units are

$[M]$ kilogram, kg

$[L]$ meter, m

$[t]$ second, sec

From physics, a force of 1 newton is the force that will accelerate 1.0 kg at 1.0 m/sec^2 or

$$g_0 = \frac{(1.0\text{ kg})\,(1.0\text{ m/sec}^2)}{(1\text{ N})} = 1\text{ kg m/N sec}^2$$

Now the unit of force, the newton, is a derived quantity, i.e., 1 N = 1 kg m/sec^2, and

$$g_0 = \frac{1\text{ N}}{1\text{ N}} = 1.0$$

Note that g_0 in SI is actually dimensionless with a unit magnitude. However, using g_0 in the form 1 kg m/N sec^2 in equations where g_0 appears will allow many equations to come out in correct units with ease and, therefore, g_0 will be carried throughout this text.

In Engineers units, the primary dimensions and units are

[*F*] pound force, lbf

[*M*] pound mass, lbm

[*L*] foot, ft

[*t*] second, sec

Again, from physics, 1 lbf will accelerate 1.0 lbm at 32.1740 ft/sec^2 or, in this instance,

$$g_0 = \frac{(1.0 \text{ lbm})\,(32.1740 \text{ ft/sec}^2)}{(1 \text{ lbf})}$$

$$= 32.1740 \text{ lbm ft/lbf sec}^2$$

Often, many problems can be understood when fundamental physical characteristics of matter, or *properties,* are stated in either dimensional or dimensionless form. In this text, capital letters generally denote extensive properties, lowercase letters imply intensive properties, and a superscript bar indicates molar properties. To illustrate, recall the concept of pressure *P,* which is force/ unit area or

$$[P] = [F]/[L]^2$$

Density ρ is mass/unit volume or

$$[\rho] = [M]/[L]^3$$

Specific volume v is volume/unit mass or

$$[v] = [L]^3/[M] = 1/[\rho]$$

Specific weight γ is weight/unit volume and, since "weight" is a force due to gravity, Newton's second law can then be used to express γ as

$$[\gamma] = [F]/[L]^3 = [M] \cdot (g/g_0)/[L]^3$$

or

$$\gamma = \rho \cdot (g/g_0) \tag{1.2}$$

Specific gravity, S.G., is the density of a fluid to that of a reference fluid and is dimensionless.

EXAMPLE 1.2 Oxygen is the principal oxidant in most combustion processes. For O_2 at 25°C and 101 kPa absolute pressure, calculate (a) the specific gas constant R, J/K kg, (b) the specific volume v, m³/kg; (c) the density ρ, kg/m³; and (d) the specific weight γ, N/m³.

Solution

1. Specific gas content:

$$R = \frac{\overline{R}}{MW} = \frac{8314.34 \text{ J/kg mole K}}{32 \text{ kg/kg mole}}$$

a) $R_{O_2} = 259.82$ J/kg K

2. Specific volume:

$$v = \frac{RT}{P} = \frac{(259.82 \text{ N m/kg K})(25 + 273\text{K})}{(101 \text{ kN/m}^2)(1000 \text{ N/kN})}$$

b) $v = 0.767$ m³/kg

3. Density:

$$\rho = \frac{1}{v} = \frac{1}{0.767 \text{ m}^3/\text{kg}} = 1.304 \text{ kg/m}^3$$

4. Specific weight:

$$\gamma = \rho \frac{g}{g_0} = (1.304 \text{ kg/m}^3) \frac{(9.8066 \text{ m/sec}^2)}{(1.0 \text{ kg m/sec}^2 \text{ N})}$$
$$= 12.79 \text{ N/m}^3$$

Comments Note that, in part d, if g_0 is omitted, γ = m kg/sec² m³ and, since 1 N ≡ 1 m kg/sec² γ = N/m³. By carrying g_0 through the calculation although the quantity is unity and dimensionless, the answer comes out in correct units with ease.

Many important relations will be expressed in terms of these properties. Recall, for example, the ideal gas law that relates pressure P, volume V, mass m, or moles N, and temperature T, of a substance. Several forms are shown below.

Mass basis: $P = P<m,V,T>$ (1.3)

Molar basis: $P = P<N,V,T>$

or

$$PV = mRT \qquad PV = N\overline{R}T \tag{1–3a}$$
$$Pv = RT \qquad P\bar{v} = \overline{R}T \tag{1–3b}$$
$$P = \rho RT \qquad P = \overline{\rho}\overline{R}T \tag{1–3c}$$

where the universal gas constant $\overline{R}$ is equal to

$$\overline{R} = \begin{cases} 8314.34 \text{ J/kg mole K} \\ 82.057 \text{ atm cc/g mole K} \\ 1{,}545.3 \text{ ft lbf/lbm mole °R} \\ 1.987 \text{ cal/g mole K} \\ 1.987 \text{ Btu/lbm mole °R} \end{cases}$$

Now the specific gas constant R can be determined as

$$PV = mRT = N\overline{R}T$$

or

$$R = (N/m) \cdot \overline{R}$$

From dimensional analysis, $[N]/[m]$ is moles per unit mass or molecular weight MW, and

$$R = \frac{\overline{R}}{MW} \tag{1.4}$$

1.5 ENGINEERING AND APPLIED COMBUSTION

Most professional engineering practices today can be segregated into three distinct skill areas: theory, design, and operation; see Table 1.4. Theory can be considered to be that part of engineering that focuses on precise formulation and development of basic principles needed to predict specific characteristics of producible products, a very intellectual and education-centered activity and illustrated by the material presented in Section 1.4. Design encompasses those many segments of the profession that apply and utilize standards and codes to fabricate beneficial products, basically a systems or program managerial activity. Note that design relationships most frequently have their roots in fundamentals, but they may be influenced by certain manufacturing or operational considerations. Operational engineering then deals with all those aspects and actions necessary to maintain essential products reliably, a training and people-centered activity. Again, it should be apparent that operational methods are interrelated with both

Table 1.4 Basic Skill Areas of Professional Engineering Practice

Theory	Design	Operation
Precise formulation: engineering physics and/or chemistry	General application: engineering codes and/or rules	Specific utility: engineering tools and/or practices
Basic Principles: differential and integral calculus	Complex Programs: statistical and numerical analysis	Particular problems basic math and geometry
Education-centered: requiring abstract mental skills necessary for system prediction	Educ/Train-Centered: requiring project managerial skills necessary for system production	Training-Centered requiring people motivational skills necessary for system protection

elements of theory and design. Figure 1.6 can be a useful illustration to help study how specific interactions between design, operation, and/or theory impact on a given engineering issue.

Engineering involves all those activities necessary to conceive, manufacture, and maintain useful products for the benefit of mankind. Today's world makes it essential for engineering students to grasp both fundamental practices, or technology, as well as fundamental principles, or science. Most undergraduate engineering curricula in the United States are based on four-year programs, with emphasis placed on basic theoretical subject matter. There is greater pressure now being placed on the university system to provide students with more true design experience. Operation and/or industrial experience occur in only a few institutions that have co-op studies. Some debate exists as to whether there is sufficient time and even a need for introducing these three skills at the college level.

The romantic notion is that engineers in the past and today are individuals whose abilities must span all three skill areas. The reality and complexity of today's engineering needs, however, force individuals to specialize in one of the three areas. Table 1.5 illustrates the path along which successfully developed combustion-related products will most frequently progress in going from an idea or concept to a useful configuration. The process begins with a physical attribute of combustion, and theoretical modeling is developed from basic laws, usually expressed in terms of simple physics and simple components. A theoretical model is translated into a design tool by passing successfully through the research laboratory, where a simple test

Table 1.5 The Classical Path of Technology Development

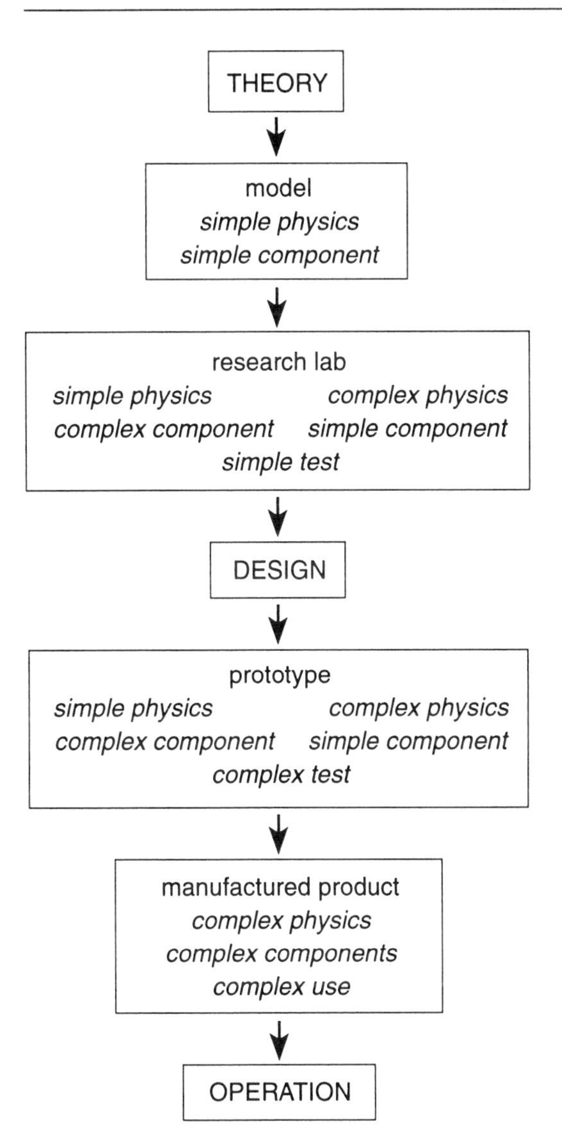

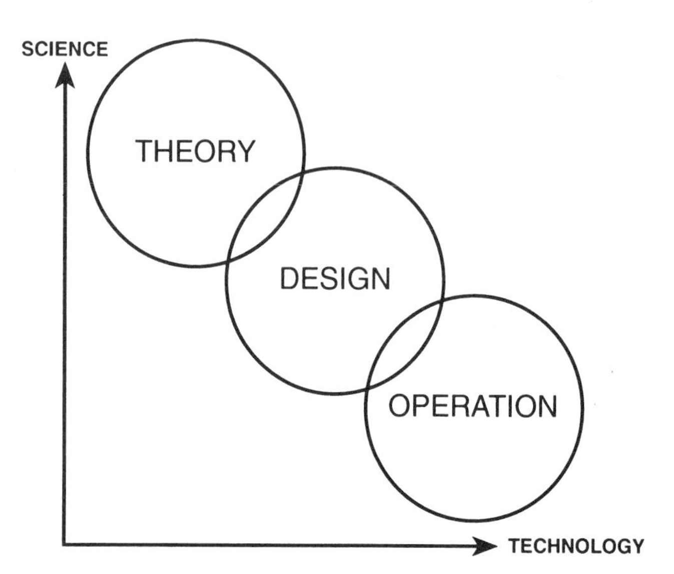

Figure 1.6 Visualization of the engineering profession.

validates or extends the initial basic theory. In the research lab, the nature of physics or component characteristics of an experiment are more complex than the original model developed from theory. Success at the research laboratory level allows a design engineer to transfer applied knowledge of combustion to industry through the development of a prototype, in which an even more complex test establishes success or failure of the concept at the design level. Success at the design level results in production of a manufactured product, which is then subjected to complex use, at which point success or failure is again redefined.

Applied Combustion will attempt to introduce basic theory, design rules, and some general operational characteristics of certain important combustion machinery. It is the author's intent that this broader purview of the subject will enable the reader to develop a better understanding of the subject for future professional development.

PROBLEMS

1.1 Energy equivalents in various dimensions and units are often useful in combustion calculations. Determine the equivalent value of 1 hp in (a) ft lbf/hr, (b) Btu/hr, and (c) kW.

1.2 Determine the equivalent temperature in Kelvin for an absolute temperature of (a) 500, (b) 1000, (c) 1500, and (d) 2000°R.

1.3 One kg of carbon will theoretically release approximately 33,000 kJ/kg of energy by complete combustion. Calculate the amount of energy that would ideally be obtained if the carbon mass could be completely converted into nuclear energy, ($E = mc^2$), kJ/kg.

1.4 The specific volume of gasoline is approximately 0.0238 ft^3/lbm. Find (a) its density, lbm/ft^3; (b) the mass of fuel in a 20-gal tank, lbm; and (c) its specific weight, N/m^3.

1.5 Standard atmospheric condition in theoretical combustion calculations is often chosen as 14.7 psia. Calculate the standard atmosphere in (a) lbf/ft^2, (b) ft H_2O, (c) mm Hg, and (d) Pa.

1.6 The specific gravity of a fuel is expressed as S.G.<60°F>, indicating that both the oil and water are measured at a temperature of 60°F. A diesel fuel has a specific gravity at 60°F of 0.8762. Determine (a) the density of the fuel, lbm/ft^3; and (b) the specific weight of the fuel, lbf/gal.

1.7 A 100-liter tank is to hold H_2 gas at 25°C and 400 kPa. Calculate (a) the mass of H_2 contained in the vessel, lbm; (b) the specific volume of H_2, ft^3/lbm; and (c) the density of H_2, lbm/ft^3.

1.8 Air is to be supplied to the combustor can of a gas turbine. Inlet conditions for the air are 740°F and 174 psia. The mass flow rate of air is 9000 lbm/min. Determine (a) the volumetric flow rate for the air, ft^3/min; and b) the molar flow rate for the air, lb mole air/min.

1.9 An automobile will consume about 0.4 lbm fuel/hr for each horsepower of power developed. Assuming that gasoline can release 19,000 Btu/lbm of energy, determine the thermal efficiency of the engine.

1.10 A stationary steam-driven electric power plant has a thermal efficiency of approximately 35%. For a 750,000-kW unit, calculate (a) the ideal required heat addition, Btu/min; (b) the ideal amount of nuclear material required to produce this energy, lbm/yr; and (c) the amount of coal necessary to produce this energy if the coal has an energy content of approximately 12,000 Btu/lbm coal.

1.11 The ideal energy content of a fuel is expressed in terms of its heating value. The approximate heating values of three important energy resources are listed below:

Natural gas	1,000 Btu/ft^3
Coal	12,000 Btu/lbm
Crude oil	138,000 Btu/gal

Determine the fuel energy equivalents of (a) natural gas to fuel oil, ft^3/gal; (b) coal to natural gas, lbm/ft^3; and (c) crude oil to coal, gal/lbm.

1.12 The fuel energy equivalence of a natural gas and coal supply is 0.78 m^3 gas/kg coal, while the fuel energy equivalence of crude oil to coal is 0.092 m^3 oil/kg coal. If the heating value of coal is approximately 29,075 kJ/kg, determine (a) the energy equivalence of natural gas to oil, m^3 gas/m^3 oil; (b) the heating value of natural gas, kJ/m^3; and (c) the heating value of the fuel oil, kJ/m^3.

1.13 Refer to the three fuels given in Problem 1.11 and their heating values to determine the fuel economic equivalence of (a) coal, $/ton; and (b) gas, $/$ft^3$. Assume that oil is priced at $50/bbl (42 gal).

2

Combustion and Energy

2.1 INTRODUCTION

Combustion engineering requires an ability to analyze energetics of chemically reactive mixtures. Most thermodynamics texts address the general subject of energy conversion and conservation as well as cover principles of mass conservation, property relationships, equations of state, and process relationships. Classical thermodynamics provides an overall bookkeeping technique useful when describing different engineering components and processes. The particular discipline termed *thermochemistry* deals with global energy analysis associated with combustion processes. In this chapter, a basic comprehension of thermodynamics will be assumed, and a general energy balance for a chemically reactive medium will be developed.

In most cases, the conservation of energy principle is stated for changes, or *processes,* that occur between some initial stable, i.e., *equilibrium* condition, or *state,* and some final equilibrium condition; see Figure 2.1. Since thermodynamics does not address the rate at which these processes occur, it will be necessary in later chapters to introduce materials that will describe the true dynamic nature of combustion.

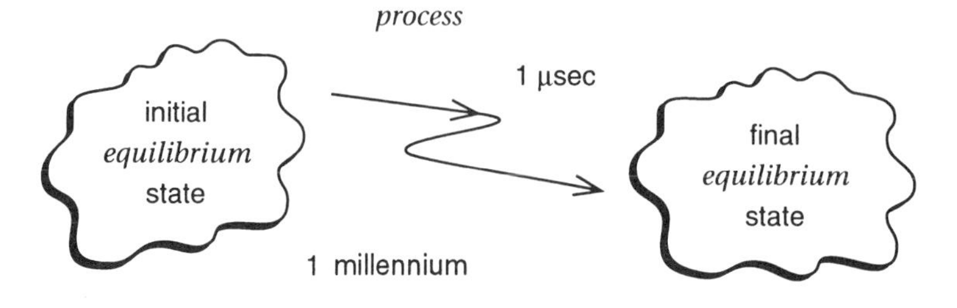

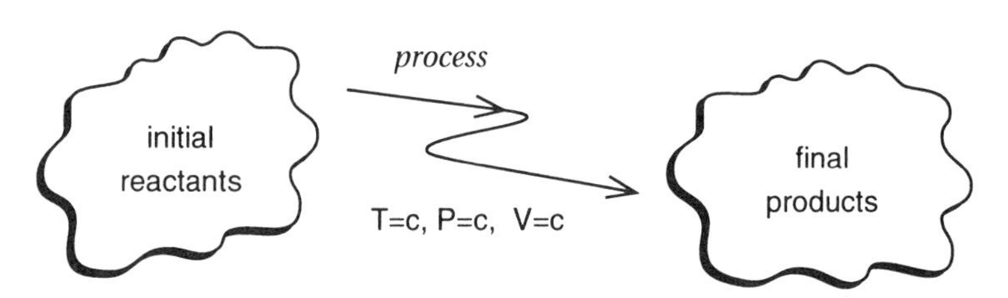

Figure 2.1 A generalized thermochemical process.

2.2 THE CONSERVATION OF MASS

The conservation of mass principle is a fundamental engineering concept. Consider a fixed total mass of a chemically reactive mixture, using a *control mass* or *closed system*. In general, the chemical composition of such a reactive mixture will change with time. The initial equilibrium condition is referred to as the *reactant* state, while the final equilibrium condition is termed the *product* state.

A mathematical means of expressing composition change and mass conservation for reactive systems is provided by the *chemical reaction* equation, written in general form as

$$\Sigma \nu_i R_i \rightarrow \Sigma \nu_j P_j \tag{2.1}$$

where

ν_i = mass or moles of species i

R_i = reactant species i

ν_j = mass or moles of species j

P_j = product species j

All initial reactants are placed on the left-hand side of the general reaction equation, while any final product appears on the right-hand side. Particular processes are represented by an arrow, with indication of the type of process shown above, i.e., $P = c$, $V = c$, $T = c$.

A description based on mass or weight for a mixture of compounds, existing in either reactant or product state, termed a *gravimetric analysis*, expresses the total mass in terms of each pure constituent. For a mixture of k total chemical species, a mass fraction mf_i for each i component species can be written as

$$mf_i \equiv \frac{m_i}{m_{\text{tot}}} \qquad \frac{\text{kg}_\text{i}}{\text{kg}_{\text{tot}}} \left(\frac{\text{lbm}_i}{\text{lbm}_{\text{tot}}} \right) \tag{2.2}$$

and the total mass m_{tot} is then equal to

$$\sum_{i=1}^{k} m_i = m_{\text{tot}} \qquad \text{kg (lbm)} \tag{2.2a}$$

and

$$\sum_{i=1}^{k} mf_i = 1.0 \tag{2.2b}$$

Even though the total mass of a combustion process may remain constant, concentration of constituents such as oxygen or carbon dioxide may change during a reaction. The mass fraction of a species i as a reactant may be quite different from that as a product species j.

Often, it is more convenient to describe chemically reactive mixtures on a *molar* basis. Molar analysis utilizes a relative chemical mass scale for compounds based on the periodic chart. For a mixture of k total chemical species, a mole fraction $\bar{x}_i$ for each i species can be expressed as

$$\bar{x}_i = \frac{N_i}{N_{\text{tot}}} \frac{\text{kg mole}_\text{i}}{\text{kg mole}_{\text{tot}}} \left(\frac{\text{lb mole}_\text{i}}{\text{lb mole}_{\text{tot}}} \right) \tag{2.3}$$

and the total moles N_{tot} is equal to

$$\sum_{i=1}^{k} N_i = N_{\text{tot}} \qquad \text{kg mole (lb mole)} \tag{2.3a}$$

$$\sum_{i=1}^{k} \bar{x}_i = 1.0 \tag{2.3b}$$

In Chapter 1, the bridge between molar and mass analysis was shown to be molecular weight. The mole-mass relationship for species i is given as

$$m_i = N_i \times MW_i$$

where

m_i = mass species i, kg (lbm)
N_i = moles species i, kg mole (lb mole)
MW_i = molecular weight species i, kg/kg mole (lbm/lb mole)

For a mixture of k species, the total mass can be written

$$m_{\text{tot}} = N_{\text{tot}}\, MW_{\text{tot}} = \sum_{i=1}^{k} N_i MW_i \tag{2.4}$$

or

$$MW_{\text{tot}} = \sum_{i=1}^{k} \left(\frac{N_i}{N_{\text{tot}}}\right) MW_i = \sum_{i=1}^{k} \bar{x}_i MW_i \tag{2.4a}$$

EXAMPLE 2.1 Standard atmospheric conditions for many combustion calculations can be represented by the following mole fractions:

N_2 78%
O_2 21%
Ar 1%

For these conditions, determine (a) the mass fractions of N_2, O_2, and Ar; (b) the mixture molecular weight, kg/kg mole; and (c) the specific gas constant R for air, J/kg K.

Solution

1. Mole fractions:

$\bar{x}_{N_2} = 0.78 \qquad \bar{x}_{O_2} = 0.21 \qquad \bar{x}_{Ar} = 0.01$

2. Mass fractions:

	$\bar{x}_i$	MW_i	$\bar{x}_i MW_i$	mf_i[a]
O_2	0.21	32.0	6.72	0.232
N_2	0.78	28.0	21.84	0.754
Ar	0.01	40.0	0.40	0.014
	$\frac{\text{kg mole}_i}{\text{kg mole}_{\text{tot}}}$	$\frac{\text{kg}_i}{\text{kg mole}_i}$	$\frac{\text{kg}_i}{\text{kg mole}_{\text{tot}}}$	$\frac{\text{kg}_i}{\text{kg}_{\text{tot}}}$

[a] a. $mf_{O_2} = \dfrac{\bar{x}_{O_2}}{\Sigma \bar{x}_i MW_i} = \dfrac{6.72}{6.72 + 21.84 + 0.40} = 0.232$

$$mf_{N_2} = \frac{21.84}{28.96} = 0.754, \qquad mf_{Ar} = \frac{0.46}{28.96} = 0.014$$

b. $MW_{air} = \Sigma \bar{x}_i MW_i = 6.72 + 21.84 + 0.40 = 28.96$ kg/kg mole

c. Specific gas constant:

$$R_{air} = \frac{\overline{R}}{MW} = \frac{8314 \text{ J/kg mole K}}{28.96 \text{ kg/kg mole}} = 287 \text{ J/K kg}$$

The general principle for mass conservation for a chemical reaction can be stated in terms of an atomic mass balance, i.e., total number of *atoms* of an element in the final product state, such as hydrogen atom, must equal the initial number of atoms of the element in the reactant state. The conservation of atomic species will yield a set of equations equal to the unique number of elements in the reaction, i.e., H, N, O, C.

The most prominent chemical reaction associated with power production and heat transfer is the *oxidation* reaction, i.e.,

"Fuel" + oxidizer → Products

Reactants and/or products of combustion may exist as a solid, liquid, and/or vapor. In addition, the contributing species may be single species, pure compounds, complex mixtures, stable molecules, and/or radicals. Traditionally, most fuels of engineering relevance have been hydrocarbon compounds, represented as C_xH_y. Furthermore, the most commonly used oxidizing agent is air. For engineering purposes, ideal air is assumed to be approximately 21% O_2 and 79% N_2 by volume. Thus, for 1 mole of O_2, there are corresponding 3.76 moles of N_2 or 4.76 moles of air. This theoretical air, termed *dry air,* has a molecular weight of approximately 28.96 in this text.

For a given fuel-oxidant reaction, all actual potential combining proportions are infinite, i.e., they can be a function of (1) the fuel, (2) the oxidant, (3) the combustion process, and/or (4) the combustion device. Complete ideal combustion of a fuel assumes formation of the most fully oxidized products. The most stable oxidized product of carbon is carbon dioxide, CO_2, while the most stable oxidized product of hydrogen is water, H_2O. Therefore, the most stable oxidized products formed from complete combustion of any hydrocarbon fuel would be only CO_2 and H_2O.

Sufficient oxidant theoretically to support complete combustion of a particular fuel and ideally to form its most stable products can be expressed in terms of the *stoichiometric* equation. For example, consider the two stoichiometric oxidation reactions below.

$$C_S + O_2 \xrightarrow{P} CO_2$$

$$CH_4 + 2(O_2 + 3.76N_2) \xrightarrow{V} CO_2 + 2H_2O + 7.52N_2$$

For a fuel-air mixture, stoichiometric conditions are often referred to as 100% *theoretical air.* It is important to note that an actual reaction having stoichiometric fuel-air proportions will not actually produce complete combustion.

Since actual oxidation reactions are incomplete, many combustion systems operate using *excess air,* i.e., a percentage of air over and above stoichiometric proportions. For instance, 150% theoretical air is equivalent to 50% excess air. A fuel-air mixture having excess air is termed a *fuel-lean* mixture, while a mixture that has excess fuel is called a *fuel-rich* mixture.

Practical engineering considerations give rise to the use of a *fuel-to-air ratio, FA,* or an *air-to-fuel ratio, AF.* These ratios can be expressed on a mass or molar basis as follows:

Mass basis:

$$AF = \frac{\text{mass of air}}{\text{mass of fuel}} \qquad \frac{\text{kg air}}{\text{kg fuel}}\left(\frac{\text{lbm air}}{\text{lbm fuel}}\right) \tag{2.5}$$

$$FA = \frac{\text{mass of fuel}}{\text{mass of air}} \qquad \frac{\text{kg fuel}}{\text{kg air}}\left(\frac{\text{lbm fuel}}{\text{lbm air}}\right) \tag{2.5a}$$

Mole basis:

$$\overline{AF} = \frac{\text{moles of air}}{\text{moles of fuel}} \qquad \frac{\text{kg mole air}}{\text{kg mole fuel}}\left(\frac{\text{lb mole air}}{\text{lb mole fuel}}\right) \tag{2.6}$$

$$\overline{FA} = \frac{\text{moles of fuel}}{\text{moles of air}} \qquad \frac{\text{kg mole fuel}}{\text{kg mole air}}\left(\frac{\text{lb mole fuel}}{\text{lb mole air}}\right) \tag{2.6a}$$

A useful dimensionless fuel-oxidant ratio, or *equivalence ratio Φ*, can be defined as

$$\Phi \equiv \frac{\left(\dfrac{\text{moles fuel}}{\text{moles oxidant}}\right)_{\text{actual}}}{\left(\dfrac{\text{moles fuel}}{\text{moles oxidant}}\right)_{\text{stoichiometric}}} \tag{2.7}$$

where

$\Phi < 1$ fuel-lean mixture

$\Phi = 1$ stoichiometric mixture

$\Phi > 1$ fuel-rich mixture

EXAMPLE 2.2 The reaction of gaseous hydrogen and oxygen to form water is well known. The combustion of carbon with oxygen forming carbon dioxide is also an important chemical reaction. Write the chemical reactions for these two oxidation reactions, and calculate (a) the reactant's molar analysis, (b) the reactant's gravimetric analysis, and (c) mass of oxygen required per mass of fuel.

Solution

1. Stoichiometric hydrogen combustion:

$$H_2 + aO_2 \rightarrow bH_2O$$

Hydrogen atom balance $\quad 2 = 2b \quad b = 1$

Oxygen atom balance $\quad 2a = 1 \quad a = 0.5$

$$H_2 + \tfrac{1}{2}O_2 \rightarrow H_2O$$

2. Mole fractions:

$$\bar{x}_{H_2} = \frac{1}{1 + 0.5} = 0.667$$

$$\bar{x}_{O_2} = \frac{0.5}{1.5} = 0.333$$

3. Mass fractions:

	$\bar{x}_i$	MW_i	$\bar{x}_i MW_i$	mf_i^a
H_2	0.667	2.00	1.333	0.111
O_2	0.333	32.00	10.667	0.889
	$\dfrac{\text{lb mole}_i}{\text{lb mole}_{tot}}$	$\dfrac{\text{lbm}_i}{\text{lb mole}_i}$	$\dfrac{\text{lbm}_i}{\text{lb mole}_{tot}}$	$\dfrac{\text{lbm}_i}{\text{lbm}_{tot}}$

$$MW = \Sigma \bar{x}_i MW_i = 1.333 + 10.667 = 12.00 \text{ lbm/lb mole}$$

[a] $mf_{H_2} = \dfrac{1.333}{12.00} = 0.111, \quad mf_{O_2} = \dfrac{10.667}{12.00} = 0.889$

4. Mass O_2/H_2:

$$OF = \frac{Z \text{ lbm } O_2}{1.0 \text{ lbm } H_2} = \frac{88.9 \text{ lbm } O_2}{11.1 \text{ lbm } H_2} \qquad Z = 8.01 \frac{\text{lbm } O_2}{\text{lbm } H_2}$$

5. Stoichiometric carbon combustion:

$$C_S + aO_2 \rightarrow bCO_2$$

Carbon atom balance $\quad 1 = b$

Oxygen atom balance $\quad 2a = 2 \quad a = 1$

$$C_S + O_2 \rightarrow CO_2$$

6. Mole fractions:

$$\bar{x}_{C_s} = \frac{1}{1 + 1} = 0.50$$

$$\bar{x}_{O_2} = 1 - 0.5 = 0.5$$

7. Mass fractions:

	x_i	MW_i	$\bar{x}_i MW_i$	mf_i^b
C_S	0.50	12.00	6.00	0.273
O_2	0.50	32.00	16.00	0.727
	$\frac{\text{lb mole}_i}{\text{lb mole}_{tot}}$	$\frac{\text{lbm}_i}{\text{lb mole}_{tot}}$	$\frac{\text{lbm}_i}{\text{lb mole}_{tot}}$	$\frac{\text{lbm}_i}{\text{lbm}_{tot}}$

$$MW = \Sigma \bar{x}_i MW_i = 6.00 + 16.00 = 22.00 \text{ lbm/lb mole}_{tot}$$

[b] $mf_i = \frac{6.00}{22.00} = 0.273, \quad mf_{O_2} = \frac{16.00}{22.00} = 0.727$

8. Mass O_2/C_S:

$$\text{OF} = \frac{Z \text{ lbm } O_2}{1.0 \text{ lbm C}} = \frac{72.7 \text{ lbm } O_2}{27.3 \text{ lbm C}} \qquad Z = 2.663 \frac{\text{lbm } O_2}{\text{lbm } C_S}$$

Comments: This problem illustrates the variations in air-fuel ratios for hydrocarbon compounds. Not only will the stoichiometric proportions vary with each particular fuel, but the ratio will differ depending on whether the results are based on a mass or molar analysis.

2.3 THERMODYNAMIC PROPERTIES

An analytical treatment of the energetics of combustion will require the use of thermodynamic material characteristics, or *properties*. Several of the more familiar include: temperature T, total pressure P, total volume V, moles N_i (or $\bar{x}_i$), mass M_i (or mf_i), total internal energy U, and/or total enthalpy H. The state of a simple pure homogeneous substance can be described by specifying two independent and substance-dependent, or *intrinsic,* properties. Internal energy and pressure are examples of intrinsic properties, while kinetic and potential energies are examples of nonintrinsic, or *extrinsic,* properties.

Combustion reactions involve multiphase, multicomponent mixtures that undergo chemical transformation in both time and space. Obviously, such a complex medium cannot be treated as a pure homogeneous compound. Even for most nonreactive mixtures, two independent intrinsic properties, such as temperature and total pressure, plus the chemical composition, are necessary to define a state. Considering the many possible mixture combinations pertinent to combustion analysis, simple tabulations, or graphical representation of all mixture thermodynamic properties is prohibitive.

Thermochemical calculations should be based on fundamental relations expressed collectively in terms of pure component species properties. Furthermore, it will be convenient to think of reactant and product conditions as being two distinct nonreactive equilibrium mixtures. Many reaction mixtures can be treated as ideal-gas mixtures in part because of the high product temperatures as well as the chemical composition of most major combustion components and uncertainties of heat release of the actual combustion process.

The Gibbs–Dalton law describes the thermodynamic nature of an ideal-gas mixture in terms of each pure component species. The rule states that, *in an equilibrium mixture of ideal gases, each component acts as if it were alone in the system at* V_{tot}, T. By careful application of this law, thermodynamic state and useful thermochemical properties for a mixture can be developed.

For an equilibrium mixture of nonreactive gases A, B, C, . . . , and using the Gibbs-Dalton law, one can write the following facts:

$$V_A = V_B = V_C = \cdots = V_i = V_{tot} \tag{2.8}$$

$$T_A = T_B = T_C = \cdots = T_i = T \tag{2.9}$$

From the conservation of mass principle,

$$N_{tot} = N_A + N_B + N_C + \cdots = \sum_{i=1}^{k} N_i \tag{2.3a}$$

From Chapter 1, the ideal-gas equation of state was given as

$$P_{tot}V_{tot} = N_{tot}\overline{R}T \tag{1.4}$$

Combining Equations (1.4), (2.3), (2.8), and (2.9) yields

$$P_{tot}V_{tot} = (N_A + N_B + N_C + \cdots)\bar{R}T = \Sigma N_i\bar{R}T$$

or

$$P_{tot} = \frac{N_A\bar{R}T}{V_A} + \frac{N_B\bar{R}T}{V_B} + \frac{N_C\bar{R}T}{V_C} + \cdots = \Sigma\frac{N_i\bar{R}T}{V_i} \tag{2.10}$$

The pressure exerted by a pure component species i if it were alone in a closed system at the total volume and mixture temperature is termed its *partial pressure* P_i or

$$P_i = N_i\bar{R}T/V \tag{2.11a}$$

and

$$\frac{P_i}{P_{tot}} = \frac{N_i\bar{R}T/V}{N_{tot}\bar{R}T/V} = \frac{N_i}{N_{tot}} = \bar{x}_i$$

$$P_i = \bar{x}_iP_{tot} \tag{2.11b}$$

where

$$\Sigma P_i = \sum_{i=1}^{k}\bar{x}_iP_{tot} = P_{tot} \tag{2.11c}$$

P = pressure

i = chemical species i

k = total number of species

The total pressure is equal to the sum of partial pressures where partial pressure of a species i is equal to total pressure multiplied by its mole fraction.

Additional thermodynamic properties, such as internal energy and enthalpy can also be evaluated by applying the Gibbs–Dalton law.

Internal energy U, for example, generally is expressed as a function of temperature, volume, and composition, i.e., $U = U<T, V, N_i>$. Since, for ideal gases, $U = U<T, N_i>$, the total internal energy of a mixture can be written as

Extensive:

$$U_{tot} = m_{tot}u_{tot} = \Sigma m_iu_i \qquad \text{kJ (Btu)} \tag{2.12a}$$

or

$$U_{tot} = N_{tot}\bar{u}_{tot} = \Sigma N_i\bar{u}_i \qquad \text{kJ (Btu)} \tag{2.12b}$$

Intensive:

$$u_{\text{tot}} = \sum_{i=1}^{k} \left(\frac{m_i}{m_{\text{tot}}} \right) u_i = \sum_{i=1}^{k} mf_i u_i \qquad \frac{\text{kJ}}{\text{kg mixt}} \left(\frac{\text{Btu}}{\text{lbm mixt}} \right) \tag{2.12c}$$

or

$$\bar{u}_{\text{tot}} = \sum_{i=1}^{k} \left(\frac{N_i}{N_{\text{tot}}} \right) \bar{u}_i = \sum_{i=1}^{k} \bar{x}_i \bar{u}_i \qquad \frac{\text{kJ}}{\text{kg mole mixt}} \left(\frac{\text{Btu}}{\text{lbmole mixt}} \right) \tag{2.12d}$$

Enthalpy H is generally expressed as a function of temperature, pressure, and composition, i.e., $H = H < T, P, V_i >$. Again, for ideal gases $H = H<T, N_i>$, and the total enthalpy of a mixture is stated as

Extensive:

$$H_{\text{tot}} = m_{\text{tot}} h_{\text{tot}} = \Sigma m_i h_i \qquad \text{kJ (Btu)} \tag{2.13a}$$

or

$$H_{\text{tot}} = N_{\text{tot}} \bar{h}_{\text{tot}} = \Sigma N_i \bar{h}_i \qquad \text{kJ (Btu)} \tag{2.13b}$$

Intensive:

$$h_{\text{tot}} = \sum_{i=1}^{k} mf_i h_i \qquad \frac{\text{kJ}}{\text{kg mixt}} \left(\frac{\text{Btu}}{\text{lbm mixt}} \right) \tag{2.13c}$$

or

$$\bar{h}_{\text{tot}} = \sum_{i=1}^{k} \bar{x}_i \bar{h}_i \qquad \frac{\text{kJ}}{\text{kg mole mixt}} \left(\frac{\text{Btu}}{\text{lb mole mixt}} \right) \tag{2.13d}$$

A constant-volume specific heat C_v is defined as

$$C_v \Big)_{\text{tot}} = \frac{\partial u_{\text{tot}}}{\partial T} \Big)_v \qquad \frac{\text{kJ}}{\text{kg K}} \left(\frac{\text{Btu}}{\text{lbm °R}} \right) \tag{2.14}$$

or for a nonreactive ideal-gas mixture using Equation (2.12c),

$$= \frac{\partial}{\partial T} \left(\frac{1}{m_{\text{tot}}} \sum_{i=1}^{k} m_i u_i \right)_v$$

$$= \frac{1}{m_{\text{tot}}} \Sigma\, m_i \frac{\partial u_i}{\partial T} \Big)_v = \Sigma mf_i C_{v_i}$$

$$C_v \Big)_{\text{tot}} = \sum_{i=1}^{k} mf_i C_{v_i} \qquad \frac{\text{kJ}}{\text{kg K}} \left(\frac{\text{Btu}}{\text{lbm °R}} \right) \tag{2.14a}$$

or

$$\overline{C}_v\Big)_{tot} = \sum_{i=1}^{k} \bar{x}_i \overline{C}_{v_i} \qquad \frac{\text{kJ}}{\text{kg mole K}} \left(\frac{\text{Btu}}{\text{lb mole °R}} \right) \tag{2.14b}$$

A constant-pressure specific heat C_p can be defined in similar fashion for a nonreactive ideal-gas mixture as

$$C_p\Big)_{tot} = \frac{\partial h_{tot}}{\partial T}\Big)_p \qquad \frac{\text{kJ}}{\text{kg K}} \left(\frac{\text{Btu}}{\text{lbm °R}} \right) \tag{2.15}$$

$$= \frac{\partial}{\partial T} \left(\frac{1}{m_{tot}} \sum_{i=1}^{k} m_i h_i \right)_p$$

$$= \frac{1}{m_{tot}} \Sigma m_i \frac{\partial h_i}{\partial T}\Big)_p = \Sigma m f_i C_{p_i}$$

$$C_p\Big)_{tot} = \sum_{i=1}^{k} m f_i C_{p_i} \qquad \frac{\text{kJ}}{\text{kg K}} \left(\frac{\text{Btu}}{\text{lbm °R}} \right) \tag{2.15a}$$

or

$$\overline{C}_p\Big)_{tot} = \sum_{i=1}^{k} \bar{x}_i \overline{C}_{p_i} \qquad \frac{\text{kJ}}{\text{kg mole K}} \left(\frac{\text{Btu}}{\text{lb mole °R}} \right) \tag{2.15b}$$

Enthalpy and internal energy are related by the equation $h = u + Pv$. For an ideal gas, this can be expressed as

$$h = u + RT \tag{2.16}$$

and

$$dh = du + RdT \tag{2.16a}$$

combining Equations (2.14), (2.15), and (2.16a) and noting that specific heats for an ideal gas depend only on temperature,

$$C_p dT = C_v dT + RdT$$

or

$$R_{tot} = C_{p_{tot}} - C_{v_{tot}} \qquad \frac{\text{kJ}}{\text{kg}} \left(\frac{\text{Btu}}{\text{lbm}} \right) \tag{2.17}$$

and

$$R_{tot} = \sum_{i=1}^{k} mf_i C_{p_i} - \sum_{i=1}^{k} mf_i C_{v_i}$$

$$= \sum_{i=1}^{k} mf_i \, (C_{p_i} - C_{v_i})$$

$$R_{tot} = \sum_{i=1}^{k} mf_i R_i$$

The specific heat ratio γ is defined as

$$\gamma \equiv \frac{C_p}{C_v} = \frac{\overline{C}_p}{\overline{C}_v} \tag{2.18}$$

Using Equations (2.17) and (2.18), one can also write

$$C_p = \frac{R\gamma}{\gamma - 1} \quad \text{or} \quad \overline{C}_p = \frac{\overline{R}\gamma}{\gamma - 1} \tag{2.19a}$$

and

$$C_v = \frac{R}{\gamma - 1} \quad \text{or} \quad \overline{C}_p = \frac{\overline{R}}{\gamma - 1} \tag{2.19b}$$

Therefore,

$$\gamma_{tot} = \frac{\Sigma \bar{x}_i \overline{C}_{p_i}}{\Sigma \bar{x}_i \overline{C}_{v_i}} = \frac{\Sigma mf_i C_{p_i}}{\Sigma mf_i C_{v_i}} \tag{2.18a}$$

EXAMPLE 2.3 Consider a homogeneous mixture of ethylene, C_2H_4, and 130% theoretical air. Calculate as a function of temperature (a) the mixture gas constant R, kJ/kg; (b) the mixture constant-pressure specific heat C_p, kJ/kg K; (c) the mixture constant-volume specific heat C_v, kJ/kg K; and (d) the specific heat ratio γ.

Solution

1. Stoichiometric equation:

$$C_2H_4 + aO_2 \rightarrow bCO_2 + cH_2O$$

Carbon atom balance	$b = 2$	
Hydrogen atom balance	$c = 2$	
Oxygen atom balance	$2a = 2(2) + 2$	$a = 3$

2. Actual reaction:

$$C_2H_4 + (1.3)(3)[O_2 + 3.76N_2] \rightarrow$$
$$2CO_2 + 2H_2O + 0.9O_2 + (3.9)(3.76)N_2$$

3. Mixture molecular weight:

$$\bar{x}_{C_2H_4} = \frac{1.0}{1.0 + (3.9)(4.76)} = 0.0511$$

$$\bar{x}_{air} = 1.0000 - 0.0511 = 0.9489$$

$$MW_{tot} = \Sigma \bar{x}_i MW_i = (0.0511)(28) + (0.9489)(28.97)$$

$$MW_{tot} = 28.92 \text{ kg/kg mole mixt}$$

4. Thermodynamic properties:

$$\text{a. } R = \frac{\overline{R}}{MW} = \frac{8.314 \text{ kJ/mole K}}{28.92 \text{ kg/kg mole}} = 0.2875 \text{ kJ/kg K}$$

and

$$C_p = \overline{C}_p/MW = \overline{C}_p/28.92 \qquad C_p - C_v = R \qquad C_p/C_v = \gamma$$

$$\overline{C}_p = \Sigma \overline{x}_c \overline{C}_{p_c} = 0.9489\, \overline{C}_{p_{air}} + 0.0511\, \overline{C}_{p_{C_2H_4}}$$

5. From Appendixes B.4 and B.10,

	Air		C_2H_4	
T	$\overline{C}_p$	$\overline{C}_p$	$\overline{C}_p$	$\overline{C}_p$
298	6.947	29.087	10.250	42.917
400	7.010	29.351	12.679	53.087
600	7.268	30.431	16.889	70.714
800	7.598	31.813	20.039	83.903
1000	7.893	33.048	22.443	93.969
1200	8.109	33.952	24.290	101.70
1400	8.289	34.706	25.706	107.63
1600	8.437	35.326	26.794	112.19
K	cal/g mole K	kJ/kg mole K	cal/g mole K	kJ/kg mole K

(*cont.*)

T	$\overline{C}_{p_{\text{mixt}}}$	C_p	C_v	γ
298	29.794	1.030	0.7425	1.387
400	30.564	1.057	0.7695	1.374
600	32.489	1.123	0.8355	1.344
800	34.475	1.192	0.9045	1.318
1000	36.161	1.250	0.9625	1.298
1200	37.414	1.294	1.0065	1.286
1400	38.432	1.329	1.0415	1.276
1600	39.254	1.357	1.0695	1.269
	kJ/kg mole K	kJ/kg K	kJ/kg K	—

Comments This problem illustrates that the assumption of constant specific heats and specific heat ratio may, in certain problems with a wide temperature variation, lead to serious error. Note that the specific heats C_p and C_v both increase with increasing temperature while γ decreases with increasing temperature.

The enthalpy of a single compound can be expressed on a unit mole basis as a function of temperature and pressure as

$$\bar{h}_i = \bar{h}_i\langle T, P\rangle \qquad \frac{\text{kJ}}{\text{kg mole}_i}\left(\frac{\text{Btu}}{\text{lb mole}_i}\right) \tag{2.20}$$

Differential changes in enthalpy are equal to

$$d\bar{h}_i = \left.\frac{\partial \bar{h}_i}{\partial T}\right)_P dT + \left.\frac{\partial \bar{h}}{\partial P}\right)_T dP \tag{2.20a}$$

Since, for an ideal gas, enthalpy is independent of pressure

$$\bar{h}_i = \bar{h}_i\langle T\rangle \tag{2.21}$$

and Equation (2.15) becomes

$$d\bar{h}_i = \left.\frac{\partial \bar{h}_i}{\partial T}\right)_P dT = \overline{C}_{p_i}\langle T\rangle\, dT \tag{2.21a}$$

integrating Equation (2.21a) between limits

$$\Delta\bar{h}_i = \int_{h_0}^{h} d\bar{h}_i = \int_{T_0}^{T} \overline{C}_{p_i}\langle T\rangle\, dT \tag{2.21b}$$

where

T = system temperature

T_0 = reference datum temperature = 25°C (77°F)

The temperature dependency of ideal-gas specific heats can be expressed in virial form. Consider, for example, the constant-pressure molar specific heat of a pure component i, expressed as a fitted power series in temperature as

$$\overline{C}_{p_i}\langle T\rangle = a_1 + a_2T + a_3T^2 + a_4T^3 + a_5T^4 \cdots \tag{2.22}$$

Tabulated values for $\overline{C}_{p_i}$ versus T for several gases can be found in Appendix B.

Using the virial form for $\overline{C}_p$, Equation (2.22), and substituting into Equation (2.21b) yields an expression for changes in enthalpy for a compound as

$$\bar{h}_i\langle T_2\rangle - \bar{h}_i\langle T_0\rangle = \int_{T_0}^{T} [a_1 + a_2T + a_3T^2 + a_4T^3 + a_5T^4]dT \tag{2.23}$$

$$= [a_1T + \frac{a_2}{2}T^2 + \frac{a_3}{3}T^3 + \frac{a_4}{4}T^4 + \frac{a_5}{5}T^5]_{T_0}^{T} \tag{2.23a}$$

Tabulated values for $\Delta\bar{h}_i\langle T\rangle$ for several gases are also found in Appendix B.

Changes in internal energy for an ideal gas can be obtained from an appropriate expression in terms of enthalpy

$$\bar{h}_i = \bar{u}_i + P_i\bar{v}_i = \bar{u}_i + \overline{R}T \tag{2.24}$$

$$\bar{u}_i = \bar{h}_i - \overline{R}T \frac{\text{kJ}}{\text{kg mole}} \left(\frac{\text{Btu}}{\text{lb mole}}\right)$$

and

$$\Delta\bar{u}_i = \bar{u}_i\langle T\rangle - \bar{u}_i\langle T_0\rangle = \Delta\bar{h}_i - \overline{R}\Delta T = \Delta\bar{h}_i - \overline{R}(T-T_0) \tag{2.24a}$$

EXAMPLE 2.4 A 10-kg mixture of 25% CO and 75% N_2 by weight is cooled from 1800 to 800K. For these conditions, determine (a) Δh, kJ/kg mixture, (b) $\Delta\bar{h}$, kJ/kg mole; (c) $\Delta\bar{u}$, kJ/kg mole; and (d) Δu, kJ/kg.

Solution:

1. Mixture properties from Appendix B:

	$\bar{h}\langle T\rangle - \bar{h}\langle T_0\rangle$[a]		$\bar{h}\langle 1800\rangle - \bar{h}\langle 800\rangle$
	1800K	800K	
N	49,017.2	15,056.5	33,960.7
CO	49,557.3	15,186.2	34,371.1
		kJ/kg mole	

[a]Values are equal to 4.187 × $[\bar{h}\langle T\rangle - \bar{h}\langle T_0\rangle]$ from Appendix B.

where

$$\Delta h)_{N_2} = \left.\frac{\Delta \bar{h}}{MW}\right)_{N_2} = \frac{33{,}960.7 \text{ kJ/kgmole}}{28 \text{ kg/kg mole}} = 1212.9 \frac{\text{kJ}}{\text{kg}}$$

$$\Delta h)_{CO} = \left.\frac{\Delta \bar{h}}{MW}\right)_{CO} = \frac{34{,}371.1 \text{ kJ/kgmole}}{28 \text{ kg/kg mole}} = 1227.5 \frac{\text{kJ}}{\text{kg}}$$

2. Mixture enthalpy:

$$\Delta h = \Sigma m f_i \Delta h_i = \left(0.25 \frac{\text{kg CO}}{\text{kg mixt}}\right)\left(1227.5 \frac{\text{kJ}}{\text{kgCO}}\right) + \left(0.75 \frac{\text{kg } N_2}{\text{kg mixt}}\right)\left(1212.9 \frac{\text{kJ}}{\text{kg } N_2}\right)$$

a. $\Delta h = 1{,}216.6$ kJ/kg mixt

3. Mixture molar enthalpy:

$$\Delta \bar{h}_{\text{tot}} = \Delta h_{\text{tot}} \times MW_{\text{tot}} = (1{,}216.6 \text{ kJ/kg mixt}) \ (28 \text{ kg/kg mole})$$

b. $\Delta \bar{h} = 34{,}064.8$ kJ/kg mole

4. Mixture molar internal energy:

$$\bar{u}_i = \bar{h}_i - \bar{R}T_i$$

and

$$\Delta \bar{u}_i = \Delta \bar{h}_i - \bar{R}\Delta T$$

or

$$\Delta \bar{u} = 34{,}064.8 \text{ kJ/kg mole} - (8.314 \text{ kJ/kg mole K}) \ (1000\text{K})$$

c. $\Delta \bar{u} = 25{,}750.8$ kJ/kg mole

5. Mixture internal energy

$$\Delta u_{\text{tot}} = \Delta \bar{u}_{\text{tot}}/MW_{\text{tot}}$$

$$\Delta u_{\text{tot}} = \frac{25{,}750.8 \text{ kJ/kg mole}}{28 \text{ kg/kg mole}}$$

d. $\Delta u = 919.7$ kJ/kg mixt

Comments: This problem illustrates the method used in preparing thermodynamic tables for gas mixtures. Tables such as for air and for products of combustion of fuel with 200% theoretical air and 400% theoretical air, as well as those for particular water vapor-air mixtures, are familiar to the engineering student of thermodynamics. For complex calculations, it is convenient to store this type of information in computer codes for access and easy solution to many thermochemical problems.

Several additional properties, including entropy and Gibbs and Hemholtz functions, will be covered in Chapter 3.

2.4 HEATS OF REACTION

Thermodynamic processes will usually result in a change in temperature, pressure, and/or volume. Thermochemical processes produce additional effects, including compositional changes and release or absorption of a form of energy due to chemical reactions. For the purposes of analysis, it will be convenient to model the actual energetics of combustion reactions in terms of an ideal hypothetical process.

Chemical processes can result in energy being liberated by a reaction, termed an *exothermic reaction,* or the process may require energy to be absorbed, termed an *endothermic reaction;* see Figure 2.2. Furthermore, any energy absorbed or released at the reference datum by reaction would cause temperature to change and, therefore, make it necessary to transfer energy, usually as heat, to return the final temperature to its datum value.

Consider, for example, a closed system in which a chemical reaction occurs at constant total pressure; see Figure 2.3. Experience indicates that the temperature and composition will change as reactants are converted to products. To analyze this problem, it will be assumed that all chemical reactions, i.e., mass conversions and energy release, will occur only at a datum state, i.e., conditions of constant temperature and total pressure; T_0,

Exothermic reaction: *reaction in which energy is* liberated.

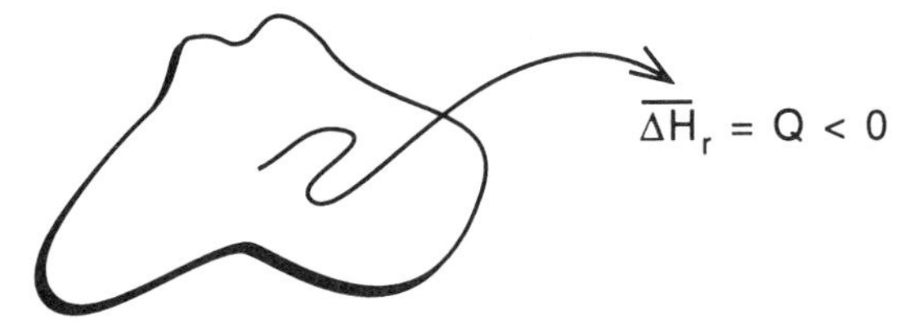

Endothermic reaction: *reaction in which energy is* absorbed.

Figure 2.2 Energy of reaction.

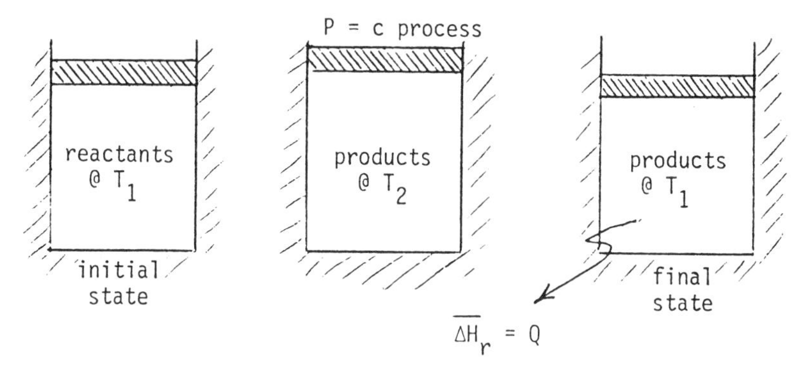

Figure 2.3 Heat of reaction.

P_0. This idea is similar to boiling and/or condensation energy transfer, in which T_{sat} and P_{sat} do not change but rather the energy transfer results in phase changes from liquid $\rightleftarrows$ vapor.

This energy transfer is called the *heat of reaction*, ΔH_R, and is defined as the *energy absorbed or liberated from a chemical reaction when products are brought back to the initial temperature of the reactants.* Since chemical reactions involve rearrangement of chemical bonds and formation and destruction of species, it is necessary to develop a means of comparing the ideal chemical energetics of various compounds, i.e., a kind of a chemical potential energy of a species. Thus, for instance, certain compounds such as fuels have a great capacity to react with oxygen, i.e., have high chemical potential energies, whereas product species such as CO_2 and H_2O have low chemical potential energies. This new species property, an *energy of formation,* will be seen to have direct bearing on the heat of reaction.

Two standard states have been used to define the datum in thermochemistry: the *gaseous monatomic species datum* and the *natural elemental species datum.* In both standards, the datum pressure and temperature are

$$P_0 = 1 \text{ atm} = 101 \text{ kPa} = 14.7 \text{ psi}$$

$$T_0 = 25°\text{C} = 298\text{K} = 77°\text{C} = 537°\text{R}$$

Standard state compounds are selected compounds that are assigned zero chemical potential energy at standard temperature and pressure (*STP*). Since our datum is for a constant-pressure process, this energy of formation is termed the *enthalpy of formation,* $\bar{h}^0_{f_i}$.

In the monatomic standard, all monatomic species, i.e., O, H, N, etc., have $\bar{h}^0_{f_i} = 0$ at *STP.* In the natural elemental standard, all natural elemental compounds have $\bar{h}^0_{f_i} = 0$ at *STP.* A natural elemental compound is a species in which only one element appears and in a form in which it occurs naturally at *STP,* i.e., H_2, N_2, O_2, and *solid* carbon. Note that, for example, CO is not a natural elemental species since it contains more than one atomic species and O is also not a natural elemental species since oxygen occurs naturally as O_2 at *STP.* The natural elmental standard will be used in this text.

Two important heats of reaction warrant special attention: the heat of formation and the heat of combustion. The *heat of formation,* ΔH_f, is the *energy absorbed or liberated by a chemical reaction when the product* (a single chemical compound) *is brought back to the initial temperature of the reactants* (only natural elemental species). The heat of formation reaction for several compounds is shown below.

Elemental compounds		Single compounds
$H_2 + \frac{1}{2}0_2$	$\overset{STP}{\rightarrow}$	H_2O
$C_S + 2H_2$	$\overset{STP}{\rightarrow}$	CH_4
$C_S + O_2$	$\overset{STP}{\rightarrow}$	CO_2
$8C_s + 9H_2$	$\overset{STP}{\rightarrow}$	C_8H_{18}

Another useful heat of reaction is the *heat of combustion,* ΔH_c, which is the *energy liberated when a compound reacts with an oxidant to form the most oxidized form of the reactants, and is brought back to the initial reactant temperature.* Recall that, for hydrocarbon chemistry, the most common oxidant is O_2 and the most stable oxidized products are CO_2 and H_2O; i.e.,

$$\text{Fuel} + \text{oxidant} \overset{STP}{\rightarrow} aCO_2 + bH_2O$$

If the water formed exists in the vapor state, the heat of combustion is termed *lower heating value.* If the water in the product state exists as a liquid, the heat of combustion is termed *higher heating value of the fuel.* This implies that the product mixture gave up its additional latent heat in going from a vapor to a liquid state.

Obviously, actual combustion processes do not occur at *STP,* and real combustion processes may not go to completion. It is therefore necessary to develop a means of expressing the energetics of chemical reactions in the more general case of combustion, which will be done in the following sections.

2.5 FIRST LAW FOR REACTIVE SYSTEMS

The first law of thermodynamics leads directly to an energy conservation principle. An extensive development of this concept can be found in most undergraduate thermodynamics texts. This fundamental relationship provides a statement, in general terms, of conservation and conversion of various energy forms.

From classical thermodynamics and the first law, one can obtain a *differential* form of the general energy equation for an *open system* (also termed a fixed or control volume) as

$$\frac{dE}{dt} = \dot{Q} - \dot{W} + \sum_{i_{\text{in}}} \dot{m}_i e_i - \sum_{j_{\text{out}}} \dot{m}_j e_j \qquad \text{kW} \left(\frac{\text{Btu}}{\text{min}} \right) \tag{2.25a}$$

or

$$\frac{dE}{dt} = \dot{Q} - \dot{W} + \sum_{i_{\text{in}}} \dot{N}_i \dot{e}_i - \sum_{j_{\text{out}}} \dot{N}_j \bar{e}_j \tag{2.25b}$$

where

E = total system energy in C.V.

$\dot{Q}$ = rate of heat transfer to/from (+ in/− out) the C.V.

$\dot{W}$ = rate of work transfer to/from (− in/+ out) the C.V.

$m_{i,j}$ = mass flow rate of material transferred across the C.V. boundaries

$\dot{N}_{i,j}$ = molar flow rate of material transferred across the C.V. boundaries

$e_{i,j}$ = intensive total energy per unit mass of material transferred across the C.V. boundaries

$\bar{e}_{i,j}$ = intensive total energy per unit mole of material transferred across the C.V. boundaries

This relationship can be expressed on a mass or molar basis but, in this text, will most frequently be written on a molar basis.

Equation (2.25) shows that the total energy E of a control volume may change with time, i.e., increase, decrease, and/or remain constant. Furthermore, it indicates that the *rate of change* of total control volume energy can be the result of several influences, including heat and/or work transfer rates to and/or from an open system, as well as various energy fluxes associated with material transfer across the boundaries. Figure 2.4 and Table 2.1 illustrate the significance of this equation to certain applications.

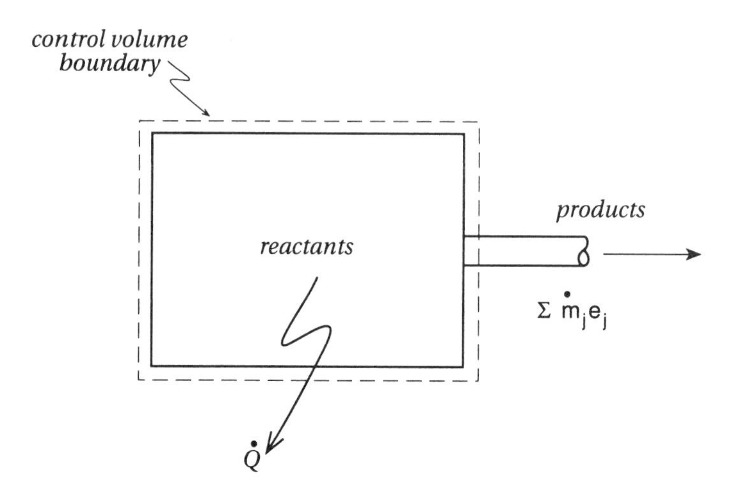

Figure 2.4 Open-system energy transfers.

Table 2.1 General Energy Equation and Applications

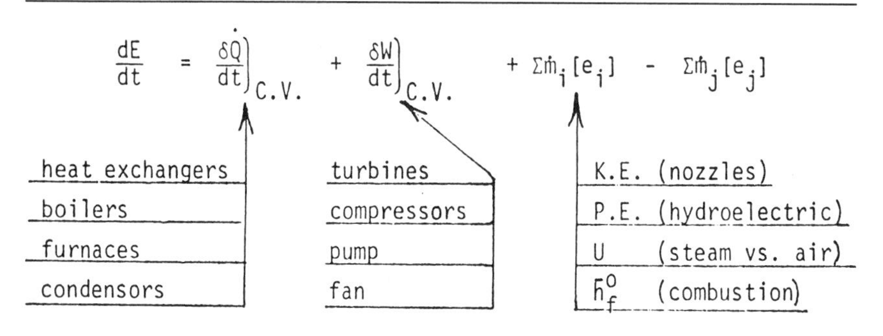

$$\frac{dE}{dt} = \left.\frac{\delta \dot{Q}}{dt}\right)_{C.V.} + \left.\frac{\delta W}{dt}\right)_{C.V.} + \Sigma \dot{m}_i [e_i] - \Sigma \dot{m}_j [e_j]$$

$\delta \dot{Q}/dt$	$\delta W/dt$	$\Sigma \dot{m}_i [e_i]$
heat exchangers	turbines	K.E. (nozzles)
boilers	compressors	P.E. (hydroelectric)
furnaces	pump	U (steam vs. air)
condensors	fan	$\bar{h}_f^o$ (combustion)

Many specific forms of the energy equation can easily be obtained by a proper application of the general energy equation as expressed by Equation (2.25). To do so, one must: (1) understand the physics of a particular problem under consideration, (2) formulate the proper mathematical statement of the appropriate energy balance, and (3) obtain the necessary properties required for solution.

To illustrate a technique for reducing the general energy equation to a suitable specific form, consider the case in which all the properties within a C.V., as well as those transferring across the boundary, do not change with time, i.e., *steady state and steady flow.* Under these restrictions, the rate of change in total C.V. energy must remain constant or

$$\frac{dE}{dt} = 0$$

Thus, the net energy flux into C.V. must equal the net energy flux out of C.V. or, when one equates transient energy forms of heat transfer and power to energy as properties crossing the boundaries due to mass transfer,

$$\dot{Q} - \dot{W} = \sum_{j_{\text{out}}} \dot{m}_j e_j - \sum_{i_{\text{in}}} \dot{m}_i e_i = \sum_{j_{\text{out}}} \dot{N}_j \bar{e}_j = \sum_{i_{\text{in}}} \dot{N}_i \bar{e}_i \tag{2.26}$$

Unsteady applications that allow mass transfer from an open system, such as in rocket propulsion systems, are instances in which the rate of decrease in total C.V. energy is written as

$$\frac{dE}{dt} = \dot{Q} - \sum_{j_{\text{out}}} \dot{m}_j e_j = \dot{Q} - \sum_{j_{\text{out}}} \dot{N}_j \bar{e}_j \tag{2.27}$$

Consider another illustrative case in which no material is transferred across the boundary, i.e., a *closed system* (often termed a fixed or control mass). In this instance,

$$\dot{m}_i = \dot{m}_j = 0$$

and

$$\frac{dE}{dt} = \dot{Q} - \dot{W}$$

Multiplying both sides by *dt,*

$$dE = \delta Q - \delta W$$

In many engineering applications, closed systems can produce *boundary expansion work,* such as with a piston-cylinder geometry. In this instance, the force through a distance relationship for work takes the familiar form,

$$\delta W = P dV \tag{2.28}$$

and

$$dE = \delta Q - P dV \tag{2.29}$$

For a device that operates on a thermodynamic cycle, the initial and final states must be identical, requiring that the system must function as a closed system and integrating around the cyclic loop

$$\oint dE = 0 = \oint \delta Q - \oint \delta W$$

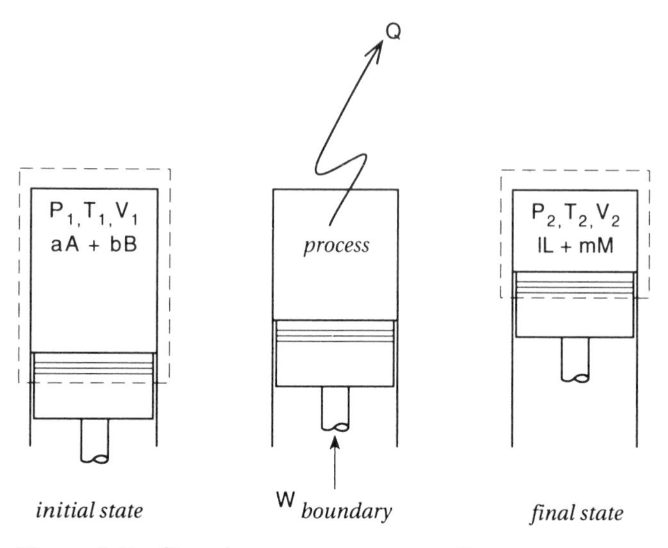

Figure 2.5 Closed-system energy transfer.

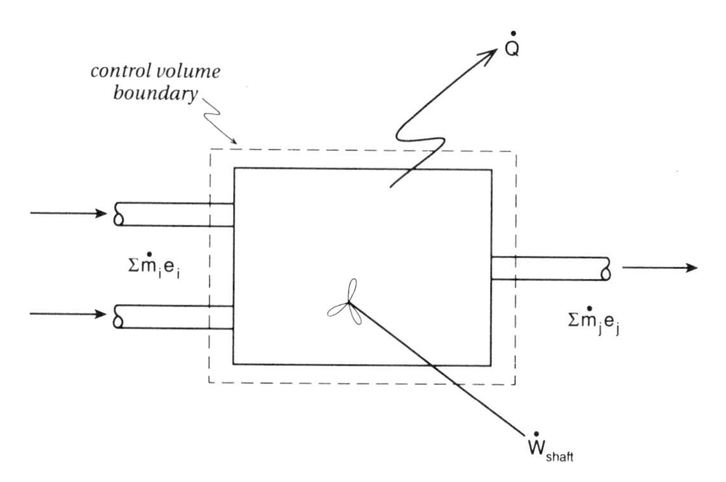

Figure 2.6 Unsteady open-system energy transfer.

$$Q_{net} = W_{net} \tag{2.30}$$

The total energy E of an ideal-gas mixture according to the Gibbs–Dalton law can be written as

$$E_{tot} = \Sigma m_i e_i = \Sigma N_i \bar{e}_i \qquad \text{kJ (Btu)} \tag{2.31}$$

or

$$e_{tot} = \Sigma m f_i e_i \qquad \text{kJ/kg (Btu/lbm)} \tag{2.31a}$$

and

$$\bar{e}_{tot} = \Sigma \bar{x}_i \bar{e}_i \qquad \text{kJ/kg mole (Btu/lb mole)} \tag{2.31b}$$

A component species i can store energy in several ways, including potential, kinetic, internal, and chemical potential energy or

$$e_i = \frac{g}{g_0} z_i + \frac{V_i^2}{2g_0} + u_i + h_{f_i}^0 \tag{2.32}$$

Neglecting kinetic and potential energy terms for the moment, the total energy of a mixture of fixed total mass equals

$$E_{tot} = U_{tot} = \Sigma N_i(\bar{u}_i + \bar{h}_{f_i}^0) \tag{2.33}$$

Since enthalpy data are available in this text, Equation (2.33) can also be written as

$$E_{tot} = U_{tot} = \Sigma N_i[\bar{h}_i\langle T\rangle - \bar{R}T + \bar{h}_f^0] \tag{2.34}$$

A plot of total mixture energy versus temperature for a fixed composition will yield a curve having a positive slope since both u and h increase with increasing temperature; see Figure 2.7. In addition, mixture lines representing reactants have a greater total energy value at a given temperature than corresponding curves for products because fuels have larger chemical potential energies.

In open-system applications, there is an additional energy flux term associated with pressure gradients in the flowfield termed *flow work* or *flow energy.* Flow work or energy per unit mass crossing the control volume as a result of mass transfer is equal to the local product $P_i v_i$. The total energy term associated with mass that can cross a control volume is then equal to

$$e_i = \frac{g}{g_0} z_i + \frac{V_i^2}{2g_0} + u_i + h_f^0 + p_i v_i \tag{2.35}$$

Since $h_i \equiv u_i + p_i v_i$, Equation (2.35) becomes

$$e_i = \frac{g}{g_0} z_i + \frac{v_i}{2g_0} + h_i + h_f^0 \tag{2.35a}$$

The general energy equation equates heat- and work-transfer rates to changes in energy levels of materials that undergo state changes or processes. Therefore, it is the difference between energy terms rather than

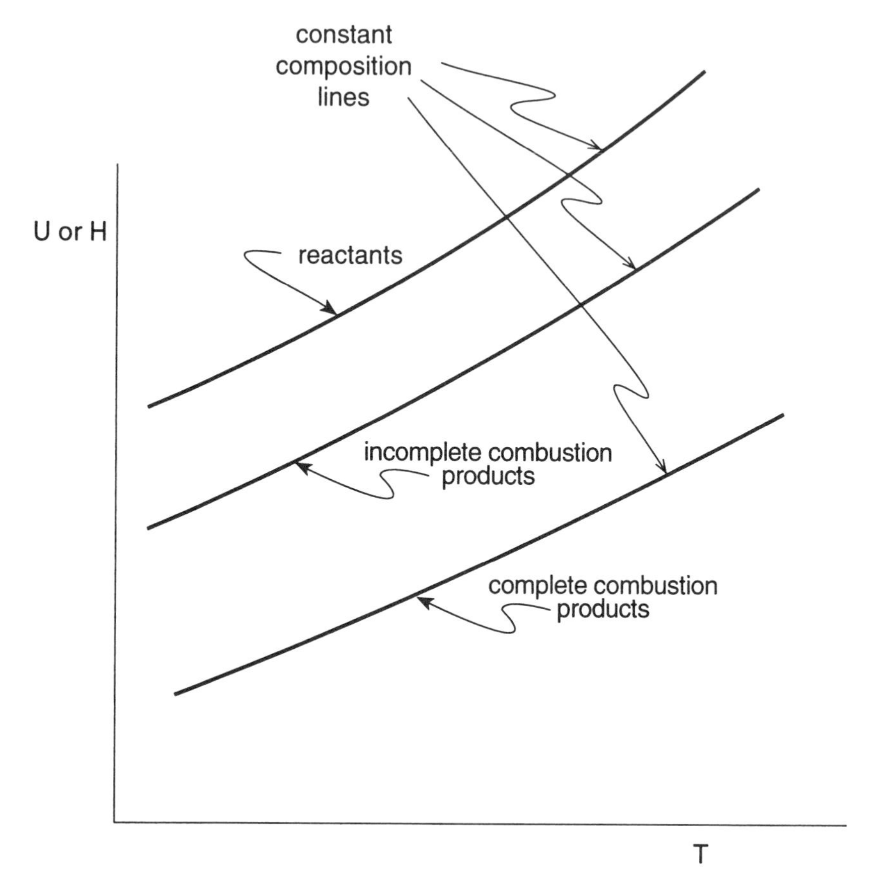

Figure 2.7 Total mixture energy lines vs. temperature.

their absolute values that is most important to the engineer. Since thermochemical reactants in this text assume all chemical changes occur at *STP*, it will be convenient to express internal energy and enthalpy values referenced to the *STP* state as a datum (Figure 2.8)

When this approach is used, the total molar enthalpy of a species i may be expressed as

$$\left.\bar{h}_i\right)_{\text{tot}} \equiv \bar{h}_i<T> - \bar{h}_i<T_0> + \bar{h}^0_{f_i} \tag{2.36}$$

and the change in total enthalpy is equal to

$$\begin{aligned} \Delta H_{\text{tot}} &= H_{\text{tot}}<T_2> - H_{\text{tot}}<T_1> \\ &= \Sigma N_j\{\bar{h}_j<T_2> - \bar{h}_j<T_0> + \bar{h}^0_{f_j}\} \\ &\quad - \Sigma N_i\{\bar{h}_i<T_1> - \bar{h}_i<T_0> + \bar{h}^0_{f_i}\} \end{aligned} \tag{2.37}$$

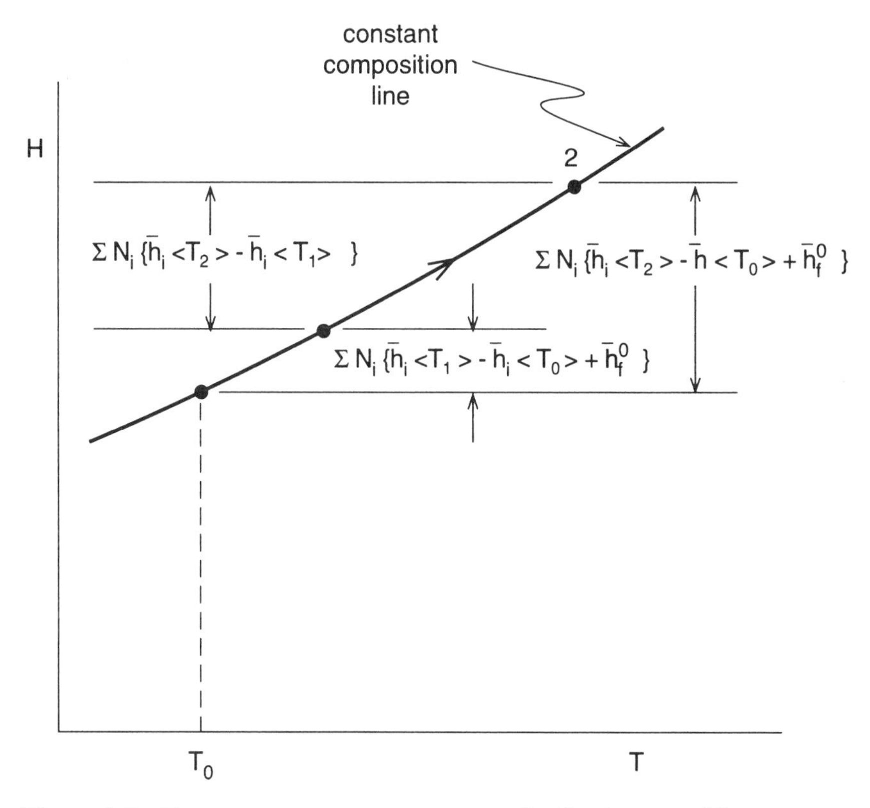

Figure 2.8 Energy vs. temperature process for fixed compositions.

Consider the case of a nonreactive gas mixture undergoing a process, i.e., $N_i = N_j$, $T_2 \neq T_1$, and

$$\begin{aligned} \Delta H_{\text{tot}} &= \Sigma N_i\{\bar{h}_i{<}T_2{>} - \bar{h}_i{<}T_0{>} + \bar{h}^0_{f_i}\} \\ &= \Sigma N_i\{\bar{h}_i{<}T_1{>} - \bar{h}_i{<}T_0{>} + \bar{h}^0_{f_i}\} \\ &= \Sigma N_i\,\{\bar{h}_i{<}T_2{>} - \bar{h}_i{<}T_1{>}\} \end{aligned} \tag{2.38}$$

Equation (2.38) is the familiar form found in most introductory thermodynamics texts.

Using the energy equation as developed in this chapter, an expression for various thermochemical processes pertinent to combustion can now be written. Consider the energy analysis for the heat of formation of a particular species occurring at *STP.* In this instance, $P = P_0 = \text{const}$, $T = T_0 = \text{const}$, the initial state constituents are standard elemental species, and the final state consists of a single product species. For a fixed total mass system analysis, the closed-system form of the energy equation is applicable, or

$$dE = dU_{tot} = \delta Q - PdV \tag{2.39}$$

$$\delta Q_{STP} = dU_{tot} + PdV + \cancel{VdP} = dH_{tot} \tag{2.40}$$

Integrating Equation (2.40) yields

$$Q_{STP} = \Delta H_{tot} = \sum_{prod} N_j[\bar{h}_f^0 + \Delta \bar{h}]_j - \sum_{react} N_i[\bar{h}_f^0 + \Delta \bar{h}]_i \tag{2.40a}$$

where

$$\Delta \bar{h}_{i,j} = 0 \text{ (reaction at } STP) \qquad \text{i.e., } T_2 = T_1 = T_0$$

and

$$\bar{h}_f^0)_{reactants} = 0 \quad \left\{ \begin{array}{l} \text{product is formed from natural} \\ \text{elemental compounds at } STP \end{array} \right.$$

$$N_j = 1.0 \qquad \text{one product species}$$

The heat of formation ΔH_f is therefore given as

$$Q_{S.T.P.} = 1.0\, \bar{h}_{f_i}^0 \equiv \Delta H_{f_i} \tag{2.41}$$

Appendix B lists the heat of formation at STP for various compounds.

Consider the energy analysis for heat of combustion of a stoichiometric hydrocarbon fuel and air mixture reaction at STP. Again, this process requires that $P = P_0$ = const and $T = T_0$ = const, with the initial state a stoichiometric fuel and air mixture and the final state consisting of most oxidized products, i.e., CO_2 and H_2O. Equation (2.40a) again applies and

$$Q = \Delta H_{tot} = \sum_{prod} N_j[\bar{h}_f^0 + \Delta \bar{h}]_j - \sum_{react} \text{N}_i[\bar{h}_f^0 + \Delta \bar{h}]_i \tag{2.40a}$$

$$\begin{aligned} Q = &\Big\{ N_{CO_2}[\bar{h}_f^0 + \Delta \bar{h}]_{CO_2} + N_{H_2O}[\bar{h}_f^0 + \Delta \bar{h}]_{H_2O} \\ &+ N_{N_2}[\bar{h}_f^0 + \Delta \bar{h}]_{N_2} \Big\}_{prod} - \Big\{ - \Big\} N_{fuel}[\bar{h}_f^0 + \Delta \bar{h}]_{fuel} \\ &+ N_{O_2}[\bar{h}_f^0 + \Delta \bar{h}]_{O_2} + N_{N_2}[\bar{h}_f^0 + \Delta \bar{h}]_{N_2} \Big\}_{react} \end{aligned}$$

where

$$N_{i,j} = \text{determined from the stoichiometric equation}$$

$$\Delta \bar{h}_{i,j} = \text{O (reaction at } STP)$$

$$Q_{STP} = N_{CO_2}[\bar{h}_f^0]_{CO_2} + N_{H_2O}[\bar{h}_f^0]_{H_2O} - N_{fuel}[\bar{h}_f^0] = \Delta H_c \tag{2.42}$$

EXAMPLE 2.5 Carbon monoxide, CO, is an important product, resulting from incomplete combustion of many hydrocarbon fuels. Consider the stoichiometric reaction of carbon monoxide with oxygen. Determine (a) the constant-pressure heat of combustion of carbon monoxide at *STP*, Btu/lbm CO; (b) the constant-volume heat of combustion, Btu/lbm CO; and (c) the constant-pressure heat of combustion at 1800°R and 14.7 psia, Btu/lbm CO.

Solution:

1. Stoichiometric equation:

$$CO + aO_2 \overset{STP}{\rightarrow} bCO_2$$

Carbon atom balance: $b = 1$

Oxygen atom balance: $1 + 2a = 2b \qquad a = 0.5$

$$CO + \tfrac{1}{2}O_2 \rightarrow CO_2$$

2. Energy equation: $P = c \ (STP)$

$$\cancel{dE} = \delta Q - \cancel{\delta W} + \Sigma N_i \bar{e}_i - \Sigma N_j \bar{e}_j$$

$$Q = \sum_{\text{prod}} N_i[\bar{h}_f^0 + \{\bar{h}<T> - \bar{h}<T_0>\}]_i - \sum_{\text{react}} N_j[\bar{h}_f^0 + \{\bar{h}<T> - \bar{h}<T_0>\}]_j$$

$$Q = 1.0[\bar{h}_f^0 + \cancel{\Delta\bar{h}}]_{CO_2} - 0.5[\bar{h}_f^0 + \cancel{\Delta\bar{h}}]_{O_2} - 1.0[\bar{h}_f^0 + \Delta\bar{h}]_{CO}$$

$$Q = \left[-94{,}054 + 26416 \text{ cal/g mole}\right]\left[1.8001 \frac{\text{Btu/lb mole}}{\text{cal/g mole}}\right]$$

$$= -121{,}755 \text{ Btu/lb mole CO}$$

or

$$\text{a. } Q = \frac{-121{,}755 \text{ Btu/lb mole}}{28 \text{ lbm/lb mole}} = -4348.4 \text{ Btu/lbm CO}$$

3. Energy equation: $V = c \quad (STP)$

$$Q = \sum_{\text{prod}} N_i[\bar{h}_f^0 + \{\bar{h}<T> - \bar{h}<T_0>\} - \bar{R}T]_i - \sum_{\text{react}} N_j[\bar{h}_f^0 + \{\bar{h}<T> - \bar{h}<T_0>\} - \bar{R}T]_j$$

$$Q = 1.0[\bar{h}_f^0 + \cancel{\Delta\bar{h}}]_{CO_2} - 0.5[\cancel{\bar{h}_f^0} + \cancel{\Delta\bar{h}}]_{O_2} - 1.0[\bar{h}_f^0 + \cancel{\Delta\bar{h}}]_{CO} - (1.0 - 0.5 - 1.0)\,\bar{R}T$$

$$= -121{,}755 + (0.5)(1.987)(537)$$

$$= -121{,}221 \text{ Btu/lb mole CO}$$

$$\text{b. } Q = \frac{-121{,}221}{28} = -4329.3 \text{ Btu/lbm CO}$$

4. Energy equation: $P = c \qquad T = 1800°\text{R}$

$$\begin{aligned} Q &= 1.0[-94{,}054 + 7{,}984]_{CO_2} \\ &\quad - 0.5[0 + 5{,}427]_{O_2} \\ &\quad - 1.0[-26{,}416 + 5183]_{CO} \\ &= -67{,}550.5 \text{ cal/g mole CO} \end{aligned}$$

$$Q = \frac{(-67{,}550.5)(1.8001)}{28}$$

$$\text{c. } Q = -4342.8 \text{ Btu/lbm CO}$$

Comments: This problem illustrates that the heating value of a fuel at *STP* in many engineering problems is a good approximation for the heat released by combustion at conditions other than at *STP.*

Since formation and combustion reactions both occur at *STP,* one can simply add the appropriate heats of formation via stoichiometric equations to determine heats of combustion. Note also that addition of N_2 or excess O_2 does not change the value of heat of combustion since both those compounds are elemental species. The heat of combustion ΔH_c can be visualized on an $H-T$ diagram as the vertical distance at $T = T_0$ between the reactant mixture curve and the product curve; see Figure 2.9. The heat of combustion at *STP* for various species is found in Appendix B.

Finally, consider the energy balance for control volume analysis of a burner; see Figure 2.10. For the purposes of analysis, it will be assumed that the burner system operates at steady-state, steady-flow conditions and that, in this case, no work or power transfer occurs. Also assuming negligible changes in kinetic and potential energy fluxes, one can write the energy balance for the burner as

$$\cancel{\frac{dE}{dt}} = \frac{\delta Q}{dt} - \cancel{\frac{\delta W}{dt}} + \sum_{i_{in}} \frac{dN_i}{dt} \left[\bar{h}_{tot}<T_i> \right] - \sum_{j_{out}} \frac{dN_j}{dt} \left[\bar{h}_{tot}<T_j> \right] \tag{2.43}$$

$$\begin{aligned} \dot{Q} &= \sum_{j_{out}} \dot{N}_j [\bar{h}_f^0 + \{\bar{h}<T_j> - \bar{h}<T_0>\}]_j \\ &\quad - \sum_{i_{in}} \dot{N}_i [\bar{h}_f^0 + \{\bar{h}<T_i> - \bar{h}<T_0>\}]_i \end{aligned} \tag{2.44}$$

Energy transfer *from* the gases as heat due to combustion is influenced by (1) particular fuel ($\bar{h}_{f_i}^0$), (2) particular *AF* ratio (N_i and N_j), (3) reactant state (fuel and oxidant temperatures), and (4) product state (completeness of combustion and temperature).

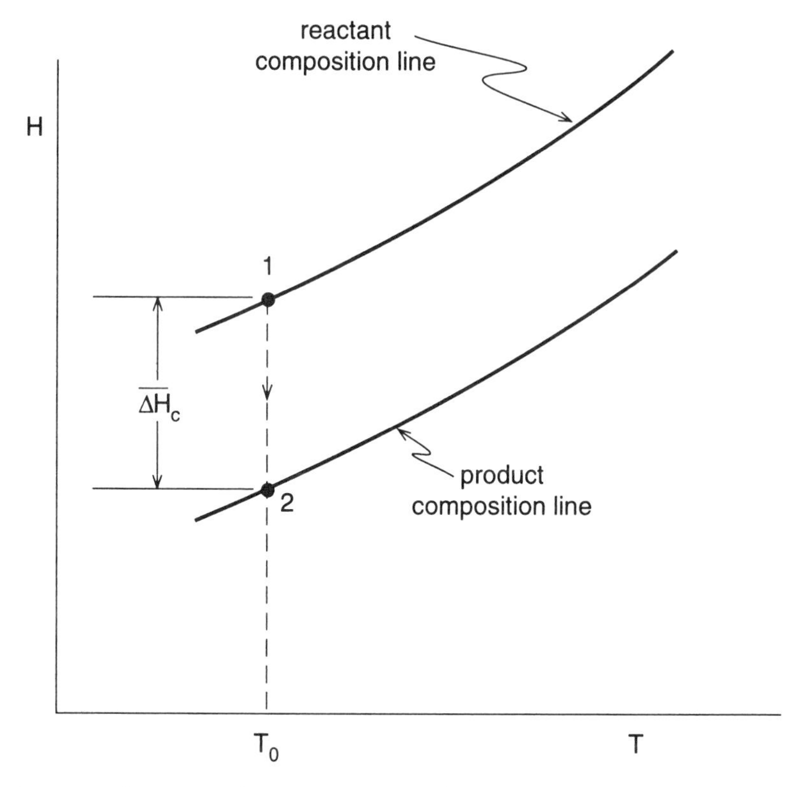

Figure 2.9 Heat of combustion H-T diagram.

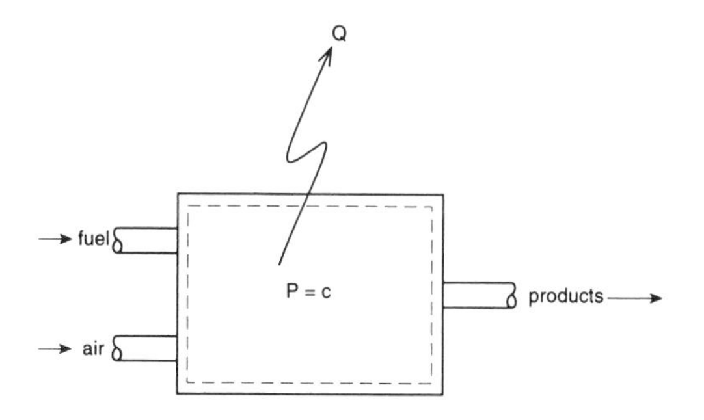

Figure 2.10 Combustor or burner schematic.

EXAMPLE 2.6 A rigid vessel filled with diatomic hydrogen and air is initially at 25°C and 1 atm total pressure. Calculate the adiabatic flame temperature for ideal complete combustion for an equivalence ratio of (a) 1.0, (b) 1.4, (c) 1.8, (d) 0.6, and (e) 0.2.

Solution:

1. Stoichiometric equation:

$$H_2 + 0.5[O_2 + 3.76N_2] \xrightarrow{V} H_2O + 1.88N_2$$

2. Equivalence ratio:

$$\Phi = \frac{FA)_{\text{actual}}}{FA)_{\text{stoich}}}$$

Assume that the moles of air remain constant and vary the moles of hydrogen, giving

$$\Phi\, H_2 + 0.5[O_2 + 3.76\, N_2] \xrightarrow{V} \text{products}$$

3. Rich combustion:

$$\Phi > 1.0 \qquad\qquad (\text{excess } H_2)$$

$$\Phi H_2 + 0.5[O_2 + 3.76\, N_2] \rightarrow aH_2 + bH_2O + 1.88N_2$$

$$\text{Hydrogen atom balance} \qquad \Phi = a + b$$

$$\text{Oxygen atom balance} \qquad 1 = b$$

and

Φ	a	b
1.0	0	1.0
1.4	0.4	1.0
1.8	0.8	1.0

4. Energy balance:

$$\delta Q - \delta W = \cancel{\delta Q}^{\,0} - \cancel{P\,dV}^{\,0} = dU$$

or

$$U<T_2> = U<T_1>$$

$$\sum_{\text{prod}} N_i\{\bar{h}^0_{f_i} + [\bar{h}<T_2> - \bar{h}<T_0>]_i - \bar{R}T_2\}$$
$$= \sum_{\text{react}} N_j\{\bar{h}^0_{f_j} + [\bar{h}<T_1> - \bar{h}<T_0>]_j - \bar{R}T_1\}$$

5. Energy balance—rich combustion:

$$a[\cancel{\bar{h}_f^0} + \Delta\bar{h}]_{H_2} + [\bar{h}_f^0 + \Delta\bar{h}]_{H_2O} + 1.88[\cancel{\bar{h}_f^0} + \Delta\bar{h}]_{N_2}$$
$$- (2.88 + a)\,\bar{R}T_2 = \Phi[\cancel{\bar{h}_f^0} + \cancel{\Delta\bar{h}}]_{H_2}$$
$$+ 2.38[\cancel{\bar{h}_f^0} + \cancel{\Delta\bar{h}}]_{air} - (\Phi + 2.38)\,\bar{R}T_1$$

where $\left.\bar{h}_f^0\right)_{H_2O_g} = -57{,}798$ cal/g mole (Appendix B)

6. For $\Phi = 1.0$, using values for $\Delta\bar{h}_{(j)}$ from Appendix B;

$$\Delta\bar{h})_{H_2O} + 1.88\Delta\bar{h})_{N_2} - 5.723\ T_2 = 55{,}797$$

by trial and error,

$$T_2 \simeq 3045\ \text{K}$$

For $\Phi = 1.4$,

$$0.4\Delta\bar{h})_{H_2} + \Delta\bar{h})_{H_2O} + 1.88\overline{\Delta h})_{N_2} - 6.517\ T_2 = 55{,}560$$

by trial and error,

$$T_2 \simeq 2805\ \text{K}$$

For $\Phi = 1.8$,

$$0.8\overline{\Delta h})_{H_2} + \Delta\bar{h})_{H_2O} + 1.88\overline{\Delta h})_{N_2} - 7.312\ T_2 = 55{,}320$$

by trial and error

$$T_2 \simeq 2605\ \text{K}$$

7. Lean combustion: $\Phi < 1.0$ (excess air)

$$\Phi H_2 + 0.5[O_2 + 3.76\ N_2] \xrightarrow{V} cO_2 + dH_2O + 1.88\ N_2$$

Hydrogen atom balance $\Phi = d$

Oxygen atom balance $1 = 2c + d$

Φ	c	d
0.6	0.2	0.6
0.2	0.4	0.2

8. Energy balance—lean combustion:

$$c[\bar{h}_f^0 + \Delta\bar{h}]_{O_2} + d[\bar{h}_f^0 + \Delta\bar{h}]_{H_2O} + 1.88[\bar{h}_f^0 + \Delta\bar{h}]_{N_2}$$
$$- (c + d + 1.88)\bar{R}T_2 = \Phi[\bar{h}_f^0 + \Delta\bar{h}]_{H_2} + 2.38[\bar{h}_f^0 + \Delta\bar{h}]_{air}$$
$$- [\Phi + 2.38]\bar{R}T_1$$

9. For $\Phi = 0.6$,

$$0.2\Delta\bar{h})_{O_2} + 0.6\Delta\bar{h})_{H_2O} + 1.88\Delta\bar{h})_{N_2} - 5.325\ T_2 = 32{,}914$$

by trial and error,

$$T_2 \cong 2240\ \text{K}$$

For $\Phi = 0.2$,

$$0.4\Delta\bar{h})_{O_2} + 0.2\Delta\bar{h})_{H_2O} + 1.88\Delta\bar{h})_{N_2} - 4.928\ T_2 = 10{,}030$$

by trial and error

$$T_2 \cong 800\ \text{K}$$

Comments: This problem illustrates the fact that, for most oxidation reactions, the peak flame temperature occurs at or near stoichiometric conditions. In practice, the maximum adiabatic flame temperature may occur just on the rich side of stoichiometric proportions. This slight shift can be attributed to the reduction in specific heat of the rich combustion products with a corresponding increase in temperature. Also, in actual flame systems, incomplete combustion and heat loss to surroundings will reduce the temperature, but the above calculated trend of temperature versus Φ will still hold.

The general combustion process can also be visualized by use of the $H-T$ diagram; see Figure 2.11. Any actual heat release can be modeled using three fictitious processes. First, if reactant conditions are offset from T_0 on the reactant line, an amount of energy is needed to "cool" reactants to *STP* conditions, process $1 \rightarrow a$. Next, the vertical path $a \rightarrow b$ represents the heat of combustion in going from reactant to products. Finally, the product state 2 indicates that an additional amount of energy is needed to heat the product mixture from *STP* to the final state, process $b \rightarrow 2$.

A *combustion efficiency* η_c can be defined in similar fashion to that done in thermodynamics for turbines and compressors; see Figure 2.12, as

$$\eta_c = \frac{\text{actual heat released}}{\text{ideal heat released}} = \frac{\Delta H}{\Delta H_c} \leq 1.0 \tag{2.45}$$

For heat-transfer applications, such as in furnaces, heaters, and boilers, high efficiency requires that the devices result in product temperatures approaching T_0, i.e., large heat transfer *from* the gases; see Figure 2.12. In the limit, the ideal heat release by reaction would equal the heat of combustion. For high-power output applications, such as in internal and external combustion engines, the need to produce a maximum combustion temperature leads to the concept of an *adiabatic flame temperature,* i.e., a combustion process in which there is no heat transfer, i.e., $\Delta H = \text{O}$.

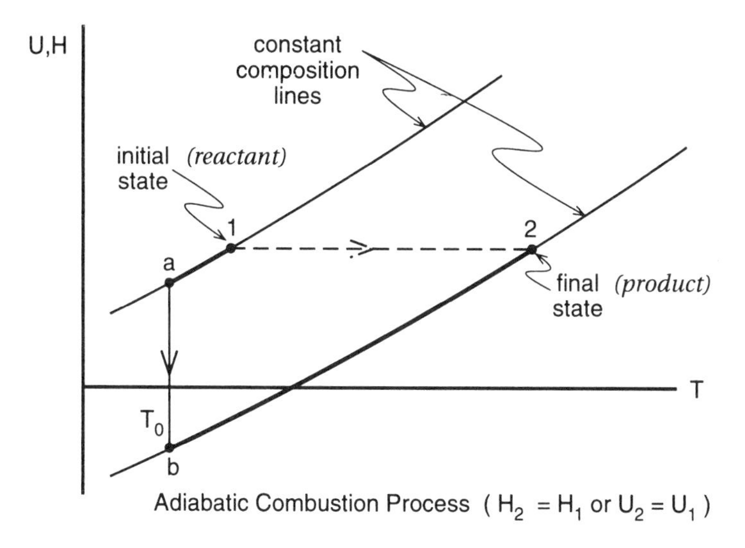

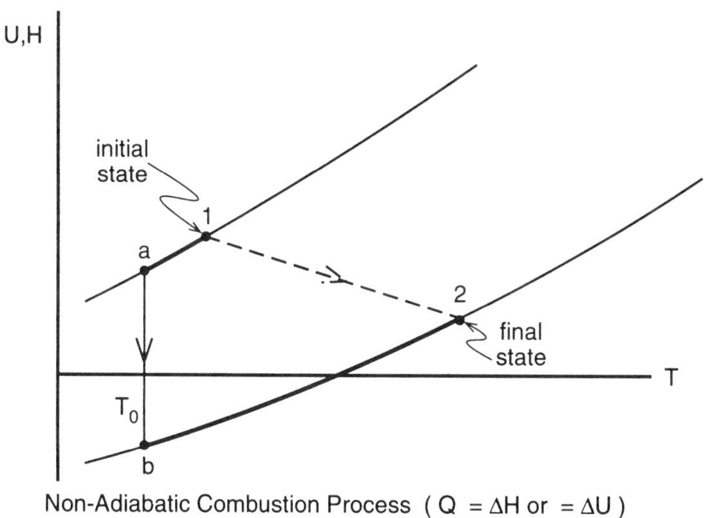

Figure 2.11 *U-T* or *H-T* diagram for chemical reaction.

$$\eta_c = \frac{\overline{\Delta H}}{\overline{\Delta H}_c}$$

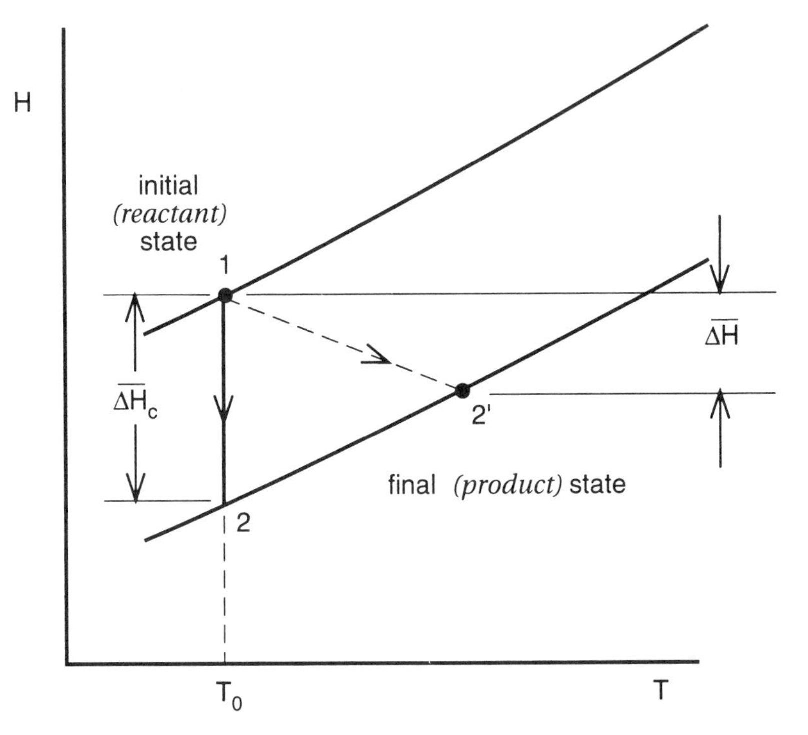

Figure 2.12 H-T diagram and combustion efficiency.

PROBLEMS

2.1 Consider a stoichiometric mixture of hydrogen and air. Calculate for these conditions the (a) reactant mass fractions, *mf;* (b) reactant mixture molecular weight, kg/kg mole; (c) molar $\overline{AF}$ ratio, kg mole air/kg mole fuel; and (d) mass *FA* ratio, kg fuel/kg air.

2.2 Repeat Problem 2.1 for methane, CH_4.

2.3 Repeat Problem 2.1 for propane, C_3H_8.

2.4 Repeat Problem 2.1 for iso-octane, C_8H_{18}.

2.5 Repeat Problem 2.1 for dodecane, $C_{12}H_{26}$.

2.6 Repeat Problem 2.1 for ethanol, C_2H_5OH.

2.7 An unknown hydrocarbon fuel is burned with air. The molar $\overline{AF}$

ratio for the combustion process is equal to 21.43 lbm air/lbm fuel, while the corresponding mass *FA* ratio is given as 0.0451 lbm fuel/lbm air. Calculate (a) fuel molecular weight, lbm/lb mole; (b) fuel structure $CxHy$, %; and (c) reaction excess air, % for the reaction.

2.8 A gaseous fuel having a volumetric analysis of 65% CH_4, 25% C_2H_6, 5% CO, and 5% N_2 is burned with 30% excess air. Determine (a) mass *AF* ratio, lbm air/lbm fuel; (b) mass of CO_2 produced, lbm CO_2/lbm fuel; (c) mass of water formed, lbm H_2O/lbm fuel; and (d) mass of products formed, lbm products/lbm reactants.

2.9 The dry products of combustion have the following molar percentages:

CO 2.7% O_2 5.3% H_2 0.9%
CO_2 16.3% N_2 74.8%

Find, for these conditions: (a) mixture gravimetric analysis; (b) mixture molecular weight, lbm/lbm mole; and (c) mixture specific gas constant *R*, ft lbf/lbm °R.

2.10 A hydrocarbon fuel, $CxHy$, is burned at constant pressure in air. The volumetric analysis of the products of combustion on a dry basis are:

CO 1.1% 0_2 8.3%
CO_2 7.8% N_2 82.8%

Determine (a) the molar carbon to hydrogen ratio for the fuel; (b) fuel mass fraction composition, %C and %H; (c) theoretical air, %; (d) mass *AF* ratio, kg air/kg fuel; and (e) dew point temperature for 1 atmospheric combustion, °C.

2.11 A homogeneous mixture of methane, CH_4, and air at *STP* has a specific heat ratio of 1.35. Obtain for this mixture (a) the reactant mixture mole fractions, %; (b) the reactant mixture molecular weight kg/kg mole; (c) the mixture constant-pressure specific heat, kJ/kg K; (d) the mixture constant volume specific heat, kJ/kg K; and (e) the mixture density, kg/m^3.

2.12 Using the JANAF data for O_2 and N_2 and assuming that air consists of 1 kg mole O_2 and 3.76 kg mole, N_2 find, for the mixture: (a) molar specific heats C_p and C_v, kJ/kg mole K; (b) mass specific heats, kJ/kg K; and (c) specific heat ratio γ for 1 atm and 300K.

2.13 Repeat Problem 2.12 for a temperature of 1000K.

2.14 A fuel composed of 50% C_7H_{14} and 50% C_8H_{18} by weight is burned using 10% excess air. For 100 lbm of fuel, determine: (a) the volume of *STP* air required for ideal combustion, ft^3; and (b) the mass *AF* ratio, lbm air/lbm fuel.

2.15 Consider the reaction of propane, C_3H_8, and 120% theoretical air. Assuming ideal combustion, determine (a) product mole fractions,

%; (b) product mass fractions, %; (c) product mixture molecular weight, kg/kg mole; and (d) product mixture specific gas constant, kJ/kg K.

2.16 For a reactive mixture, the dew point temperature is defined as the saturation temperature of water corresponding to the water vapor partial pressure. Calculate (a) product mole fractions, %; (b) product water vapor partial pressure; and (c) dew point temperature, K, for the ideal constant-volume combustion of acetylene, C_2H_2, and 20% excess air. Assume that the reactants are at 14.5 psi and 77°F, while the products are at 960°R.

2.17 The lower heating value of gaseous butene, C_4H_8, is 19,483 Btu/lbm. For reaction conditions at *STP*, calculate (a) molar enthalpy of formation of butene, Btu/lb mole; (b) mass enthalpy of formation, Btu/lbm; and (c) the enthalpy of hydrogenation associated with the reaction $C_4H_8)g + H_2)g \rightarrow C_4H_{10})g$, Btu/lb mole H_2.

2.18 A propane torch, C_3H_8, burns lean with 120% excess air. Temperature of the flame is 1400°R. Find (a) molar *AF* ratio, lb mole air/lb mole fuel; (b) ideal combustion product mole fractions, %; (c) volume of exhaust gas to volume of combustion air, ft^3 gas/ft^3 air; and (d) heat released by an ideal combustion process, Btu/lbm fuel.

2.19 A rigid vessel contains a mixture of one mole of methane, CH_4, and three moles of oxygen, O_2. The reactants initially are at 14.7 psia and 537°R. After complete combustion, the products are cooled to 1080°R. Find (a) the vessel volume, ft^3; (b) the final pressure, psia; (c) the product dew point temperature, °F; (d) the heat transfer, Btu/ft^3, and (e) the product gravimetric analysis, %.

2.20 An internal combustion engine runs on liquid octane, C_8H_{18}, and 150% theoretical air. The fuel-air mixture enters the engine at steady-state and steady-flow conditions of 25°C and 101 kPa. The heat loss from the engine is equal in magnitude to 20% of the work output. Assuming complete combustion of the mixture, calculate (a) the equivalence ratio; (b) the product gas dew point temperature, °C; (c) the net work, kJ/kg mole of fuel; (d) the engine thermal efficiency, %; and (e) the fuel consumption required to produce 400 kW, kg/sec.

2.21 A furnace burns 2500 ft^3/hr of a gas having the following volumetric analysis: 90% CH_4, 7% C_2H_6, and 3% C_3H_8. Both gas and air are supplied at 25°C and atmospheric pressure. The flue gas in the exhaust has a temperature of 1300K and passes through a stack with an inside diameter of 30 cm. Calculate (a) volumetric analysis of the ideal dry flue gas, %; (b) velocity of stack gas, cm/sec; (c) heat release for ideal combustion, Btu/hr; and (d) combustion efficiency, %.

2.22 Calculate the (a) adiabatic flame temperature, K; (b) total moles of products; (c) final total pressure, kPa; and (d) product dew point temperature, °C, for the constant-volume combustion of methane, CH_4, and oxygen with a 0.75 equivalence ratio. Assume homogeneous gas-phase reactant conditions of 101 kPa and 298K.

2.23 A mixture of hydrogen and air initially at 25°C and 1-atm pressure is burned in a constant-pressure adiabatic process in which the adiabatic flame temperature is 2800K. Determine (a) volume of products to the volume of reactants; (b) molar *AF* ratio, kg mole air/kg mole fuel; (c) excess air required, %; and (d) equivalence ratio for the process. Assume complete combustion in all calculations.

2.24 Dodecane, $C_{12}H_{26}$, and 120% theoretical air are supplied to a constant-pressure atmospheric combustor. Fuel is supplied at 77°F, and the product adiabatic flame temperature is 4500°R. Calculate (a) required air preheat temperature, °R; (b) heat release, Btu/lbm fuel; and (c) dry exhaust gas volumetric flow rate for a fuel mass flow rate of 1 lbm/hr.

3

Combustion and Entropy

3.1 INTRODUCTION

In Chapter 2, rudiments of a general energy analysis for thermochemical reactions were developed. Ideal complete combustion of any particular hydrocarbon fuel-air mixture assumes that only CO_2 and H_2O are produced as products of reaction. In such instances, one would need knowledge of both reactant and product compositions via the appropriate stoichiometric equation. The inability to achieve complete combustion in practice is due directly to both thermochemical and combustion machinery influences and interactions, i.e., the *fuel–engine interface.* Incomplete combustion products, or *pollutants,* generated by internal and external combustion systems such as automotive engines, jet aircraft, fossil-fueled stationary power plants, as well as residential space heating systems, are a direct result of the irreversible nature of high-temperature chemical reactions and their nonequilibrium occurrence within various combustion devices.

A knowledge of effluent exhaust gas composition and an ability to control or limit the concentrations of critical chemical species in such streams are major concerns facing engineers today. In this chapter, the product composition for incomplete combustion will be modeled thermodynamically using the concept of an equilibrium product state. In addition, second law analysis and property entropy will be written for chemically reactive mixtures, and new properties, including the entropy, Gibbs, and Hemholtz functions of reaction, as well as an equilibrium constant, will be presented.

3.2 EQUILIBRIUM AND CHEMICAL REACTIONS

Since actual combustion reactions do not go to completion, it will be useful to develop characteristics that will specify an equilibrium product composition. Recall from classical thermodynamics that an isolated system, one having no mass or energy transfer with its surroundings, is in an equilibrium state if it has time-independent properties and is not capable of any spontaneous change of state. Nature tends spontaneously toward equilibrium, and systems that are not in equilibrium will try to relax to a stable state. Systems relax by energetically interacting with their surroundings, a rate process that is a strong function of both time and space.

Thermodynamic equilibrium implies a lack of certain potential differences, or gradients, which are expressed in terms of local thermodynamic properties. For instance, the absence of force imbalances or pressure gradients implies mechanical equilibrium, a condition that, if not true, would require a work transfer to achieve an equilibrium state (see Figure 3.1). Thermal equilibrium requires that no temperature gradients exist and, hence, no heat transfer. Furthermore, for a pure homogeneous species, phase equilibrium means that no phase change will spontaneously occur when different phases of a pure species are in direct contact, i.e., liquid $\rightleftarrows$

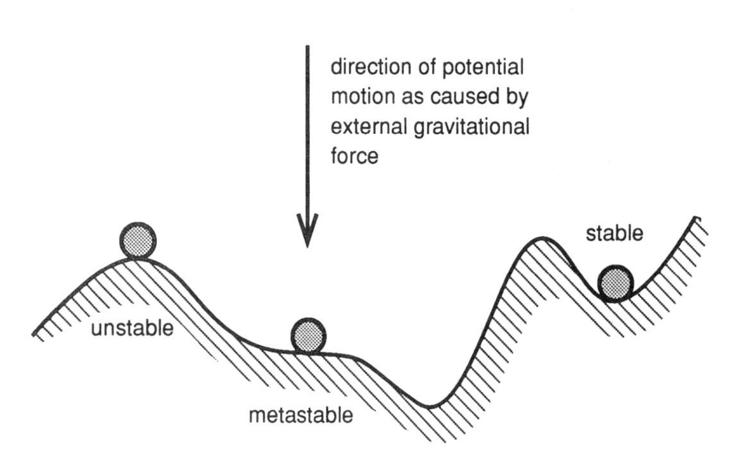

Figure 3.1 Mechanical illustration of processes and equilibrium.

vapor. Finally, chemical equilibrium requires that no spontaneous change in chemical composition occur when different species are present together in a mixture. Full equilibrium for reactive media, such as products of incomplete combustion, would require that all the above criteria be true. Often, an assumption of partial equilibrium can be justified in many engineering calculations in that such an approach yields reasonable results.

In Chapter 2, thermochemical calculations assumed a partial equilibrium approach to combustion that is *frozen* complete combustion product composition.

Frozen thermochemistry:

Reactants in initial *equilibrium* state → *Assumed* products in final state

$$CH_4 + 2O_2 \rightarrow CO_2 + 2H_2O$$

Since the actual path for combustion involves molecular rearrangement of fuel and oxidant atoms to form products, the potential exists for forming a wide variety of stable and unstable products. Since the number of possible product species may exceed the number of independent atom balances that can be written, the equilibrium product state appears indeterminate.

Equilibrium thermochemistry:

Reactants in initial *equilibrium* state	→	*Actual* products in final state

$$CH_4 + 2O_2 \rightarrow aCO_2 + bCO + cC_s + dO_2 + eH_2O + fO_3 + \cdots$$

where

$$a = ?, \qquad b = ?, \qquad c = ?, \cdots$$

To predict an ideal equilibrium product composition for an incomplete combustion case, additional principles beyond that of mass and energy conservation will be needed. In the next section, consequences of the second law of thermodynamics will be explored in regard to chemically reactive systems.

3.3 ENTROPY

Energy conversion processes have both a sense of magnitude and direction. Classical thermodynamics and the first law introduced the concept of energy conservation and the associated property energy. The second law of thermodynamics and its related property, entropy, lead to the concept of a continual entropy increase of the universe, that is, that all real processes must maintain or increase the entropy level of the universe.

The idea of perpetual motion machines, i.e., devices that violate thermodynamic principles, is first encountered in physics. Some devices, or perpetual motion machines of the first kind, do not conserve energy, while others are classified as perpetual motion machines of the second kind, i.e., entropy violators.

Natural experience dictates that combustion processes are highly irreversible and, therefore, must conserve energy but increase the total entropy of the universe. A process is assumed to be *reversible* if it can be retraced in such a way that the complete universe can be restored to its original condition without increasing the entropy of the universe. The Clausius inequality expresses the differential charge in entropy for a closed system in terms of a corresponding heat transfer as

$$T\,dS \geq \delta Q \qquad \text{kJ (Btu)} \tag{3.1}$$

which, for reversible heat transfer, becomes

$$T\,dS = \delta Q_{\text{rev}} \tag{3.1a}$$

and, for irreversible heat transfer, yields

$$T\,dS > \delta Q_{\text{irrev}} \tag{3.1b}$$

Consider the closed system shown in Figure 3.2 undergoing a general but differential change of state. From conservation of energy principles and applying Equation (2.29), the change in internal energy, for negligible kinetic and potential energy changes, must be balanced by required heat and boundary expansion work transfers as

$$\delta Q = dU + P\,dV \tag{2.29}$$

If the process is assumed to be reversible, Equations (3.1a) and (2.29) can be combined to give one important $T\,dS$ equation,

$$T\,dS = dU + P\,dV \tag{3.2a}$$

or, rearranging and noting for an ideal gas $PV = mRT$,

$$dS = \frac{dU}{T} + \frac{R}{V}\,dV \qquad \text{kJ/K (Btu/°R)} \tag{3.2b}$$

Using the definition of *enthalpy,*

$$H \equiv U + PV \tag{3.3}$$

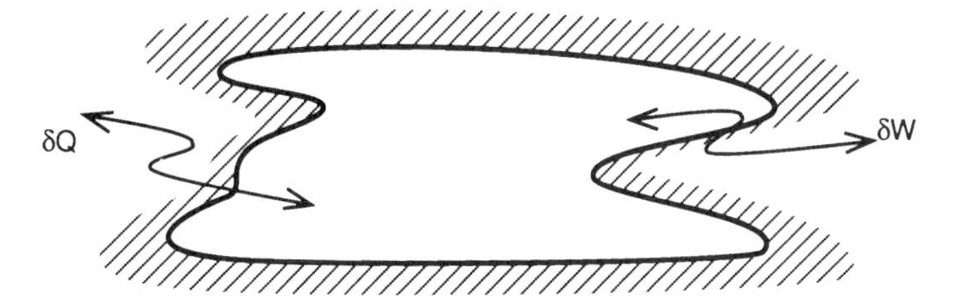

Figure 3.2 Closed system undergoing a differential change of state.

and differentiating gives

$$dH \equiv dU + P\,dV + V\,dP \tag{3.3a}$$

Substituting Equation (3.3a) into Equation (3.2a) yields a second important TdS equation

$$T\,dS = dH - P\,dV - V\,dP + PdV$$
$$T\,dS = dH - V\,dP \tag{3.4a}$$

Rearranging and assuming that the substance obeys the ideal-gas law,

$$dS = \frac{dH}{T} - \frac{R}{P}\,dP \qquad \text{kJ/K (Btu/°R)} \tag{3.4b}$$

Equations (3.2b) and (3.4b) are differential equations that can be used to evaluate entropy changes for a pure component species as a function of V and T or P and T. Since thermochemical data for pure gases in Appendix B are tabulated in terms of enthalpies, it will be convenient to evaluate entropy in terms of the second of the TdS equations, Equation (3.4b).

Consider evaluation of entropy for an ideal-gas species i having the molar equation of state, $P_i\bar{v}_i = \bar{R}T$, undergoing a differential change of state. Equation (3.4b) can be expressed in molar form as

$$d\bar{s}_i = \frac{d\bar{h}_i}{T} - \bar{R}\frac{dP_i}{P_i} \qquad \frac{\text{kJ}}{\text{kg mole K}}\left(\frac{\text{Btu}}{\text{lb mole °R}}\right) \tag{3.5}$$

Recall that, for a pure gaseous species undergoing a process without composition change, enthalpy changes are only functions of temperature or

$$d\bar{h}_i = d[\bar{h}_f^0 + \bar{h}<T> - \bar{h}<T^0>] = d\bar{h}<T>$$

and

$$d\bar{h}_i = \bar{C}_{p_i}<T>dT \tag{2.36}$$

For a nonreactive ideal-gas species, entropy changes can now be evaluated by substituting Equation (2.36) into Equation (3.5), giving

$$d\bar{s}_i\langle T\rangle = \frac{\overline{C}_{p_i}\langle T\rangle\, dT}{T} - \overline{R}\,\frac{dP_i\langle T, P\rangle}{P_i} \tag{3.6}$$

Entropy changes are evaluated with respect to the reference datum as was done for $\bar{u}$ and $\bar{h}$ in Chapter 2 as

$$\Delta\bar{s}_i = \int_{T_0}^{T} \frac{\overline{C}_{p_i}\langle T\rangle}{T}\, dT - \overline{R}\int_{P_0}^{P} \frac{dP_i}{P_i} \tag{3.7}$$

where, for entropy,

P_i = compound *partial* pressure $< P_{tot}$
P_0 = reference datum pressure = 1 atm.
T_0 = reference datum temperature = 25°C (77°F)

The first term in Equation (3.7) is defined as the *absolute standard state entropy of species i,* $\bar{s}_i^0\langle T\rangle$ and is evaluated as a function of temperature as

$$\bar{s}_i^0\langle T\rangle \equiv \int_{T_0}^{T} \frac{\overline{C}_{p_i}\langle T\rangle}{T}\, dT \tag{3.8}$$

Recall from Chapter 2 that molar specific heat $\overline{C}_{p_i}$ of a pure species is temperature-dependent, and this relation was represented in virial form in Equation (2.22) as

$$\overline{C}_{p_i}\langle T\rangle = a_1 + a_2T + a_3T^2 + a_4T^3 + a_5T^4 \cdots \tag{2.22}$$

Substituting Equation (2.22) into Equation (3.8) gives an expression for absolute standard state entropy of species as

$$\bar{s}_i^0\langle T\rangle = \int_{T_0}^{T} [a_1 + a_2T + a_3T^2 + a_4T^3 + a_5T^4 \cdots]\frac{dT}{T} \tag{3.8a}$$

and integrating yields

$$\bar{s}_i^0\langle T\rangle = a_1 \ln T + a_2T + \frac{a_3}{2}T^2 + \frac{a_4}{3}T^3 + \frac{a_5}{4}T^4 + \cdots + c_1 \tag{3.9}$$

Equation (3.9) is constrained to satisfy the requirement in keeping with the third law of thermodynamics that entropy at absolute zero must be zero, i.e., $\bar{s}_i^0\langle 0\rangle = 0$. Tabulated values of $\bar{s}_i^0\langle T\rangle$ for several gases can be found in Appendix B.

A general expression for changes in entropy for an ideal-gas species can now be evaluated by combining Equations (3.9) and (3.7), giving

$$\Delta \bar{s}_i = \Delta \bar{s}_i^0 - \bar{R} \ln \left(\frac{P_i}{P_0} \right) \frac{\text{kJ}}{\text{kg mole K}} \left(\frac{\text{Btu}}{\text{lb mole °R}} \right) \tag{3.9a}$$

Recall from Chapter 2 that for a closed system, reversible boundary expansion work can be evaluated by integrating Equation (2.28) or

$$\delta \bar{w}_{\text{rev}} = P<\bar{v}>d\bar{v} \tag{2.28}$$

A P-$\bar{v}$ diagram can be used in such instances to visualize the process, and the area under the reversible process line is equal to work transfer.

Equation (3.1a) can be used to determine reversible heat transfer for a closed system as

$$\delta \bar{q}_{\text{rev}} = T<\bar{s}>d\bar{s} \tag{3.1a}$$

and, furthermore, this relationship suggests an additional two-dimensional thermodynamic coordinate system, the T-$\bar{s}$ diagram. The T-$\bar{s}$ diagram can be used to visualize reversible heat transfer associated with a closed system. Figure 3.3 illustrates how several process lines for ideal gases appear on both the P-$\bar{v}$ and T-$\bar{s}$ planes.

Equations (3.2b) and (3.4b) also can be used to predict the shape of particular process lines on a T-$\bar{s}$ diagram. For example, a constant-pressure process diagram for a gas can be evaluated using Equation (3.4b), written in molar form as in Equation (3.6), as

$$d\bar{s}_i = \frac{\bar{C}_{p_i}<T>dT}{T} - \bar{R}\frac{dP_i}{P_i} = \frac{\bar{C}_{p_i}<T>\,dT}{T} \tag{3.6a}$$

Rearrangement yields an expression for the slope of a constant-pressure process curve as

$$\left.\frac{dT}{d\bar{s}_i}\right)_P = + \frac{T}{\bar{C}_{p_i}<T>} > 0 \tag{3.10}$$

Using Equation (3.2b) and similar arguments, the slope of a constant-volume process line on a T-$\bar{s}_i$ diagram for a gas having a constant molar specific heat $\bar{C}_{v_i}$ can be shown to equal

$$\left.\frac{dT}{d\bar{s}_i}\right)_{\bar{v}} = + \frac{T}{\bar{C}_{v_i}<T>} > 0 \tag{3.11}$$

Since $\bar{C}_p - \bar{C}_v = \bar{R}$, then $\bar{C}_p > \bar{C}_v$ and, at a particular temperature, the slope of a constant-pressure process line will be greater than the corresponding

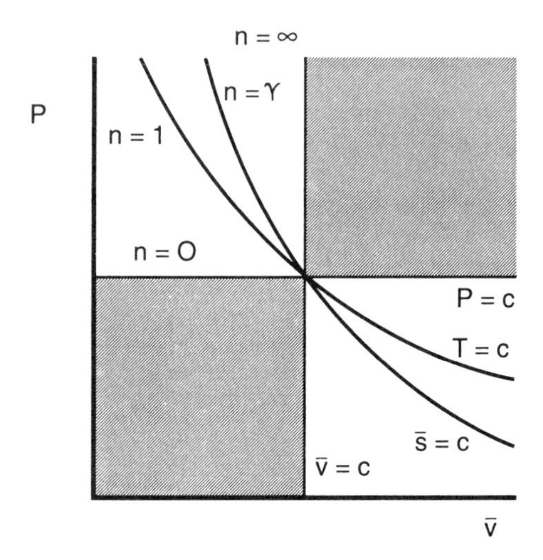

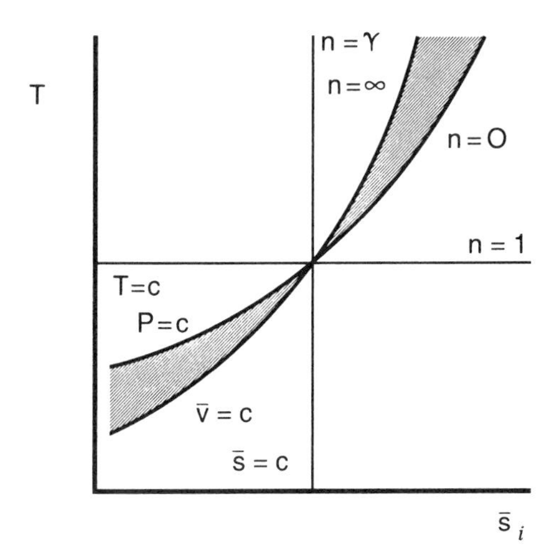

Figure 3.3 $P-\bar{v}$ and $T-\bar{s}$ ideal gas polytropic process diagrams.

constant-volume process line; see Figure 3.3. Furthermore, since specific heats of gases in general are temperature-dependent, the actual $P = c$ and/or $V = c$ process lines may differ from the constant molar specific heat case.

EXAMPLE 3.1 The reversible adiabatic, or *isentropic,* process is an important idealization utilized in many combustion calculations. Consider an ideal gas with constant specific heats. For a closed system, use the polytropic equation

$$Pv^n = \text{const}$$

and (a) find an expression for n in terms of $\overline{C}_p$ and $\overline{C}_v$, (b) express the change in pressure in terms of temperature and n, (c) express the change in volume in terms of temperature and n, and d) calculate n for air at 298 and 1200K.

Solution

1. Boundary expansion work:

$$ {}_1w_2 = \int_1^2 P\,dv = \int_1^2 \left(\frac{c}{v^n}\right) dv = c \int_1^2 v^{-n}\,dv $$

$$ = \frac{c}{1-n}(v_2^{\,1-n} - v_1^{\,1-n}) = \frac{Pv^n}{1-n}\,(v_2^{\,1-n} - v_1^{\,1-n}) $$

$$ {}_1w_2 = \frac{1}{1-n}(P_2v_2 - P_1v_1) = \frac{R}{1-n}(T_2 - T_1) $$

2. Energy balance: $\delta Q - \delta W = dU$

$$ -\delta w = C_v<T>\,dT = C_v\,dT $$

$$ -{}_1w_2 = C_v[T_2 - T_1] $$

3. Combining items 1 and 2,

$$ \left(\frac{-R}{1-n}\right)(T_2 - T_1) = C_v(T_2 - T_1) $$

$$ R = C_v(n-1) $$

$$ n - 1 = \frac{R}{C_v} = \frac{C_p - C_v}{C_v} = \frac{C_p}{C_v} - 1 $$

$$ n - 1 = \gamma - 1 $$

a. $n = \gamma$

4. Isentropic process:

$$P_1 v_1^{\gamma} = P_2 v_2^{\gamma}$$

$$P_1 \left(\frac{RT_1}{P_1} \right)^{\gamma} = P_2 \left(\frac{RT_2}{P_2} \right)^{\gamma}$$

$$\left(\frac{P_1}{P_2} \right)^{1-\gamma} = \left(\frac{T_2}{T_1} \right)^{\gamma}$$

or

$$\text{b. } \frac{P_2}{P_1} = \left(\frac{T_2}{T_1} \right)^{\gamma/(\gamma-1)}$$

5. Isentropic process:

$$P_1 v_1^{\gamma} = P_2 v_2^{\gamma}$$

$$\frac{RT_1}{v_1} (v_1)^{\gamma} = \frac{RT_2}{v_2} (v_2)^{\gamma}$$

$$\left(\frac{v_1}{v_2} \right)^{\gamma-1} = \left(\frac{T_2}{T_1} \right)$$

$$\text{c. } \frac{v_2}{v_1} = \left(\frac{T_2}{T_1} \right)^{1-\gamma}$$

6. From Appendix B for air,

$$\overline{C}_p<298\text{K}> = 6.947 \qquad \overline{C}_p<1200\text{K}> = 8.109$$

where

$$\gamma = \frac{C_p}{C_p - R} = \frac{\overline{C}_p}{\overline{C}_p - \overline{R}}$$

$$\text{d. } \gamma<298\text{K}> = \frac{6.947}{6.947 - 1.987} = 1.40$$

$$\gamma<1200\text{K}> = \frac{8.109}{8.109 - 1.987} = 1.32$$

Using the Gibbs–Dalton law, the entropy of a mixture of ideal gases can be expressed in terms of pure component entropy values as

$$S_{tot} = \sum_{i=1}^{k} m_i s_i \quad \text{kJ/K (Btu/°R)} \tag{3.12}$$

or

$$s_{tot} = \sum_{i=1}^{k} mf_i s_i \quad \text{kJ/kg mixt K (Btu/lbm mixt °R)} \tag{3.12a}$$

and

$$S_{tot} = \sum_{i=1}^{k} N_i \bar{s}_i \quad \text{kJ/K (Btu/°R)} \tag{3.13}$$

or

$$\bar{s}_{tot} = \sum_{i=1}^{k} \bar{x}_i \bar{s}_i \quad \text{kJ/kg mole mixt K (Btu/lb mole mixt °R)} \tag{3.13a}$$

An important process associated with the analysis of many combustion processes and devices is the *isentropic* or constant-entropy process. For a closed system containing an ideal-gas mixture undergoing an isentropic process, it is necessary that 1) no chemical change in composition occur, i.e., N_i = const, and 2) no change in total composition occur, i.e., N_{tot} = const. Thus, for an isentropic process, see equation (3.14) on p. 73.

$$S<T_2> = \Sigma N_i \bar{s}_i<T_2, P_2> = \Sigma N_i \bar{s}_i<T_1, P_1> = S<T_1> \tag{3.14}$$

Expressing the *STP* reference pressure P_0 in units of atmospheres, $P_0 \equiv 1$ atm, and substituting Equation 3.9 yields

$$\Sigma N_i\{\bar{s}_i^0<T_2> - \overline{R}\ 1n(P_{i_2}/P_0)\} = \Sigma N_i\{\bar{s}_i^0<T_1> - \overline{R}\ 1n(P_{i_1}/P_0)\}$$

or

$$\Sigma N_i \overline{R}\{1n(P_{i_2}) - 1n(P_{i_1})\} = \Sigma N_i\{\bar{s}_i^0<T_2> - \bar{s}_i^0<T_1>\} \tag{3.14a}$$

Dividing by N_{tot} and using the definition of partial pressure

$$P_i = \bar{x}_i\ P_{tot} \tag{2.11b}$$

and rearranging gives an isentropic process relationship in terms of pressure and temperature as

$$\begin{aligned}\Sigma \bar{x}_i \overline{R}\{1n(\bar{x}_i P_{tot_2}/\bar{x}_i P_{tot_1}) &= \Sigma \bar{x}_i \overline{R}\ 1n(P_{tot_2}/P_{tot_1}) \\ &= \overline{R}\ 1n(P_{tot_2}/P_{tot_1}) = \Sigma \bar{x}_i\{\bar{s}_i^0<T_2> - \bar{s}_i^0<T_1>\}\end{aligned}$$

or

$$\left(\frac{P_2}{P_1}\right) = \frac{\exp\{\Sigma\bar{x}_i\bar{s}_i^0<T_2>/\overline{R}\}}{\exp\{\Sigma\bar{x}_i\bar{s}_i^0<T_1>/\overline{R}\}} \tag{3.15}$$

Using the ideal-gas equation $P\bar{v} = \overline{R}T$, Equation (3.15) can also be written as

$$\left(\frac{\bar{v}_2}{\bar{v}_1}\right) = \left(\frac{T_2}{T_1}\right) \times \left[\frac{\exp\{\Sigma\bar{x}_i\bar{s}_i^0<T_1>/\overline{R}\}}{\exp\{\Sigma\bar{x}_i\bar{s}_i^0<T_2>/\overline{R}\}}\right] \tag{3.16}$$

For the case of a single species, i.e., $\bar{x}_i = 1.0$ or, if one is able to calculate the expression $\bar{s}^0<T>$ for mixtures using Equation (3.6b), the isentropic process relationship Equation (3.15) reduces to the form

$$\frac{P_2}{P_1} = \frac{\exp(\bar{s}^0<T_2>/\overline{R})}{\exp(\bar{s}^0<T_1>/\overline{R})} \tag{3.17}$$

Since Equation (3.17) is only a function of temperature, it is convenient to define a *relative pressure* P_r as

$$P_r<T> \equiv \exp(\bar{s}^0<T>/\bar{R}) \tag{3.18}$$

which yields the familiar isentropic process equation for variable specific heats,

$$\frac{P_2}{P_1} = \frac{P_r<T_2>}{P_r<T_1>} \tag{3.19}$$

Also, defining a *relative volume* $\bar{v}_r$ as

$$\bar{v}_r \equiv \frac{\overline{R}T}{P_r} \tag{3.20}$$

and substituting into Equation (3.19) gives the alternate isentropic process relation.

$$\frac{v_2}{v_1} = \frac{v_r<T_2>}{v_r<T_1>} \tag{3.21}$$

If one further assumes that $\overline{C}_p$ and $\overline{C}_v$ are temperature-independent or that they may be treated as constants, the above isentropic relations can be written as follows. Combining Equations (3.8) and (3.18),

$$P_r<T> = \exp\{\frac{\overline{C}_p}{\overline{R}} \int_{T_0}^{T} \frac{dT}{T}\} \tag{3.22}$$

$$= \exp\{\frac{\overline{C}_p}{\overline{R}} \ln(T/T_0)\}$$

$$= (T/T_0)^{\overline{C}_p/\overline{R}} \tag{3.22a}$$

Equation (3.19), the isentropic process for the constant specific heat case, then can be expressed as

$$\frac{P_2}{P_1} = \frac{P_r<T_2>}{P_r<T_1>} = \frac{(T_2/T_0)^{\overline{C}_p/\overline{R}}}{(T_1/T_0)} = \left(\frac{T_2}{T_1}\right)^{\overline{C}_p/\overline{R}} \tag{3.23}$$

Recall from Chapter 2 the expression for $\overline{C}_p$ and $\overline{R}$ in terms of γ, where $\gamma = \overline{C}_p/\overline{C}_v$, is equal to

$$\frac{\overline{C}_p}{\overline{R}} = \frac{\gamma}{\gamma - 1} \tag{2.19}$$

and, then,

$$\frac{P_2}{P_1} = \left(\frac{T_2}{T_1}\right)^{\gamma/(\gamma-1)} \tag{3.24}$$

Using the ideal-gas law $P\bar{v} = \overline{R}T$, Equation (3.24) can also be written as

$$\frac{v_2}{v_1} = \left(\frac{T_2}{T_1}\right)^{-1/(\gamma-1)} \tag{3.25}$$

and

$$\frac{P_2}{P_1} = \left(\frac{v_1}{v_2}\right)^{\gamma} \tag{3.26}$$

Table 3.2 summarizes the various isentropic process relations applicable to nonreactive gas mixtures.

EXAMPLE 3.2 As indicated in Example 3.1, air has variable specific heats for the range of temperatures encountered in combustion processes. Using the JANAF air tables found in Appendix B, calculate (a) the relative pressure P_r as a function of temperature and (b) the relative volume v_r as a function of temperature. Find (c) the final temperature of air, initially at 14.7 psi and 77°F when isentropically compressed at 140 psi.

Solution:

1. Relative pressure $P_r<T>$:

$$P_r<T> = \exp\left\{\frac{\bar{s}^0<T>}{\overline{R}}\right\} \tag{3.18}$$

2. Relative volume $v_r<T>$:

$$v_r<T> = \frac{\overline{R}T}{P_r<T>} \tag{3.20}$$

3. From the air data found in Appendix B:

$$\text{at } T = 536°\text{R (298K)}$$
$$\bar{s}^0<T> = 46.255 \text{ cal/g mole K}$$

and

$$P_r<298\text{K}> = \exp\left\{\frac{46.255 \text{ cal/g mole K}}{1.987 \text{ cal/g mole K}}\right\}$$
$$= 1.288 \times 10^{10}$$

with

$$v_r<298\text{K}> = \frac{(82.057 \text{ cc atm/g mole K})(298\text{K})}{1.288 \times 10^{10}}$$
$$= 1.899 \times 10^{-6}$$

Using the data for $\bar{s}^0<T>$, a table can be generated for $P_r<T>$ and $v_r<T>$. Results are given in Table 3.1.

4. Isentropic process—variable specific heats:

$$\frac{P_2}{P_1} = \frac{P_r<T_2>}{P_r<T_1>}, \qquad P_r<T_2> = \left(\frac{140}{14.7}\right) 1.288 = 12.267$$

and by interpolation

T	P_r
900	8.286
992.5	12.267
1080	16.028

c. $T_2 = 992.5°\text{R}$

The Clausius inequality, Equation (3.1), can also be expressed in a rate form as

$$T\frac{dS}{dt} \geq \frac{\delta Q}{dt} \qquad \text{kW}\left(\frac{\text{Btu}}{\text{hr}}\right) \tag{3.27}$$

Table 3.1 Isentropic Properties for Air

T		$\bar{s}^{\circ}\langle T\rangle$	$Pr\langle T\rangle \times 10^{-10}$	$v_r\langle T\rangle \times 10^{10}$
536	298	46.255	1.288	18,990
540	300	46.372	1.366	18,020
720	400	48.378	3.749	8,755
900	500	49.954	8.286	4,952
1080	600	51.265	16.028	3,072
1260	700	52.397	28.334	2,027
1440	800	53.400	46.938	1,399
1620	900	54.305	74.017	997.8
1800	1000	55.129	112.06	732.3
1980	1100	55.887	164.10	550.0
2160	1200	56.588	233.52	421.7
2340	1300	57.241	324.37	328.9
2520	1400	57.852	441.16	260.4
2700	1500	58.427	589.21	208.9
2880	1600	58.969	773.98	169.6
3060	1700	59.482	1002.0	139.2
3240	1800	59.970	1280.9	115.3
3420	1900	60.434	1617.8	96.37
3600	2000	60.877	2021.9	81.17
3780	2100	61.301	2502.8	68.85
3960	2200	61.707	3070.2	58.80
4140	2300	62.097	3736.1	50.52
4320	2400	62.471	4509.8	43.67
4500	2500	62.831	5405.6	37.95
4680	2600	63.178	6437.0	33.14
4860	2700	63.513	7619.1	29.08
5040	2800	63.837	8968.5	25.62
5220	2900	64.150	10,499	22.67
5400	3000	64.454	12,234	20.12
°R	K	cal/g mole K	—	—

Table 3.2 Isentropic Process Relationships

Constant specific heats

$$\frac{P_2}{P_1} = \left(\frac{T_2}{T_1}\right)^{\gamma/(g-1)} \qquad \frac{v_2}{v_1} = \left(\frac{T_2}{\mathrm{T}_1}\right)^{1/(1-\gamma)}$$

Variable specific heats—single species

$$\frac{P_2}{P_1} = \frac{P_r\langle T_2\rangle}{P_r\langle T_1\rangle} \qquad \frac{v_2}{v_1} = \frac{v_r\langle T_2\rangle}{v_r\langle T_1\rangle}$$

Variable specific heats—mixture

$$\left(\frac{P_2}{P_1}\right) = \frac{\exp\{\Sigma \bar{x}_i \bar{s}_i^0\langle T_2\rangle/\bar{R}\}}{\exp\{\Sigma \bar{x}_i \bar{s}_i^0\langle T_1\rangle/\bar{R}\}} \qquad \left(\frac{\bar{v}_2}{\bar{v}_1}\right) = \left(\frac{T_2}{T_1}\right) \times \left[\frac{\exp\{\Sigma \bar{x}_i \bar{s}_i^0\langle T_1\rangle/\bar{R}\}}{\exp\{\Sigma \bar{x}_i \bar{s}_i^0\langle T_2\rangle/\bar{R}\}}\right]$$

or

$$\frac{dS}{dt} \geq \frac{1}{T}\frac{\delta Q}{dt}$$

A relationship, valid for closed systems, i.e., systems of fixed mass or a control mass, can be written using open-system or control-volume parameters as

$$\left.\frac{dS}{dt}\right)_{\text{C.M.}} = \left.\frac{\partial S}{\partial t}\right)_{\text{C.V.}} + \sum_{\text{out}} \dot{m}_j s_j - \sum_{\text{in}} \dot{m}_i s_i \geq \frac{\dot{Q}}{T} \tag{3.28a}$$

or

$$\left.\frac{dS}{dt}\right)_{\text{C.M.}} = \left.\frac{\partial S}{\partial t}\right)_{\text{C.V.}} + \sum_{\text{out}} \dot{N}_j \bar{s}_j - \sum_{\text{in}} \dot{N}_i \bar{s}_i \geq \frac{\dot{Q}}{T} \tag{3.28b}$$

For steady-state, steady-flow conditions and no entropy production within the control volume, Equation (3.28b) reduces to

$$\sum_{\text{out}} \dot{N}_j \bar{s}_j - \sum_{\text{in}} \dot{N}_i \bar{s}_i \geq \frac{\dot{Q}}{T} \quad \text{kW}\left(\frac{\text{Btu}}{\text{hr}}\right) \tag{3.29a}$$

and

$$\sum_{\text{out}} N_j \bar{s}_i - \sum_{\text{in}} N_k \bar{s}_i \geq \frac{Q}{T} \quad \text{kJ (Btu)} \tag{3.29b}$$

For an adiabatic process,

$$\sum_{\text{out}} N_j \bar{s}_j \geq \sum_{\text{in}} N_i \bar{s}_i \tag{3.30}$$

and for reversible adiabatic or isentropic processes for an open system,

$$S_{\text{out}} = S_{\text{in}} \tag{3.31}$$

which, in fact, is identical in form to the previous expressions developed in this section for ideal-gas mixture entropy analysis.

Combustion charts for a unit mass of selected reactant and equilibrium product mixtures are available; see Figures 3.4 and 3.5. Charts for important hydrocarbon fuel–air mixtures graphically show values of temperature, pressure, specific volume, internal energy, enthalpy, and entropy for easy use in engineering calculations. When working with combustion charts, one must

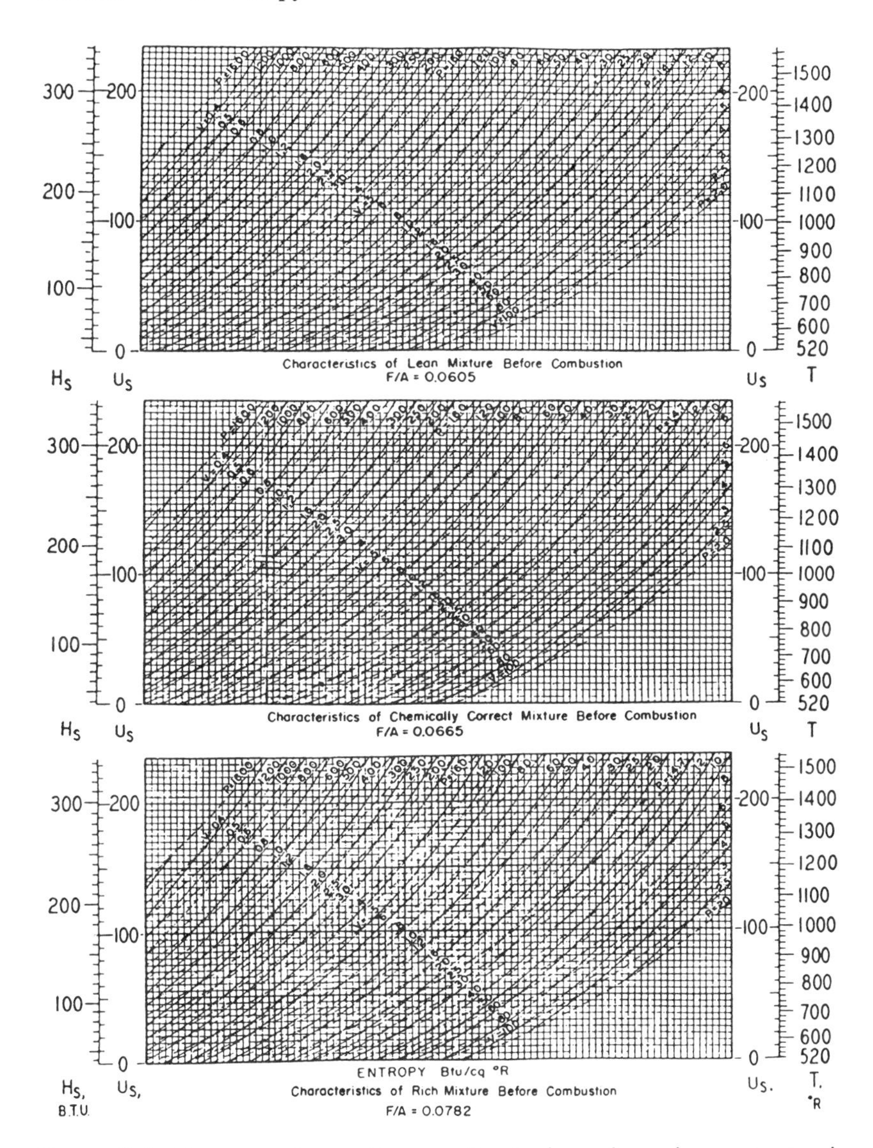

Figure 3.4 Energy-entropy diagram for hydrocarbon-air reactant mixtures.

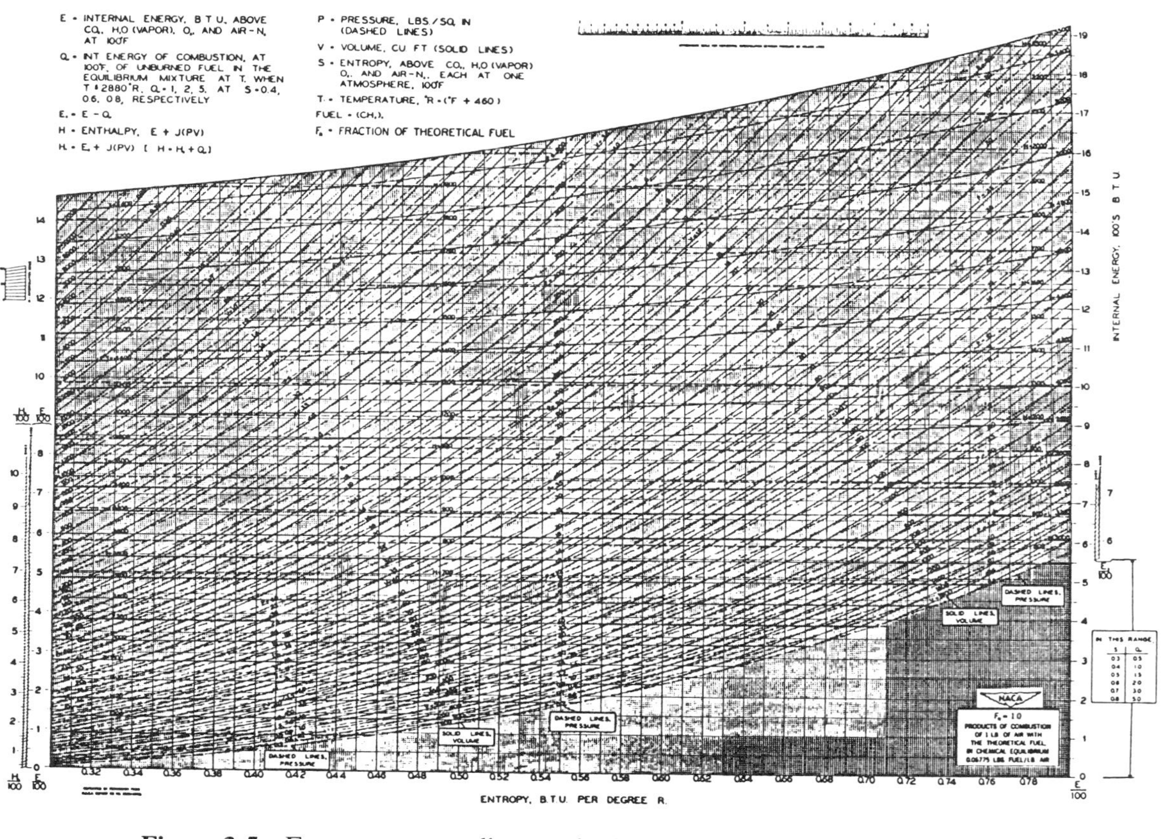

Figure 3.5 Energy-entropy diagram for hydrocarbon-air product mixture.

carefully note the standard state datum used. The Hottel charts, for example, are based on a 1 atm, 520°R datum with zero internal energy and entropy being assigned to CO_2, H_2O, O_2, and N_2. The Starkman–Newhall charts use a 1 atm, 537°R datum with zero energy and entropy assigned to C_{solid}, H_2, O_2, and N_2. Powell selected 1 atm and 0°R as the reference, with zero enthalpy assigned to C_{solid}, H_2, O_2, and N_2 and zero entropy in compliance with the third law. Various processes, i.e., $T = C$, $P = C$, $U = C$, etc., can be drawn on these charts and appropriate thermodynamic parameters can then be read directly instead of being calculated using the above relationships.

Several computer routines, such as the program written by Gordon and McBride at NASA Lewis, now exist and allow routine calculation of changes in energy, enthalpy, entropy, pressure, temperature, and composition for complex mixtures over a much wider range of fuels and compositions than are available in graphical form. This approach will become more prevalent with greater use of the minicomputer in engineering.

3.4 GIBBS AND HEMHOLTZ FUNCTIONS

The actual exhaust gases emitted from an internal or external combustion device are reactive and will not be fully oxidized. Local conditions surrounding the product gases as well as environmental influences, in fact, prevent complete reaction from occurring. Space-time factors such as heat transfer, free expansion, dissociation, and other irreversibilities will influence the reactive mixture. A useful concept for describing an incomplete product state in such instances is that of an exhaust gas mixture equilibrium condition.

Consider a fixed sample of a reactive gas mixture such as the exhaust gases described above. Assume further that this mixture is maintained at a constant temperature and total pressure. If this mixture is not in complete equilibrium, composition adjustments will be necessary to achieve such a state. An energy balance including the potential for boundary expansion work is

$$dU = \delta Q - \delta W = \delta Q - P\,dV \tag{2.29}$$

Using Equation (3.3a) for dH and after rearranging,

$$\delta Q = dH - VdP \tag{3.32}$$

A new and useful property, the *Gibbs function* G, can now be introduced and defined as

$$G = H - TS \qquad \text{kJ (Btu)} \tag{3.33}$$

and

$$g = h - Ts \qquad \text{kJ/kg (Btu/lbm)} \tag{3.33a}$$

or

$$\bar{g} = \bar{h} - T\bar{s} \qquad \text{kJ/kg mole (Btu/lb mole)} \tag{3.33b}$$

In addition, the *Hemholtz function A* can be expressed as

$$A = U - TS \qquad \text{kJ (Btu)} \tag{3.34}$$

and

$$a = u - Ts \qquad \text{kJ/kg (Btu/lbm)} \tag{3.34a}$$

or

$$\bar{a} = \bar{u} - T\bar{s} \qquad \text{kJ/kg mole (Btu/lb mole)} \tag{3.34b}$$

Differentiating Equation (3.33) for the Gibbs function G yields

$$dG = dH - T\,dS - S\,dT \tag{3.35}$$

Substituting Equation (3.32) for dH into Equation (3.35) gives an expression for the differential change in G as

$$dG = \delta Q - V\,dP - T\,dS - s\,dT \tag{3.36}$$

Equation (3.36) suggests that the Gibbs function for a reactive mixture can be changed by changes in T, P, and/or composition. Recall that it was assumed for the reactive mixture that T and P_{tot} were held constant, so that $dT = dP_{\text{tot}} = 0$ or

$$dG_{T,P} = \delta Q - V\cancel{dP} - T\,dS - S\cancel{dT} = \delta Q - T\,dS$$

From the Clausius inequality, Equation (3.1),

$$T\,dS \geq \delta Q \tag{3.1}$$

or

$$T\,dS - \delta Q \geq 0$$

and

$$\delta Q - T\,dS \leq 0$$

Substituting Equation (3.1) into Equation (3.36) gives the result

$$dG_{T,P} \leq 0 \tag{3.37}$$

Figure 3.6 shows a plot of Gibbs function versus changing composition at fixed temperature and total pressure. Equation (3.37) would require the equilibrium point to exist at the minimum of the curve. In other words, for a reversible differential change, i.e., at the equilibrium point, $dG = 0$ and, for all irreversible changes, dG must be <0. Equilibrium composition for a

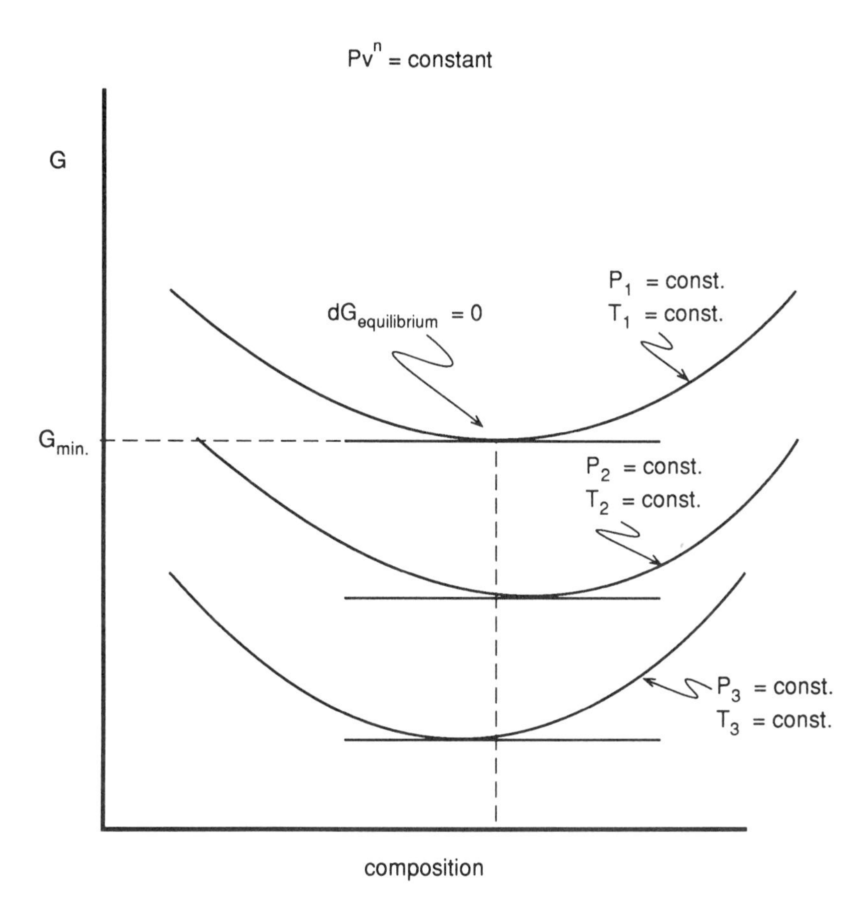

Figure 3.6 Gibbs function for a reactive mixture.

reactive mixture is therefore seen to be that composition which, at a particular temperature and total pressure, defines its minimum Gibbs function.

Consider now a specific sample of a reactive mixture that is maintained at constant temperature and total volume. Differentiating the expression for the Hemholtz function, Equation (3.34) gives

$$dA = dU - T\,dS - S\,dT \tag{3.38}$$

Substituting Equation (2.29) for dU yields

$$dA = \delta Q - P\,dV - T\,dS - S\,dT \tag{3.39}$$

Equation (3.39) indicates that the Hemholtz function for a reactive mixture can change by changing T, V, and/or composition. For a reactive mixture in which it is assumed that T and V are constant, $dT = dV = 0$, or

$$dA_{T,V} = \delta Q - P\,dV - T\,dS - S\,dT = \delta Q - T\,dS \tag{3.40}$$

Again, from the Clausius inequality, Equation (3.1),

$$T\,dS \geq \delta Q$$

or

$$T\,dS - \delta Q \geq 0$$

and

$$\delta Q - T\,dS \leq 0$$

Substituting Equation (3.1) into (3.40) yields the result that, at constant temperature and total volume,

$$dA_{T,V} \leq 0 \tag{3.41}$$

Figure 3.7 shows a plot of Hemholtz function versus changing composition at fixed T and V for a reactive system. Equation (3.41) requires the equilibrium point, similar to the case for the Gibbs function, to occur at a position of minimum Hemholtz value.

EXAMPLE 3.3 Consider the dissociation reaction

$$\tfrac{1}{2}O_2 \xrightarrow{STP} O$$

Using JANAF data, calculate for the reaction at STP: (a) the enthalpy of reaction ΔH_r^0, cal; (b) the entropy of reaction ΔS_r^0, cal/K; (c) the Hemholtz function of reaction ΔA_r^0, cal; and (d) the Gibbs function of reaction ΔG_r^0, cal.

Solution:

1. Enthalpy of reaction at STP

$$\Delta H_r^0<25°C> = \sum_{i\,\text{prod}} N_i[\bar{h}_f^0 + \Delta\bar{h}]_i - \sum_{j\,\text{react}} N_j[\bar{h}_f^0 + \Delta\bar{h}]_j$$

using the data found in Appendix B for $\bar{h}_f^0$.

a. $\Delta H_r^0 = 1[59{,}559 \text{ cal/g mole}]_O - 0.5[0]_{O_2} = 59{,}559$ cal

2. Entropy of reaction at STP

$$\Delta S_r^0<25°C> = \sum_{i\,\text{prod}} N_i[\bar{s}_i^0<T> - \bar{R}\ln(P_i/P_0)]_i - \sum_{j\,\text{react}} N_j[\bar{s}_j^0<T> - \bar{R}\ln(P_j/P_0)]_j$$

using the data found in Appendix B for $\bar{s}^0$.

b. $\Delta S_r^0 = 1[38.468 \text{ cal/g mole K}]_O - 0.5[49.004 \text{ cal/g mole K}]_{O_2}$
$= 13.966 \text{ cal/K}$

3. Hemholtz function of reaction at *STP*:

$$\bar{a}_i = \bar{u}_i - T\bar{s}_i$$

and

$$\bar{u}_i = \bar{h}_i - P\bar{v}_i = \bar{h}_i - \bar{R}T$$

or

$$\bar{a}_i = \bar{h}_i - T(\bar{s}_i + \bar{R})$$

$$\begin{aligned}\Delta A_r^0\langle 25°\text{C}\rangle &= \sum_{i\,\text{prod}} N_i[\bar{h}_{f_i}^0 - T^0(\bar{s}_i^0\langle T^0\rangle + \bar{R})]_i \\ &\quad - \sum_{j\,\text{react}} N_j[\bar{h}_{f_j}^0 - T^0(\bar{s}_i\langle T^0\rangle + \bar{R})]_j \\ &= 1[59{,}559 - (298)(38.468 + 1.987)]_O \\ &\quad - 0.5[0 - (298)(49.004 + 1.987)]_{O_2}\end{aligned}$$

c. $\Delta A_r^0\langle 25°\text{C}\rangle = 55{,}101 \text{ cal}$

4. Gibbs function of reaction at STP:

$$\begin{aligned}\Delta G_r^0\langle 25°\text{C}\rangle &= \sum_{i\,\text{prod}} N_i[\bar{h}_{f_i}^0 - T^0\bar{s}_i^0\langle T^0\rangle]_i \\ &\quad - \sum_{j\text{react}} N_j[\bar{h}_{f_j}^0 - T^0\bar{s}_j^0\langle T^0\rangle]_j \\ &= 1\,[59{,}559 - (298)(38.468)]_O \\ &\quad - 0.5[0 - (298)(49.004)]_{O_2}\end{aligned}$$

d. $\Delta G_r^0\ \langle 25°\text{C}\rangle = 55{,}397 \text{ cal}$

or, alternately,

$$\begin{aligned}\Delta G_r^0\langle 25°\text{C}\rangle &= \Sigma H_r^0\langle 25°\text{C}\rangle - T^0\Delta S_r^0\langle 25°\text{C}\rangle \\ &= 59{,}559 - (298)(13.966)\end{aligned}$$

$$\Delta G_r^0\ \langle 25°\text{C}\rangle = 55{,}397 \text{ cal}$$

3.5 EQUILIBRIUM CONSTANTS

Consider now a reactive gas mixture sample that contains species A, B, . . . , L, and M, all assumed to be in equilibrium at a specified temperature and total pressure but unknown mole fractions. Furthermore, assume

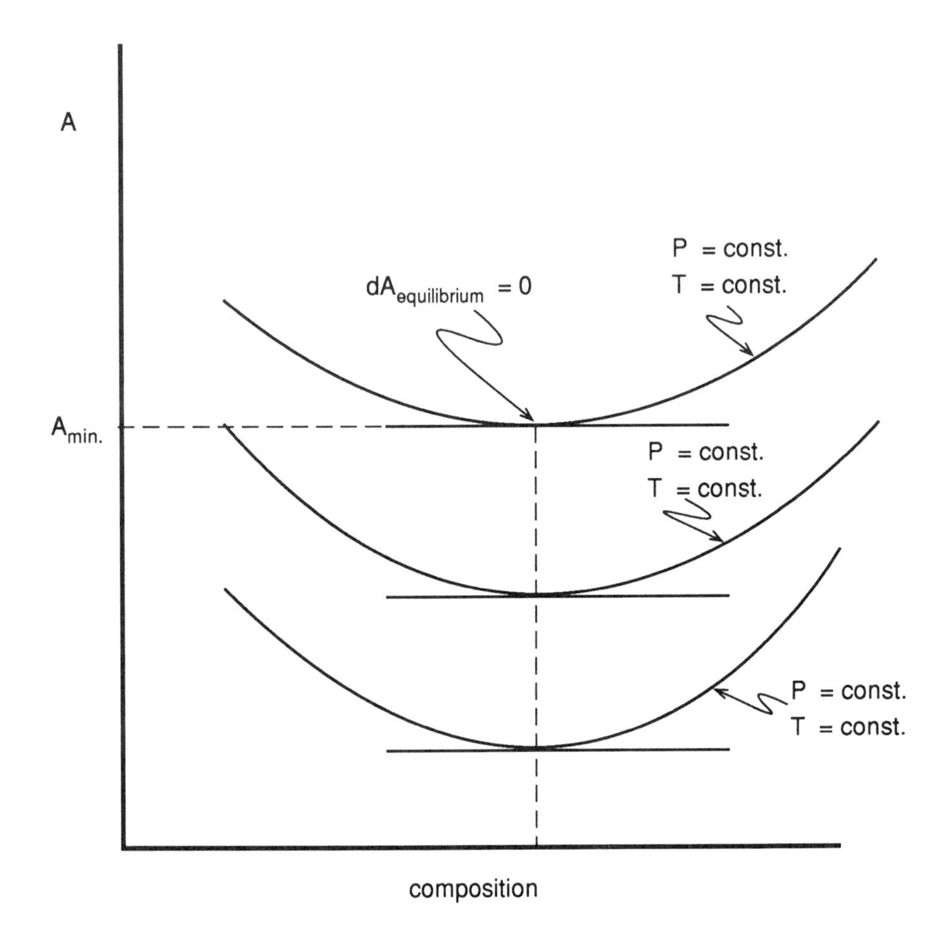

Figure 3.7 Hemholtz function for a reactive mixture.

that these compounds were produced by the following incomplete combustion reaction

$$\text{Reactants} \rightarrow N_1\text{A} + N_2\text{B} + N_3\text{C} + \cdots + N_i\text{L} + N_{i+1}\text{M} \tag{3.42}$$

In Section 3.4, it was shown that, for a reactive gas mixture at equilibrium, the Gibbs function is a minimum. Using the Gibbs–Dalton law for the mixture

$$G_{\text{tot}} = \Sigma N_i \bar{g}_i\langle T,P\rangle \qquad \text{kJ (Btu)} \tag{3.43}$$

Specific molar Gibbs function for a pure species $\bar{g}_i$ can be written as

$$\bar{g}_i<T, P> = \bar{h}_i<T, P> - T\bar{s}_i<T, P> \qquad (3.33b)$$

while specific molar enthalpy $\bar{h}_i$ was expressed in Chapter 2 as

$$\bar{h}_i<T, P> = \bar{h}_f^0 + [\bar{h}_i<T> - \bar{h}_i <T^0>] \qquad (2.36)$$

An expression for specific molar entropy $\bar{s}_i$ was developed in this chapter and given as

$$\bar{s}_i<T, P> = \bar{s}_i^{\,0}<T> - \bar{R} \ln(P_i/P_0) \qquad (3.9)$$

Substituting Equations (2.36) and (3.9) into Equation (3.33b) and recalling that the units of atmospheres for pressure make $P_0 = 1$ yield

$$\bar{g}_i<T, P> = \bar{h}_f^0 + [\bar{h}_i<T> - \bar{h}_i<T^0>] - T\,[\bar{s}_i^{\,0}<T> - \bar{R} \ln P_i] \qquad (3.33c)$$

Defining a standard state molar Gibbs function $\bar{g}_i^0<T>$ as

$$\bar{g}_i^0<T> \equiv \bar{h}_{f_i}^0 + [\bar{h}_i<T> - \bar{h}_i<T^0>] - T\bar{s}_i^0<T> \qquad (3.33d)$$

and

$$\bar{g}_i <T, P> = \bar{g}_i^0 <T> - \bar{R}T \ln P_i$$

The total mixture Gibbs function can now be expressed as

$$G = \Sigma N_i[\bar{g}_i^0<T> - \bar{R}T \ln P_i] \qquad (3.43a)$$

and, at equilibrium,

$$dG_{T,P} = 0 = \Sigma N_i d\bar{g}_i<T,P> + \Sigma \bar{g}_i<T,P>dN_i$$

or

$$0 = \Sigma[\bar{g}_i^0<T> - \bar{R}T \ln P_i]dN_i \qquad (3.44)$$

For a fixed amount of reactive-gas mixture, one cannot independently or arbitrarily specify the concentration of each constituent and maintain equilibrium. Near an equilibrium point, the concentration relationship is expressed using the *equilibrium reaction* written in general form for the stated case above as

$$a\mathrm{A} + b\mathrm{B} \overset{T,P}{\rightleftarrows} l\mathrm{L} + m\mathrm{M} \qquad (3.45)$$

Note that the equilibrium coefficients a, b, l, and m are not concentration coefficients N_i given by Equation (3.42).

To illustrate, recall from chemistry the familiar equilibrium relationship for a mixture of H_2, O_2, and H_2O,

$$2H_2 + O_2 \rightleftarrows 2H_2O \qquad (3.46)$$

Equation (3.46) shows that H_2 and O_2 can combine to produce H_2O, whereas H_2O can concurrently dissociate to yield H_2 and O_2. The equilibrium concentrations of H_2, O_2, and H_2O depend on the temperature and total pressure of the gas mixture and will be defined by minimum Gibbs function for the mixture.

At high temperatures, such as those that occur in fuel–oxidant reactions, additional equilibrium reactions can be written between stable molecules and radicals. In fact, it is the endothermic dissociation effects at high temperature that work against achieving complete combustion. Examples of these types of reactions include

$$O_2 \rightleftarrows 2O$$
$$CH_4 \rightleftarrows CH_3 + H$$
$$N_2 \rightleftarrows 2N$$

For particular components, equilibrium equations can be written in several different but consistent forms. Consider, for example, the relationship between CO, O_2, and CO_2.

$$CO + \tfrac{1}{2}O_2 \rightleftarrows CO_2$$
$$2CO + O_2 \rightleftarrows 2CO_2$$
$$4CO_2 \rightleftarrows 4CO + 2O_2$$

In the above equations, one sees that any adjustment in CO, O_2, and CO_2 concentrations must be in proportion to their equilibrium coefficients. A general means of expressing the concentration adjustments required for CO, O_2, and CO_2 in terms of an *extent of reaction* $d\bar{x}$ is equal to

$$dN_{CO} = \mp 2d\bar{x}$$
$$dN_{O_2} = \mp d\bar{x}$$
$$dN_{CO_2} = \pm 2d\bar{x}$$

or

$$dN_i = \pm n_i d\bar{x}$$

Returning to the generalized equilibrium reaction equation between species A, B, . . . , L, and M, Equation (3.45), and substituting for minimum Gibbs function, Equation (3.44), one obtains an expression for the reaction $aA + bB \rightleftarrows lL + mM$,

$$0 = [\bar{g}^0_A\langle T\rangle - \overline{R}T \ln(P_A)]dN_A + [\bar{g}^O_B\langle T\rangle - \overline{R}T \ln(P_B)]dN_B + [\bar{g}^O_L\langle T\rangle - \overline{R}T \ln(P_L)]dN_L + [\bar{g}^O_M\langle T\rangle - \overline{R}T \ln(P_M)]dN_M \quad (3.47)$$

Expressing the concentration changes in terms of the extent of reaction

$$\begin{aligned} dN_A &= \mp a\, d\bar{x} \\ dN_B &= \mp b\, d\bar{x} \\ dN_L &= \pm l\, d\bar{x} \\ dN_M &= \pm m\, d\bar{x} \end{aligned} \tag{3.48}$$

or

$$0 = \Big\{[\bar{g}_A^O<T> - \bar{R}T \ln(P_A)](-a) + [\bar{g}_B^O<T> - \bar{R}T \ln(P_B)](-b) + [\bar{g}_L^O<T> - \bar{R}T \ln(P_L)](+l) + [\bar{g}_M^O<T> - \bar{R}T \ln(P_M)](+m)\Big\} d\bar{x}$$

Rearranging the above expression and noting that $d\bar{x} \neq 0$,

$$\begin{aligned} &+(a\bar{g}_A^O<T> + b\bar{g}_B^O<T> - l\bar{g}_L^O<T> - m\bar{g}_M^O<T>) \\ &= +a\bar{R}T \ln(P_A) + b\bar{R}T \ln(P_B) - l\bar{R}T \ln(P_L) - m\bar{R}T \ln(P_M) \end{aligned} \tag{3.49}$$

A *standard state Gibbs function of reaction* ΔG_r^0 can be written for the reaction as

$$\Delta G_r^0<T> = l\bar{g}_L^O<T> + m\bar{g}_M^O<T> - a\bar{g}_A^O<T> - b\bar{g}_B^O<T> \tag{3.50}$$

and substituting into Equation (3.49) gives

$$-\Delta G_r^0<T> = \bar{R}T[\ln(P_A)^a + \ln(P_B)^b - \ln(P_L)^l - \ln(P_M)^m] \tag{3.51}$$

or

$$-\Delta G_r^0<T> = \bar{R}T \ln\left[\frac{(P_A)^a(P_B)^b}{(P_L)^l(P_M)^m}\right] \tag{3.51a}$$

and

$$\exp\left\{\frac{-\Delta G_r^0<T>}{\bar{R}T}\right\} = \frac{(P_A)^a(P_B)^b}{(P_L)^l(P_M)^m} \tag{3.51b}$$

An *equilibrium constant,* $K_p<T>$, for the equilibrium reaction between ideal-gas species

$$a\text{A} + b\text{B} \rightleftarrows l\text{L} + m\text{M}$$

can now be defined and written as

$$K_p<T> = \frac{\left(P_i/P_0\right)_L^l \left(P_i/P_0\right)_M^m}{\left(P_i/P_0\right)_A^a \left(P_i/P_0\right)_B^b} \tag{3.52}$$

where

P_0 = standard state pressure = 1 atm
P_i = equilibrium partial pressure, in atmospheric units

or, for the case in which $P_{0_i} \equiv 1$ atm,

$$K_p\langle T\rangle = \frac{(P_L)^l (P_M)^m}{(P_A)^a (P_B)^b} \tag{3.53}$$

Comparing the equilibrium constant expression, Equation (3.53), to minimum Gibbs function, Equation (3.51c), one can show that the standard Gibbs free energy of reaction ΔG_r^0 and equilibrium constant K_p are related as

$$\Delta G^0{}_r\langle T\rangle = -\bar{R}T \ 1\text{n} \ K_p\langle T\rangle$$

or

$$K_p\langle T\rangle = \exp\{-(\Delta G_r^0\langle T\rangle/\bar{R}T)\} \tag{3.54}$$

EXAMPLE 3.4 Dissociation of stable chemical compounds into active radical species occurs in many high-temperature combustion reactions. For example, diatomic oxygen O_2 can dissociate to form monatomic oxygen O, while monotomic oxygen can recombine and form diatomic oxygen. Consider the equilibrium reaction between O_2 and O

$$O_2 \overset{P,\,T}{\rightleftarrows} 2O$$

Using JANAF data calculate (a) $\Delta G_r^0\langle STP\rangle$, cal/g mole; (b) $K_p\langle STP\rangle$; (c) log $k_p\langle T\rangle$, and (d) the equilibrium mole fractions of O_2 and O at *STP*. Repeat part (d) for 1000, 1500, and 2000K and 1 atm total pressure.

Solution

1. Gibbs function of reaction:

$$\Delta G_r^0\langle STP\rangle = 2\Delta G_f^0\langle STP\rangle\Big)_O - \Delta G_f^0\langle STP\rangle\Big)_{O_2} = (2)(55{,}395) - 0$$

a. $\qquad = 110{,}790$ cal

2. Equilibrium constant:

$$K_p\langle STP\rangle = \exp\left\{\frac{-\Delta G_r^0\langle STP\rangle}{\bar{R}T^0}\right\} = \exp\left(\frac{-(110{,}790)}{(1.987)(298)}\right)$$

$$= 5{,}5098 \times 10^{-82}$$

b. $K_p\langle 298\text{K}\rangle = 5{,}5098 \times 10^{-82}$

3. Now, JANAF value for K_p for O based on forming one mole of compound from standard state species

$$\tfrac{1}{2}O_2 \overset{T}{\rightleftarrows} O \qquad \text{(I)}$$

and

$$\log K_p)_{O_I} = \log K_p)_{O_{JANAF}}$$

Original equilibrium reaction:

$$O_2 \overset{T}{\rightleftarrows} 2O \qquad \text{(II)}$$

and

$$\log K_p)_{O_{II}} = 2 \log K_p)_{O_I} = 2 \log K_p)_{O_{JANAF}}$$

T	$\log K_p)_I$	$\log K_p)_{II}$	K_{p_0}
298	−40.604	−81.208	6.194×10^{-82}
1000	−9.807	−19.614	2.432×10^{-20}
1500	−5.395	−10.790	1.6218×10^{-11}
2000	−3.178	−6.356	4.4055×10^{-7}

4. Mole fraction of equilibrium concentrations:

$$K_p\langle T\rangle = \frac{(P/P_0)^2{}_O}{(P/P_0)_{O_2}} = \frac{(\bar{x}P_0/P_0)^2{}_O}{(\bar{x}P_0/P_0)_{O_2}} = K_p\langle T\rangle = \frac{\bar{x}^2{}_O}{\bar{x}_{O_2}}$$

Let $y = x_O$, $1 - y = x_{O_2}$, or

$$K_p\langle T\rangle = \frac{y^2}{1-y}$$

$$y^2 + K_p y - K_p = 0$$

$$y = \frac{-K_p}{2} \pm \frac{1}{2}\sqrt{(K_p)^2 + 4K_p}$$

Solving for y yields

$$y = \frac{-K_p}{2} + \frac{1}{2}\sqrt{(K_p)^2 + 4K_p}$$

and

T	y	$\bar{x}_{O}$	$\bar{x}_{O_2}$
298	2.49×10^{-41}	0	1.0
1000	1.56×10^{-10}	1.56×10^{-10}	~1.0
1500	4.03×10^{-6}	4.03×10^{-6}	0.99999579
2000	6.64×10^{-4}	6.64×10^{-4}	0.999336

Equilibrium constants for particular reactions, see Figure 3.8, can be calculated from Gibbs free energy of reaction. Values for K_p for several compounds are found in Appendix B. In this instance, the equilibrium equations are written using natural elemental species on the left-hand side with a single compound on the right-hand side such as shown at the top of p. 95.

$$C_s + 2H_2 \rightleftarrows CH_4$$
$$C_s + O_2 \rightleftarrows CO_2$$
$$\tfrac{1}{2}O_2 \rightleftarrows O$$

EXAMPLE 3.5 The equilibrium reaction for the combustion of carbon and oxygen can be written as

$$C_s + O_2 \overset{P_0,T}{\rightleftarrows} CO_2$$

The corresponding value of the equilibrium constant K_p for this reaction is a function of temperature. For the equilibrium reaction above at 1000K, evaluate (a) K_p, (b) the derivative $d(\ln K_p/dT)$, and (c) the value of K_p at 1500K using parts a and b.

Solution:

1. Equilibrium constant:

$$K_p<T> = \exp\left\{\frac{-\Delta G^0<T>}{\bar{R}T}\right\}$$

and

$$\ln K_p<T> = \frac{-\Delta G^0<T>}{\bar{R}T}$$

2. Gibbs free energy of reaction:

$$\Delta G_r^0<T> = \Delta H_r^0<T> - T\Delta S_r^0<T>$$

or

$$\ln K_p<T> = \frac{-\Delta H^0<T> + T\Delta S^0<T>}{\bar{R}T}$$

$$\frac{d}{dT}\left\{ \ln K_p<T> \right\} = \frac{d}{dT}\left\{ \frac{-\Delta H^0<T>}{\bar{R}T} \right\} + \frac{d}{d}\left\{ \frac{T\Delta S^0<T>}{\bar{R}T} \right\}$$

$$= \frac{-\bar{R}T\dfrac{d\Delta H^0}{dT} - \Delta H^0\dfrac{d\bar{R}T}{dT}}{\bar{R}^2T^2} + \frac{\bar{R}T^2\dfrac{d\Delta S^0}{dT} + T\Delta S^0\dfrac{d\bar{R}T}{dT}}{\bar{R}^2T^2}$$

$$= -\frac{1}{\bar{R}T}\left\{ \frac{d\Delta H^0}{dT} \frac{-\Delta H^0}{\bar{R}T^2} \right\} + \frac{1}{\bar{R}}\left\{ \frac{d\Delta S^0}{dT} + \frac{\Delta S^0}{\bar{R}T} \right\}$$

3. *TdS* equation:

$T\Delta S = \Delta H - V\Delta P$ since $P_0 = \text{const}$

or

$$\frac{d}{dT}\left\{ \ln K_p<T> \right\} = \frac{-1}{\bar{R}T}\frac{d}{dT}\left\{ T\Delta S^0 \right\} - \frac{\Delta H^0}{\bar{R}T^2} + \frac{1}{\bar{R}}\frac{d\Delta S^0}{dT} + \frac{\Delta S^0}{\bar{R}T}$$

$$\frac{d}{dT}\left\{ \ln k_p<T> \right\} = \frac{-1}{\bar{R}}\frac{d\Delta S^0}{dT} - \frac{\Delta S^0}{\bar{R}T}\frac{dT}{dT} - \frac{\Delta H^0}{\bar{R}T^2}$$

$$\frac{1}{\bar{R}}\frac{\Delta S^0}{dT} + \frac{\Delta S^0}{\bar{R}T}$$

or

$$\frac{d}{dT}\left\{ \ln K_p<T> \right\} = \frac{-\Delta H_r^0<T>}{\bar{R}T^2}$$

and

$$\frac{d}{d(1/T)}\left\{ \ln K_p<T> \right\} = \frac{-\Delta H_r^0<T>}{\bar{R}}$$

4. For the reaction $C_S + O_2 \rightleftarrows CO_2$:

$$\frac{d}{dT}\left\{ \ln K_p<\mathrm{T}> \right\} = \frac{-\left[(\bar{h}_f^0 + \Delta\bar{h})_{CO_2} - (\bar{h}_f^0 + \Delta\bar{h})_{C_S} - (\bar{h}_f^0 + \Delta\bar{h})_{O_2} \right]}{\bar{R}T^2}$$

5. Thermochemical data:

T	$\bar{h}_f^0)_{CO_2}$	$\bar{h})_{CO_2}$	$\bar{h}_f^0)_{C_S,O_2}$	$\bar{h})_{O_2}$	$\bar{h})_{C_S}$	$\Delta H_r^0\langle T\rangle$
500	−94,054	1,987	0.0	1455	569	−94,091
1000	−94,054	7,984	0.0	5427	2024	−93,521
1500	−94,054	14,750	0.0	9706	5552	−94,562
K	$\frac{\text{cal}}{\text{g mole}}$	$\frac{\text{cal}}{\text{g mole}}$	$\frac{\text{cal}}{\text{g mole}}$	$\frac{\text{cal}}{\text{g mole}}$	$\frac{\text{cal}}{\text{g mole}}$	cal

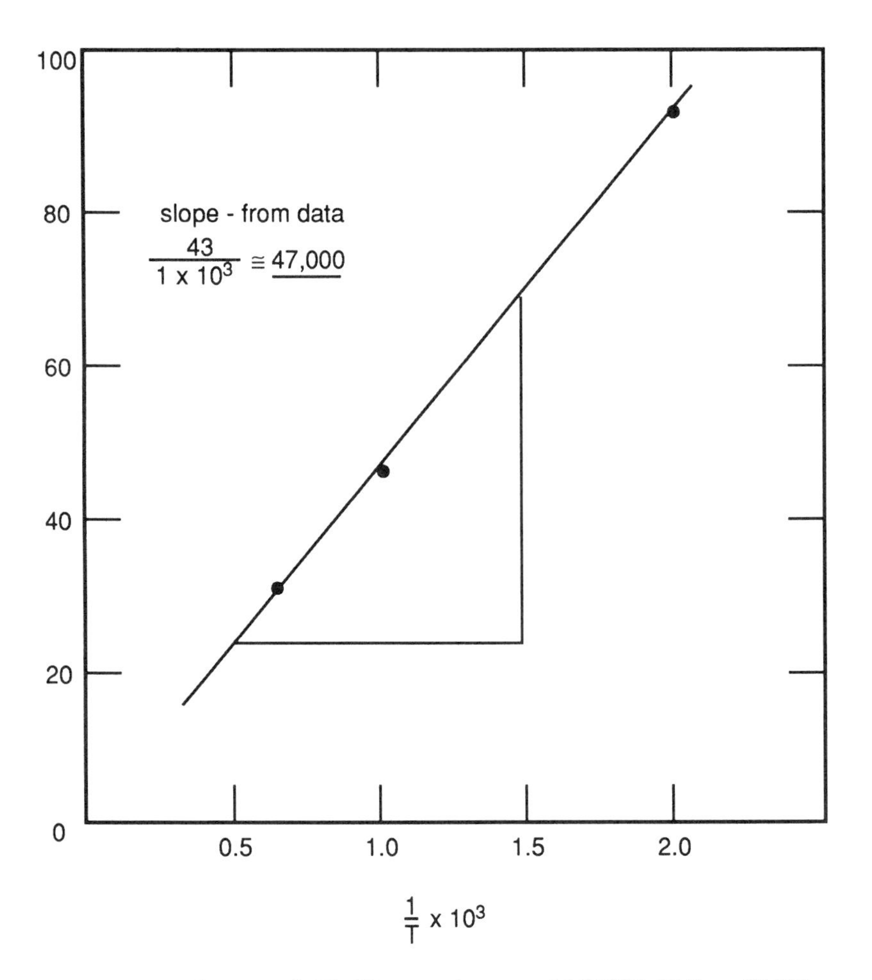

Slope, from van't Hoff's equation = +94,000/(1.986) ≅ 47,331

T	K_p	ln K_p
500	1.8197×10^{41}	95.005
1000	4.7863×10^{20}	47.617
1500	6.3241×10^{13}	31.779

The general nature of reactive gas mixtures and chemical equilibrium can best be illustrated using a device termed a van't Hoff's equilibrium box; see Figure 3.9. Inside the box, the species A, B, . . . , L, and M are assumed to be in an equilibrium state. Energy transfers to and/or from the box as heat and/or work cause temperature and total pressure in the box to change. After any energy transfer, a new state will be established, and it may be necessary to adjust the species concentrations using mass transfer components shown to re-establish equilibrium. A complete analysis of this process and determination of a new equilibrium state, i.e., temperature, pressure, and composition, would require an application of both the first and second laws. Since thermodynamics does not tell the rate of such processes, Chapter 8 will introduce the dynamic, i.e., time factors, or reaction kinetics, necessary to describe chemical equilibrium more completely.

3.6 THE FUEL CELL

The van't Hoff's equilibrium box shown in Figure 3.9 suggests the plausibility of a device that can convert reactants isothermally into products while producing power rather than heat. Chemical potential energy stored in a fuel converted directly into electricity would provide an efficient energy conversion process, i.e., a direct energy conversion system. By using a reversible electrochemical process, electrical power could be produced without a need for traditional internal or external combustion heat engines. Two familiar examples of controlled chemical reactions used to generate electrical energy rather than heat are: (1) the chemical battery and (2) the fuel cell. The battery contains a fixed amount of chemicals that produce electricity until all reactive material is depleted, whereas a fuel cell will generate electricity continuously as long as reactants are supplied to the unit.

In the 1830s, Sir William Grove experimentally demonstrated that electrolysis of water was a reversible process and that H_2 and O_2 could be reacted in an electrochemical cell, or *fuel cell,* to produce power. Practical fuel cell development began with the work of Francis Bacon in the 1930s but was interrupted by World War II. By 1959, Bacon was able to develop a 6-kW H_2–O_2 demonstration fuel cell. This unit was the forerunner of fuel cells

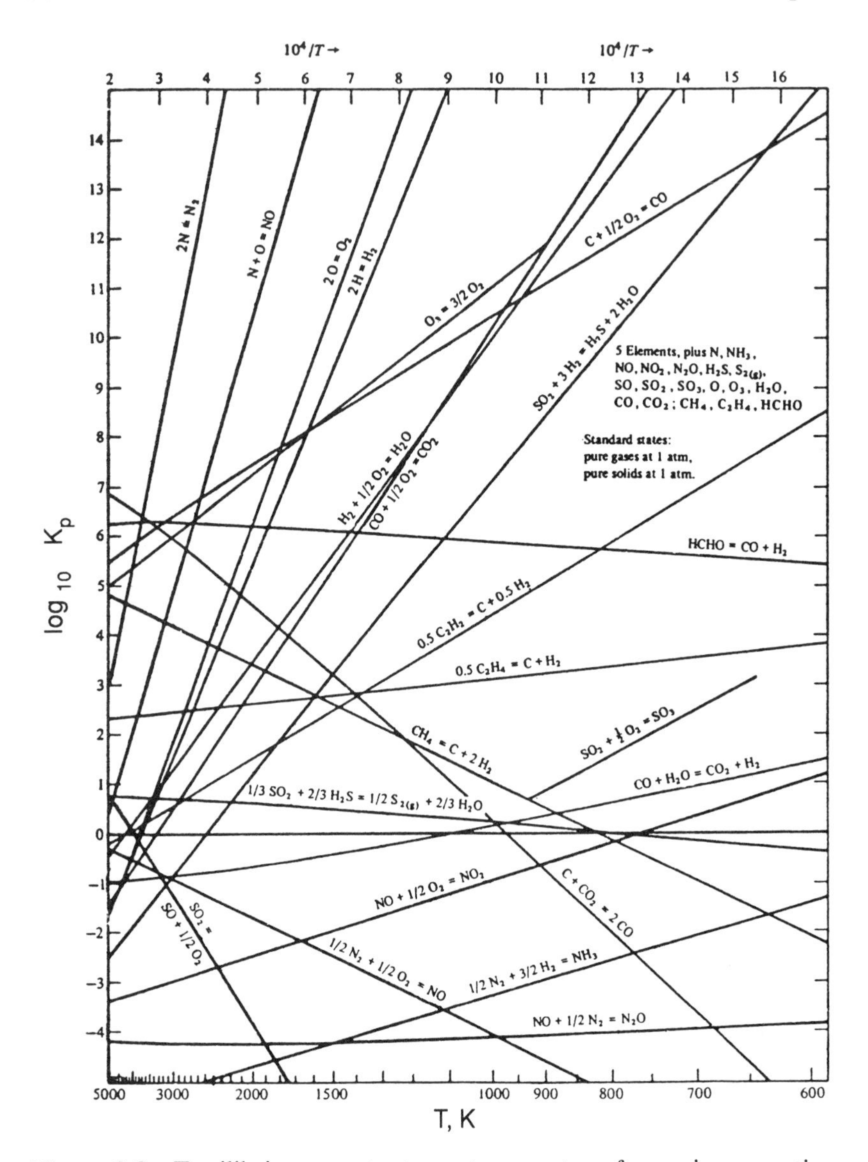

Figure 3.8 Equilibrium constants vs. temperature for various reactions.

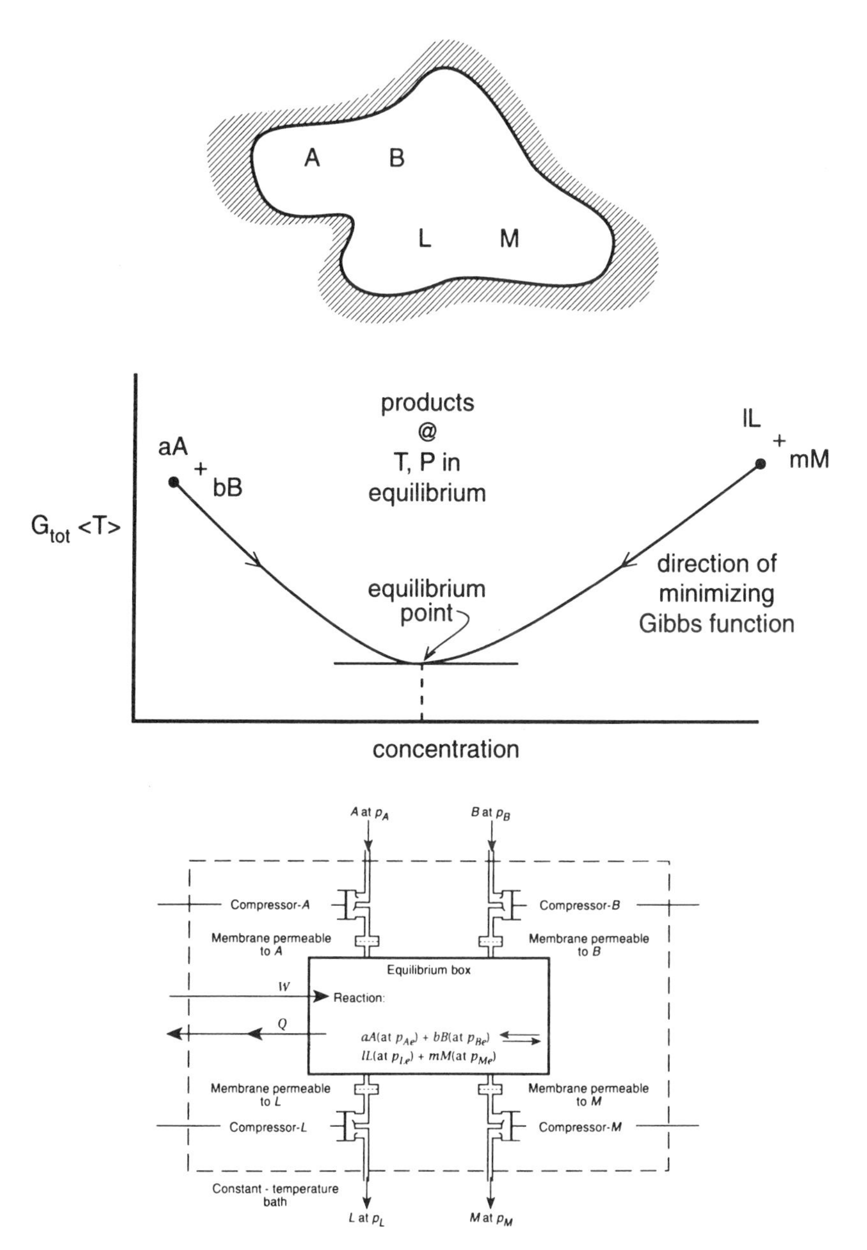

Figure 3.9 Van't Hoff's equilibrium box.

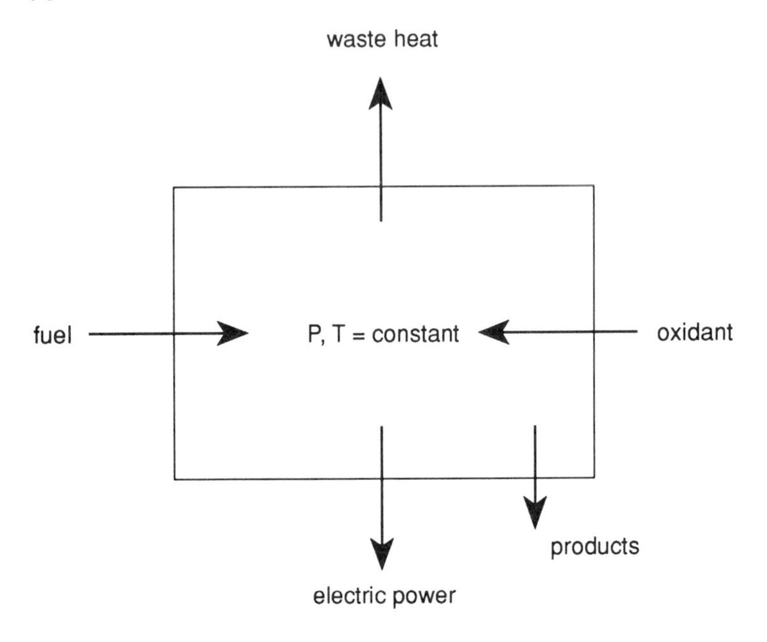

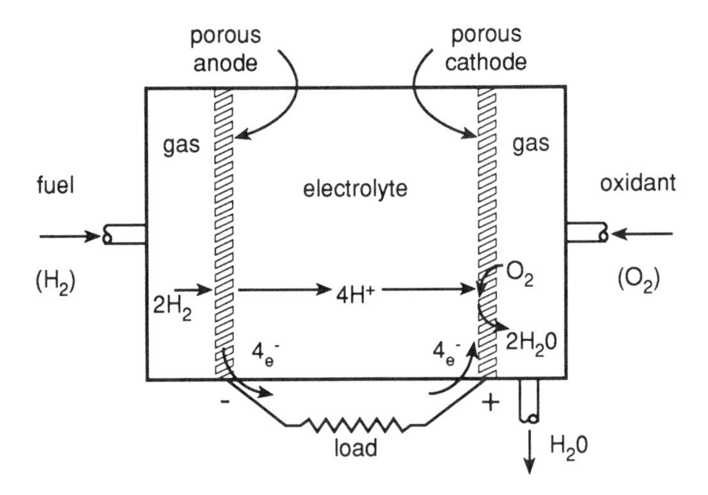

Figure 3.10 The fuel cell.

used in the U.S. space program. Many configurations and variations of fuel cells are currently being developed for potential commercial application.

General operation of a fuel cell, Figure 3.10, requires that an easily ionized fuel, such as H_2, be fed continuously to the unit. Hydrogen is bubbled through a porous electrode to an electrolytic solution, such as phosphoric acid, where it is reduced at an anode surface. The anode

Table 3.3 Ideal Fuel Cell Reactions and Efficiency

REACTANT Half-Cell Reaction Complete-Cell Reaction	Ideal Efficiency[a] (Percent)
HYDRAZINE (aq)	
N_2H_4 (aq) + 4 OH^- ⇄ N_2 + 4 H_2O + 4 e^-	
N_2H_4 (aq) + O_2 ⇄ N_2 + 2 H_2O	99.4
ETHANOL (aq)	
C_2H_5OH (aq) + 3 H_2O ⇄ 2 CO_2 + 12 H^+ + 12 e^-	
C_2H_5OH (aq) + 3 O_2 ⇄ 2 CO_2 + 3 H_2O	97.5
BENZENE	
C_6H_6 + 12 H_2O ⇄ 6 CO_2 + 30 H^+ + 30 e^-	
C_6H_6 + 7½ O_2 ⇄ 6 CO_2 + 3 H_2O	97.2
METHANOL (aq)	
CH_3OH (aq) + H_2O ⇄ CO_2 + 6 H^+ + 6 e^-	
CH_3OH (aq) + 1½ O_2 ⇄ CO_2 + 2 H_2O	97.1
PROPYLENE	
C_3H_6 + 6 H_2O ⇄ 3 CO_2 + 18 H^+ + 18 e^-	
C_3H_6 + 4½ O_2 ⇄ 3 CO_2 + 3 H_2O	95.1
ACETYLENE	
C_2H_2 + 4 H_2O ⇄ 2 CO_2 + 10 H^+ + 10 e^-	
C_2H_2 + 2½ O_2 ⇄ 2 CO_2 + H_2O	95.0
PROPANE	
C_3H_8 + 6 H_2O ⇄ 3 CO_2 + 20 H^+ + 20 e^-	
C_3H_8 + 5 O_2 ⇄ 3 CO_2 + 4 H_2O	95.0
ETHYLENE	
C_2H_4 + 4 H_2O ⇄ 2 CO_2 + 12 H^+ + 12 e^-	
C_2H_4 + 3 O_2 ⇄ 2 CO_2 + 2 H_2O	94.3
ETHANE	
C_2H_6 + 4 H_2O ⇄ 2 CO_2 + 14 H^+ + 14 e^-	
C_2H_6 + 3½ O_2 ⇄ 2 CO_2 + 3 H_2O	94.1
METHANE	
CH_4 + 2 H_2O ⇄ CO_2 + 8 H^+ + 8 e^-	
CH_4 + 2 O_2 ⇄ CO_2 + 2 H_2O	91.9

Table 3.3 Continued

REACTANT Half-Cell Reaction Complete-Cell Reaction	Ideal Efficiency[a] (Percent)
CARBON MONOXIDE	
$CO + H_2O \rightleftarrows CO_2 + 2\,H^+ + 2\,e^-$	
$CO + \frac{1}{2}\,O_2 \rightleftarrows CO_2$	90.9
HYDROGEN	
$H_2 \rightleftarrows 2\,H^+ + 2\,e^-$	
$H_2 + \frac{1}{2}\,O_2 \rightleftarrows H_2O$	83.0

[a]Standard-state free energy of reaction at 25°C divided by standard-state heat of reaction at 25°C.
Source: Carl Berger, ed., *Handbook of Fuel Cell Technology* (Englewood Cliffs, NJ: Prentice-Hall, Inc., 1968).

(usually made from a noble metal such as platinum) catalytically causing hydrogen to give up electrons forming hydrogen ions via an anode reaction,

$$2H_2 \rightarrow 4H^+ + 4e^- \tag{3.55}$$

Oxygen must also be continuously fed to the cell but is bubbled through a porous cathode electrode. Hydrogen ions produced at the anode pass through the electrolyte and combine with O_2 and electrons at the cathode to produce water via an anode reaction

$$4H^+ + re^- + O_2 \rightarrow 2H_2O \tag{3.56}$$

The electrons produced at the anode travel through an external circuit and an electrical load to the cathode and can be used to provide electric power.

Ideally, cell output is governed by change in Gibbs function for the fuel cell chemical reaction which, in turn, is affected by mixture composition, reaction pressure, and cell temperature. Since hydrogen is not naturally available, additional reactions are being considered for commercial fuel cell applications. Table 3.3 lists ideal cell reactions and energy conversion efficiencies for several prospective fuels.

The environmental impact of a fuel cell would be less than that of a conventional heat engine. Fuel cells operate at temperatures below typical hydrocarbon combustion reaction temperatures. As such, there should be less waste heat produced, i.e., lower thermal pollution of the environment. Even water produced can be recovered and used as in space flight applications. In addition, there would be little or no high-temperature combustion pollutants such as nitric oxides, carbon monoxide, and unburned hydrocarbons generated by the low-temperature fuel cell operation.

Development and successful commercialization of fuel cell technology now is a question of economics of scale. The need to optimize various components, such as electrolyte and electrode performance, as well as to minimize losses and irreversibilities, is critical to growth of the fuel cell industry. Several programs are currently in progress that attempt to produce fuel cells having high power densities for mobility applications. Other designs are being investigated that would use available fuels yet have high thermal efficiencies for use in electric power operation.

PROBLEMS

3.1 An ideal-gas mixture has a molecular weight of 40 kg/kg mole and a constant-pressure specific heat of 0.525 kJ/K kg. At 480-kPa absolute pressure and 14.5-m^3 volume, the temperature of the gas is 422 K. Determine for the mixture (a) the specific gas constant R, kJ/K kg, (b) the constant-volume specific heat Cv, kJ/kg-K; (c) the specific heat ratio γ; and (d) the mass of gas in the cylinder, kg.

3.2 Air initially at 14.7 psi and 77°F is isentropically compressed until the original volume is reduced by 90%. Using constant specific heats, find (a) the final temperature, °F; and (b) the final pressure, psi. Repeat parts a and b using the isentropic tables for air based on variable specific heat analysis found in Example 3.2

3.3 The modern compression-ignition engine injects fuel into hot compressed air. Fuel ignition occurs by contacting air that has been raised above the self-ignition temperature of the fuel. Consider air initially at 101 kPa and 30°C and a fuel with a self-ignition temperature of 600°C. Assuming an isentropic compression process for air and using constant specific heat analysis, find (a) the final pressure, kPa; (b) the final specific volume, m^3/kg; and (c) the required compression work, kJ/kg. Repeat parts (a)–(c) using the isentropic tables for air based on variable specific heat analysis found in Example 3.2

3.4 An axial compressor is used to compress a mixture of methane, CH_4, and ethylene, C_2H_4. The mixture of 50% methane on a molar basis and compressor inlet conditions are 101 kPa, 21°C, and 0.012 m^3/sec. Discharge conditions from the unit are 550 kPa and 163°C. For these conditions and assuming constant specific heats, find (a) the mixture mass flow rate, kg/sec; (b) the ideal isentropic discharge temperature; (c) the entropy increase across the compressor, kJ/K kg; (d) the actual compressor power, kW; and (e) the compressor adiabatic efficiency, i.e., isentropic work/actual work, kW.

3.5 Gasoline can be approximated in many combustion calculations using *n*-octane. Using the JANAF data for C_8H_{18} found in Appendix B, determine the specific heat ratio at 25°C for (a) stoichiometric fuel-air mixture, (b) a fuel-rich mixture having an equivalence ratio of 0.55, and (c) a fuel-lean mixture having an equivalence ratio of 0.55. Repeat parts (a)–(c) for an average temperature between 25°C and the isentropic compression temperature for an 18:1 compression ratio.

3.6 Repeat 3.5 using methanol, CH_3OH, instead of C_8H_{18}.

3.7 Consider the reaction of formation of carbon dioxide from natural elemental species. For reaction at *STP*, determine (a) the entropy of reaction, Btu/°R lb mole; (b) the Gibbs function of reaction, Btu/lb mole; and (c) the Hemholtz function of reaction, Btu/°R lb mole.

3.8 Repeat Problem 3.7 for a reaction temperature at 1800°R.

3.9 Consider the ideal *STP* stoichiometric combustion reaction of acetylene. For these conditions, determine (a) the change in enthalpy for the reaction, kJ/K kg mole; (b) the change in entropy for the reaction, kJ/kg mole; and (c) the change in Gibbs free energy for the reaction, kJ/kg mole.

3.10 Consider the ideal *STP* stoichiometric combustion reaction of methane. For these conditions, determine (a) the change in internal energy for the reaction, Btu/°R lb mole; (b) the change in entropy for the reaction, Btu/lb mole; and (c) the change in Hemholtz function for the reaction, Btu/lb mole.

3.11 Consider the one-atmospheric equilibrium reaction

$$N_2 \overset{P_0,T}{\rightleftarrows} 2N$$

at 4000K. For this condition, determine (a) the equilibrium mole fractions of N_2 and N, %; (b) the mixture Gibbs function, Btu/lb mole; and (c) whether this condition gives a minimum value to the Gibbs function.

3.12 Repeat Problem 3.11 using O_2 instead of N_2.

3.13 Many combustion processes involve products of combustion at temperatures not in excess of 2000°F. Often, the effects of dissociation of stable species are ignored for these temperatures. Is this assumption truly justified for (a) O_2, (b) N_2, (c) CO_2, or (d) H_2O?

3.14 H_2 can be commercially generated by oxidizing CO using steam via the water–gas reaction

$$H_2Og + CO \overset{T,P}{\rightleftarrows} H_2 + CO_2$$

For this reaction, (a) write an expression for K_p for the equilibrium reaction in terms of the partial pressures of H_2, CO_2, CO, and H_2O; (b) evaluate log $10K_p<T>$ for the reaction using JANAF data for an equilibrium temperature of 2000K; and (c) the mole fractions of H_2, CO_2, CO, and H_2O at 2000K using parts (a) and (b).

3.15 Hydrogen can be produced by steam-cracking methane according to the reaction

$$CH_4 + 2H_2Og \quad \overset{T,P}{\rightleftarrows} \quad CO_2 + 4H_2$$

An initial mixture at 1-atm total pressure and 800K consists of 50% excess steam. Determine (a) the equilibrium constant for the conversion and (b) the equilibrium mole fractions of CH_4, H_2O, CO_2, and H_2. Repeat part b for (c) a temperature of 1200K and (d) a total pressure of 5 atm.

3.16 The equilibrium composition for CO_2 dissociation to form CO and O_2 at high temperatures can be described by the reaction

$$2CO_2 \quad \overset{T,P}{\rightleftarrows} \quad 2CO + O_2$$

For this equilibrium reaction, (a) write the JANAF equilibrium reaction and K_p for CO in terms of appropriate partial pressures. (b) Repeat part a for CO_2, and (c) show that, for the equilibrium reaction above,

$$\log Kp\Big)_{\text{reaction}} = \log Kp\Big)_{CO} - \log Kp\Big)_{CO_2}$$

and (d) calculate the equilibrium constant K_p, for $T = 3800$K.

3.19 Solid carbon reacts with stoichiometric air at 600K and 1-atm total pressure. Assume ideal complete combustion, and determine the adiabatic flame temperature. Repeat the calculation assuming that the products consists only of CO, CO_2, O_2, and N_2.

3.20 Methane, CH_4, reacts with 90% theoretical in a steady-flow reactor. The reactants enter at 1-atm pressure and 77°F. Assuming complete ideal combustion, determine the adiabatic flame temperature for a constant-pressure combustion process. Repeat the calculation, but assume that CO_2 and H_2O dissociation satisfies the equilibrium reaction

$$H_2Og + CO \quad \overset{T,P}{\rightleftarrows} \quad H_2 + CO_2$$

Furthermore assume that no O_2 exists in the product composition.

3.21 A fuel cell thermochemical efficiency can be defined in terms of the

Gibbs function of reaction and the enthalpy of reaction of the cell equilibrium reaction as

$$\eta_{FC} = \frac{\Delta G}{\Delta H}$$

Consider the fuel cell reaction $H_2 + 0.5O_2 \rightleftarrows H_2O$ at *STP* conditions, and calculate (a) the enthalpy of reaction, kJ/kg mole; (b) the Gibbs function of reaction, kJ/kg mole; and (c) the fuel cell efficiency.

3.22 Repeat Problem 3.21 for a fuel cell reaction of 800K.

3.23 Repeat Problem 3.21 for the *STP* fuel cell reaction $CO + 0.5O_2 \rightleftarrows CO_2$

3.24 Repeat Problem 3.21 for the *STP* fuel cell reaction

$$CH_4 + 2O_2 \rightleftarrows CO_2 + 2H_2O$$

4

Solid Fuels

4.1 INTRODUCTION

Any general engineering discussion of combustion should, in addition to presenting principles of thermochemistry, review several specific topics including energy characteristics of various important fuel resources. Many reasons can be cited for selecting solid fuels as the first fuel science subject covered in this text, such as the historical role fuels such as coal played in the birth of the Industrial Revolution, recent growth and viability of these solid fuels as alternatives to declining world oil supplies, and the increasing need to control pollution associated with many solid waste management systems. Also, solid materials may be used as future feedstocks for producing various synthetic liquid and gaseous fuels.

A significant amount of material has been written and published that addresses energy issues relating to the world's supplies and reserves of solid fuels, including their origins, economics, and potential political impact. This discussion, however, will focus mainly on those particular solid fuels that are compatible with industrial combustion-generated power or heat transfer applications. The following material will be limited, therefore, to a consideration of certain energy aspects, general emissions characteristics, and the combustion machinery necessary to burn these selected reactants.

4.2 SOLID FUEL THERMOCHEMISTRY

Most naturally occurring solid fuel resources are primarily hydrocarbon-based compounds. The prominent solid phase component in these fuels is carbon. From Appendix B, the heat of combustion of solid carbon is equal to

$$\Delta\bar{H}_c = 393{,}790 \quad \text{kJ/kg mole} \qquad (169{,}300 \quad \text{Btu/lb mole}) \tag{4.1}$$

or

$$\Delta H_c = 32{,}785 \quad \text{kJ/kg carbon} \qquad (14{,}095 \quad \text{Btu/lbm}) \tag{4.2}$$

The heating value of hydrocarbon-based solid fuels should therefore be approximately equal to or less than 32,780 kJ/kg fuel (14,000 Btu/lbm).

Since the composition of a solid fuel such as coal varies depending on such factors as mine location and particular samples of fuel burned, we need a more practical method than those ideal concepts indicated in Chapter 2 for determining the heating value of such complex reactants. A constant-volume reaction vessel is often used to measure experimentally the energy released by complete combustion of a given solid or liquid fuel sample; see Figure 4.1. The techniques used to determine experimentally a heat of combustion are based on principles of *calorimetry*. In order to ensure accurate measurements of energy in a fuel sample, bomb calorimetry requires the use of standard procedures, such as those established by the American Society for Testing Materials (ASTM), skillful calibration, and repeatable operation of facilities.

A reaction vessel with the fuel and excess pure oxygen mixture is immersed in a water bath initially at ambient conditions, usually *STP* temperature of 25°C. Energy released by complete combustion of the reactants within the vessel is absorbed by the surrounding water bath, which causes the water temperature in the jacket to rise. By burning a pure compound having a known heating value, such as benzoic acid, a thermal response, or *water equivalence*, of the apparatus to a particular heat release can be established. After calibrating the unit, any combusted fuel that releases approximately the same total energy as the calibration fuel would also cause the water jacket temperature to rise and, by using its water equivalence, the fuel heating value can thus be determined. Using a constant-volume oxygen bomb calorimeter, the heating value of solid and certain liquid fuels can be experimentally determined, using Equation (4.3) as

$$HV = \frac{W\,\Delta T - E_1 - E_2 - E_3 - E_4}{m} \tag{4.3}$$

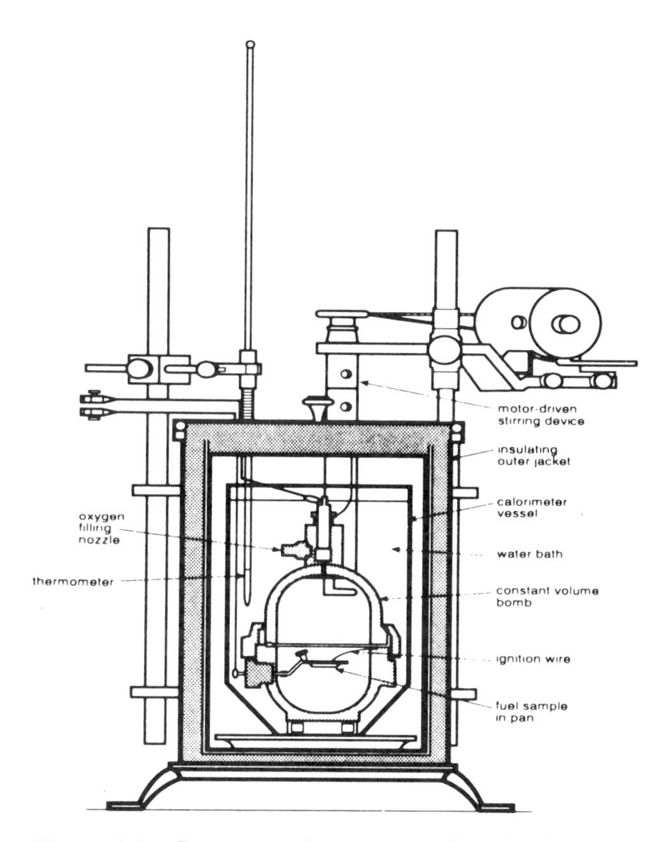

Figure 4.1 Constant-volume oxygen bomb calorimetry.

where

W = water equivalent, kJ/°C (Btu/°F)

ΔT = temperature rise, °C (°F)

E_1 = correction for heat of formation of nitric acid

E_2 = correction for heat of formation of sulfuric acid

E_3 = correction for combustion of gelatin capsule (used with liquid fuel testing)

E_4 = correction for heat of combustion of firing wire

m = weight of fuel sample, kg (lbm)

HV = heating value of fuel, kJ/kg (Btu/lbm)

Accuracy requires mass measurement of the fuel sample to within ± 0.001 g and water jacket temperature to within ± 0.002°C. Ignition of premeasured fuel and excess pure oxygen is accomplished by passing current through an ignition wire. Several corrections are included in Equation (4.3) to account for extraneous experimental factors such as energy released by burning any ignition wire and energy associated with the formation of any nitric and/or sulfuric acid. Experimental errors can arise because of improper fuel sizing or contamination with impurities during preparation. Note that, to ensure equal water equivalents, calibration and experimental runs require that fuel sample weight be carefully controlled. Improper ignition, for example, from an insufficient ignition current or misalignment or grounding of the ignition wire, or insufficient oxygen for complete combustion would also contribute to experimental error. Improper calibration or testing can result in an erroneous water equivalence and predicted heating values when a constant-volume oxygen bomb calorimeter is used.

Most natural solid fuels exist in the environment as complex organic chemical and mineral compounds that have undergone various degrees of aging and chemical conversion. Solid hydrocarbon fuel resources, beginning with wood through coal, are actually vegetation in various stages of decay and consist chiefly of carbon, hydrogen, oxygen, moisture, sulfur, carbon monoxide, methane, ethane, sulfuric acid, tars, and ammonia.

Specification of the precise energy characteristics of these solid fuels would require a complete knowledge of all constituents expressed in terms of each constituent mass fraction, i.e., an *ultimate analysis*. Often a *proximate* or *fixed carbon* analysis is sufficient for engineering consideration to define the fuel's combustion characteristics expressed as a *fixed carbon*, C, analysis or

$$\%\ C = 100\% - \%\ \text{moisture} - \%\ \text{volatiles} - \%\ \text{ash} \tag{4.4}$$

Moisture is water trapped within a fuel sample; volatiles are constitutents that evaporate during low-temperature heating of the fuel; while ash is inorganic mineral impurity or residue left when a fuel has been completely burned. The fixed carbon is a cokelike residue fuel resource left after water vapor, volatiles, and ash are removed.

EXAMPLE 4.1 A constant-volume bomb calorimeter is to be used to determine the heating value of a solid fuel. To calibrate the vessel, 0.848 g of benzoic acid, having a heating value of 26,452 kJ/kg, is completely burned by reaction with oxygen, causing the water jacket, stirrer, bomb thermometer, etc., to be heated. Precise measurement of the temperature rise of the water gives a reading of +2.49°C. After calibration, 1.05 g of the solid fuel is then completely burned, giving a temperature increase of 4.882°C. Neglecting the correction terms, calculate (a) the water equivalent for the bomb, kJ/°C rise, and (b) the heating value of the fuel, kJ/kg.

Solution

1. Water equivalence:

$$W = \frac{(HV)(m)}{\Delta T} = \frac{(26{,}452\ \text{kJ/kg})(0.848\ \text{g})}{(1{,}000\ \text{g/kg})\ (2.49\ ^\circ\text{C})}$$

a. $W = 9.009$ kJ/°C rise

2. Heating value:

$$HV = \frac{(W)(\Delta T)}{m} = \frac{(9.009\ \text{kJ/}^\circ\text{C})(1000\ \text{g/kg})(4.882^\circ\text{C})}{(1.05\ \text{g})}$$

$$= 41{,}888\ \text{kJ/kg}$$

b. $HV = 41{,}890$ kJ/kg of fuel

Empirical equations, such as DuLong's formula, can be used to predict a higher heating value of solid fuels when an ultimate analysis is known.

$$HHV = 33{,}960\ C + 141{,}890\left[H - \frac{O}{8}\right] + 9420\ S \qquad \text{kJ/kg} \tag{4.5a}$$

$$HHV = 14{,}600\ C + 61{,}000\left[H - \frac{O}{8}\right] + 4050\ S \qquad \text{Btu/lbm} \tag{4.5b}$$

where

HHV = higher heating value

C = mass fraction of carbon – fuel ultimate analysis

H = mass fraction of hydrogen – fuel ultimate analysis
O = mass fraction of oxygen – fuel ultimate analysis
S = mass fraction of sulfur – fuel ultimate analysis

DuLong's formula expresses heating value in terms of major solid fuel constituents, their heating value, and corresponding mass fractions.

The lower heating value of a solid fuel can be determined from higher heating value of the fuel and a knowledge of the mass of water formed by combustion to the mass of fuel as

$$LHV = HHV - Wh_{fg} \qquad \text{kJ/kg (Btu/lbm)} \tag{4.6}$$

where

LHV = lower heating value
W = mass of water formed/mass of fuel burned
h_{fg} = latent heat of vaporization for water at heating value conditions

4.3 COAL AND OTHER SOLID FUEL RESOURCES

The world's most prominent natural solid fuel resource is coal. Coal, remnants of plants and other vegetation that have undergone varying degrees of chemical conversion in the biosphere, is not a simple homogeneous material but rather is a complex substance having varying chemical consistency. Energy characteristics of coal obviously depend on many factors, including particular composition of the original vegetation, amount and type of inorganic materials present in the debris, and the specific history that these local materials have undergone during regional deposition and conversion.

Plant life first begins to decay by anaerobic, or bacterial, action, often in swamps or other aqueous environments, producing a material known as *peat.* The decomposing material is next covered and folded into the earth's crust via geological action that provides extreme hydrological pressure and heating required for the coal conversion process, as well as an environment that drives off volatiles and water. This complex transformation, or *coalification process,* results in change, or *metamorphosis,* over great periods of time and in a variety of fuels ranging from peat, which is principally cellulose, to hard, black coal, which is principally amorphous carbon.

Standards for coal classification have been established by the ASTM and allow coal to be identified by rank and grade. *Rank* is a measure of the time period for coal conversion; i.e., a low rank has less carbon, whereas a high rank coal is mostly carbon. *Grade* is a measure of the degree of fixed carbon in a coal sample. Four grades of coal are commonly specified, starting from lowest to highest, are: lignite, subbituminous, bituminous, and anthracite. Table 4.1 gives typical energy characteristics of coal.

EXAMPLE 4.2 Cellulose, $C_6H_{10}O_5$, is a major constituent of both organic and inorganic waste material. The higher heating value of cellulose is approximately 7500 Btu/lbm. (a) Determine the stoichiometric reaction for the ideal combustion of cellulose; (b) estimate the heat of formation at STP, Btu/lb mole; (c) find the heat released from burning cellulose with 140% theoretical air in an atmospheric burner, assuming the stack temperature to be 1520°F; and (d) find the ratio of cellulose to that of carbon required to produce the same heat ideal release.

Solution:

1. Stoichiometric equation:

$$C_6H_{10}O_5 + aO_2 \rightarrow bCO_2 + cH_2O_g$$

Carbon atom balance:

$$b = 6$$

Hydrogen atom balance:

$$c = 5$$

Oxygen atom balance:

$$5 + 2a = 12 + 5$$

$$a = 6$$

or

$$C_6H_{10}O_5 + 6[O_2 + 3.76\ N_2] \rightarrow 6CO_2 + 5H_2O + (6)(3.76)N_2$$

2. Energy balance (STP):

$$\begin{aligned} Q &= \sum_{\text{prod}} N_j\,[\bar{h}_f^0 + \Delta\bar{h}]_j - \sum_{\text{react}} N_i[\bar{h}_f^0 + \Delta\bar{h}]_i \\ &= [6(-94{,}054)(1.8)_{CO_2} + 5(-68{,}317)(1.8)_{H_2O}] \\ &\quad -\bar{h}_f^0)_{C_6H_{10}O_5} \end{aligned}$$

and

$$Q = [(6)(12) + (10)(1) + (5)(16)][-7500] \text{ Btu/lb mole fuel}$$

or

$$\bar{h}_f^0)_{C_6H_{10}O_5} = -1{,}215{,}000 + 6(94{,}054)(1.8) + 5(68{,}317)(1.8)$$

b. $$= -415{,}640 \text{ Btu/lb mole}$$

3. 140% theoretical air combustion with $T_2 = 1980°R$:

$$C_6H_{10}O_5 + (1.4)(6)[O_2 + 3.76\ N_2] \xrightarrow{P}$$

$$6CO_2 + 5H_2O + (0.4)(6)O_2 + (1.4)(6)(3.76)N_2$$

$$\begin{aligned} Q = {} & [\{6(-94{,}054 + 9{,}296)_{CO_2} \\ & + (5)(-57{,}798 + 7{,}210)_{H_2O_g} + (0.4)(6)(O + 6{,}266)_{O_2} \\ & + (1.4)(6)(3.76)(O + 5{,}917)_{N_2}\}\ (1.8) \\ & - 1.0\ (-415{,}640 + O)_{C_6H_{10}O_5}] \end{aligned}$$

$$Q = -5.92 \times 10^5 \text{ Btu/lb mole cellulose}$$

or

c. $$|q| = \frac{5.92 \times 10^5 \text{ Btu/lb mole}}{162 \text{ lbm/lb mole}} = 3654 \text{ Btu/lbm cellulose}$$

4. Heat flux—using the higher heating value:

Cellulose:

$$\dot{Q}<STP> = \dot{m} \times 7500 \text{ Btu/lbm cellulose}$$

Carbon:

$$\dot{Q}<STP> = \dot{m} \times 14095 \text{ Btu/lbm carbon}$$

or

d. $$\frac{m<\text{lbm cellulose}>}{m<\text{lbm carbon}>} = \frac{14095}{7500} = 1.88$$

Lignite, the lowest grade of coal, takes its name from the Latin word for wood, *lignum.* Since this material is in the early stages of coalification, it often has a woodlike structure and is high both in moisture content and volatile components. Lignite has a low heat of combustion, nearly equal that of cellulose.

Subbituminous, the next highest grade, is often referred to as black lignite. U.S. subbituminous coal reserves are large; and, in general, this type of coal has a higher fixed carbon content than lignite but may still have a high degree of moisture as well as a considerable ash content. Subbituminous is often not usable as a natural fuel resource, but its low sulfur content; relatively free-burning, noncaking characteristics, and relatively low energy density make it a good candidate for combustion in coal-burning systems in which sulfur content may be critical.

Table 4.1 Properties of Coal

Coal grade	Approximate heating value	
	kJ/kg coal	Btu/lbm coal
Anthracite	30,240–33,730	13,000–14,500
Bituminous	27,910–34,420	12,000–14,800
Subbituminous	19,310–23,260	8,300–10,000
Lignite	13,260–17,450	5,700– 7,500

		Proximate analysis, %				Nominal heating value	
Coal grade	Source	Moisture	Volatile matter	Fixed carbon	Ash	kJ/kg	Btu/lbm
Anthracite	PA	4.4	4.8	81.8	9.0	30,540	13,130
Bituminous							
Low-volatile	MD	2.3	19.6	65.8	12.3	30,750	13,220
Medium-volatile	AK	3.1	23.4	63.6	9.9	31,470	13,530
High-volatile	OH	5.9	43.8	46.5	3.8	30,590	13,150
Subbituminous	WA	13.9	34.2	41.0	10.9	24,030	10,330
	CO	25.8	31.1	38.4	4.7	19,960	8,580
Lignite	ND	36.8	27.8	30.2	5.2	16,190	6,960

Source: U.S. Bureau of Mines.

EXAMPLE 4.3 The ultimate analysis of a coal as fired is reported to be as listed below.

As-fired constituent	*Percentage*	*As-fired constituent*	*Percentage*
C_S	72.8	H_2O	3.5
H_2	4.8	S	2.2
O_2	6.2	ash	9.0
N_2	1.5		

Assuming ideal complete combustion of the coal at *STP*, calculate (a) the molar air–fuel ratio; (b) the mass air–fuel ratio; (c) the higher heating value of the coal using DuLong's formula, kJ/kg fuel; and (d) the higher heating value of the coal using JANAF data, kJ/kg fuel.

Solution:

1. Mole fraction fuel sample, ashless

	mf_i	MW_i	mf_i/MW_i	$\bar{x}_i$
C_S	0.8000	12.0	0.06667	0.6760
H_2	0.0527	2.0	0.02635	0.2672
O_2	0.0681	32.0	0.002128	0.02158
N_2	0.0165	28.0	0.000589	0.005972
H_2O	0.0385	18.0	0.002139	0.02169
S	0.0242	32.0	0.000756	0.007665
	$\dfrac{kg_i}{kg_{tot}}$	$\dfrac{kg_i}{kg_{mole}}$	$\dfrac{kg_{mole_i}}{kg_{tot}}$	$\dfrac{kg_{mole_i}}{kg_{mole_{tot}}}$

where, for example, for carbon

$$mf_i = \frac{72.8}{100 - 9.0} = 0.8000$$

$$\Sigma \left.\frac{mf}{MW}\right)_i = 0.06667 + 0.02635 + 0.002128 + 0.000589$$
$$+ 0.002139 + 0.000756 = 0.09863 \text{ kg mole/kg)}_{tot}$$

$$\bar{x}_i = \frac{0.06667 \text{ kg mole}_i/\text{kg}_{tot}}{0.09863 \text{ kg mole}_{tot}/\text{kg}_{tot}} = 0.6760 \frac{\text{kg mole}_i}{\text{kg mole}_{tot}}$$

2. Stoichiometric equation:

$$[0.6760\ C_S + 0.2672\ H_2 + 0.0216\ O_2 + 0.0060\ N_2$$
$$+ 0.02169\ H_2O + 0.00767\ S]$$
$$+ a[O_2 + 3.76\ N_2] \rightarrow bCO_2 + cH_2O + dSO_2 + eN_2$$

Carbon atom balance:

$$b = 0.676$$

Hydrogen atom balance:

$$2(0.2672 + 0.02169) = 2c \qquad c = 0.28889$$

Sulfur atom balance:

$$d = 0.00767$$

Oxygen atom balance:

$$2(0.0216) + 0.02169 + 2a = 2(0.676) + 0.28889 + 2(0.00767)$$

$$a = 0.7957$$

Nitrogen atom balance:

$$e = (0.7957)(3.76) + 0.0060 = 2.998$$

3. Air–fuel ratio:

Molar:

$$\text{a. } \overline{AF} = \frac{(0.7957)(4.76) \text{ kg mole air}}{1.0 \text{ kg mole fuel}} = 3.7875$$

Mass:

$$AF = \frac{(3.7875 \text{ kg moles air})(28.97 \text{ kg/kg mole air})}{(1.0 \text{ kg mole fuel})(0.09863 \text{ kg mole fuel/kg})^{-1}}$$

$$\text{b. } AF = 10.822$$

4. DuLong's formula:

$$HHV = 33{,}960\ C + 141{,}890 \left[H - \frac{O}{8} \right] + 9420\ S$$

or

$$HHV = 33{,}960(0.728) + 141{,}890[0.048 - (0.062/8)] + 9420(0.0222)$$

$$\text{c. } HHV = 30{,}640 \text{ kJ/kg coal}$$

5. Energy balance:

$$Q = \sum_{\text{prod}} N_i\,[\bar{h}_f^0 + \Delta\bar{h}]_i - \sum_{\text{react}} N_j[\bar{h}_f^0 + \Delta\bar{h}]_j$$

where

$$\sum_{\text{prod}} N_i\bar{h}_{f_i}^0 = [0.676(-94{,}054)_{CO_2} + 0.28889(-68{,}317)_{H_2O}$$

$$+ 0.00767(-70{,}947)_{SO_2} + 2.998(0)_{N_2}] \text{ cal/g mole}$$

$$= -83{,}860 \text{ cal/g mole}$$

$$\sum_{\text{react}} N_j \bar{h}^0_{f_j} = [0.676(0)_{C_S} + 0.2672(0)_{H_2} + 0.0216\,(0)_{O_2} + 0.006(0)_{N_2} + 0.02169(-68{,}317)_{H_2O} + 0.00767(0)_S] = -1482 \text{ cal/g mole}$$

$$Q = [-83{,}860 - (-1482) \text{ cal/g mole}]\left[4.187\,\frac{\text{kJ/kg mole}}{\text{cal/g mole}}\right] = 344{,}917 \text{ kJ/kg mole}$$

d. $$HHV = (344{,}917 \text{ kJ/kg mole fuel})(0.09863 \text{ kg mole fuel/kg}) = (34{,}020 \text{ kJ/kg ashless})\left(\frac{91 \text{ kg ashless}}{100 \text{ kg coal}}\right) = 30{,}960 \text{ kJ/kg coal}$$

Bituminous coal, the next highest grade, is a soft coal having a high sulfur content. It will produce significant smoke when burned and is therefore also not usable as a natural fuel. The largest U.S. coal reserves are bituminous; and, in general, this class of coal will have a lower fixed carbon but higher volatile concentration than the highest grade of coal. Bituminous coal, because of its high volatile content, is a good candidate solid fuel for use as a feedstock for producing synthetic liquid and gaseous fuels. It will burn easily in a pulverized or powdered form since, when heated, it tends to reduce to a cohesive, sticky mass.

Anthracite, the highest grade of coal, is a brittle, homogeneous and hard, black substance having a high fixed carbon content (usually above 90%) and a low percentage of volatiles (less than 15%). It is a dense material, bordering on graphite, having a high energy density, low sulfur content, but the lowest known U.S. reserves of the four grades of coal. Anthracite is a slow-burning coal that is difficult to ignite and tends not to cake. The composition and location of anthracite have made it an energy resource heavily used in the steel industry.

The most important renewable solid fuel resource, which has been used in many cultures for millennia, is wood. Timber is a natural solar energy storage medium, having a reproduction rate that can vary anywhere from 25 to 100 years, depending on the type of tree. Wood was a major energy source in the early history of the United States, and at the turn of the century, it still accounted for one-fourth of the nation's total energy supply. U.S. timber resources could yield about the same energy as that available in proven U.S. natural gas reserves, about a third more energy than that in proven oil reserves, and less than 10% of energy potential in proven coal reserves. Estimates further indicate that a region of approximately 30% of the total farming acreage in the United States could theoretically meet energy input requirements of all installed electric generating capacity in the United States.

The large water requirements for biomass production tend to limit use of timber today as a major fuel resource. Some consideration, however, has been given to producing liquid and gaseous fuels from biomass growing in aqueous environments such as the ocean. Biomass fuels generated from vegetation will be discussed in later chapters. Proper wood usage could contribute to development of forested regions as long as its burning rate does not ecologically and aesthetically exceed the growth rate in the region. History tells of what occurs when this principle is violated. Mediterranean forests were cut down during the years of the Roman Empire to provide fuel for smelting iron. Toward the end of the Roman Empire, local forests were so depleted that the Empire had to resort to moving their iron works north to German forests and importing iron from the provinces.

Although the industrial and commercial use of trees and other forms of vegetation as a solid biomass fuel is not as extensive today as in the past, its use in local applications and as a general-purpose or supplementary source of energy for domestic heating and cooking in specific areas is still significant. In certain regions, steam power plants purchase chipped wood as a fuel. Commercial use of wood and wood wastes is more common in the lumber industry. Wood waste left from manufacturing of lumber can be as high as 50% of the original logging and consists of unusable side slabs, bark, shavings, and sawdust. The larger pieces of this waste material are reduced in size for proper burning using a device termed a *hog*. Sized chips, shavings, and sawdust are mixed to yield a wood fuel commonly referred to as *hog fuel*. Sawmills burn these materials as a means of disposal and a readily available source of energy. Paper mills also burn their wet wood byproducts as a means of disposal.

The worldwide energy crisis of the 1970s motivated many Americans again to use wood for domestic heating. Burning wood in suitable wood-burning stoves may provide an auxiliary source of heat when an appropriate high energy density, slow-burning, low-creosote timber supply is available. Not all wood-burning stoves and fireplaces, however, are efficient space heaters. Many homes heated using wood actually lose more energy than is released by combustion since even good wood-burning stoves have efficiencies on the order of only 50%. Ash left from burning can be used in a garden, and burning wood is also a good means of utilizing diseased timber, trees downed by wind, as well as *cull* trees, i.e., those so twisted that they have no lumber value.

Table 4.2 Typical Analyses of Dry Wood

	Wt %						*HHV*[a]	
Type	C	H_2	S	O_2	N_2	Ash	kJ/kg	Btu/lbm
Cedar, white	48.80	6.37	—	44.46	—	0.37	19,540	8,400
Cypress	54.98	6.54	—	38.08	—	0.40	22,960	9,870
Fir, Douglas	52.3	6.3	—	40.5	0.1	0.8	21,050	9,050
Hemlock, western	50.4	5.8	0.1	41.4	0.1	2.2	20,050	8,620
Pine, pitch	59.00	7.19	—	32.68	—	1.13	26,330	11,320
white	52.55	6.08	—	41.25	—	0.12	20,700	8,900
yellow	52.60	7.02	—	40.07	—	0.31	22,350	9,610
Redwood	53.5	5.9	—	40.3	0.1	0.2	21,030	9,040
Ash, white	49.73	6.93	—	43.04	—	0.30	20,750	8,920
Beech	51.64	6.26	—	41.45	—	0.65	20,380	8,760
Birch, white	49.77	6.49	—	43.45	—	0.29	20,120	8,650
Elm	50.35	6.57	—	42.34	—	0.74	20,490	8,810
Hickory	49.67	6.49	—	43.11	—	0.73	20,170	8,670
Maple	50.64	6.02	—	41.74	0.25	1.35	19,960	8,580
Oak, black	48.78	6.09	—	44.98	—	0.15	19,030	8,180
red	49.49	6.62	—	43.74	—	0.15	20,210	8,690
white	50.44	6.59	—	42.73	—	0.24	20,490	8,810
Poplar	51.64	6.26	—	41.45	—	0.65	20,750	8,920

[a]Calculated from reported high heating value of kiln-dried wood, assumed to contain 8% moisture.
Source: Forest Products Laboratory, U.S. Dept of Agriculture.

Wood has a variable consistency and, like coal, is actually a heterogeneous fuel source; see Table 4.2. The major fuel component in wood is cellulose $(C_6H_{10}O_5)_n$, with trace amounts of lignin, resins, gum, water, and ash. Water content in wood, a critical parameter in its ability to burn properly, ranges from 30 to 50% by weight in green wood to 18–25% in air-dried lumber. Even burning various types of dried timber will release a different energy per unit mass because of the nonuniform concentrations of resins, gums, and other trace combustible components in wood. Since Dulong's formula does not include heat released by these trace constituents, predicting heating value with this relation may predict improper heats of combustion for wood.

Coal or wood can be upgraded as a fuel suitable for certain applications by simply driving off volatiles and moisture content of the raw materials. Destructive distillation, accomplished by limited combustion in an oxygen-deficient atmosphere, or *pyrolysis,* produces a solid carbonaceous residue, or *char,* that can subsequently be burned. Char produced using coal resources is called *coke* which, because it is purer than raw coal, is a useful fuel in the metallurgical industry in the production of iron and steel. Char produced using wood is referred to as *charcoal* and is used, among other things, as a filtering medium.

Certain solid by-products of the food processing industry can also produce a useful fuel. As examples, nut shells, rice hulls, coffee grounds, and corn husks are burnable and can be used to provide some of the necessary heating in fruit and vegetable processing plants. Sugarcane grows abundantly in Central America and the tropics and, on a moisture- and ash-free basis, consists of 83–90% sugar and 17–10% cellulose by weight. After extracting the sugar by shredding, rolling, and steam heating, the wasteful fibrous cane by-product, termed *bagasse,* can be burned in a suitable boiler used to generate steam required to extract sugar from the cane.

Modern industrialized societies tend to produce an abundance of both organic and inorganic solid wastes. The traditional means for disposing of these high-technology waste by-products is by burial in sanitary landfills. Many constituents of *municipal solid waste,* or MSW, are not inert and, when buried, will begin a process of biodegradation that can produce potentially explosive pockets of gas consisting of approximately 50% methane–50% carbon dioxide by volume. In addition, toxic compounds such as chlorine, fluorine, and other trace substances present in raw MSW can leach into the water table surrounding a landfill site. The continued introduction of non-inert MSW into a limited number of suitable landfill sites is therefore a serious environmental issue, which many growing urban centers must now face.

An alternate means of MSW disposal, since those materials by weight consist of 50–80% combustibles, is incineration. After burning, original waste material is transformed into a sterile, solid burned-out residue that is reduced to 5% of the original volume or 17% of its original mass. Incineration has been used chiefly as a means of volume reduction of MSW prior to burial in landfills. Mass burning of trash does release energy in a form as heat that could be recovered if appropriate steam-generating systems are utilized. Incinerators with waste heat boilers have been used to provide steam for heating purposes since the turn of the century in both the United States and Europe. Direct combustion of 10–20% MSW, 90–80% coal mixtures by weight, has been successfully burned using particular coal-burning equipment. Use of MSW as a suitable solid fuel does raise some engineering concerns and constraints, such as generation of unhealthy levels of hazardous air pollution, including airborne ash; presence of highly explosive constituents, such as gas cans, fireworks, TNT; unacceptably high moisture content; long-term instability due to the biodegradable nature of MSW; and production of pungent odors. Incineration-fired steam generators tend to have abnormally corrosive flue gases that attack heat transfer surfaces in a unit because of the presence of chlorine, fluorine, sulfur, and plastics in the waste stream. Incinerators for MSW reduction have been curtailed as emission control requirements have become more stringent. Modern refuse disposal combustion facilities that are environmentally safe are under design and construction in the United States as well as overseas in Europe and Japan. The Europeans and the Japanese, because of their limited land, energy, and economic resources, were concerned with resource recovery systems before major U.S. efforts began after the energy crisis of the 1970s.

Efforts are also being made to use MSW as a feedstock for commercially produced synthetic solid, liquid, and/or gaseous fuels. Liquid and gaseous fuel resource characteristics will be considered in Chapters 5 and 6, and so the following material will deal only with the technology necessary for producing a solid *refuse-derived fuel,* or RDF. Currently operating RDF facilities demonstrate that the concept is viable. However, since such facilities are complex, design and operation of an RDF-producing plant must be in the hands of experts knowledgeable in both the characteristics of waste-generated fuel combustion and the particular heat engines necessary to utilize these materials. Operational RDF-producing facilities consist of the following major components:

1. Staging/sorting area
2. Shredders
3. Air classifiers
4. Metal magnetic separators
5. Sizing devices
6. Bag houses
7. Fuel densifiers
8. Transfer mechanisms

Arrangement of these components can be varied to optimize a facility for local conditions of refuse, fuel form desired, and local environmental and social requirements. The following discussion describes a general flow of MSW through such an RDF plant; see Figure 4.2.

Solid RDF production facilities should generate minimal airborne pollutants. Dust and potential explosion of certain waste items are hazards when RDF facilities are in operation. Mechanical shredding, waste stream separation or classifying, and suitable processing of raw MSW yield a paperlike substance that can be burned directly. Additional steps will allow long-term storage and use of RDF as a clean-burning, renewable, low-sulfur fuel that

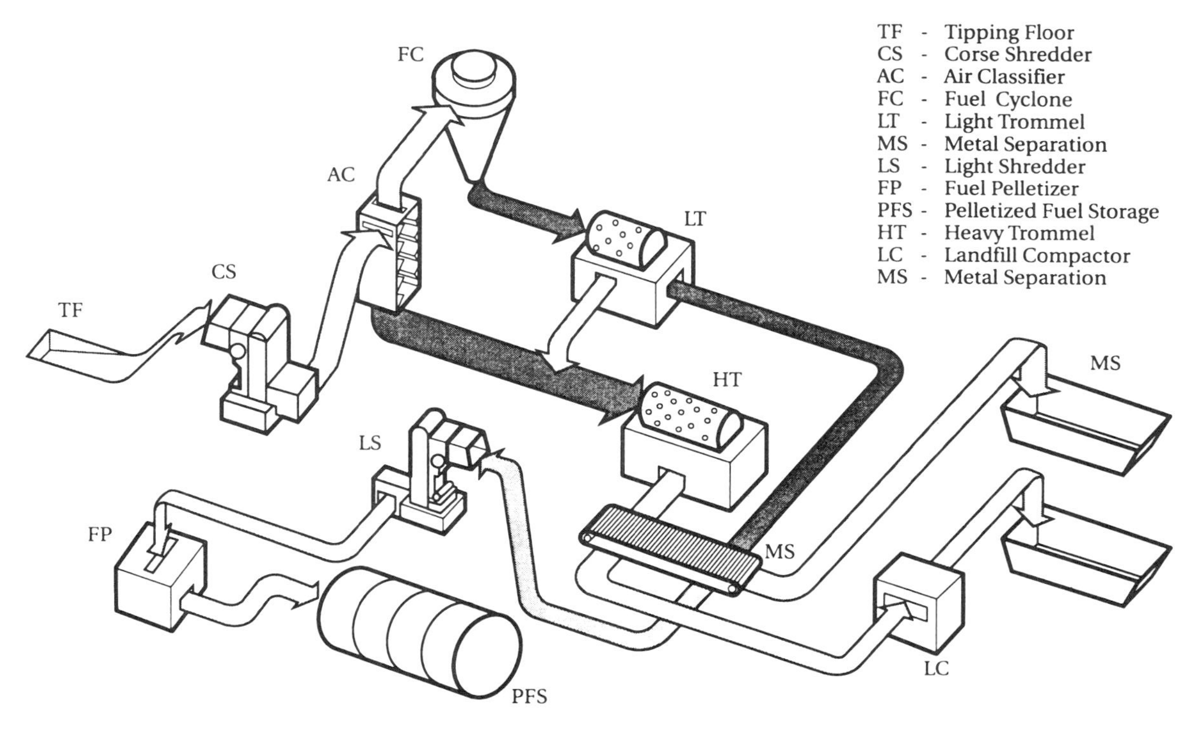

Figure 4.2 Schematic of a municipal solid waste (MSW) fuel processing plant.

can supplement coal in the generation of steam and/or electricity. Removal of metals and glass from MSW prior to RDF production purifies as well as energetically upgrades the fuel on a mass basis.

Municipal waste is initially delivered to a plant at the staging/sorting area. This area should be enclosed for aesthetic, environmental, and operational benefits. Visual inspection of MSW being dumped allows plant operators to remove any massive and/or hazardous objects prior to its being fed to the RDF plant. Continuous feed and control of this supply stream occurs at the staging area.

Two successful techniques of staging/sorting are used: tipping floor transfer or a recessed floor transfer configuration. With the tipping floor, refuse delivery trucks drive onto a large, flat floor dump their waste, and then leave. Front-end loaders transfer and sort material being fed to the plant. With the recessed floor configurations, refuse trucks back up to a large on-grade floor, dump their waste, and then leave. Hydraulic pickers, overhead cranes, or front-end loaders then transfer material to the RDF-producing plant feed lines. Long-term storage of material at the staging/sorting area is not desirable.

Typically, MSW is continually fed to a shredder, mill, or hammer system that reduces the waste stream to small pieces. The operation, wear, and energy input required by any shredder are a strong function of a specific waste-stream composition, i.e., the presence of abrasive and potentially explosive constitutents. Separation of metals, glass, stones, and ceramics from the flow prior to shredding could prolong the life of the mill hammers. Removal of potentially explosive materials, such as gas cans, TNT, and fireworks, would also increase safe operation of the shredder. The required specific energy input per mass of fuel produced is greatly affected by this separation process. This size reduction step, however, does not segregate, separate, or select waste stream components.

The air classifier is the major component that separates light organic and inorganic combustible materials from reamaining heavy objects. This unit, which is often several stories tall, uses high-velocity air supplied by fans to separate out pulverized waste. Light materials, consisting mainly of paper, plastics, wood chips, and other combustibles, are drawn out the top while heavier particles, chiefly metal, glass, and ceramics, fall to the bottom.

A magnetic system can extract ferrous metals from the waste stream for resale as scrap. An "aluminum magnet" using eddy current separation techniques can also be used to remove aluminum for recycling. Glass, stones, and ceramics are removed from the heavy fraction by a sizing device. This unit can be either a trommel or a vibrating screen, which allows smaller materials to pass through a fine mesh while restricting the passage of larger pieces. Abrasives, such as fine sand, glass, and grit, are separated by this technique either for further processing or transfer to a sanitary landfill.

Dust generated by the operation of an RDF-producing facility is collected by the bag house. The collection system consists of pickup hoods that draw dusty air from various points around the plant, i.e., shredders and conveyor transfer points. A series of cyclone separators removes dust, after which the air is filtered, cleaned, and discharged to the environment.

The light combustible stream can be transferred for direct combustion or further processing. Compacting and extrusion of processed RDF product to form cubettes, sheets, or pellets heat the material to temperatures that kill bacteria, which normally would cause MSW to decompose. The long-term storage of RDF is thereby enhanced as long as the material is kept dry. The compactor is the plant component that densifies this stream and produces the final form of storable RDF. Also, if the light stream is chemically treated with an embrittling agent, this material can be ground to a powder or dust in a mill to make a fine mesh pulverized RDF.

When properly processed, highly refined RDF can be transferred, stored, and burned with high efficiency in a variety of compatible systems. In addition, potential recovery of metals and glass, as well as reduced landfill volumes, makes RDF processing more acceptable than other solid waste utilization schemes. Conversion of MSW into fuel-grade RDF offers the following benefits:

1. A long-term energy source since industrial and urban waste will continue to be generated.
2. Established and successful technology beyond the pilot plant stage.
3. Minimal environmental impact on air pollution standards of all waste recovery techniques.
4. Maximum utilization of waste materials, i.e., fuel production, as well as proven recovery of metals and glass.
5. Minimum required landfilling of inert material.
6. Commercial viability and replicability throughout the United States in the near term.
7. RDF is a low-ash, low-sulfur, low-slagging, odorless, higher-Btu fuel source than raw MSW.
8. RDF technology can produce a variety of different fuels for a variety of applications:
   a. Densified RDF
   b. Pulverized RDF
   c. Coal-RDF mixture
   d. Waste oil–impregnated RDF
   e. RDF-derived alcohol

Table 4.3 Typical Solid Fuel Analysis

		Stoker coal	Wood (bark)	RDF	RDF (densified)
HHV	kJ/kg	32,100	18,610–20,930	12,790–13,960	12,790–13,960
(dry basis)	(Btu/lbm)	(13,800)	(8,000–9,000)	(5,500–6,000)	(5,500–6,000)
Volatiles	%	37	70–75	50–60	50–60
Moisture	%	1.8	50	15–25	15–20
Ash	%	6.6	3–5	15–20	15–20
Density	kg/m^3	800–961	160–240	48–97	480–641
	(lbm/ft^3)	(50–60)	(10–15)	(3–6)	(30–40)
Particle size	cm (in.)	90% < 3.175 (1.25)	100% < 10.16 (4.0)	10.16 (4) nominal	100% 5.08 (2)
		30% < 0.635 (0.25)	95% < 5.08 (2.0)		
			50% < 1.25 (0.5)		

9. Compatibility with existing boilers and demonstrated efficient combustion therein.

The technology of refuse-derived fuel will achieve the goals set forth above if successes and failures of previous fuel processing and combustion systems are fully understood. Producing fuels using solid wastes in the future is as much an economic as an engineering issue. Economics and energy engineering dictate that each plant design be unique for local conditions. Municipal solid waste is a seasonally changing, nonhomogeneous mixture containing a variety of combustible and recyclable materials in various sizes and shapes. As such, a knowledge of local area waste stream time-dependent composition is essential. An emerging RDF technology does raise several negative economic points, including: social cost required to produce these fuels in a usable and socially acceptable manner; economics of scale necessary to ensure successful operation of an RDF processing plant; and the requirement that resources other than gas and oil be utilized in RDF production. Positive economic benefits can result from RDF technology commercialization, including the cost benefit of producing RDF over the cost of MSW landfill disposal; the cost benefit of replacing oil and gas with RDF for use in certain suitable applications; and the cost benefit of revenue-producing recycling of scarce materials as recovered during RDF production.Various solid fuels are compared in Table 4.3.

4.4 SOLID FUEL COMBUSTION

Burning solid fuels, such as coal, requires both proper machinery and sufficient preparation of reactants in order to achieve and control the chemical energy conversion processes during combustion. An understanding of major solid fuel–heat engine interface characteristics is essential in discussions of direct combustion of chemical energy resources, as presented in earlier sections of this chapter. For example, because of the structure, rank, chemical composition, caloric value, porosity, and caking nature of various grades of coal, different methods of firing are required. Also, the nonuniform nature of these reactive materials influences, among other factors, the required fuel–air ratio necessary to sustain complete combustion, the amount of energy released, and the type of critical emissions present in the effluent product stream. Coal generally burns as either a *caking* or *free-burning* solid fuel. Caking coals, when heated, tend to fuse and form a semicoke substance that is impervious to the passage of air and combustion gases. Free-burning coals (FBCs), however, tend to crumble into smaller pieces when heated, which then burn separately.

Efficient solid fuel combustion systems will be successful only if they provide a proper distribution and flow rate of reactants, proper ignition, and a proper firing temperature. Solid fuel combustion systems therefore must (1) prepare fuel for burning, (2) transfer energy released by the complex chemical reactions as either heat or work, (3) minimize incomplete combustion and losses, as well as (4) control formation and removal of emissions. Heat engines used with solid fuels are generally large and traditionally have been steam generators that are used for heat and power applications. Coal-burning steam-generating facilities, in addition to firing, must store, handle, and crush or pulverize coal. Coal-fired units are designed to burn a reasonable range of statistically varying coal supplies. Dust and bottom ash collection, as well as stack systems that remove soot and fly ash, is also needed to lessen the environmental impact of these byproducts.

Three general methods for burning coal have been used: stoker systems, pulverized coal systems, and cyclone-firing systems. Stoker systems, limited to small capacity boilers of less than 181,440 kg/h (400,000 lbm/h) steam flow rates, were developed in the late nineteenth century and used predominantly until the 1920s. In the 1920s, pulverized coal firing was introduced, allowing coal to be burned in boilers having greater capacities, with steam flow rate of 181,440 kg/h (400,000 lbm/h), and is still the chief means of burning coal to produce power commercially. Cyclone-fired furnaces, developed in the 1940s, provide an improvement in efficient coal burning over either stoker or pulverized coal–firing systems. Considerable efforts are now also being made to commercialize a fourth coal-burning technique—fluidized bed combustion. Coal-fired fluidized bed combustors are a promising alternative that could improve combustion efficiency of coal-fired steam generators, lower sulfur oxide emissions of commercial power plants, and eliminate ash slagging and fouling of steam-generating system.

The first mechanical coal-firing technique to replace hand firing was the *stoker* coal-fired unit. Three general classes of mechanical stokers, overfeed, spreader, and underfeed systems, have been used. Mechanical stokers are limited to low-capacity steam generation but are relatively simple, are economical, and can be designed to accommodate a wide range of unsized solid fuels ranging from all ranks of coal to wood, bagasse, and even RDF.

Coal combustion using stokers involves three phases of reaction: drying, distillation of volatiles, and burning of fixed carbon on a burning bed; see Figure 4.3. When firing, crushed coal is admitted from a storage hopper through a regulating coal feeder and introduced onto a moving grate. Various overfeed grate designs have been utilized, such as the chain and

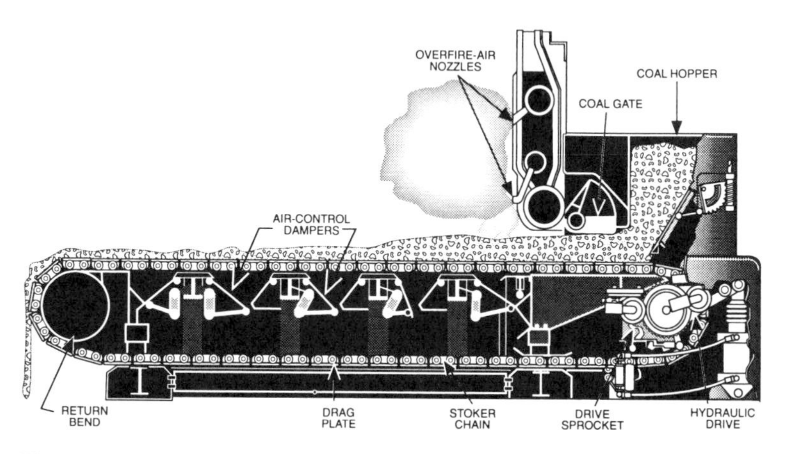

Figure 4.3 Stoker-fed coal combustion system schematic.

traveling grate systems and the vibrating and oscillating grate systems. Stoker feed furnaces burn both fine and coarse coal in a burning bed that moves away from the coal injection point. The larger particles lie on the grate and react with the underfire primary air supply while smaller particles are held in suspension over the grate and are ignited by the overfire secondary air and hot product gases. Size distribution is critical in that large or coarse coal does not provide as great a total surface area for complete combustion as smaller pieces would. Too fine a coal bed would require a very fine grate opening, which would make it difficult for combustion air to pass through. The layer of coal on a moving grate is heated, dried, pyrolyzed, and ignited by radiation from furnace gases directly above the bed. Fine coal in suspension, hydrocarbon, and other combustible gases driven off the coal bed by distillation, as well as products of incomplete combustion such as CO, burn together above the grate in overfire gases. The moving bed becomes thinner as it proceeds until, ideally, at the end of travel, all that remains is ash, which is dumped as bottom ash into an ash hopper.

Spreader stokers fire coal in much the same way as a traveling grate overfeed stoker. The essential difference between spreader and moving grate stokers is that a spreader system uses a combination of suspension burning and a thin, fast-burning fuel bed, whereas overfeed systems utilize chiefly a mass-bed burning approach. In spreader firing, coal is uniformly spread over a burning coal bed by means of a distributor; see Figure 4.4. Spreader coal firing can be used with either a stationary or moving bed.

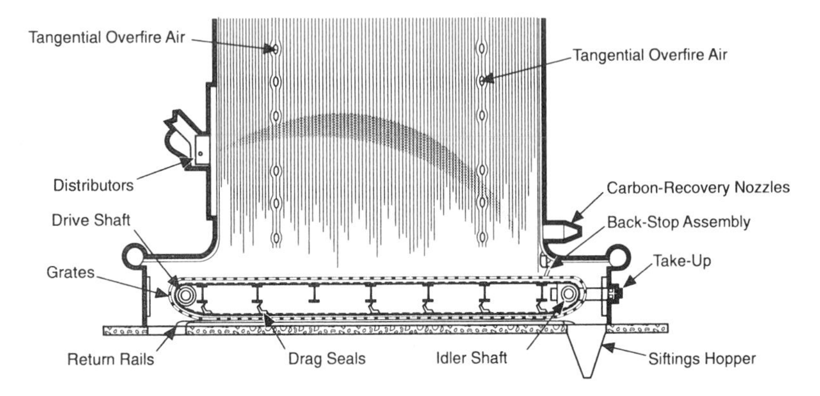

Figure 4.4 Spreader-fed coal combustion system schematic.

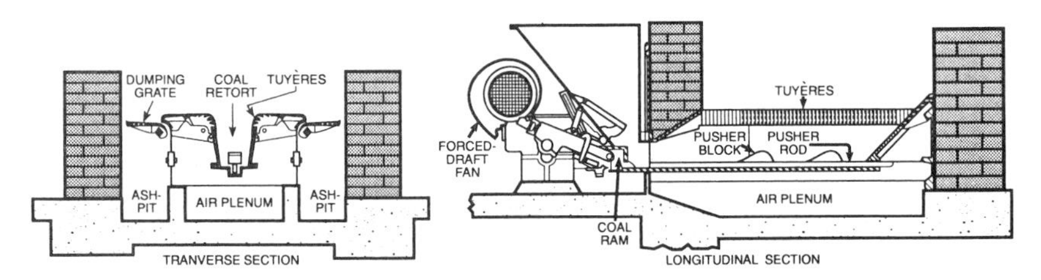

Figure 4.5 Underfeed stoker coal combustion schematic.

Underfire stokers operate by feeding crushed coal through a channel, or *retort,* below the burning bed; see Figure 4.5. When the passage is filled, the coal is pushed up and over on two sides, forming and feeding the burning fuel bed. Combustion air passes up through tuyeres with a burning bed, which is being pushed toward ash dumps by retorted coal. Underfeed stokers are best for burning caking coals, spreader stokers will burn all ranks of coal except anthracite, and traveling stokers will fire all but strongly caking bituminous coals.

Commercial operation of large coal-fired steam generators producing 453,600–1,814,400 kg/h (100,000–4,000,000 lbm/h) steam began in the 1920s with the introduction of pulverized coal firing. The added cost of operating pulverized coal plants, in terms of increased complexity and plant size, is offset by a gain in steam generation and reduction in labor in these units over stoker-fired, coal-fired boiler installations. Essential combustion elements of a pulverized coal facility are shown in Figure 4.6.

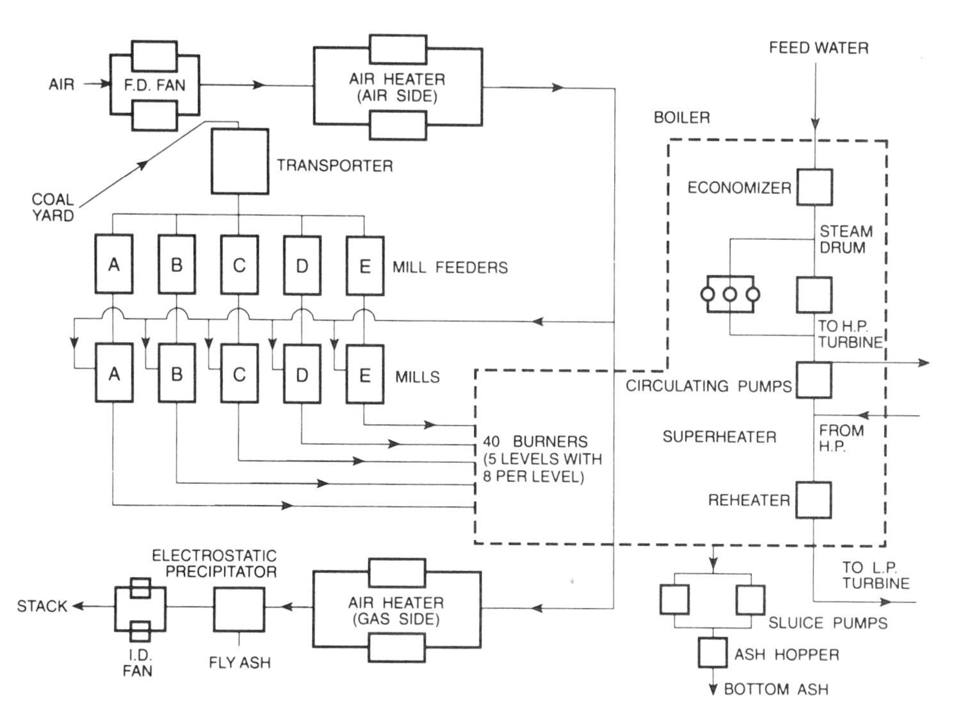

Figure 4.6 Pulverized coal combustion system schematic.

Coal is moved from a supply yard to the furnace on a *transporter* that fills *mill feeders,* i.e., containers that are used to store and meter proper amounts of coal for burning. Gravity-fed coal is delivered from a silo, hopper, or bin mill feed system to coal grinding components, or *mills.* Coarse coal is pulverized into a fine dust having particle dimensions of a few thousandths of a centimeter, or approximately 200 mesh size. Environmental air is supplied for combustion using a *forced-draft fan.* Hot exhaust gases from the furnace are used to preheat air in a heat exchanger, termed an *air preheater.* Approximately 20% of the air flow, or *primary air,* is used to dry and blow pulverized coal from mills into the furnace, while the remaining, or *secondary,* air is supplied directly to the furnace to promote complete combustion.

Pulverized coal particles burn in suspension within the furnace. The smaller size, increased surface area, and long residence time of pulverized coal particles confined within the fireball make this technique of firing very fuel-insensitive. The homogeneous fine coal–air stream burns much like a gaseous torch. Radiant distillation of coal particles blown into the furnace leaves a coke particle that then burns by diffusion of oxygen through the CO, CO_2, and N_2 atmosphere surrounding the particles. An *induced-draft fan* extracts products of combustion, i.e., *flue* or *stack* gases, from the furnace and pulls exhaust gases through the hot side of the air preheater, by means of appropriate fly ash removal machinery, and discharges remaining material to the environment.

The cyclone-fired coal burner, introduced in the 1940s, improved certain negative features of pulverized coal combustion systems; see Figure 4.7. Coal feed is similar to pulverized coal units except that coal is not crushed as fine, only four mesh, thereby requiring less power for grinding. Cyclone coal firing uses a small horizontal water-cooled burner, which requires less furnace volume than that used in pulverized coal plants. Crushed coal is admitted to a horizontal burner, with combustion air entering tangentially, inducing swirl and greater turbulence, which promotes better burning. Cyclone burners allow higher gas temperatures, 1650°C (3000°F), enabling quick firing and complete combustion to occur. In addition, high firing rates melt 30–50% of ash, forming a liquid slag that can be removed as a liquid bottom. Pulverized coal units generate 10–30% ash as bottom ash and 90–70% as fly ash. Cyclone units, however, generate only 50–70% of the ash as fly ash. Also, cyclone furnaces do not require the combustion

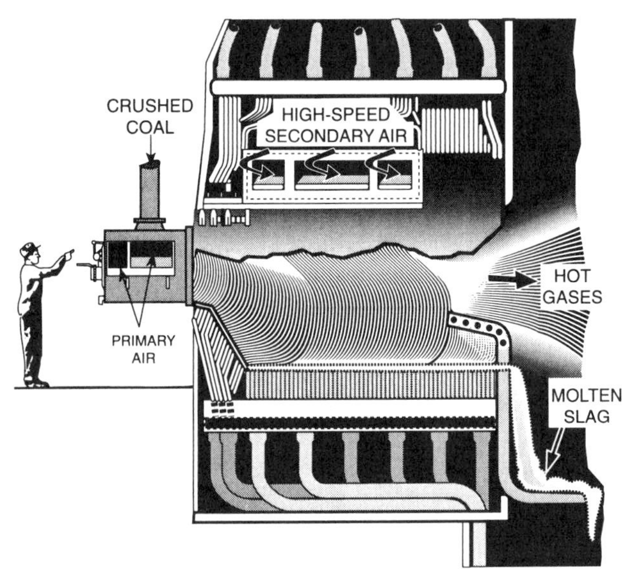

Figure 4.7 Cyclone combustor schematic.

space that pulverized coal combustion systems do, so that cyclone coal combustion plants need not be as large.

The *fluidized bed combustor,* or FBC, is a more recent solid fuel combustion technology, currently being developed for commercial use around the world; see Figure 4.8. Combustion occurs within a unique bed consisting chiefly of sand, with lesser amounts of limestone, ash, and only about 1% fuel. Air with a nominal velocity of 122–366 cm/sec (4–12 ft/sec) passes up through the bed to provide sufficient momentum for the bed to overcome its static weight; generate turbulent action, causing the bed to behave much like a boiling liquid; and supply sufficient oxygen for complete combustion. Too high a flow rate would cause the reacting bed to blow away, whereas too low a velocity would make the bed slump and not burn properly. Potentially, FBC can burn coal as well as a variety of the solid fuels described earlier in this chapter. In addition, control of critical by-products of solid fuel combustion, such as ash, sulfur dioxide, and nitric oxide, may be possible with this technology.

Turbulent mixing maintains a fairly uniform operating temperature of approximately 680–870°C (1150–1600°F) throughout the burning bed. These reaction temperatures are generally below ash melting temperatures, so that the slagging and fouling normally associated with coal-fired steam generators is absent. Furthermore, at these temperatures, hot corrosion problems associated with steam-generating tubes are minimal; this allows tubes to be buried directly in the combustion bed. Steam generation using a solid bed provides greater heat transfer rates, which translates directly into greater steam generation rates, less required surface area,

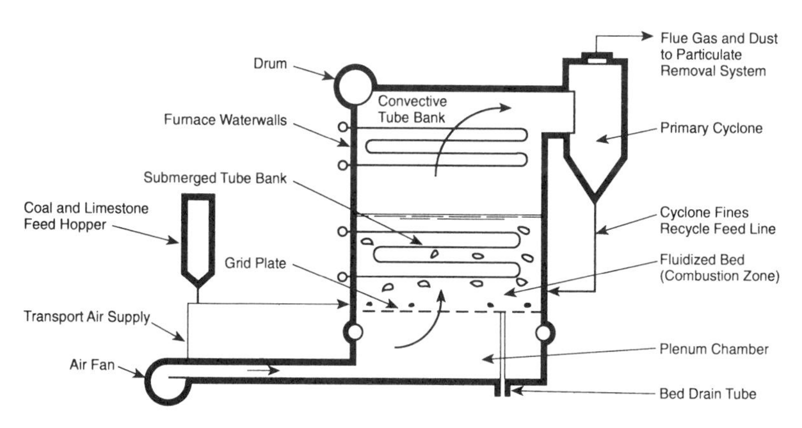

Figure 4.8 Fluidized bed combustor (FBC) schematic.

Table 4.4 Usual Amount Excess Air Supplied to Fuel-Burning Equipment

Fuel	Type of furnace or burner	Wt. % excess air
Pulverized coal	Completely water-cooled furnace for slag tap or dry ash removal	15–20
	Partially water-cooled furnace for dry ash removal	15–40
Crushed coal	Cyclone furnace pressure or suction	10–15
Coal	Stoker-fired, forced-draft, B&W chain grate	15–50
	Stoker-fired, forced-draft, underfeed	20–50
	Stoker-fired, natural draft	50–65
Fuel oil	Oil burners, register type	5–10
	Multifuel burners and flat flame	10–20
Natural, coke oven, and refinery gas	Register-type burners	5–10
	Multifuel burners	7–12
Blast furnace gas	Intertube nozzle-type burners	15–18
Wood	Dutch oven (10–23% through grates) and Hofft type	20–25
Bagasse	All furnaces	25–35

Source: Steam Its Generation and Use, B&W Co. (1963).

lower unit weight, and reduced capital cost for FBC systems. Since coal combustion occurs in the FBC, most of the ash is also entrained within the active bed, thus reducing fly ash. This method of burning is compatible with high-sulfur coal. Fuel sulfur oxidizes to form sulfur dioxide, SO_2, and limestone, dolomite, or other absorbent particles present in the bed react with generated sulfur dioxide to form calcium sulfate. This stable material can then be removed from the bed and be either sent to a landfill or used as a feedstock to produce sulfur.

Many FBC design problems remain to be solved before successful commercialization can be realized, including better coal feed and bed removal systems, suitable scaling of pilot plants to commercial size, and optimizing sorbent concentrations for sulfur removal, to name but a few. Cleanup of gases from FBC would open up the possibility of operating a pressurized fluidized bed combustor (PFBC), which could allow hot gases to provide additional expansion work in combined cycle applications.

The increased use of solid fuels and solid fuel combustion systems makes their environmental impact and pollution control essential. The resurgence of coal-fired electric power generation and cogeneration, i.e., combined electric and steam production, escalating amounts of MSW, and its incineration, landfilling, or processing to produce RDF have made improvements in combustion efficiency and emissions control both a local and a federal matter. Engineers in the future will need to make major fuel engine design choices for solid fuel combustion units, choices that will concurrently influence selection of fuel options, method of firing, and means of controlling emissions. Table 4.4 lists some excess air requirements.

The carbon-to-hydrogen atom ratio of any hydrocarbon fuel is one of many important parameters that affect energy content, density, and other combustion properties. Coal and many of the solid fuel resources discussed in this chapter have a high carbon-to-hydrogen ratio and therefore will produce a greater proportion of carbon dioxide than other hydrocarbon fuels. In fact, CO_2 produced per unit of thermal energy release is greater for coal than for other liquid and/or gaseous hydrocarbon fuels. In the limit, ideal combustion of pure carbon yields only CO_2 or

$$C_S + O_2 \rightarrow CO_2 \tag{4.7}$$

and

1 kg mole carbon → 1 kg mole carbon dixoide

(1 lb mole C) → (1 lb mole CO_2)

or

1 kg carbon → 44 kg carbon dioxide

(1 lbm C) → (44 lbm CO_2)

It has been documented by groups such as the National Oceanic and Atmospheric Administration and the National Academy of Sciences that the world's atmospheric CO_2 level has been steadily increasing. Although only a part of atmospheric CO_2 results from combustion sources, increased coal combustion would accelerate this growth at a greater rate. Certain gases, such as CO_2 and water vapor, are known to absorb high-temperature thermal radiation in the infrared region of the electromagnetic spectrum. These gases are opaque to low-temperature radiation in the infrared region, while gases such as oxygen and nitrogen are transparent throughout the spectrum. This ability of CO_2 and H_2O in the atmosphere to pass solar energy to the earth but block environmental radiation back into space is commonly referred to as the *greenhouse effect.* Predictions suggest that a continued increase in atmospheric CO_2 could raise the earth's air temperature an average of 2–4°C by the next century. The delicate balance in the earth's biosphere requires a greater understanding of, and concern for the generation of CO_2 and its consumption by vegetation and the ocean, as well as geothermal impact of these parameters on the earth's weather.

Other solid fuel combustion products, in addition to CO_2, may also have potential long-term biological and geological impact on the earth's ecosystem. As an example, certain exhaust gas species can interact with moisture and, given proper conditions, produce a dilute acidic solution; i.e., CO_2 can form carbonic acid, sulfur dioxide can form sulfuric acid, and nitric oxide can form nitric acid. Acid corrosion within coal-fired steam generators is well known, but concern is growing about the effects these species have on the environment when they precipitate in the form of acid rain.

The Clean Air Act Amendments of 1977 set air-quality standards for the United States that limit pollution from combustion of coal and other solid fuels. Solid fuels, when burned, generate three major pollutants that are to be regulated and controlled by this law: sulfur oxides, SO_x; nitrogen oxides, NO_x; and particulates. Fossil-fueled power plants annually produce approximately 65% of the total sulfur oxides, 29% of the nitric oxides, and 24% of the particulate emissions, with almost two-thirds of these pollutants coming from coal-fired plants alone.

Sulfur present in solid fuels, such as coal, will form sulfur oxides, or SO_x, during combustion, or

$$S + aO_2 \rightarrow bSO_2 + cSO_3$$

These oxides and other pollutants, along with the products of combustion CO_2 and H_2O, are released into the atmosphere by a power plant. Sulfur oxides can precipitate in rain or snow as acid rain or as dry SO_x particles that settle on trees and soil, which then interact with rain or snow to produce sulfuric acid:

$$SO_x + aH_2O \rightarrow bH_2SO_4 + cH_2SO_3$$

Sulfur dioxide control can be accomplished prior to combustion by choosing to burn only low-sulfur coal. Burning low-sulfur coals may involve an engineering design tradeoff since most low-sulfur coals are low-energy coals. In addition, most U.S. coal reserves are high-sulfur fuels. Benefication, or coal cleaning, to alter sulfur, ash, and moisture content prior to burning, can produce harmful by-products and waste that are as difficult to dispose of as the original sulfur. Controlling sulfur dioxide emissions during combustion may be possible using a fluidized bed combustion system. Flue gas desulfurization techniques are necessary to control SO_x after combustion, as in pulverized coal units.

Current practice for desulfurizing flue gases utilizes a scrubbing agent such as water and/or an alkaline material like lime or limestone. Most commercial units are wet systems, and a variety of geometries, including spray towers, packed towers, and spray chambers, are in use. Wet scrubbing provides several benefits, including the ability to scrub flue gases efficiently from a variety of solid fuels, the ability also to remove some particulates, little or no restrictions on flue gas temperatures or moisture, and desulfurization within a small region of space. There are also some disadvantages of wet systems such as problems associated with disposal of the slurry, acidic misting, and high power requirements. Efforts currently focus on dry absorption systems and regenerative techniques that would allow material to recover sulfur reversibly.

The NO_x emissions associated with solid fuel combustion are a direct result of several fuel-engine parameters, such as actual reaction temperatures, means of preparing and sizing solid fuel particles being burned, percent nitrogen originally present in the fuel, and percent excess air used in firing. Any technique used and any alteration in design or operation of solid fuel burners that favorably shifts temperature and partial pressures of O and N via the $O_2 \rightleftarrows 2O$ and $N_2 \rightleftarrows 2N$ reactions will contribute to a reduction in NO_x emissions. Deposits from reaction, such as ash, make heat-absorbing surfaces in boilers ineffective, increase flue gas temperatures, and contribute to formation of NO_x. Burning solid fuels, such as coal, can generate both thermal NO_x because of high-temperature N-O thermochemistry and fuel NO_x because of the presence of fuel nitrogen. Reducing reaction temperature is a means of controlling thermal NO_x, and several techniques are currently used, including a reduction in excess air, flue gas recirculation, and water injection. Control of the highly oxidizing primary combustion, reduction in air preheat, and off-stoichiometric burning also can be used to control fuel NO_x.

Solid fuel combustion also generates a high level of fly ash and uncontrolled particulates. These fine particles, or *soot,* consisting of inert materials, trace toxic metals such as cadmium and arsenic, and solid carbon, are of concern currently because:

1. They are principal carriers of toxic and carcinogenic trace metals.
2. They are more easily inhaled.
3. Their greater surface area increases reactive or absorptive capacity.
4. Particles in the size range 0.1–1.0 μ scatter or absorb sunlight, lowering visibility.
5. Increases in consumption of coal could significantly increase the amount of fine particles in the atmosphere.

Particulate levels are a result of the type of fuel being burned and the method of firing. For example, RDF is a greater source of fly ash than coal. Also, stoker and spreader firing produces greater particulates than pulverized burners which, in turn, produce more particulates than cyclone or FBC systems. Mechanical collectors, or *cyclones,* use a vortex motion and inertia to collect particulates. The performance of cyclone systems is a function of gas volume handled, particulate loading, cyclone inlet conditions, gas velocity, and cyclone geometry. Cyclones are simple, low-cost devices having no moving parts, but they are ineffective with small particulates and are subject to erosion, reducing the effectiveness of the units. Electrostatic precipitators are used frequently with pulverized coal units and utilize a high-voltage DC discharge between two plates to charge and collect particles on a collecting plate. These collectors are highly efficient and reliable and can be scraped periodically to remove collected material, but they are costly to operate. Bag houses use fabric filters to collect the particulate.

4.5 BOILER ENERGY BALANCE

The combustion of most solid fuels described in this chapter is compatible with steam-generating systems used for industrial heating and/or production of power; see Figure 4.9. Many important energy characteristics of such external combustion machinery can be determined from a knowledge of the following items:

1. Energy available in the fuel as fired
2. Energy absorbed by generated steam
3. Energy loss due to dry flue gas conditions
4. Energy loss due to moisture in fuel
5. Energy loss due to moisture in combustion air
6. Energy loss due to moisture formed from burning hydrogen in fuel
7. Energy loss due to incomplete combustion
8. Energy loss due to radiation heat transfer and other unaccounted effects

In order to quantify the performance of a steam generator or boiler, several measurements are required to be made during any standard boiler test. It is important first that the unit achieve steady-state conditions and that these conditions be maintained throughout the entire test period. Often test codes, such as the ASME Boiler Test Codes, are used to determine a *boiler heat balance.* These procedures have been developed and published in engineering literature. Based on published procedures and practices included in these codes, the following items are found to be needed for predicting combustion aspects of a steam generator.

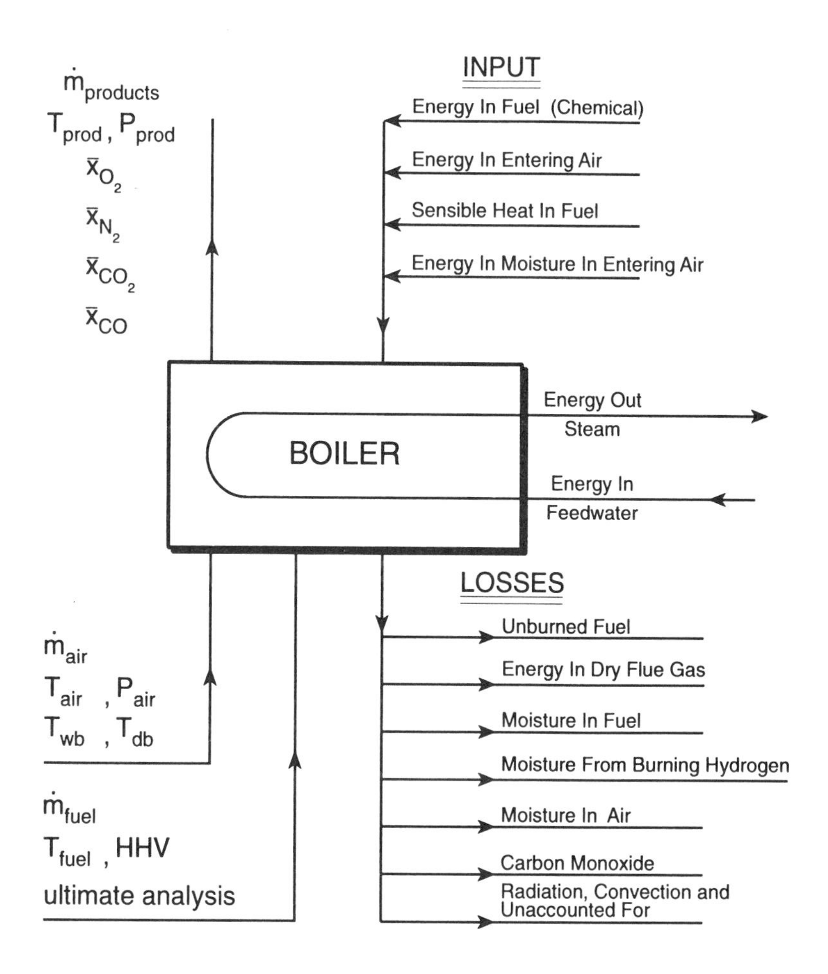

OUTPUT = INPUT - LOSSES

$$\text{EFFICIENCY} = \frac{\text{OUTPUT}}{\text{INPUT}} = \frac{Q_b}{\text{HHV}}$$

Figure 4.9 Elements of a boiler energy balance.

1. Ultimate analysis of the fuel along with its higher heating value
2. Mass of fuel fired during the test
3. Fuel temperature entering the burner
4. Relative humidity, barometric pressure, and dry bulb temperature of combustion air
5. Total mass of steam generated during the test
6. Average temperature and pressure of both the feedwater and the generated steam.
7. Flue gas temperature measured just at the exit of the heat exchanger.
8. Average flue gas analysis as sampled just at the heat exchanger exit, i.e., mole fractions of CO, CO_2, O_2, and N_2.

Recall from Chapter 2 that, ideally, energy released by a fuel–air mixture can be predicted using the heating value of the fuel. Therefore, if a boiler could operate such that (1) ideal complete combustion occurred, (2) reactants and stack temperature were both at *STP*, and (3) water in the exhaust gases formed by combustion was a liquid, then the heat released would be a maximum and equal to the higher heating value of the fuel. If the generated steam could then absorb all the chemical potential energy of the fuel released by combustion, the necessary heat transfer for making steam would equal

$$\dot{Q} = \dot{m}_{H_2O}(h_2 - h_1) \qquad \text{kW (Btu/h)} \tag{4.8a}$$

and

$$\dot{Q} = \dot{m}_f \, HHV \qquad \text{kW (Btu/h)} \tag{4.8b}$$

where

$\dot{m}_{H_2O}$ = feedwater mass flow rate
$\dot{m}_f$ = fuel mass flow rate
HHV = fuel higher heating value
h_2 = steam boiler or superheater generated outlet enthalpy
h_1 = feedwater enthalpy

A *boiler combustion efficiency* η_b can be defined as the actual energy absorbed by the feedwater to the ideal chemical potential energy available in the fuel, or

$$\eta_b = \frac{\dot{m}_{H_2O}(h_2 - h_1)}{\dot{m}_f \, HHV} < 1.0 \tag{4.9}$$

Since combustion, heat transfer, and other irreversible events take place in a boiler, losses occur, and boiler combustion efficiency will not be 100%.

It is useful to calculate these losses in terms that then will allow a direct comparison to be made to the heating value of a fuel being fired.

The energy actually absorbed by the feedwater as it is converted into steam measured per mass of fuel is equal to

$$Q_1 = M_s(h_2 - h_1) \qquad \text{kJ/kg fuel (Btu/lbm)} \tag{4.10}$$

where M_s = the mass of steam generated per mass of fuel fired, and

$$M_s = \frac{\dot{m}_{H_2O}}{\dot{m}_f} \tag{4.11}$$

Frequently, actual air temperature as supplied to the burner is not at standard state or even ambient conditions. This is due to air having passed through auxiliary machinery, such as fans and heaters, that are required to supply air to the boiler and that preclude corrosion. In actual operation, then, the boiler stack temperature will be considerably in excess of ambient conditions. It is also reasonable to assume that, in the boiler, complete combustion will not have occurred. All these factors will cause boiler efficiency to be less than 100%.

The amount of energy loss due to dry flue gas can now be expressed as

$$Q_2 = M_g C_p(T_4 - T_3) \qquad \text{kJ/kgfuel (Btu/lbm)} \tag{4.12}$$

where M_g = mass of dry flue gas per mass of fuel fired,

and

$$M_g = \frac{\dot{m}_g}{\dot{m}_f} \tag{4.13}$$

C_p = mean constant pressure specific heat of dry flue gases evaluated at conditions between air supply and flue gas temperatures

T_4 = flue gas temperature measured at the heat exchanger exit

T_3 = air supply dry bulb temperature

Any water formed in the exhaust gases should be above its corresponding dew point temperature and therefore vaporized by the flue gases. The latent heat of vaporization associated with this process will not then be available to generate steam. Several factors contribute to the presence of moisture in the combustion process and the resulting energy loss, including the combustion of hydrogen in the fuel, moisture present in the fuel, and moisture existing in the combustion air supply.

Recall from Chapter 2 that, ideally, 2 moles H_2 form 2 moles H_2O or 4 mass units H_2 produce 36 mass units H_2O. Therefore, 9 mass units of water are produced by each mass unit of H_2. The loss due to moisture produced from hydrogen in the fuel is then given as

$$Q_3 = \frac{9H}{100}[h_6 - h_5] \qquad \text{kJ/kg fuel (Btu/lbm)} \tag{4.14}$$

where

H = H_2 ultimate analysis in fuel, %

h_6 = enthalpy of superheated steam at low pressure (10 kPa, 1 psi) and flue gas temperature measured at the heat exchanger exit

h_5 = enthalpy of saturated liquid at fuel supply temperature.

The loss associated with moisture in the fuel is determined in similar fashion to hydrogen loss above as

$$Q_4 = \frac{M_l}{1 - M_l}(h_6 - h_5) \qquad \text{kJ/kg fuel (Btu/lbm)} \tag{4.15}$$

where M_l = mass of fuel moisture per mass of fuel fired.

The loss due to moisture in the air supplied for combustion can also be evaluated and is equal to

$$Q_5 = M_a M_v(h_6 - h_7) \qquad \text{kJ/kg fuel (Btu/lbm)} \tag{4.16}$$

where

M_a = mass of air supplied per mass of fuel

M_v = mass of moisture per mass of dry air

$$M_a = \frac{\dot{m}_a}{\dot{m}_f} \tag{4.17}$$

$$M_v = \frac{\dot{m}_v}{\dot{m}_a} \tag{4.18}$$

h_7 = enthalpy of superheated steam at air supply temperature and water vapor partial pressure

Incomplete combustion of carbon in the fuel to form CO instead of CO_2 also will result in less energy being available to generate steam. The energy released by oxidation of CO to form CO_2, a measure of this loss, is given by the heating value of CO or

$$Q_6 = HV_{CO} < \frac{\text{energy release}}{\text{unit mass CO}} > \tag{4.19}$$

From a boiler flue gas analysis, major dry exhaust gas constituents are found, in general, to be CO, CO_2, O_2, and N_2. By measuring concentrations of these species, their mole fractions in dry gas mixture can be determined. Obviously, the presence of nitric oxides, sulfur, ash, and other trace elements influence boiler performance, but these components do not have a pronounced effect on a boiler energy analysis. The mass of CO formed per mole of dry exhaust gas is therefore equal to the molecular weight of CO multiplied by the mole fraction of CO or

$$28\, \bar{x}_{CO} < \frac{\text{mass CO}}{\text{moles dry gas}} > \tag{4.20}$$

As previously indicated, carbon, when oxidized, will form carbon monoxide and carbon dioxide, and the mass of carbon per moles of dry exhaust gas is given by the expression

$$12\, [\bar{x}_{CO} + \bar{x}_{CO_2}] < \frac{\text{moles carbon}}{\text{moles dry gas}} > \tag{4.21}$$

The mass of carbon per mass of fuel fired, C, can be obtained from the fuel gravimetric analysis. Combining Equations (4.19–4.21) yields an expression for energy loss due to incomplete combustion of carbon per unit mass of fuel as

$$Q_6 = \left[\frac{28\, C}{12}\right]\left[\frac{\bar{x}_{CO}}{\bar{x}_{CO} + \bar{x}_{CO_2}}\right] HV_{CO} \qquad \text{kJ/kg fuel (Btu/lbm)} \tag{4.22}$$

The energy loss to radiation and convective heat transfer as well as other unaccounted-for losses can now be expressed in terms of the heating value of the fuel and the losses evaluated above as

$$Q_7 = HHV - \sum_{i=1}^{6} Q_i \quad \text{kJ/kg fuel (Btu/lbm)} \tag{4.23}$$

The above items are often expressed alternatively as percentages of heating value. In general, acceptable test results require that unaccounted-for losses should be less than 2%. Additional minor losses could be considered if information is available and could include losses due to unburned carbon in the bottom, fly ash, and formation of nitric oxide and sulfur emissions.

EXAMPLE 4.4 The following data were obtained during a standard boiler test:

1.	Duration of test	1 hr
2.	Steam delivered by boiler	200,000 lbm
3.	Average steam temperature at superheater outlet	760°F
4.	Average steam pressure at superheater outlet	600 psia
5.	Feedwater temperature	240°F
6.	Feedwater pressure	700 psia
7.	Fuel-fired	15,385 lbm
8.	Flue gas temperature leaving last heat transfer passage	450°F
9.	Dry bulb temperature of air supplied for combustion	80°F
10.	Wet bulb temperature of air supplied for combustion	70°F
11.	Barometric pressure at test location	29.92 in. Hg
12.	Temperature of fuel supplied to burners	80°F
13.	Ultimate analysis of fuel on an as-fired basis:	

Carbon	0.8095 lbm/lbm fuel
Hydrogen	0.1143 lbm/lbm fuel
Nitrogen	0.0048 lbm/lbm fuel
Sulfur	0.0143 lbm/lbm fuel
Oxygen	0.0095 lbm/lbm fuel
Moisture	0.0476 lbm/lbm fuel
Ash	0.0000 lbm/lbm fuel
	1.0000

14. Volume analysis of flue gases in percent (Orsat):

CO_2 = 11.34%
CO = 00.71
O_2 = 5.06
N_2 = 82.89
100.00%

15. Higher heating value of fuels is 19,500 Btu/lbm dry fuel.

Calculate an energy balance for the tested boiler.

Solution:

1. Fuel analysis, dry basis gravimetric analysis:

	mf_i	$mf_{i)_I}$	$mf_{i)_{II}}$
C_S	0.8095	0.8095	0.8500
H_2	0.1143	0.1143	0.1200
O_2	0.0048	0.0048	0.0050
S	0.0143	0.0143	0.0150
N_2	0.0095	0.0095	0.0100
Moisture	0.0476	—	—
	$\frac{lbm_i}{lbm_{tot}}$	$\frac{lbm_i}{lbm_{tot}}$	$\frac{lbm_i}{lbm_{tot}}$

$\Sigma\, mf_{i)_I} = 0.9524$

2. Flue gas analysis, gravimetric:

	$\bar{x}_i$	MW	$\bar{x}_i MW_i$	mf_i
CO_2	0.1134	44	4.9896	0.1662
CO	0.0071	28	0.1988	0.0066
O_2	0.0506	32	1.6192	0.0539
N_2	0.8289	28	23.2092	0.7733
	$\frac{lb\ mole_i}{lb\ mole_{tot}}$	$\frac{lbm_i}{lb\ mole_i}$	$\frac{lbm_i}{lb\ mole_{tot}}$	$\frac{lbm_i}{lbm_{tot}}$

$MW = \Sigma \bar{x}_i MW_i = 30.0168$ lbm/lbmole

3. Energy absorbed by water:

$$Q_1 = M_S(h_2 - h_1)$$
$$= \left(\frac{200{,}000 \text{ lbm steam}}{15{,}385 \text{ lbm fuel}}\right)\left[(1385.1 - 209.9)\frac{\text{Btu}}{\text{lbm steam}}\right]^*$$
$$= 15{,}277.2 \text{ Btu/lbm fuel}$$

4. Energy loss to dry flue gases:

$$Q_2 = M_g C_p(T_2 - T_1)$$

where

*Keenan, Keyes, Hill, and Moore *Steam Tables*.

$$M_g = \frac{0.85 \text{ lbm carbon/lbm fuel}}{[(12/44)(0.1662) + (12/28)(0.0066)] \text{ lbm carbon/lbm gases}}$$
$$= 17.65 \text{ lbm gas/lbm fuel}$$
$$Q_2 = (17.65 \text{ lbm gas/lbm fuel})(0.24 \text{ Btu/lbm gas°R})(450 - 80°\text{R})$$
$$= 1567.3 \text{ Btu/lbm fuel}$$

5. Energy lost to moisture from burning hydrogen:

$$Q_3 = \frac{9H}{100}(h_4 - h_3)$$

$$= \left(\frac{9 \text{ lbm } H_2O}{\text{lbm } H_2}\right)\left(\frac{12}{100}\frac{\text{lbm } H_2}{\text{lbm fuel}}\right)[1265.1 - 48.09] \text{ Btu/lbm } H_2O$$
$$= 1314.3 \text{ Btu/lbm fuel}$$

6. Energy lost to moisture accompanying fuel:

$$Q_4 = \frac{M_l}{1 - M_l}(h_4 - h_3) = \frac{0.0476}{1 - 0.0476}[1265.1 - 48.09]$$
$$= 60.8 \text{ Btu/lbm fuel}$$

7. Energy loss to moisture in air:

$$Q_5 = M_a M_v (h_4 - h_5)$$

$$M_a = \frac{\left(0.7732 \frac{\text{lbm } N_2}{\text{lbm gas}}\right)\left(0.8500 \frac{\text{lbm C}}{\text{lbm fuel}}\right)\left(1.3040 \frac{\text{lbm air}}{\text{lbm } N_2}\right)}{\left[\left(\frac{12}{44}\right)(0.1662) + \left(\frac{12}{28}\right)(0.0066)\right]\frac{\text{lbm C}}{\text{lbm gas}}}$$
$$= 18.10 \text{ lbm air/lbm fuel}$$

$M_v = 0.0134$ lbm moisture/lbm dry air from psychometric chart @ 80°F dry bulb and 70°F wet bulb

$$h_5 \simeq h_g(T = 80°\text{F}) = 1096.4 \text{ Btu/lbm}$$

$$Q_5 = (18.10 \text{ lbm air/lbm fuel})(0.0134 \text{ lbm moisture/lbm air})(1265.1 - 1096.4)$$
$$= 40.92 \text{ Btu/lbm fuel}$$

8. Energy loss to incomplete combustion:

$$Q_6 = \left[\frac{28\ C}{12}\right]\left[\frac{\bar{x}_{CO}}{\bar{x}_{CO} + \bar{x}_{CO_2}}\right] HV_{CO}$$

$$= (28/12)(0.85)\left[\frac{0.0071}{0.1134 + 0.0071}\right]\left[\frac{(67{,}636)(1.8001)}{(28)}\right]$$

$$= 508.1 \text{ Btu/lbm fuel}$$

9. Energy loss to radiation and unaccounted-for losses:

$$Q_7 = HHV - (Q_1 + Q_2 + Q_3 + Q_4 + Q_5 + Q_6)$$
$$= 19{,}500 - (15{,}277.2 + 1567.3 + 1314.4$$
$$+ 60.8 + 40.9 + 508.1)$$
$$= 731.3 \text{ Btu/lbm dry fuel}$$

10. Boiler energy balance summary:

No.	Item	Btu/lbm	%
1	Gain by water and steam	15,277.2	78.34
2	Loss to dry flue gases	1,567.3	8.04
3	Loss to moisture from burning H_2	1,314.4	6.74
4	Loss to moisture in fuel	60.8	0.31
5	Loss to moisture in air	40.9	0.21
6	Loss to incomplete combustion	508.1	2.61
7	Radiation and unaccounted losses	731.3	3.75
8	Higher heating value of fuel	19,500.0	100.00

Certain terminology and definitions are commonly used when discussing boiler combustion systems. *Capacity* C_b is the heat absorption rate of a boiler and its feedwater and generated steam

$$C_b = \dot{m}_s(h_2 - h_1) \qquad \text{kW (mBtu/hr)} \tag{4.24}$$

where $\dot{m}_s$ = steam mass flow rate, kg/sec (lbm/hr). *Heat rate HR* compares the boiler heat addition to the net power plant output, an inverse of thermal efficiency η or

$$HR = \eta^{-1} \tag{4.25}$$

and

$$HR = \frac{3413}{\eta} \qquad \frac{\text{Btu}}{\text{kW hr}}$$

$$HR = \frac{2544}{\eta} \qquad \frac{\text{Btu}}{\text{hp hr}}$$

The heat rate per unit volume of furnace space is termed the *furnace rating* *FR* or

$$FR = \frac{\dot{m}_f \, HHV}{\text{Furnace volume}} \qquad \text{kW/m}^3 \ (\text{Btu/ft}^3 \text{ hr}) \tag{4.26}$$

A more complete discussion of boiler equipment and the general subject of steam generation can be found in the literature.

EXAMPLE 4.5 Municipal solid waste (MSW) is to be utilized as an alternate fuel for steam generation. The fuel composition on a weight basis is given as

C_S 25.0% H_2 3.3% N_2 0.5%
H_2O 28.0% O_2 21.1% S 0.1%
Glass, metals, ash 22.0%

The material stream is treated to remove the noncombustibles for recycling. Air enters the burner at 25°C and 104 kPa, and the dry exhaust gases leave the boiler at 500K and 101 kPa, with the following stack analysis.

CO_2 13.6% O_2 4.4%
CO 3.4% SO_2 0.02%

Find: (a) excess air, %; (b) stack dew point temperature, °C; (c) sulfur dioxide production, kg SO_2/kg fuel; (d) boiler efficiency, %.

Solution:

1. Fuel mole fractions:

	mf_{orig}	mf_i	MW_i	$(mf/MW)_i$	$\bar{x}_i$
C_S	0.250	0.3205	12	0.026708	0.3490
H_2O	0.280	0.3590	18	0.019944	0.2606
H_2	0.033	0.0423	2	0.021150	0.2764
O_2	0.211	0.2705	32	0.008453	0.1105
N_2	0.005	0.0064	28	0.000229	0.0030
S	0.001	0.0013	32	0.000041	0.0005
Waste	0.220	—	—	—	—

$\Sigma(mf/MW)_i = 0.0765251$
or $MW = 13.068$ kg/kgmole

2. Stoichiometric equation:

$$[0.349C_S + 0.2606H_2O + 0.2764H_2 + 0.1105O_2 + 0.0030N_2 + 0.0005S] + aO_2 \rightarrow bCO_2 + cH_2O + dSO_2$$

Carbon: $b = 0.349$

Hydrogen: $c = (0.2606 + 0.2764) = 0.537$

Sulfur: $d = 0.0005$

Oxygen: $0.2606 + (2)(0.1105) + 2a = (2)(0.349) + 0.537 + (2)(0.0005)$

$$a = 0.3772$$

3. Actual equation:

$$[0.349C_S + 0.2606H_2O + 0.2674H_2 + 0.0030N_2 + 0.1105O_2 + 0.0005S] + a[0.3772][O_2 + 3.76N_2] \rightarrow b\,[0.136CO_2 + 0.034CO + 0.044O_2 + 0.002SO_2 + 0.7858N_2] + cH_2O$$

Carbon: $0.349 = b(0.136 + 0.034)$

$$b = 2.053$$

Hydrogen: $0.2606 + 0.2764 = c = 0.537$

Nitrogen: $0.003 + (a)(0.3772)(3.76) = (2.053)(0.7858)$

$$a = 1.135$$

4. Excess air, %:

113.5% theoretical air or 13.5% excess air

5. Dew point T:

$$\bar{x}_{H_2O} = \frac{0.537}{2.053} = 0.2616$$

$$P_{H_2O} = (0.2616)(101) = 26.4 \text{ kPa}$$

$$T_{sat} <26.4 \text{ kPa}> = T_{db} \cong 66°C$$

6. SO_2 production:

$$\frac{m_{SO_2}}{m_{fuel}} = \frac{(2.053)(0.0002)(32) \text{ lbm } SO_2}{(1.0)(13.068) \text{ kg fuel}} = 1.0 \frac{\text{kg } SO_2}{\text{kg fuel}}$$

but there is 0.78 lbm fuel/1.0 lbm of fuel + waste, or

$$\frac{m_{SO_2}}{m_{fuel}} = \left(0.78 \frac{\text{kg fuel}}{\text{kg MSW}}\right)\left(1.0 \frac{\text{kg } SO_2}{\text{kg fuel}}\right) = 0.78 \frac{\text{kg } SO_2}{\text{kg MSW}}$$

7. Boiler efficiency:

$$Q = \sum_{i\,\text{prod}} N_i[\bar{h}_f^o + \Delta\bar{h}] - \sum_{\text{react}} N_j[\bar{h}_f^o + \Delta\bar{h}]$$

$$\begin{aligned} Q = {} & (2.053)(0.136)[\bar{h}_f^o + \Delta\bar{h}]_{CO_2} + (2.053)(0.034)[\bar{h}_f^o + \Delta\bar{h}]_{CO} \\ & + (2.053)(0.044)\,[\bar{h}_f^o + \Delta\bar{h}]_{O_2} + (2.053)(0.0002)[\bar{h}_f^o + \Delta\bar{h}]_{SO_2} \\ & + (0.7858)(2.053)[\bar{h}_f^o + \Delta\bar{h}]_{N_2} + (0.537)[\bar{h}_f^o + \Delta\bar{h}]_{H_2O_g} \\ & - (1.135)(0.3772)[\bar{h}_f^o + \Delta\bar{h}]_{O_2} - (1.135)(3.76)(0.3772)[\bar{h}_f^o + \Delta\bar{h}]_{N_2} \\ & - (0.0005)[\bar{h}_f^o + \Delta\bar{h}]_{S} - (0.1105)[\bar{h}_f^o + \Delta\bar{h}]_{O_2} - (0.003)[\bar{h}_f^o + \Delta\bar{h}]_{N_2} \\ & - (0.2764)[\bar{h}_f^o + \Delta\bar{h}]_{H_2} - (0.349)[\bar{h}_f^o + \Delta\bar{h}]_{C_s} - (0.2606)[\bar{h}_f^o + \Delta\bar{h}]_{H_2O} \end{aligned}$$

$$\begin{aligned} Q = {} & (2.053)(0.136)(-94{,}054 + 958)_{CO_2} + (2.053)(0.034)[-26{,}417 + 711]_{CO} \\ & + (2.053)(0.044)(724)_{O_2} + (2.053)(0.0002)(-20{,}947 + 1016)_{SO_2} \\ & + (0.7858)(2.053)(710)_{N_2} + (0.537)(-57{,}798 + 825)_{H_2O} \\ & - (0.2606)(-68{,}317)_{H_2O} = -39396.5 \text{ cal/g mole of fuel} \end{aligned}$$

$LHV = Q$ <above without the $\Delta\bar{h}$>

$$\begin{aligned} & = (2.053)(0.136)(-94{,}054)_{CO_2} + (2.053)(0.034)(-26417)_{CO} \\ & + (2.053(0.0002)(-70{,}947)_{SO_2} + (0.537)(-57{,}798)_{H_2O_g} \\ & - (0.2606)(-68{,}317) \\ & = -41368 \text{ cal/g mole fuel} \end{aligned}$$

$$\eta_{\text{boiler}} = \frac{-39396.5 \text{ cal/g mole fuel}}{-41368.0 \text{ cal/g mole}} = 0.953$$

$$= 95.2\%$$

PROBLEMS

4.1 An anthracite coal sample has the following dry-basis ultimate analysis:

C_S	83.73%	S	0.48%	N_2	0.80%
H_2	2.14%	O_2	1.92%	Ash	10.93%

For this solid fuel, (a) write the stoichiometric equation for the ideal combustion reaction; (b) determine the mass *AF* ratio on a dry basis, lbm air/lbm fuel; and (c) calculate the amount of carbon dioxide formed per unit of fuel fired, lbm CO_2/lbm coal.

4.2 Two hundred pounds of coal/min having the following ultimate analysis is burned in a boiler:

C_S	74.17%	S	1.95%	N_2	2.40%
H_2	5.64%	O_2	9.02%	Ash	6.82%

Determine, for the unit: (a) the rating for a forced-draft fan designed to supply 35% excess air at 90°F and 14.7 psia, ft^3/min; and (b) the mass flow rate of dry flue gas resulting from combustion if the Orsat analysis yields 11.95% CO_2, 6.20% O_2, and 1.33% CO, lbm/min.

4.3 The ultimate analysis of a bituminous coal sample is reported to be:

C_S	76.3%	S	1.4%	N_2	2.3%
H_2	4.8%	O_2	4.1%	H_2O	4.1%
		Ash	7.0%		

Determine the following: (a) fuel analysis on a dry basis, %; (b) fuel analysis on a dry and ashless basis, %; and (c) the amount of refuse from combustion to be handled if this fuel is burned at a rate of 200 tons/hr as received and the refuse is found to be 10% carbon in ash, lbm refuse/hr.

4.4 Ten pounds mass of coal/min having an ultimate analysis as given below is completely burned in a small power boiler.

C_S	79.5%	S	1.2%	N_2	2.7%
H_2	4.1%	O_2	6.6%	Ash	5.9%

Twenty-five percent excess air is supplied at 70°F and 28-in. Hg. Flue gas leaves the unit at 155°F and 28-in. Hg. For complete combustion, find (a) the ideal *AF* ratio, lbm air/lbm fuel; b) the mass flow rate of CO_2, lbm/min; (c) the flue gas volumetric flow rate, ft^3/min; (d) the percent O_2 in dry flue gas analysis, %; and (e) exhaust dew point temperature, °F.

4.5 Fifty kilograms of coal/sec having the ultimate analysis as given below is burned in a boiler.

C_s	78.2%	O_2	6.4%	H_2	5.2%
N_2	1.6%	S	1.3%	Ash	7.3%

One hundred and twenty percent theoretical air is supplied at 70°C and 101 kPa. The flue gases leave the unit at 327°C and 101 kPa. Calculate (a) the *MW* of the coal, kg/kg mole; (b) Orsat analysis if 10% of the ideal CO_2 is CO, %; (c) the dry flue gas *MW*, kg/kg mole; (d) the dry flue gas volumetric flow rate, m^3/sec; and (e) the exhaust dew point temperature, °C.

4.6 Pulverized coal is fired at a rate of 100 tons/hr in a stationary power plant. The analysis of the coal is reported to be:

C_S	72.6%	H_2O	4.1%	N_2	3.2%
H_2	6.9%	O_2	8.7%	Ash	4.5%

The stack gas analysis on a volumetric basis yields the following:

CO_2	10.38%	O_2	7.55%
CO	1.83%	N_2	80.24%

Refuse pit analysis on a mass basis yields:

C_S	23%
Ash	77%

Determine the following: (a) volume of air supplied measured at 14.7 psia and 90°F, ft^3/hr; (b) refuse collection rate, lbm/hr; and (c) the volume of dry products measured at a total pressure of 14.7 psia and stack temperature of 310°F.

4.7 A sample of dry wood has the following ultimate analysis:

C_S	53.6%	S	0.1%	N_2	29.8%
H_2	6.4%	O_2	7.2%	Ash	2.9%

Calculate (a) the stoichiometric equation; (b) the *AF* ratio, kg air/kg wood; and (c) the higher heating value, kJ/kg wood.

4.8 Consider the dry wood in Problem 4.7. For 0, 20, 40, and 80% moisture, calculate (a) the net heating value of the wood, Btu/lbm; (b) the mass of water to mass of dry wood, %; (c) the heat required to vaporize the water, Btu/lbm of dry wood; and (d) the percentage of water at which the ideal net heat of combustion would be zero, %.

4.9 A processed coal, consisting *only* of carbon and ash, is burned at a rate of 30,000 kg/hr. The combustion occurs with 20% excess air, and the entire ash is collected after combustion at a rate of 4500 kg/hr. Determine (a) the ultimate analysis of the coal, (b) the *AF* ratio, kg air/kg fuel; (c) the volume of air supply required at 101 kPa and 27°C; and (d) the higher heating value of the coal, kJ/kg.

4.10 Municipal solid waste (MSW) is being utilized as an alternate fuel for steam generation. The fuel composition on a weight basis is:

C_S	25.0%	O_2	21.1%	H_2	3.3%
H_2O	28.0%	N_2	0.5%	S	0.1%
	Glass, metals, ash		22.0%		

Dry exhaust gases leave the boiler at 500K and 101 kPa with the following stack analysis:

CO_2	13.6%	O_2	4.4%
CO	3.4%	SO_2	0.02%

Air enters the burner at 25°C and 104 kPa. Determine (a) the excess air, %; (b) the stack dew point temperature, °C; (c) the sulfur dioxide production rate, kg SO_2/kg fuel; and (d) the boiler efficiency, %.

4.11 A small one-room cabin is to be heated during the winter using a wood-burning stove having a 60% combustion efficiency. The room will require on the average about 8000 Btu/hr of heat from the stove. Specifications for a given wood supply on a weight basis are given below:

C_S	44.00%	S	0.10%	N_2	0.25%
H_2	5.60%	O_2	36.15%	H_2O	12.00%
Inerts	1.90%	(Density = 30 lbm/ft^3)			

Estimate the following: (a) minimum required cords of wood burned/hr (1 cord = 4×4×8 ft = 128 ft^3), cords/hr; (b) equivalent gallons of fuel oil saved if fuel oil has a higher heating value of 19,000 Btu/lbm, gal/hr; and (c) the economic equivalency of these two fuels on a Btu basis for a fuel price of \$1.50/gal, \$/cord.

4.12 The stoichiometric combustion of an unknown ashless coal sample yields a 17.44% CO_2, 1.94% CO, and 0.97% O_2 dry flue gas analysis. The molar $\overline{AF}$ ratio for this reaction is 4.286 lb mole air/lb mole fuel while producing 0.2815 lbm H_2O/lbm ashless coal. The product molecular weight is 30.351 lbm/lb mole. For this ashless coal, find (a) the stoichiometric equation; (b) mass fractions of the coal sample, %; (c) molecular weight of the coal, lbm/lb mole; and (d) the mass *FA* ratio, lbm coal/lbm air.

4.13 A power plant, rated at 500 MW and 39% thermal efficiency, is to be switched over to the low-sulfur coal having the ultimate analysis shown below.

	Original coal	*Low-sulfur coal*
C_s	68.87%	83.84%
H_2	20.41%	4.65%
S	1.38%	0.47%
O_2	2.32%	5.36%
N_2	0.59%	1.17%
Ash	6.43%	4.51%

Assume that the volumetric flow rate and air supply conditions to the burners remain constant and that the original combustion requires 130% theoretical air. Find (a) the change in coal mass flow rate, %; (b) the reduction in SO_2 emissions, %; (c) the change in moisture in the exhaust, %; and (d) the change in CO_2 level in the stack, %.

4.14 The fuel for the preliminary design of a 500,000-kW steam power plant with a 40% thermal efficiency is to be coal at 5 \$/ton. The as-fired gravimetric analysis of the coal for the initial design analysis is given below:

C_S	74.79%	S	3.42%	N_2	1.20%
H_2	4.98%	O_2	6.42%	H_2O	1.55%
		Ash	7.82%		

Calculate, for the unit: (a) the specific fuel consumption, kg/kW hr; and (b) the fuel cost for operation, $/hr.

4.15 A western U.S. coal has the following ultimate analysis:

C_S	72.3%	S	0.5%	N_2	1.3%
H_2	5.8%	O_2	14.9%	Ash	5.2%

Estimate (a) the higher heating value of the coal, kJ/kg; (b) the dry SO_2 in the stack if the flue gas is 3% O_2, %; (c) the NO_x concentration as NO, assuming all the chemically bound nitrogen is converted to NO, %; and (d) repeat parts (b) and (c) product analysis, kg product/ million kJ fuel.

4.16 A fuel plant is designed to convert municipal solid waste (MSW) into a refuse-derived fuel (RDF). The plant is supplied with MSW, having the ultimate analysis shown below at a rate of 1200 tons/day.

C_S	25.2%	S	0.1%	H_2O	28.0%
H_2	3.3%	O_2	21.1%	Ash	5.6%
Glass	9.3%	Metal	7.4%		

For ideal conversion, determine the following: (a) mass flow rate of recoverable glass and metal, kg/day; (b) mass flow of landfill inerts, kg/day; (c) mass flow of RDF, kg/day; (d) higher heating value of MSW, kJ/kg; and (e) higher heating value of RDF, kJ/kg.

4.17 Determine the adiabatic flame temperature of the anthracite coal in Problem 4.1 when burned at constant pressure with 130% theoretical air. Assume that both the fuel and air are initially at 1 atm and 25°C.

4.18 A 1-g sample of solid carbon is burned in 200% excess air in a bomb calorimeter. The initial fill conditions are 62°F and 15 psia. After combustion in the constant-volume vessel, determine (a) the product mole fractions for complete combustion; (b) the adiabatic flame temperature, R; and (c) the rise in temperature of a water jacket from this process if the mass of the water is 5 lbm.

4.19 The following data were obtained during the operation of a coal-fired steam generator:

Fuel analysis		*Stack analysis*	
C_S	63.70%	CO_2	14.8%
H_2	4.78%	CO	0.0%
N_2	1.40%	O_2	4.5%
O_2	9.42%	N_2	80.7%
H_2O	8.10%		
Ash	12.60%		

Barometer	1 atm	Coal feed	3400 lbm/hr
Flue gas T	500°F	Steam rate	30,600 lbm/hr
Air supply	82°F	Refuse	445 lbm/hr
Feedwater	208°F		
Sat vap at	150 psia		

Determine, for the unit: (a) the excess air supplied for combustion, %; and (b) steam generator efficiency, %. Assume the refuse is composed of ash and carbon alone.

5

Liquid Fuels

5.1 INTRODUCTION

In Chapter 4, energy characteristics of certain solid fuel resources as well as their required combustion systems were considered in some detail. In this chapter, however, attention will be given to general energy properties of important liquid fuel resources. The specific nature of particular internal combustion engines and the nature of their liquid fuel–engine interface will be treated separately in later chapters. Availability of abundant crude oil reserves, ease of storing and handling distillate liquid fuels, and development of the internal combustion engine have each contributed to making liquid fuels a major energy resource in today's world economy. The critical dependency of mobility propulsion systems, such as spark or compression engines as well as gas turbines, on petroleum-based fuels is fundamental to their present state of development. With the projected downturn in future oil reserves, power and propulsion engineering will begin to require a greater emphasis on the utilization of alternate and synthetic liquid fuels. An understanding of liquid fuel science requires some background in chemistry and fluid mechanics.

In this chapter, various liquid fuel resources derived from crude oil reserves, as well as those produced from synthetic stocks, will be reviewed. The use of these reactants to produce heat and power depends on certain properties, but additional characteristics do result from the specific or unique nature of each fuel–engine interface. For example, the explosive nature or detonation of spark-ignition engine fuels is defined by a fuel-engine parameter, the *octane rating*.

EXAMPLE 5.1 Kerogen, or shale oil, is proposed as a potential fuel source. The ultimate analysis of a typical 1-ton sample of raw shale is given below.

Raw shale constituent	Percentage
Ash	65.7
CO_2	18.9
Organic carbon	12.4
H_2	1.8
N_2	0.4
Sulfur	0.6
H_2O	0.2

Using these percentages, determine (a) the dry and ashless gravimetric analysis of the shale; (b) the gravimetric analysis for part a on an oxygen-free basis; (c) the mass of part a as derived from 1 ton of raw shale, and (d) the mass of part b as derived from one ton of raw shale.

Solution:

1. Gravimetric analysis, dry and ashless:

$$mf_i = \frac{\text{lbm component } i}{\text{lbm mixture}}$$

a. Mass fraction, where lbm mixt = lbm total − lbm H_2O − lbm ash

$$mf_{CO_2} = \frac{0.189}{1.00 - 0.657 - 0.002} = 0.5543 \text{ or } 55.43\%$$

$$mf_{C_S} = \frac{0.124}{0.341} = 0.3636 \text{ or } 36.36\%$$

$$mf_{H_2} = \frac{0.018}{0.341} = 0.0528 \text{ or } 5.28\%$$

$$mf_{N_2} = \frac{0.004}{0.341} = 0.0117 \text{ or } 1.17\%$$

$$mf_S = \frac{0.006}{0.341} = 0.0176 \text{ or } 1.76\%$$

2. Oxygen-free basis:

$$12 \text{ lbm } C_S + 32 \text{ lbm } O_2 \rightarrow 44 \text{ lbm } CO_2$$

or for each lbm CO_2 on a mass basis:

$$\% \; C_S = \frac{12 \text{ lbm carbon}}{44 \text{ lbm } CO_2}$$

$$\% \; O_2 = \frac{32 \text{ lbm } O_2}{44 \text{ lbm } CO_2}$$

Mass fraction of dry, ashless sample:

$$mf_{C_S} = \left(\frac{12}{44}\right)(0.5543) = 0.1512$$

$$mf_{O_2} = \left(\frac{32}{44}\right)(0.5543) = 0.4031$$

$$mf_{C_S} = 0.3636$$
$$mf_{H_2} = 0.0528$$
$$mf_{N_2} = 0.0117$$
$$mf_S = 0.0176$$

b. Mass fraction, dry, ashless, oxygen-free:

$$mf_{C_S} = \frac{0.1512 + 0.3636}{1.0000 - 0.4031} = 0.8625 \text{ or } 86.25\%$$

$$mf_{H_2} = \frac{0.0528}{0.5969} = 0.0885 \text{ or } 8.85\%$$

$$mf_{N_2} = \frac{0.0117}{0.5969} = 0.0196 \text{ or } 1.96\%$$

$$mf_S = \frac{0.0176}{0.5969} = 0.294 \text{ or } 2.94\%$$

3. Total mass of dry, ashless sample now from 1.

$$m_{tot} = (0.341 \frac{\text{lbm dry, ashless sample}}{\text{lbm raw shale}}) \times (2000 \text{ lbm raw shale})$$

c. $m_{tot} = 682$ lbm (0.341 ton)

4. Total mass of dry, ashless, oxygen-free sample now from 2.

$$m_{tot} = (0.5969 \frac{\text{lbm dry, ashless, } O_2\text{-free}}{\text{lbm dry, ashless}}) \times (682 \text{ lbm})$$

d. m_{tot} = 407 lbm (0.2035 ton)

Comments: This problem illustrates the fact that potential oil from shale sources, such as kerogen, is high in waste and low in potential fuel. To remove the waste material can require high-energy inputs to produce low-energy output in terms of a viable fuel source.

5.2 LIQUID FUEL PROPERTIES

In the following section, a few major liquid fuel properties that influence combustion are covered. Standard tests for liquid fuels are established by engineering societies such as the American Society for Testing Materials (ASTM), American Society of Mechanical Engineering (ASME), and the Society of Automotive Enginers (SAE).

Liquid hydrocarbon fuels are basically mixtures of specific compounds from the following inorganic families:

Paraffins
Olefins
Naphthenes
Aromatics

Many liquid fuel chemical and combustion characteristics are a direct result of a particular nature of those hydrocarbon groups. General naming and classifying of particular species follow rules established in inorganic chemistry. For example, each hydrocarbon compound is named in terms of the number of carbon atoms per molecule by using the basic carbon atom prefixes given below, see Figure 5.1.

No. C atoms	Prefix
1	meth-
2	eth-
3	prop-
4	but-
5	pent-
6	hex-
7	hept-
8	oct-
9	non-
10	dec-
16	hexadec-

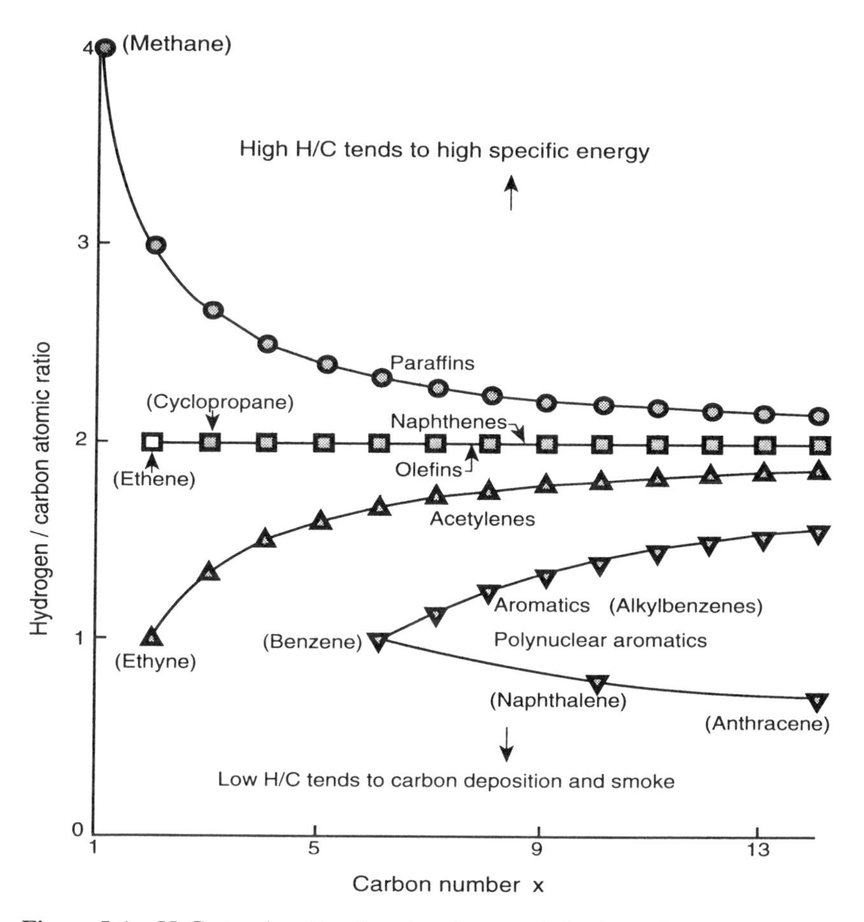

Figure 5.1 H:C atomic ratio of various inorganic hydrocarbon compounds. (From: Alternate Fuels: Chemical Energy Sources, Goodger, 1980.)

Paraffins are hydrocarbon compounds having open saturated carbon chain structures with only single carbon atom bonds. Every carbon atom has four single valence bonds available. Saturated structures are those in which each bond is made with a separate C-C or C-H atom-to-atom bond. Paraffins can be represented chemically as C_xH_{2x+2} and are recognized by their common suffix *-ane*. For example, if $x = 4$, then C_4H_{10} is termed butane; whereas, if $x = 10$ then $C_{10}H_{12}$ is called decane. Paraffins are major constituents in most crude oil base liquid fuels.

```
  |
--C--        (1st fig.)
  |
```

methane (CH_4) ethane (C_2H_6)	} natural gas components
propane (C_3H_8) butane (C_4H_{10})	} liquid petroleum gas compounds
octane (C_8H_{18})	gasoline component

Paraffins can have the same total carbon-to-hydrogen ratio but may assume considerably different structures. Open, straight chain structures, termed *normal* paraffins, are written as n-C_xH_{2x+2}, whereas branched chain structures, termed *isomers,* are designated *iso*-C_xH_{2x+2}. Radicals other than a single hydrogen atom can be bonded to the basic carbon chain, i.e., additional carbon atoms or even radicals such as the *methyl* radical CH_3.

```
    H   H   H   H   H   H   H   H
    |   |   |   |   |   |   |   |
H — C — C — C — C — C — C — C — C — H
    |   |   |   |   |   |   |   |
    H   H   H   H   H   H   H   H
                                        n-octane
```

```
                 H                  H
                 |                  |
            H —  C — H         H —  C — H
       H         |         H        |         H
       |         |         |        |         |
   H — C ——————— C ——————— C ——————— C ——————— C — H    iso-octane
       |         |         |        |         |
       H         |         H        H         H
            H —  C — H
                 |
                 H
```

Branched hydrocarbon structures can be named using alternative nomenclatures. The isooctane structure above is also termed 2, 2, 4-trimethylpentane since pentane is the principal root; i.e., the basic chain consists of five carbon atoms. Trimethyl is used because there are three methyl radicals attached to the chain, and 2, 2, 4 is used to identify that the branched radicals are attached to the second, second, and fourth carbon atoms of the basic chain.

Several general thermochemical characteristics of paraffin fuels can be cited that are a result of their structure.

Maximum H atom/carbon atom ratio
Highest heating values per unit mass of liquid hydrocarbon fuels
Lowest densities
Lowest heating values per unit volume of liquid hydrocarbon fuels
Clean burning
Stable when stored

Olefins are hydrocarbon compounds having open unsaturated carbon chain structures with both single and double carbon atom bonds. Olefins, termed alkenes, that have one double carbon bond are written chemically as C_xH_{2x} and are identifiable by their common suffix *-ene.* As an example, for $x = 5$, C_5H_{10} is pentene while, if $x = 8$, C_8H_{16} would be octene. For compounds having more than one double bond, *diolefins,* the chemical structure is C_xH_{2x-2}, and the suffix is *-diene.* Olefins are formed predominantly by cracking heavy paraffin compounds.

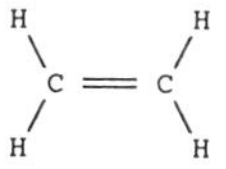

```
                                          H
                                          |
                                      H — C — H
                                          |
H       H                             H — C ══ C — H            Isobutene
 \     /                                       |
  C == C          Ethene                   H — C — H
 /     \                                       |
H       H                                      H
```

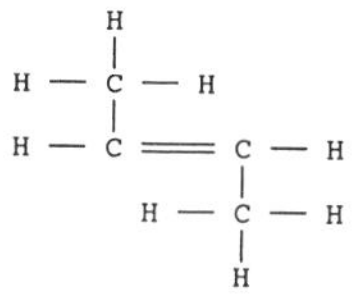

Olefins, like paraffin fuels, have certain thermochemical characteristics that are a result of their structure:

Fewer hydrogen atoms than for paraffins having the same number of carbon atoms
Unstable when formed
Can form residue or *gum* with exposed O_2
Fairly clean-burning fuel

Napthenes are hydrocarbon ring compounds having saturated carbon chain structures with only single carbon atom bonds. Chemically, napthenes are similar in character to normal paraffins and isoparaffins. Structurally, they are represented by the use of their common prefix *cyclo-* and the corresponding paraffin suffix. As an example, C_5H_{10} is termed cyclopentane.

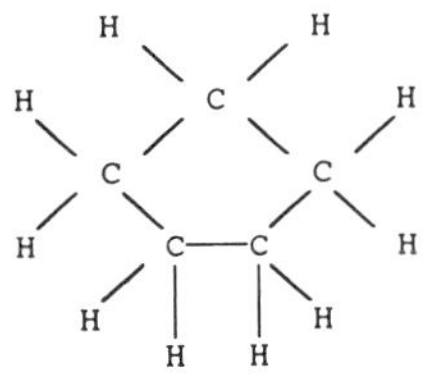

Cyclopentane

Aromatics are hydrocarbon ring structures having unsaturated carbon ring structures with both single and double carbon atom bonds. Aromatic hydrocarbons are recognized by their basic six carbon atom benzene ring structure.

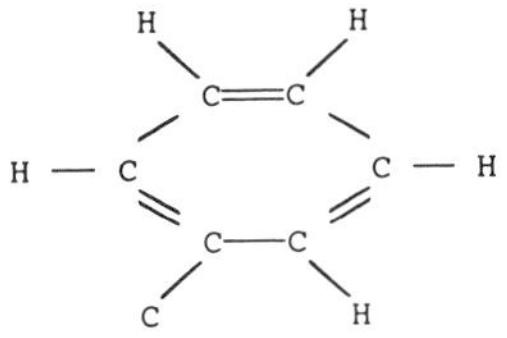

Benzene

Aromatics having single-ringed structures have the general formula C_xH_{2x-6}, and double-ringed species have the formula C_xH_{2x-12}.

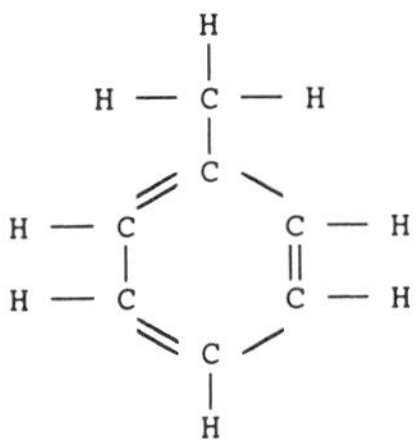

Toluene

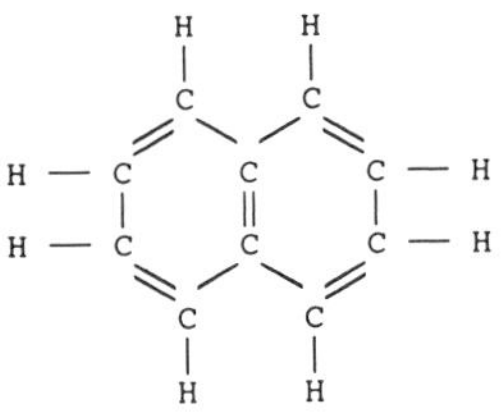

naphthalene

Aromatic hydrocarbons also have thermochemical characteristics that are a result of their structure:

Compact molecular structure
Stable when stored
Smoky combustion process
Highest fuel distillate densities
Highest heating values per *unit volume* of liquid hydrocarbon fuels
Lowest heating values per *unit mass* of liquid hydrocarbon fuels

Aromatic compounds are found in gasoline and diesel fuels and are also predominant products of coal-based distillate fuels. These particular fuel constituents are of a major concern in terms of incomplete combustion and their *carcinogenic,* or cancer-causing, formed products.

Acetylenes are unsaturated carbon chain compounds having triple carbon atom bonds. They have the same formula as diolefins and are recognized by the suffix *-yne.*

```
H — C ≡ C — H      Acetylene

        H               H
        |               |
    H — C — C ≡ C — C — H      Butyne-2
        |               |
        H               H
```

Partially oxidized hydrocarbon compounds having saturated carbon structures are known as *alcohols.* Alcohols are recognized by their common suffix *-ol* and have the chemical formula $C_xH_{y-1}OH$ where a hydrogen atom has been replaced by the hydroxyl radical OH. Alcohols have been synthesized from vegetable matter and commercially produced from wood or grain sources, i.e., ethanol. Recent attention has been given to producing methanol from natural gas or coal. The discussion of alcohol fuels and their utilization depends greatly on commercialization of technology with favorable economics and energetics; i.e., one cannot take 10 units of energy and/or monetary value to produce a fuel with 5 units of energy, or one cannot successfully market a fuel resource that costs twice that of competitive resources.

```
    H
    |
H — C — OH      Methanol
    |
    H

    H   H
    |   |
H — C — C — OH      Ethanol
    |   |
    H   H
```

Liquid fuel densities can be expressed in dimensionless form in terms of their *specific gravity,* that is, density of the liquid at a specified temperature to a reference liquid (usually water) also at the specified temperature. The measurements for specific gravity, SG, are taken at barometric conditions and are written

$$SG = \frac{\rho_{\text{fuel}}\langle T \rangle}{\rho_{\text{fuel}}\langle T \rangle} \tag{5.1}$$

Often, the fuel and water reference temperatures are taken as 15.8°C (60°F), and then specific gravity is equal to $SG\langle 60°\text{F}/60°\text{F}\rangle$. The American Petroleum Industry has developed the API gravity scale reading, which can be expressed in terms of specific gravity as

$$API \text{ gravity} = \frac{141.5}{SG\langle 60/60 \rangle} - 131.4 \tag{5.2}$$

API hydrometer readings at temperatures other than 18.5°C (60°F) can be corrected to the reference temperature as

$$API°\langle 60 \rangle = [0.002\ (60 - \text{observed°F}) + 1] \times [\text{observed } API°] \tag{5.3}$$

Note that a fuel having a low specific gravity will have a high API reading and vice versa.

Hydrocarbon liquid densities are related to their molecular structure and molecular weight by their carbon atom–to–hydrogen atom ratio. Recall that hydrogen per carbon atom is greatest for paraffin fuels; therefore, these liquids would be lighter such that specific gravity of liquid fuels will range from their lowest value (~0.8) for paraffinic compounds to (~1.0) for napthenic materials.

Specific gravity is also a convenient means for empirically expressing the heating value of liquid hydrocarbon fuels; see Figure 5.2 and Appendix D.

SI:

Fuel oil $HHV = 43{,}380 + 93\ (API - 10)$ kJ/kg (5.4a)

Kerosene $HHV = 42{,}890 + 93\ (API - 10)$ kJ/kg (5.4b)

Gasoline $HHV = 42{,}612 + 93\ (API - 10)$ kJ/kg (5.4c)

Heavy cracked fuel $HHV = 41{,}042 + 126 \times API$ kJ/kg (5.4d)

Engineers:

Fuel oil $HHV = 18{,}650 + 40\ (API - 10)$ Btu/lbm (5.5a)

Kerosene $HHV = 18{,}440 + 40\ (API - 10)$ Btu/lbm (5.5b)

Gasoline $HHV = 18{,}320 + 40\ (API - 10)$ Btu/lbm (5.5c)

Heavy cracked fuel $HHV = 17{,}645 + 54 \times API$ Btu/lbm (5.5d)

Note that the greater the percentage of hydrogen in a fuel, the higher the heating value will be since the heating value of hydrogen per unit mass is approximately four times that for carbon. Therefore, fuels with lower specific gravities, i.e., lighter fuels, will generally have greater heating values per unit mass.

EXAMPLE 5.2 The specific gravity of #2 diesel fuel measured at 60°F is found to be 32.5 *API*. For conditions of 75°F, calculate (a) the specific gravity, *API* ; (b) the density of the fuel, lbm/ft^3; (c) the higher heating value of the fuel using Appendix D; (d) the higher heating value using Equation (5.2a) and (e) the heating value using Figure 5.2.

Solution:

1. Specific gravity, Equation (5.3):

$$API <60> = 32.5$$

$$API <75> = \frac{32.5}{[0.002\,(60-75) + 1]} = 33.5$$

2. Density, from Appendix D, 33.5 *API* = 0.8576:

$$\rho_{\text{fuel}} <75°\text{F}> = SG \times \rho_{H_2O} <75°\text{F}>$$

from Appendix C, $\rho_{H_2O} <75°\text{F}> = 62.253$ lbm/ft^3

b. $\rho_{\text{fuel}} = (0.8576)(62.253) = 53.388$ lbm/ft^3

3. *HHV*—from Appendix D:

c. $HHV = 19{,}520 + 0.5(40) = 19{,}540$ Btu/lbm

4. *HHV*, Equation (5.2a):

$$\begin{aligned} HHV &= 18{,}650 + 40\,(API - 10) \\ &= 18{,}650 + 40\,(33.5 - 10) \end{aligned}$$

d. $HHV = 19{,}590$ Btu/lbm

5. *HHV*, Figure 5.2:

$API = 33.5$

e. $HHV = 19{,}530$ Btu/lbm

Every pure compound has a unique liquid-vapor relationship identified by its particular saturation temperature–pressure curve; see Figure 5.3. Recall from thermodynamics that saturation temperature and pressure are dependent properties; that is, for a pure species, any saturation pressure

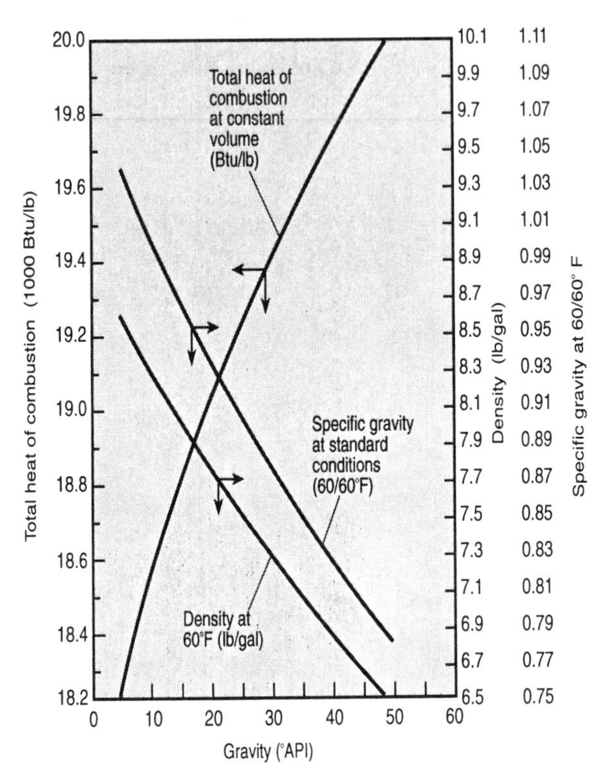

Figure 5.2 Distillate fuel properties as a function of specific gravity. (From: Steam and Its Generation and Use, 1972.)

corresponds to a unique saturation temperature. The liquid to vapor transition for a pure liquid such as benzene, methyl, or ethyl alcohol will, therefore, occur at the saturation temperature corresponding to its fuel saturation pressure. The energy required to vaporize a liquid fuel prior to combustion is equal to its *latent heat of vaporization.*

A blend of pure liquids, however, will boil off at a more complex rate, which will depend on composition as well as temperature and pressure. The vaporizing characteristic of liquid fuels, expressed as a percent of volume distilled for a fixed total pressure and variable distillation temperature, is referred to as its *volatility.* Volatility curves for several liquid fuels are shown in Figure 5.4. The ASTM has established an atmospheric distillation test to determine the volatility characteristics of liquid fuels. In this test, a sample is placed in a 100-m flask and thermally distilled by heating. Upon heating, fuel begins to vaporize at the fuel initial vaporization, or *bubble*

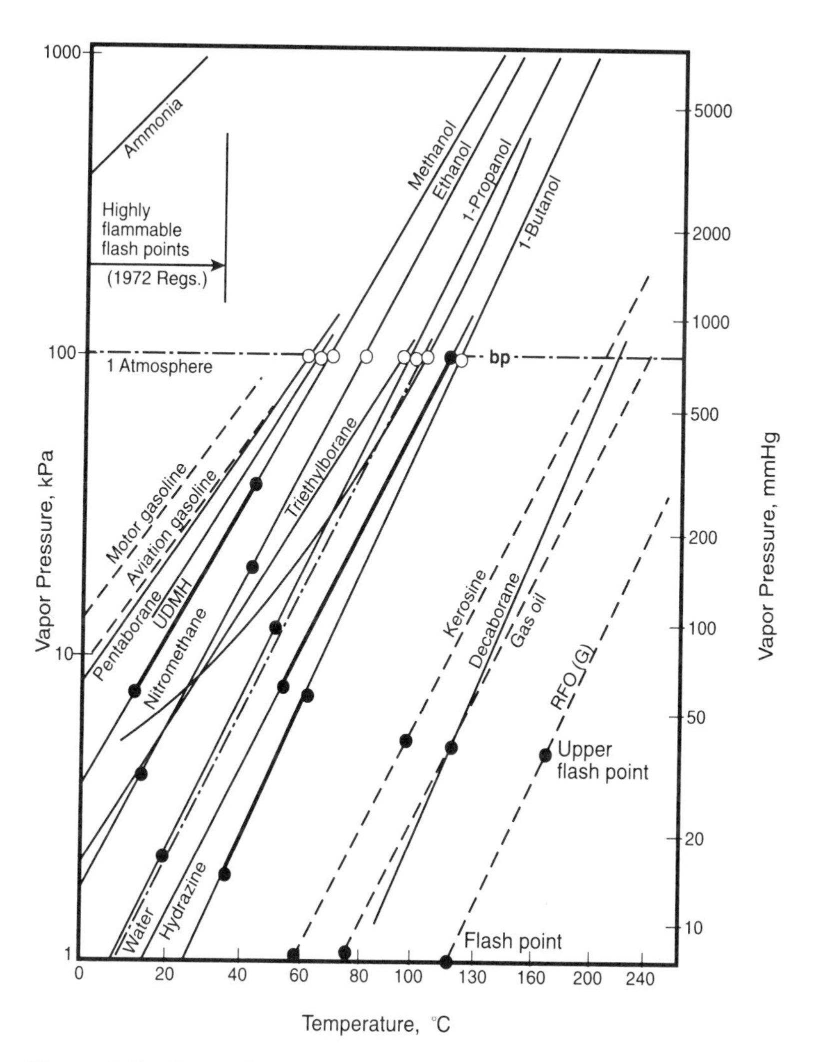

Figure 5.3 Saturation pressure-temperature curves for various petroleum and alternate liquid fuels. (From: Alternate Fuels: Chemical Energy Sources, Goodger, 1980.)

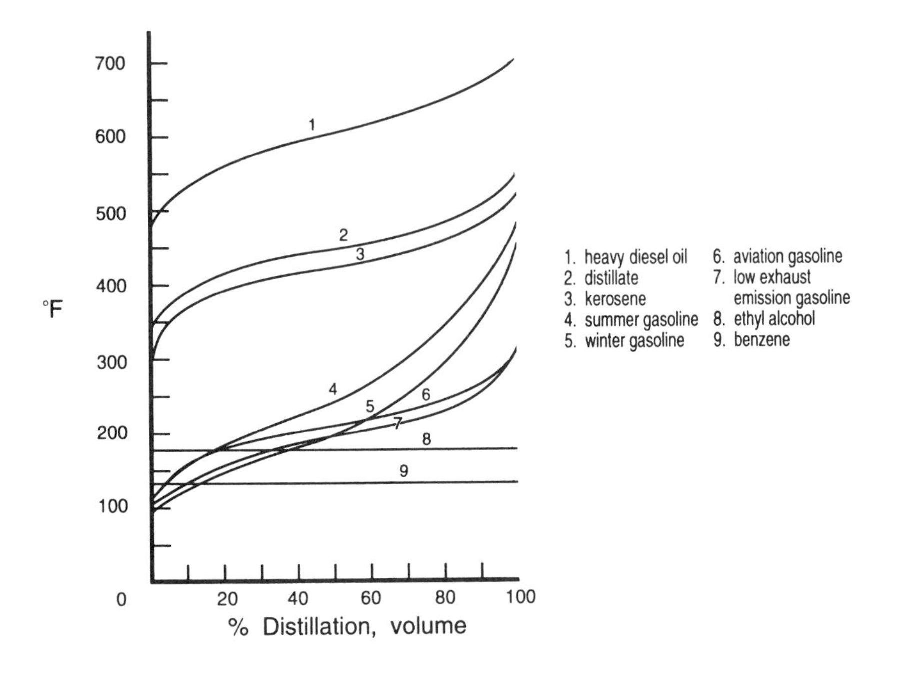

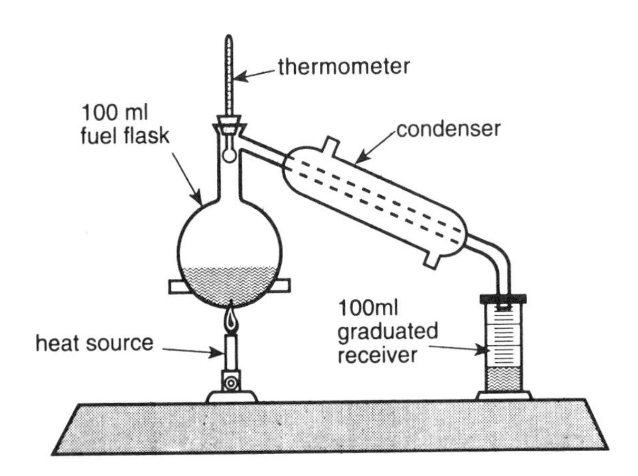

Figure 5.4 Typical ASTM petroleum distillation test facility and distillation curves. (Adapted from: *The Internal Combustion Engine,* Taylor and Taylor, 1980.)

point, temperature and, with further heating, will finally become completely vaporized at the terminating vaporization, or *dew point,* temperature. By collecting and measuring the liquid volume of recondensed vapor during this process, the volatility curve for any liquid fuel can be experimentally determined. The process will not completely distill the original fuel sample chiefly because of the loss of vapor during distillation and/or the presence of impurities, which will remain a residue.

The ASTM distillation test measures fuel volatility in a liquid-vapor fuel atmosphere. Since combustion occurs not in a fuel atmosphere but in a fuel-air environment, additional tests such as the equilibrium air distillation and Reid vapor tests have been developed to measure volatility of liquid fuels in air. In the Reid test, for example, liquid fuel is placed within a chilled fuel sample chamber, which connects below an air chamber having a volume four times that of the fuel sample. The entire unit is assembled along with a pressure gauge attached at the top of the air chamber and immersed in a 38°C (100°F) water bath. The pressure increase due to heating creates an air-water–vapor-fuel vapor mixture that is characteristic of fuel vaporization in air for this configuration.

Volatility influences the suitability of a liquid fuel for use in particular combustion systems. Some of the more general combustion factors associated with volatility include:

Potential evaporative fuel loss during storage
Potential fire hazard associated with highly volatile fuels
Proper fuel-air preparation requirements prior to combustion, i.e., carburation, atomization, and distribution
Ambient influences on fuel volatility, i.e., cold start limitations, potential vapor lock

The *flash point* temperature, a measure of the lean ignitable limit for a fuel vapor-air mixture formed above a liquid fuel, is important in relation to the potential hazards of handling flammable fuel vapors. On the other hand, the *spontaneous ignition* temperature is the minimum temperature to which a fuel vapor–air mixture must be heated in order to promote combustion without an external ignition source. This temperature is much greater than the flash point temperature of a fuel and, for hydrocarbon fuels, is generally within the ranges of 260–116°C (500–700°F). The *cloud point* is the temperature at which, when a fuel is cooled, a hazy or waxy crystallization begins to appear in the fuel. The *pour point temperature* measures the lowest temperature to which a fuel can be cooled without impeding the free flow of liquid, while the *freezing point* is the state at which a fuel is completely frozen. Liquid fuel properties are found in Table 5.1.

Viscosity, a measure of fluid resistance to internal shearing, is an important property that influences pumping, lubricity, and atomization character-

Table 5.1 Thermodynamic Properties of Various Hydrocarbon Compounds

Compound		Formula	Molecular weight	API gravity 60°F	Specific gravity 60/60°F	Freezing point °F[a]	Boiling point °F[b]	H_{fg} Btu/lbm[c]
n-*Paraffins & Isoparaffins*								
Methane	*g*	CH_4	16.04	—	—	−296.5	−258.7	—
Ethane	*g*	C_2H_6	30.07	—	—	−297.8	−127.5	—
Propane	*g*	C_3H_8	44.09	147.0	0.508	−305.8	−43.7	147.1
n-Butane	*g*	C_4H_{10}	58.12	110.8	0.584	−217.0	+31.1	155.8
Isobutane	*g*	C_4H_{10}	58.12	119.8	0.563	−255.3	10.9	141.4
n-Pentane	ℓ	C_5H_{12}	72.15	92.7	0.631	−201.5	96.9	157.5
2-Methylbutane	ℓ	C_5H_{12}	72.15	94.9	0.625	−255.8	82.1	146.6
n-Hexane	ℓ	C_6H_{14}	86.17	81.6	0.664	−139.6	155.7	157.4
2,2-Dimethylbutane	ℓ	C_6H_{14}	86.17	84.9	0.654	−147.5	121.5	138.1
n-Heptane	ℓ	C_7H_{16}	100.20	74.2	0.688	−131.1	209.2	156.8
2,2,3-Trimethylbutane	ℓ	C_7H_{16}	100.20	72.4	0.694	−12.9	177.6	137.5
n-Octane	ℓ	C_8H_{18}	114.22	68.6	0.707	−70.2	258.2	156.1
4-Methylheptane	ℓ	C_8H_{18}	114.22	68.1	0.709	−185.7	243.9	149.3
2,2,4-Trimethylpentane	ℓ	C_8H_{18}	114.22	71.8	0.696	−161.3	210.6	132.2
n-Nonane	ℓ	C_9H_{20}	128.25	64.5	0.722	−64.5	303.4	155.7
2,2,4,4-Tetramethylpentane	ℓ	C_9H_{20}	128.25	63.9	0.724	−87.8	252.1	127.8
n-Decane	ℓ	$C_{10}H_{22}$	142.28	62.3	0.730	−22	345	155
n-Tetradecane	ℓ	$C_{14}H_{30}$	198.38	54.0	0.763	+42	484	154
n-Hexadecane	ℓ	$C_{16}H_{34}$	226.43	51.5	0.774	65	536	154
n-Pentatriacontane	ℓ	$C_{35}H_{72}$	492.93	49.5	0.781	176	628	—

Olefins								
Ethylene	g	C_2H_4	28.05	—	—	−272.5	−154.7	—
Propylene	g	C_3H_6	42.08	139.6	0.522	−301.5	−53.9	—
Isobutene	g	C_4H_8	56.10	104.3	0.600	−220.6	+19.6	157.7
Pentene	ℓ	C_5H_{10}	70.13	87.5	0.646	−265.4	86.0	—
Hexene	ℓ	C_6H_{12}	84.16	77.2	0.678	−218	146.4	—
Octene	ℓ	C_8H_{16}	112.21	65.0	0.720	−152.3	250.3	158.0
Hexadecene	ℓ	$C_{16}H_{32}$	224.42	49.2	0.783	+39	527	—
Naphthenes								
Cyclopentane	ℓ	C_5H_{10}	70.13	57.2	0.750	−136.8	120.7	174.7
Cyclohexane	ℓ	C_6H_{12}	84.16	49.2	0.783	+43.8	177.3	168.8
Ethylcyclohexane	ℓ	C_8H_{16}	112.21	47.1	0.792	−168.3	269.2	155.1
Aromatics								
Benzene	ℓ	C_6H_6	78.11	28.4	0.885	+42.0	176.2	186.3
Toluene	ℓ	C_7H_8	92.13	30.8	0.872	−139.0	231.1	177.3
o-Xylene	ℓ	C_8H_{10}	106.16	28.4	0.885	−13.3	292.0	175.9
Alcohols								
Methyl	ℓ	CH_3OH	32.0	47.2	0.792	−144	149	502
Ethyl	ℓ	C_2H_5OH	46.0	48.8	0.785	−179	172	396
Propyl	ℓ	C_3H_7OH	60.0	45.6	0.799	—	208	295
Butyl	ℓ	C_4H_9OH	74.1	44.2	0.805	+26	244	254
Acetylene								
Acetylene	g	C_2H_2	26.04	—	—	+114	+114	—

Source: Most of these data were obtained from "Selected Value of Properties of Hydrocarbons," National Bureau of Standard, Circular C461, API Research Project 44.

[a]In air, 1 atm.,

[b]At 1 atm.

[c]Latent heat of fuel at constant pressure, 77°F.

Table 5.2 Major Liquid Distillate Fuels Derived from a Barrel of Crude Petroleum

Distillation product	Principal hydrocarbon families	Number of carbon atoms	% of barrel
Gasoline	Paraffins, aromatics, naphthenes	C_4–C_{10}	44
Kerosene	"	C_{10}–C_{13}	5
Distillate fuel oil	"	C_{13}–C_{18}	24
Lubricating oil	"	C_{18}–C_{45}	2
Residual fuel oil	Aromatics, naphthenes,	C_{45}	9
Asphalt and refinery fuel	other very complex compounds		16

istics of liquid fuels. Viscosity is also a major factor in selecting suitable lubricating oils. The SAE has established a worldwide viscosity numbering system for both lubricating and transmission oils.

Additional conditions of liquid fuels, such as impurities, oxidation stability, and combustion residue, are also important in asssessing their nature during combustion. For example, trace amounts of materials such as ash, sulfur, water, and metals like vanadium can influence or even limit use of liquid fuels unless they do not exceed defined concentrations. *Carbon residue* provides a means of quantifying the carbon-forming propensities of liquid fuels on fuel nozzles.

5.3 CRUDE OIL AND DISTILLATE FUELS

The world's most prominent natural liquid fuel resources are found within *crude oil,* see Table 5.2. Crude oil, or *petroleum,* was originally marine life, which has been converted over time, much like coal, into a mixture of various hydrocarbon compounds. Petroleum reserves are located within nonhomogeneous underground pools containing a variety of compounds which, depending on location, can range from heavy tarlike components to gases. Most petroleum, however, on a mass basis, consists chiefly of the following elements:

Element	Mass fraction, %
Carbon	84–87
Hydrogen	11–14
Sulfur	0–2
Nitrogen	0–0.2

The geographical location of major oil reserves, such as those found in the Middle East, have played a major role in the recent history of Western civilizations. Currently, shortfalls in petroleum supplies and a growing worldwide demand for more fuel oil and gasoline necessitate efforts to: (1) increase the yield of existing oil fields using enhanced recovery techniques; (2) explore and discover new fields, such as those being found offshore; and (3) develop alternative natural petroleum resources, such as tar sands and oil shale.

Only a small percentage of crude petroleum can be directly used as natural liquid fuels; but through further chemical processing, or *refining,* crude petroleum can yield a variety of useful fuels as well as several nonfuel products, such as greases, lubricants, and other petrochemical by-products. Any refined fuel oil or *distillate* fuel produced by a refinery will actually be a blend of many different inorganic compounds. For example, gasoline, which is often written chemically as C_xH_y, is not a single chemical compound but rather a blend of over 50 different identifiable hydrocarbon compounds. Often a single major constituent, such as isooctane C_8H_{18}, is used in thermodynamic calculations to approximate the energy characteristics of gasoline

Any hydrocarbon compound, as with all pure chemical species, has a specific boiling curve; i.e., saturation pressure is a unique function of saturation temperature. For hydrocarbons, boiling point temperatures generally increase with the number of carbon atoms in the compound; i.e., lower molecular weight components will boil at lower temperatures, followed at higher temperatures by heavier constituents. Thermal separation, a major step in refining crude oil, is achieved in the fractionating tower; see Figure 5.5. A thermal gradient is established along the column such that higher temperatures are at the bottom, with lower temperatures being maintained at the top of the unit. Major distillate fuels from crude can be classified by their volatility and, in order of increasing boiling point temperature, are:

Low molecular weight hydrocarbons
Gasolines
Napthas
Kerosenes
Diesel fuels
Lubricating oils
Residuals

EXAMPLE 5.3 Gasoline can be represented by the formula C_8H_{16}. Determine (a) *AF* ratio and *FA* ratio for the stoichiometric combustion of gasoline and air on a molar basis; (b) repeat part a on a mass basis; and (c) calculate the garvimetric and mole fractions for the stoichiometric reactants of gasoline and air.

Solution:

1. Stoichiometric reaction:

$$C_8H_{16} + aO_2 \rightarrow bCO_2 + cH_2O$$

Carbon atom balance:

$$b = 8$$

Hydrogen atom balance:

$$2c = 16 \qquad c = 8$$

Oxygen atom balance:

$$2a = 16 + 8 \qquad a = 12$$

$$C_8H_{16} + 12[O_2 + 3.76N_2] \rightarrow 8CO_2 + 8H_2O + (12)(3.76)N_2$$

2. Molar AF ratio:

$$\text{a. } \overline{AF} = \frac{(12)(4.76) \text{ moles air}}{1.0 \text{ mole fuel}} = 57.12$$

$$\overline{FA} = (57.12)^{-1} = 0.0175$$

3. Mass AF ratio:

$$\text{b. } AF = \frac{(12)(4.76 \text{ moles air})(28.97 \text{ lbm/lb mole air})}{(1.0)(96 + 16) \text{ lbm fuel}} = 14.775$$

$$FA = (14.775)^{-1} = 0.0677$$

4. Fuel-air mole fraction analysis:

$$\text{c. } \bar{x}_{C_8H_{16}} = \frac{1}{1.0 + (12)(4.76)} = 0.0172 \text{ or } 1.72\%$$

$$\bar{x}_{O_2} = \frac{12}{58.12} = 0.2065 \text{ or } 20.65\%$$

$$\bar{x}_{N_2} = \frac{(12)(3.76)}{58.12} = 0.7763 \text{ or } 77.63\%$$

5. Fuel-air gravimetric analysis:

i	$\bar{x}_i$	MW_i	$\bar{x}_i MW_i$	mf_i
C_8H_{16}	0.0172	112.0	1.9264	0.0636
O_2	0.2065	32.0	6.6080	0.2183
N_2	0.7763	28.0	21.7364	0.7181
	$\frac{\text{lb mole}_i}{\text{lb mole}_{tot}}$	$\frac{\text{lbm}_i}{\text{lb mole}_i}$	$\frac{\text{lbm}_i}{\text{lb mole}_{tot}}$	$\frac{\text{lbm}_i}{\text{lbm}_{tot}}$

$$MW = \Sigma\, \bar{x}_i MW_i = 30.271 \text{ lbm/lb mole}$$

where, for example

$$mf_i = \bar{x}_i MW_i / MW.$$

$$mf_{C_8H_{16}} = \frac{1.9264}{30.271} = 0.0636$$

Comments Since gasoline is a blend, the general chemical formula, C_xH_y, represents the overall percentage of carbon and hydrogen present in the fuel and not necessarily any single compound. Often, a single major constituent, such as isooctane, C_8H_{18}, is used to approximate the characteristics of gasoline.

The growing demands for particular fuels, such as automotive, diesel, aviation, and/or fuel heating oils, cannot be satisfied by simply using thermal distillation of crude oil. Additional refining of thermally partitioned crude is necessary to produce sufficient quantities of these fuels, for example, summer- or winter-grade gasolines. To increase the yield of these components, additional refining steps are required. Certain secondary refining processes combine smaller hydrocarbon molecules and compounds to produce larger ones, i.e., *polymerization:*

$$aC_3H_6 + bC_2H_6 + cC_2H_4 + dCH_4 \overset{T,P}{\rightleftarrows} C_3H_8$$

Lighter members of particular hydrocarbon families can be reformed by *alkalation* to yield heavier compounds of another family. For example, light olefins and isoparaffins can combine to form higher molecular weight isoparaffins. Natural crude oil vapors can be combined with kerosene or light fuel oil by *absorption* to yield additional species. Precipitation of heavier compounds formed by absorption yields natural gasoline, while compression of lighter fractions produces *liquid petroleum gas* (LPG).

Another technique for producing more specific fuels shifts hydrocarbons by breaking larger molecules into smaller species, i.e., *cracking*.

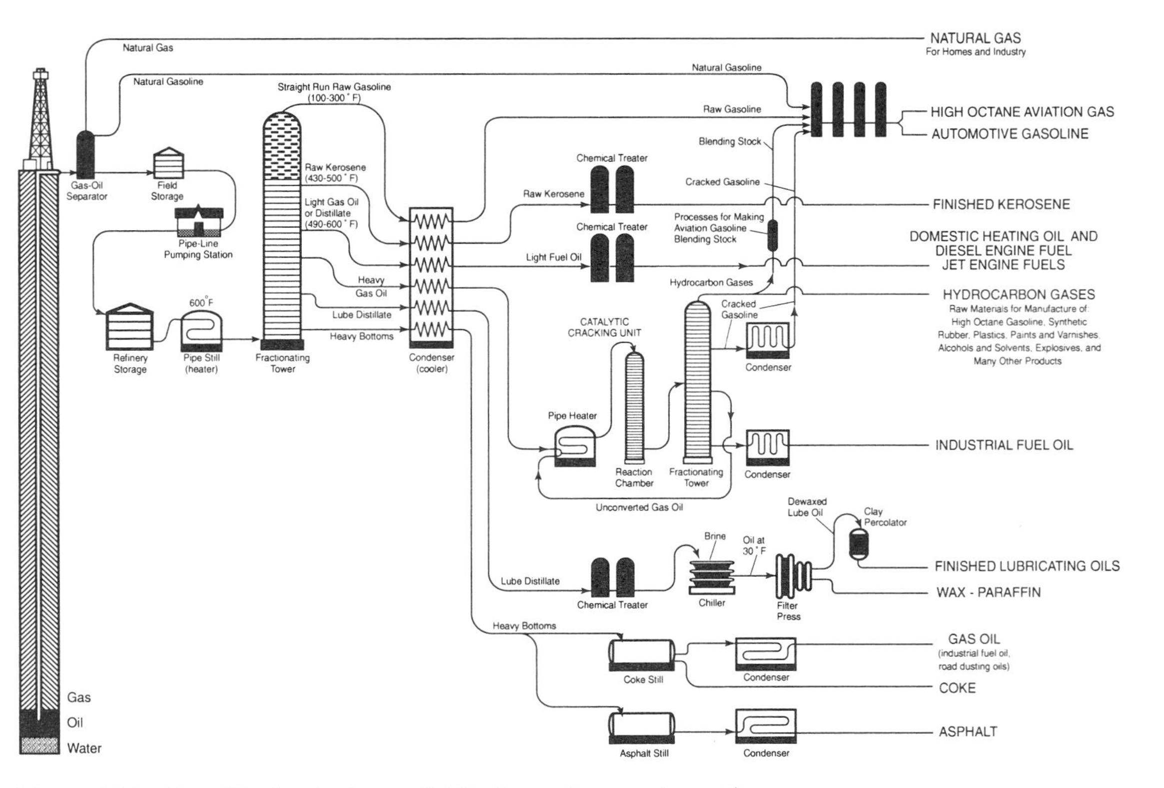

Figure 5.5 Simplified petroleum distillation refinery schematic.

$$C_{12}H_{26} \overset{T,P}{\rightleftarrows} eC_5H_{12} + fC_4H_8 + gC_3H_6$$

Thermal cracking is a high-temperature [540°C (1000°F)] and high-pressure [6870–10,306 kPa (1000–1500 psi)] conversion process. *Catalytic* cracking is a lower-temperature and lower-pressure conversion process, which uses a catalyst such as platinum. Thermal cracking done in a hydrogen gas atmosphere, or *hydrogenation,* allows unsaturated hydrocarbon compounds to pick up additional hydrogen atoms to become saturated.

By blending various refinery streams mentioned previously, specific fuel requirements can be more successfully met. Crude oil refineries have changed through the years in response to specific market demands and changes in crude resources being processed. For example, U.S. refineries initially were more concerned with producing kerosene from American reserves; gasoline was an unwanted by-product. Today, gasoline is the major product of refining, and crude oil from all parts of the world is used as feedstock. The complex modern oil refinery allows crude oil to be distilled (*separated*); to be altered structurally to produce different molecules (*converted*); to be changed molecularly while retaining the general boiling range (*upgraded*); to be processed to remove contaminants such as sulfur, nitrogen, and oxygenated species (*hydrotreated*); and to be combined (*blended*) in various refinery fractions to yield a particular product. The economics and energetics of petroleum refining is considered in detail in chemical and petroleum fuel engineering.

Gasolines are crude oil constituents that boil in the 30–200°C (85–390°F) range and are used chiefly as spark-ignition engine fuels. These refinery products must be compatible with many different types of spark-ignition engines, ranging from automotive to aviation machinery, and are required to burn in a variety of conditions, ranging from summer to winter environments. Straight run gasolines contain 5–8 carbon atoms per molecule, have a large percentage of paraffins, and are poor-quality spark-ignition engine fuels. Additional and upgraded gasoline supplies come from the refinery as a blend of straight run gasolines, thermally cracked gasolines (often formed in the presence of hydrogen), and/or reformed gasolines (often catalytically processed in the presence of platinum).

Napthas are crude oil constituents that distill in the 95–315°C (200–600°F) range. The major use of these refinery by-products, which distill between gasoline and kerosene range, is not as fuels but rather as a feedstock for producing industrial solvents.

Kerosenes, crude oil constitutents that boil in the 140–250°C (285–481°F) range, are used mainly as a burning oil in lamps, heaters, and stoves. Kerosene is a colorless hydrocarbon blend consisting of from 12–16 carbon atoms per molecule having a higher percentage of hydrogen per molecule and a greater heating value; it is lighter than gasoline. Burning oil–kerosene blends, often represented thermodynamically as dodecane,

$C_{12}H_{26}$, are high in paraffins with smaller percentages of napthanic and aromatic constituents. Power kerosenes are higher in aromatics and can be approximated by tridocane, $C_{13}H_{28}$. A gas turbine fuel, such as JP5, is an example of a straight power kerosene distillate fuel, while JP4 is an example of a gasoline-kerosene gas turbine fuel blend.

Diesel fuels, distillate fractions that boil in the range 240–370°C (465–700°F), are used primarily in compression-ignition internal combustion engines. The diesel is a broad-cut fuel combustion engine and, depending on engine size, speed, and load, can burn fuels that fall between kerosenes and heavy residuals. In general, high-speed automotive-type diesels require light kerosene-type distillates, medium-speed diesels use fuel oil–grade distillate, while low-speed marine engines can burn heavy residual fuels. Various diesel fuel grades have been established that reflect the stringent flash point, viscosity, sulfur, ash, and moisture fuel quality limits set for distillates used in compression-ignition engines.

Fuel oil, or oil burner fuel, consists of crude oil fractions that boil in the 340–420°C (640–590°F) range. Six grades of fuel oil have been established covering the requirements of atomization, smoke, and heat release of various burner fuels. The larger the fuel oil's specific gravity for a fixed boiling point temperature, the greater will be the concentration of condensables, such as napthenes, aromatics, and melting compounds, whereas a lower specific gravity implies higher percentages of open-chained paraffins. For example, fuel oils #1 and #2 are light and medium domestic fuel oils, respectively, while #6 (termed Bunker C) is a heavy residual oil that requires preheating to burn properly. Heavy fuel oils and diesel fuels are similar and consist of hydrocarbons with 12 to 16 or more carbon atoms per molecule. Some of the general thermodynamic properties of major distillate fuels are found in Table 5.3.

EXAMPLE 5.4 A fuel oil having the ash- and moisture-free analysis below is used in an oil-fired boiler. The dry analysis of the exhaust gases was measured to be as listed below:

Fuel oil constituent	Percentage
C_S	87.0%
H_2	12.0%
N_2	0.2%
S	0.2%
O_2	0.6%

Product analysis	
Gaseous component	Percentage
CO_2	11.4%
CO	2.1%
O_2	4.2%
N_2	82.3%

For these conditions, calculate (a) the mass of dry flue gas, lbm/lbm fuel; (b) the mass of air supplied, lbm/lbm fuel; (c) the air required for ideal complete combustion, lbm/lbm fuel; and (d) the percent excess air supplied.

Solution:

1. Fuel molar analysis:

i	mf_i	MW_i	mf_i/MW_i	$\bar{x}_i$
C	0.870	12.0	0.07250	0.5459
H_2	0.120	2.0	0.06000	0.4517
N_2	0.002	28.0	0.00007	0.0005
S	0.002	32.0	0.00006	0.0005
O_2	0.006	32.0	0.00019	0.0014
	$\dfrac{\text{lbm}_i}{\text{lbm}_{\text{tot}}}$	$\dfrac{\text{lbm}_i}{\text{lb mole}_i}$	$\dfrac{\text{lb mole}_i}{\text{lbm}_{\text{tot}}}$	$\dfrac{\text{lb mole}_i}{\text{lb mole}_{\text{tot}}}$

$$\sum \left(\frac{mf}{MW}\right)_i = 0.07250 + 0.0600 + 0.00007 + 0.00006 + 0.00019$$

$$= 0.13282 \text{ lb mole}_{\text{tot}}/\text{lbm}_{\text{tot}}$$

$$MW = \left(\sum \frac{mf}{MW}\right)_i^{-1} = (0.13282)^{-1} = 7.529 \text{ lbm/lb mole}$$

2. Actual combustion equation:

$$[0.5459C_s + 0.4157H_2 + 0.0005N_2 + 0.0005S + 0.0014O_2]$$
$$+ a[O_2 + 3.76N_2] \rightarrow bH_2O + cSO_2$$
$$+ d[0.114CO_2 + 0.021CO + 0.042O_2 + 0.823N_2]$$

Carbon atom balance:

$$0.5459 = (0.114 + 0.21)d \qquad d = 4.044$$

Hydrogen atom balance:

$$(2)(0.4157) = 2b \qquad b = 0.4157$$

Nitrogen atom balance:

$$(2)(0.0005) + (2)(3.76)a = (2)(0.823)(4.044)$$
$$a = 0.884$$

Sulfur atom balance: $0.005 = c$

Oxygen atom balance: $(0.0014)(2) + (2)(0.884) = 1.77$

$$0.4157 + (2)(0.0005) + 4.044\,(0.228 + 0.021 + 0.084) = 1.77$$

3. Molecular weight dry flue gas:

$$MW = (0.114)(44) + (0.021)(28) + (0.042)(32) + (0.823)(28)$$
$$= 29.99 \text{ lbm/lb mole}$$

4. Mass dry flue gas/mass fuel:

$$= \left[\frac{(4.044 \text{ lb mole gas})}{(1.0 \text{ lb mole fuel})}\right] \times \left[\frac{(29.99 \text{ lbm/lb mole gas})}{(7.529 \text{ lbm/lb mole fuel})}\right]$$

a. $\dfrac{m_g}{m_f} = 16.11$ lbm gas/lbm fuel

5. Mass air/mass fuel, actual:

$$AF = \left[\frac{(0.884)(4.76) \text{ lb moles air}}{(1 \text{ lb mole fuel})}\right] \times \left[\frac{(28.97 \text{ lbm/lb mole air})}{(7.529 \text{ lbm/lb mole fuel})}\right]$$

b. $AF = 16.19$ lbm air/lbm fuel

6. Stoichiometric reaction:

$$[0.5459C_S + 0.4517H_2 + 0.005N_2 + 0.005S + 0.0014O_2]$$
$$+ aO_2 \rightarrow bCO_2 + cH_2O + dSO_2$$

Carbon atom balance: $b = 0.5459$

Hydrogen atom balance:

$$(2)(0.4517) = (2)c \qquad c = 0.4517$$

Sulfur atom balance: $d = 0.005$

Oxygen atom balance:

$$(2)(0.0014) + 2a = (2)(0.5459) + 0.4517 + (2)(0.005)$$
$$a = 0.775$$

7. Mass air/mass fuel, ideal:

$$\text{c. } AF = \frac{(0.775)(4.76)(28.97)}{(1.0)(7.529)} = (14.19) \text{ lbm air/lbm fuel}$$

8. Percent excess air:

$$\% = \frac{(0.884 - 0.775)(4.76)}{(0.775)(4.76)} = 14.1\%$$

5.4 SYNTHETIC LIQUID FUELS

Crude oil has been the major resource base for producing a variety of primary liquid fuels for use in boilers and heaters, as well as internal and external combustion engines. Projected crude oil shortfalls and the ensuing cost of using these finite resources will cause industrial nations to look seriously in the near future for acceptable long-term liquid mobility fuel options. At present, combustion machinery such as the spark-ignition engine is being designed to burn efficiently particular liquid petroleum-based fuels. In the future, these engines will be required to burn not only those fuels more efficiently but substituted fuels as well. Synthetic fuels are manufactured, or processed, nonpetroleum-based liquid fuels that potentially can substitute for petroleum-derived fuels. Fuel alternatives are then all the various acceptable liquid fuels derived from naturally occurring or synthetically derived crudes available for use, see Table 5.4. Alternate fuels should provide performance equivalent to that by the primary fuel in a particular application. The alternate fuel may not be used at present because of logistic and/or economic considerations.

From Chapter 4, coal is seen to have several characteristics that make it somewhat unsuitable for direct use as a mobility fuel; for example, it exists naturally as a solid, it has a low H:C ratio (approximately 0.9:1 for coal versus 2:1 for crude oil), and it has high concentrations of moisture, ash, sulfur, and nitrogen. The large known U.S. reserves of coal, however, make it a logical candidate for use in producing synthetic liquid fuels, i.e., *syncrudes*. To produce a useful synoil from coal, any successful conversion must therefore show favorable economics; use minimal energy input; increase H:C ratio; reduce mineral, sulfur, and nitrogen content; and break down coal molecularly to allow carbon-hydrogen atoms to be restructured to yield a suitable liquid.

Coal is well suited for direct combustion in stationary power applications, and pulverized coal–liquid slurries have been proposed as a method of transporting coal from the mine to the utility site, where it can be

Table 5.3 Thermodynamic Properties of Major Distillate Fuels

	SG	T_{boil}		Heat of vaporization			
Gasoline	60	138	280	270	75,250	116	715
Naptha	50	171	340	240	6,690	103	670
Kerosene	40	227	440	200	5,575	86	595
Fuel oil	30	304	580	156	4,350	67	490
	API	C	°F	kJ/kg	kJ/m^3	Btu/lbm	Btu/gal

	Gasoline	Kerosene	Diesel	Fuel oil
API @ 60°	54–72	50–35	47–11	42–19
Specific gravity	0.70–0.78	0.78–0.85	0.80–0.99	0.815–0.940
Proximate analysis				
% carbon	84–86	85–88	85.7–84.9	86.1–85.0
% hydrogen	16–14	15–12	13.4–11.5	13.1–12.7
% sulfur	0.3	0.5	6.9–4.6	0.1–2.3
HHV = kJ/kg	46,520–48,265	45,125–46,985	43,030–46,520	46,100–42,565
Btu/lbm	20,000–20,750	19,400–20,200	18,500–20,000	19,818–18,300
Approximate chemical formula	C_8H_{16}	$C_{11.6}H_{23.2}$	$C_{12}H_{26}$	—

Source: NBS Misc. Pub. 97, p. 34 (1929)

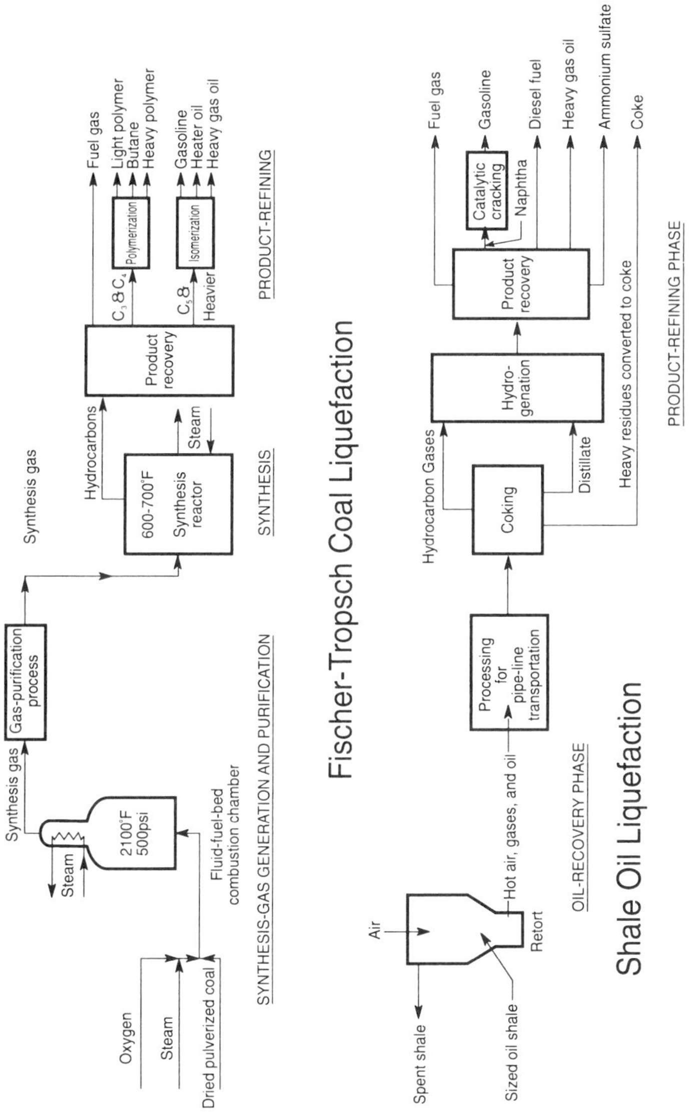

Figure 5.6 Synthetic liquid fuel recovery from coal and shale oil.

Table 5.4 Simplified Synthetic Fuel Processes
(Δ = HEAT, CAT = CATALYST, H_2 = HYDROGEN).

COAL:		
	Direct	$\text{COAL} \xrightarrow[\triangle,\ (\text{CAT})]{H_2} \text{COAL LIQUIDS} \xrightarrow[\triangle,\ \text{CAT}]{H_2} \text{SYNCRUDE} \xrightarrow[\text{REFINERY}]{} \text{PRODUCTS}$
	Indirect (Gasification)	$\text{COAL} \xrightarrow[\triangle]{H_2O} \underset{(\text{SYNGAS})}{\text{CO} + H_2} \xrightarrow[\triangle,\ \text{CAT}]{} \text{INDIRECT LIQUIDS} \xrightarrow{\text{PURIFY}} \text{PRODUCTS}$
	Pyrolysis	$\text{COAL} \xrightarrow[\triangle]{} \underset{+\text{COKE}\downarrow}{\text{LIQUIDS}} \xrightarrow[\triangle,\ \text{CAT}]{H_2} \text{SYNCRUDE} \xrightarrow[\text{REFINERY}]{} \text{PRODUCTS}$
SHALE:		$\text{OIL SHALE} \xrightarrow[\triangle]{} \text{RAW SHALE OIL} \xrightarrow[\triangle,\ \text{CAT}]{H_2} \text{SYNCRUDE} \xrightarrow[\text{REFINERY}]{} \text{PRODUCTS}$
TAR SANDS:	Near-Surface	$\underset{\text{ORE}}{\text{TAR SAND}} \xrightarrow[\text{EXTRACT}]{} \text{TAR} \xrightarrow[\text{REFINERY}]{} \text{PRODUCTS}$
	Deep	$\text{TAR SAND} \xrightarrow[\text{RESERVOIR WELLS},\ \triangle]{} \text{TAR} \xrightarrow[\text{REFINERY}]{} \text{PRODUCTS}$

Source: United States Synthetic Fuels Corporation.

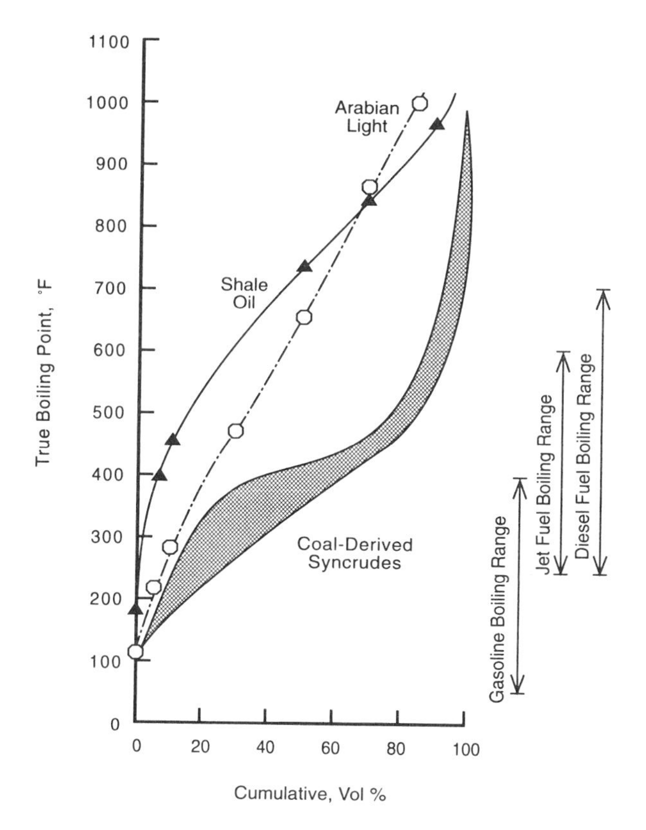

Figure 5.7 True boiling point distillation curves for Arabian light crude, shale oil, and coal-derived syncrudes.

separated and burned in coal-fired boilers. Another use for coal slurry would be to burn the entire mixture directly in residual fuel oil-fired boilers. Coal slurries such as coal-oil, coal-water and/or coal-alcohol mixtures, with as high as 40–75 wt % coal have been burned in suitable oil burners. Burning coal-oil mixtures as a simulated liquid fuel could be a means of stretching oil reserves as well as postponing replacement of certain existing oil-fired power plants with either coal or nuclear plants. However, oil-fired burner specifications and coal slurry properties do result in several adverse interface characteristics, including a need for additives due to settlement and separation of coal in nonhomogeneous slurry mixtures, increased burner abrasion, stack sooting when burning coal slurries, and reduction in heat rate of oil-fired units burning coal-liquid slurries.

The long-range future of coal-derived synfuels requires far more than supplementing residual fuel oils use with pulverized coal slurries. Marketable processes must be developed and scaled up to supply a wide range of liquid fuels. Although many methods of making synoil from coal have been attempted, at present, only a few hold promise. In fact, current successful coal liquefaction technology has its origins in work pioneered by German scientists in the 1930s prior to World War II.

Coal liquefaction techniques that can extract carbon and/or add hydrogen will increase the H:C ratio of the liquid product. *Carbonization,* or destructive thermal distillation of coal by pyrolysis, yields a carbon-rich solid char, a hydrogen-rich liquid tar, and coal gas. The low liquid yields of coal pyrolysis have properties similar to those of low-sulfur heavy crude oil, but this effluent may be processed further to yield more desirable liquid fuels. Some of the volatiles driven off by coal carbonization can also be condensed out to yield additional coal liquids.

Coal-derived syncrudes can be upgraded further and be refined in much the same manner as crude oil, to produce a wider range of useful middle-cut distillate fuels. *Hydrogenation* is a technique for direct hydrogen addition via a hydrogen-rich gas and/or a hydrogenated solvent-rich donor in the coal conversion process. In direct pyrolysis, no additional hydrogen source other than in coal itself is provided whereas, in hydropyrolysis or hydrocarbonization, tar is hydrotreated to further increase the quality of the syncrude. Pulverized coal, in this instance, is fed to a hydrogen-rich gas and recycled coal-derived liquid slurry at elevated temperature and pressure to shift the H:C ratio of the liquid. The resulting products of hydrogenated coal are solid, gas, and liquid, and the HC shift process may require catalysis, depending on the type of coal and ash content. This technique is the most probable source of future coal-derived liquid fuels.

Indirect hydrogenation uses coal-derived syngas to produce methanol and/or further conversion of methanol into gasoline. The shift of synthesis gas into low-octane gasoline and other liquids is by means of a Fisher-Tropsch catalytic converter technology developed by Franz Fischer and Hans Tropsch prior to World War II. The technology generates a large amount of intermediate- or high-Btu gas along with separated liquid products.

At present, no full-scale successful coal liquefaction industry is economically viable. Many methods that use basically the techniques discussed or combinations of various methods mentioned are being explored. The most successful coal liquefaction program was developed by the Germans during World War II, in which Fischer-Tropsch catalytic conversion, along with

Bergius hydrogeneration and catalytic hydrorefining of coal at 15–500°C (60–930°F) and 200–680-atm pressure, was used to produce gasoline and aviation fuels, see Figure 5.6.

Synthetic oil can also be produced from shale, a sedimentary rock marbled with a complex solid substance called *kerogen*. Kerogen consists chiefly of carbon, hydrogen, nitrogen, oxygen, and sulfur. Shale oil is found in rock deposits as well as in marine and coal beds and is potentially a large liquid fuel resource. Since shale oil does not exist naturally as a liquid, as does crude oil, it must be made by pyrolyzing kerogen at 480°C (900°F) to yield a synthetic liquid crude. Crude shale oil is a highly viscous liquid that is difficult to pump below 21–32°C, making transportation and storage difficult. In addition, it has a high pour point, high nitrogen but moderate sulfur content, as well as having a petroleum wax base. Shale oil can be burned as a boiler fuel or used as a refinery feedstock to yield middle-cut distillate fuels.

Shale oil is expensive, and many factors contribute to its high cost, including low organic content in shale, remoteness of the resource, as well as the capital and labor required to produce crude shale oil. Additional factors, such as air pollution from blasting, mining, crushing, transporting, and retorting shale and large quantities of polluted water associated with shale oil processing and cooling, as well as required handling of solid residue, place stringent environmental limits on shale oil production.

Table 5.5 Thermodynamic Properties of Arabian Light Crude, Shale Oil, and Coal-Derived Oil

	Arabian light crude	Shale oil	Coal-derived oil
Physical appearance	Viscous liquid	Waxy liquid	Liquid
Gravity, °*API*	33.4	20.2	18.6
Pour Point, °F	−15	+90	<-80
Sulfur, wt %	1.7	0.7	0.3
Oxygen, wt %	0.1	1.2	3.8
Total Nitrogen, wt %	0.1	2.2	0.9
Iron, ppm	<3	70	8
Arsenic, ppm	—	28	—
Hydrogen/carbon atom ratio	1.84	1.60	1.47
Boiling range, *LV* %			
St-400°F	31	4	37
400–1000°F	53	86	63
1000°F+	16	10	0

Shale oil technology generates a high level of waste and yields low amounts of oil per unit mass of raw material. Raw shale must first be mined or fractured in place, crushed, and then heated to yield a useful liquid. Both surface and in situ recovery techniques are means of retorting shale to produce syncrudes. In surface retorting, raw shale is led into a large kiln, where it is mixed and heated to produce oil, gas, and solid residue. Shale oil can be as rich as 35 wt % aromatics; however, it is still closer to crude oil in nature than coal-derived liquids and therefore has a good long-term prospect as a feedstock for producing gasoline, diesel fuel, and boiler and heating oil. Properties of various oils are found in Table 5.5.

EXAMPLE 5.5 An internal combustion engine uses liquid octane fuel. The air-fuel mixture enters the engine at 25°C and 101 kPa. The mixture, 150% theoretical air, leaves the engine at 550 K. Assuming complete combustion and that the heat loss by the engine is equal to 20% of the net work produced, calculate (a) the mass air–fuel ratio, (b) the mass flow rate of fuel required to produce 375-kw power, and (c) the corresponding mass flow rate of CO_2 produced by the engine.

Solution

1. Stoichiometric equation:

$$C_8H_{18} + aO_2 \rightarrow bCO_2 + cH_2O$$

Carbon atom balance:

$$b = 8$$

Hydrogen atom balance:

$$2c = 18 \qquad c = 9$$

Oxygen atom balance:

$$2a = (2)(8) + 9 \qquad a = 12.5$$

2. Ideal reaction (150% theoretical air):

$$C_8H_{18} + (1.5)(12.5)(O_2 + 3.76N_2) \rightarrow 8CO_2 + 9H_2O + 6.25O_2 + 70.5N_2$$

3. Air–fuel ratio (*mass basis*):

$$AF = \frac{(1.5)(12.5)(4.76 \text{ moles air})\ (28.97 \text{ kg/kg mole air})}{(1 \text{ mole fuel})\ (114 \text{ kg/kg mole fuel})}$$

a. $\qquad = 22.68$ kg air/kg fuel

4. Energy balance (*open system*):

$$Q - W = \sum_{out} N_j[\bar{h}_f^0 + \{\bar{h}<T_j> - \bar{h}<T_o>\}]_j$$
$$- \sum_{in} N_i [\bar{h}_f^0 + \{\bar{h}<T_i> - \bar{h}<T_o>\}]_i$$

where

$T_i \quad = 298\text{K} \qquad T_j = 550\text{K}$

$$\sum_{out} N_j\bar{e}_j = 8[-94{,}054 + 2537]_{CO_2}$$
$$+ 9[-57{,}798 + 2082]_{H_2O_g}$$
$$+ 6.25[1833]_{O_2} + 70.5[1769]_{N_2}$$
$$= -1097{,}409 \text{ cal} = (-1{,}097{,}409)(4.187) = -4{,}594{,}852 \text{ kJ}$$
$$\sum_{in} N_i\bar{e}_i = (1)(-59{,}740)(4.187) = -250{,}131 \text{ kJ}$$

and, since the heat loss equals 20% of the net work in magnitude,

$$-0.2W - W = -4{,}594{,}852 + 250{,}131$$
$$W = 3{,}829{,}040 \text{ kJ}$$
$$w = 3{,}829{,}040 \text{ kJ/kg fuel}$$
$$w = \frac{3{,}892{,}040 \text{ kJ/kg fuel}}{114 \text{ kg/kg mole fuel}}$$
$$w = 33{,}590 \text{ kJ/kg fuel}$$

and

$$\dot{W} = \dot{m}_f w$$

or

$$\dot{m}_f = \frac{375 \text{ kN m/sec}}{33{,}590 \text{ m kN/kg fuel}}$$

b. $= 0.0112 \text{ kg fuel/sec} = 40.3 \text{ kg fuel/hr}$

5. Carbon dioxide:
From item 1, there are 8 moles of CO_2 per mole of C_8H_{16}, or

$$\dot{m}_{CO_2} = \left[\frac{8 \text{ kgmoles } CO_2}{1 \text{ kgmole } C_8H_{16}}\right]\left[\frac{44 \text{ kg/kg mole } CO_2}{114 \text{ kg/kg mole } C_8H_{16}}\right]\left(40.3 \frac{\text{kg fuel}}{\text{hr}}\right)$$

c. $\dot{m}_{CO_2} = 124.4 \text{ kg } CO_2\text{/hr}$

Large reserves of tar sands and heavy oil are yet additional resources for long-term syncrude production. Bitumen, a hydrocarbon substance that can exist as either a solid or a heavy liquid can be found in tar sands, sandstone, limestone, and oil-impregnated rock. Because of the cohesive nature of these materials to sand and rock, they have traditionally been used to make asphalt. Petroleum techniques used to recover crude oil, such as drilling, cannot be used to recover this very viscous material. Tar sand oils have specific gravities of less than 10–15°*API* but, much like shale oil, they can be recovered, upgraded, and used to produce petroleum products. Problems associated with mining and processing of tar sands are similar to those identified with shale oil recovery. Some bitumen can be separated by hot water, while other sources require light hydrocarbon and aromatic solvent extraction. In solvent extraction systems, some solvents will remain in the solid residue. Heavy oil (10–15°*API*) and tar sands are often considered synonymous since classification and quantification of tar sands and heavy oil are difficult.

U.S. domestic synfuel resources, particularly coal and oil shale, are vast and could, therefore, supply all transportation fuel needs for centuries, see Figure 5.7. Crude oil mobility fuels will be available through the turn of the century, while synthetic fuel technology should become available at the turn of the century. With greater oil costs will come an impetus for improved fuel use, fuel economy, greater fuel tolerance, and improved combustion and engine performance. As a number of near-commercial and environmentally acceptable synfuel technologies pioneer commercial application, penetration of oil-substituted liquid fuels into the future marketplace will be a function only of oil price and availability. As long as the final products of the various synfuel options are compatible with users' needs, economics will be a driving factor behind the future of synthetic fuel.

5.5 UNCONVENTIONAL LIQUID FUELS

Recently, considerable effort has been given to developing a potentially renewable liquid biofuel alternative to supplement depleting world oil supplies. Solid biomass resources were considered in Chapter 4, and gaseous biofuels will be dealt with in Chapter 6. Animal fats, such as beef tallow and whale oil, and vegetable oils, such as peanut oil and those derived from sunflower seeds, soy, and cottonseed, have all been used as combustibles. These materials have had limited use in the past in certain specific applications, such as oil for illumination. These resources, termed *triglyceride fuels,* are hydrocarbon compounds consisting of various esters, i.e., products of alcohol and acids, such as glycerine molecules attached to long-chain fatty acids.

Photosynthesis in vegetation produces *carbohydrates* or glucose polymers, i.e., C/H/O compounds, such as sugar, starch, and/or cellulose. These materials can be converted into alcohol. A few plants, such as the milkweed, the rubber tree, the petroleum nut tree of the Philippines, and the gopher weed, actually produce hydrocarbons rather than carbohydrates. The source of liquid hydrocarbon in these plants is often concentrateed in a milky, viscous sap, called *latex,* which can be tapped and, in some cases, used directly as a fuel. Those plants that cannot be milked are first dried to remove water, reduced in size by grinding, and boiled in a solvent such as heptane to extract a liquid biocrude, which then can be processed further using current refinery technology to produce useful fuels.

Since liquid biomass fuels produced from these various feedstocks are partially oxygenated hydrocarbons, they will have a lower heating value, higher moisture content (approximately 50 wt % in fresh plants), and lower sulfur, ash, and nitrogen percentage than comparable liquid hydrocarbon fuels produced from crude oil. Liquid fuels derived from biomass are actually renewable forms of solar energy. Several major issues preclude any successful technical development of these fuel alternatives, including the food versus fuel debate, cost and energetics required to produce these fuels, and their compatibility with present and future power systems.

EXAMPLE 5.6 Methanol, CH_3OH, is being used as an automotive engine alternate fuel. An air-fuel mixture, having an equivalence ratio of 1.2 after compression is at a tool pressure of 1850 kPa abs and 680K. The combustion process occurs at constant volume, and the peak reaction temperature is 2800K. Assuming that the products of combustion contain CO, CO_2, H_2, H_2O_g, and N_2, determine (a) the combustion products' volumetric analysis and (b) the peak pressure, kPa abs.

Solution:

1. Equivalence ratio:

$$\Phi = \frac{\left.\dfrac{\text{moles fuel}}{\text{moles oxidant}}\right)_{\text{actual}}}{\left.\dfrac{\text{moles fuel}}{\text{moles oxidant}}\right)_{\text{stoichiometric}}} = 1.2$$

2. Stoichiometric equation:

$$CH_3OH + aO_2 \rightarrow bCO_2 + cH_2O$$

Carbon atom balance:

$$b = 1$$

Hydrogen atom balance:

$$2c = 4 \qquad c = 2$$

Oxygen atom balance:

$$2a + 1 = 2b + c$$
$$2a = 2 + 2 - 1$$
$$a = 1.5$$

$$CH_3OH + 1.5(O_2 + 3.76N_2) \rightarrow CO_2 + 2H_2O + (1.5)(3.76)N_2$$

3. Actual combustion:
Now, for stoichiometric reaction,

$$1 \text{ mole fuel}/(1.5)(4.76) \text{ moles air}$$

and

$$\Phi = 1.2 = \frac{x \text{ moles fuel}/(1.5)(4.76) \text{ moles air}}{1 \text{ mole fuel}/(1.5)(4.76) \text{ moles air}}$$

$$x = \left(\frac{1 \text{ mole fuel}}{(1.5)(4.76) \text{ moles air}}\right) \times (1.5)(4.76 \text{ moles air}) \times 1.2$$
$$x = 1.2$$

or

$$1.2\,CH_3OH + 1.5(O_2 + 3.76\,N_2) \xrightarrow{V} aCO + bCO_2 + cH_2 + dH_2O + eN_2$$

4. Atomic mass balances:

Nitrogen:

$$(1.5)(3.76) = e = 5.64$$

Carbon:

$$1.2 = a + b$$

Oxygen:

$$1.2 + 3.0 = 4.2 = a + 2b + d$$

Hydrogen:

$$4.8 = 2c + 2d$$

5. Equilibrium criteria: Since the products CO, CO_2, H_2, H_2O, and N_2 are *assumed to be in equilibrium,* we can write an equilibrium relation between these species. For example,

$$CO_2 + H_2 \rightleftarrows CO + H_2O$$

and

$$K_p = \frac{\left(\frac{P_i}{P_0}\right)_{CO}\left(\frac{P_i}{P_0}\right)_{H_2O}}{\left(\frac{P_i}{P_0}\right)_{CO_2}\left(\frac{P_i}{P_0}\right)_{H_2}}$$

$$= \frac{P_{CO}P_{H_2O}}{P_{CO_2}P_{H_2}}\left(P_0\right)^{2\text{-}2}$$

$$= \frac{\bar{x}_{CO}\bar{x}_{H_2O}}{\bar{x}_{CO_2}\bar{x}_{H_2}}$$

$$K_p = \frac{a\,d}{b\,c}$$

6. From the atomic mass balances in item 4.

$$a = 1.2 - b$$

$$d = 4.2 - (1.2 - b) - 2b$$
$$d = 3.0 - b$$

$$c = 2.4 - (3.0 - b)$$
$$c = b - 0.6$$

or

$$K_p = \frac{(1.2 - b)(3.0 - b)}{(b)(b - 0.6)}$$

$$K_p = \frac{3.6 - 3b - 1.2b + b^2}{b^2 - 0.6b}$$

$$K_p(b^2 - 0.6b) = 3.6 - 4.2b + b^2$$

$$(K_p - 1)b^2 + (4.2 - 0.6\,K_p)b - 3.6 = 0$$

7. Equilibrium constant K_p:

$$\log K_p = \sum_{\text{prod}} {}_i \log K_{p_i}\langle T\rangle - \sum_{\text{react}} {}_j \log K_{p_j}\langle T\rangle$$

or

$$\log K_p = [\log K_p\langle T\rangle]_{CO} + [\log K_p\langle T\rangle]_{H_2O} - [\log K_p\langle T\rangle]_{CO_2} - [\log K_p\langle T\rangle]_{H_2}$$

for $T_2 = 2800$ K

$$\log K_p = 6.649 + 1.833 - 7.664 - 0.00$$
$$\log K_p = 0.818$$
$$K_p = 6.5766$$

8. Solving for b:

$$5.5766b^2 + [(4.2) - (0.6)(6.5766)]b - 3.6 = 0$$

$$5.5766b^2 + 0.254b - 3.6 = 0$$

$$b = \frac{-0.254}{(2)(5.5766)} \pm \frac{1}{(2)(5.5766)}\sqrt{(0.254)^2 + (4)(5.5766)(3.6)}$$

$$b = 0.781$$
$$a = 1.2 - 0.781 = 0.419$$
$$d = 3.0 - 0.781 = 2.220$$
$$c = 0.781 - 0.6 = 0.181$$

Check:

$$K_p = \frac{a\,d}{b\,c} = \frac{(0.419)(2.220)}{(0.781)(0.181)} = 6.58$$

9. Mole fractions:

$$\bar{x}_{CO} = \frac{a}{a+b+c+d+e} = \frac{0.419}{9.241} = 0.04534$$

$$\bar{x}_{CO_2} = \frac{b}{a+b+c+d+e} = \frac{0.781}{9.241} = 0.08451$$

$$\bar{x}_{H_2} = \frac{c}{a+b+c+d+e} = \frac{0.181}{9.241} = 0.01959$$

$$\bar{x}_{H_2O} = \frac{d}{a+b+c+d+e} = \frac{2.220}{9.241} = 0.2402$$

$$\bar{x}_{N_2} = \frac{e}{a+b+c+d+e} = \frac{5.64}{9.241} = 0.6103$$

10. Peak pressure: $V = C$ combustion

$$P_1 V_1 = N_1 \bar{R} T_1 \qquad P_2 V_2 = N_2 \bar{R} T_2$$

$$P_2 = \frac{N_2 \bar{R} T_2}{V} = \frac{N_2 \bar{R} T_2}{N_1 \bar{R} T_1} P_1 = \frac{N_2 T_2}{N_1 T_1} P_1$$

$$P_3 = \frac{(9.241)(2800)(1850 \text{ kPa})}{(1.2 + (1.5) \times 4.76)(680)} = 8{,}441 \text{ kPa abs}$$

Fuel-grade ethanol can be produced on a commercial scale either by synthesis or fermentation reactions. The chemical industry has traditionally generated non-fuel-grade ethyl alcohol, using ethylene as feedstock. Several renewable biomass resources are currently being considered as feedstocks for commercially fermented fuel alcohol production. These materials, which are principally carbohydrates, can be chemically converted by enzyme action into alcohol, carbon dioxide, protein, and various secondary products. Potentially, any biomass material is a source of alcohol production, but the most readily available materials commercially viable on a large scale at present are sugar or starch sources. Sugar sources include sugarcane, sugar beets, and molasses, as well as fruits and their juices. Starch sources include grains, potatoes, and other root crops. The technology for cellulose enzyme conversion is not yet developed to commercial scale.

Sugar stocks are principally sucrose, $C_{12}H_{22}O_{11}$ which, under catalysis of invertase enzyme, produces glucose, $C_6H_{12}O_6$, and fructose, $C_6H_{12}O_6$, via the sucrose hydrolysis reaction.

$$\underset{\text{Sucrose}}{C_{12}H_{22}O_{11}} + H_2O \xrightarrow{\text{Invertase}} \underset{\text{Glucose}}{C_6H_{12}O_6} + \underset{\text{Fructose}}{C_6H_{12}O_6}$$

Invert sugars, i.e., glucose and fructose, are next converted by fermentation into equal moles of ethyl alcohol, C_2H_5OH, and carbon dioxide by the enzyme zymase.

$$\underset{\text{Invert sugars}}{C_6H_{12}O_6} \xrightarrow{\text{Zymase}} \underset{\text{Ethanol}}{2CH_3CH_2OH} + \underset{\text{Carbon dioxide}}{2CO_2}$$

Secondary by-products are produced from invert sugars and include aldehydes, esters such as ethyl acetate, higher molecular weight alcohols called fusel oils, fatty acids, and trace amounts of aromatics. These compounds are referred to as *congeners*.

Starch sources must be cooked to form a gel and prepared differently than sugar sources. Starch is mixed with malt, i.e., sprouted barley, which contains the diastase enzyme. This enzyme converts starch, $(C_6H_{10}O_5)_n$, into maltose sugar, $C_{12}H_{22}O_{11}$.

$$(C_6H_{10}O_5)_n \xrightarrow{H_2O} C_{12}H_{22}O_{11}$$

Starches — Malt enzyme diastase — Maltose sugar

Maltose sugar is converted to glucose, $C_6H_{12}O_6$, by the yeast enzyme.

$$C_{12}H_{22}O_{11} \xrightarrow{\text{Yeast enzyme}} C_6H_{12}O_6$$

Maltose sugar — Glucose

Glucose is converted to ethyl alcohol and CO_2 by fermentation with zymase enzymes as was the case for sugar stocks.

The technology for commercial ethanol production is available, and facilities currently in operation or near operational status demonstrate concept viability. Since the major and most stable grain feedstock in the United States is corn, the following facility will deal specifically with fermentation of cornstarch to produce fuel-grade ethanol. The production of grain alcohol fuel consists of the following major steps:

1. Corn cleaning and milling
2. Corn mill cooking
3. Saccharification
4. Fermentation
5. Distillation
6. Azeotropic drying
7. Protein by-product drying

The principle of anaerobic fermentation and distillation are specific processes currently in use in related fields such as the beverage alcohol and petrochemical industries. Operational experience and data from a growing alcohol fuel industry indicate that successful alcohol plants must be designed and operated on the basis of favorable energy balances, i.e., energy output > energy input. Successful 200-proof ethanol production requires use of nonpetroleum energy sources as well as strong regional and local availability of grain input and marketability of product output. Commercially viable ethanol-producing plants should:

1. Utilize no petroleum in the production of alcohol fuels and protein by-products
2. Reduce foreign oil energy consumption in the United States and contribute to U.S. energy independence
3. Produce a fuel-grade alcohol that can be used as a fuel or an octane-booster in unleaded gasoline
4. Produce a usable protein by-product that can meet farm-customer requirements
5. Develop the technology of alternate energy in the production of biosynthetic fuel resources

A process flow diagram for a corn-to-ethanol facility is shown in Figure 5.8. Corn is transported by rail and/or truck to a feedstock delivery where each load is recorded at a weighing station. The corn is then dumped into a transfer pit, and the delivery vehicle exits the site.

Grain is transferred from the delivery site to storage facilities and/or the main plant. The corn is first transported to a grain cleaner, where foreign material such as sand, glass, and metals are removed from the grain. Cleaned grain drops by gravity feed into a discharge hopper, where a bucket elevator transfers it to a cleaned-grain storage bin. A feed system transfers the corn to a hammermill, where corn is continuously milled to a 20-mesh screening. After milling, the grain is carried pneumatically through a flow meter for inventory control and directed to a surge hopper.

The grain delivery, storage, and milling operations are located away from alcohol-generating portions of the plant. This isolation ensures that dust from these components does not interfere with the activities required to generate anhydrous alcohol. Dust-collecting facilities are provided to remove dust from these processes and transfer the material to a baghouse area.

Starch in milled corn can be commercially converted into fermentable sugars by acid or enzymatic hydrolysis. Enzyme conversion requires cooking to gelatinize the material prior to reduction to soluble dextrins. Choice of high- or low-temperature cooking will be dictated by plant size and the energetics of its specific configuration. Cooking can be done on a batch or continuous basis, as well as by extrusion and/or direct steam injection. Large-scale facilities favor high-temperature continuous cooking using extrusion and steam injection systems in order to conserve energy, shorten cooking time, and be compatible with continuous production of alcohol. Saccharification converts dextrin into simple fermentable sugars. This process requires a lower temperature and different enzyme action than the cooking step.

Next, clean, dry, milled corn is continuously fed through the corn cooker extruder. Cooked corn leaves the process at 150°C (300°F) and is fed to an

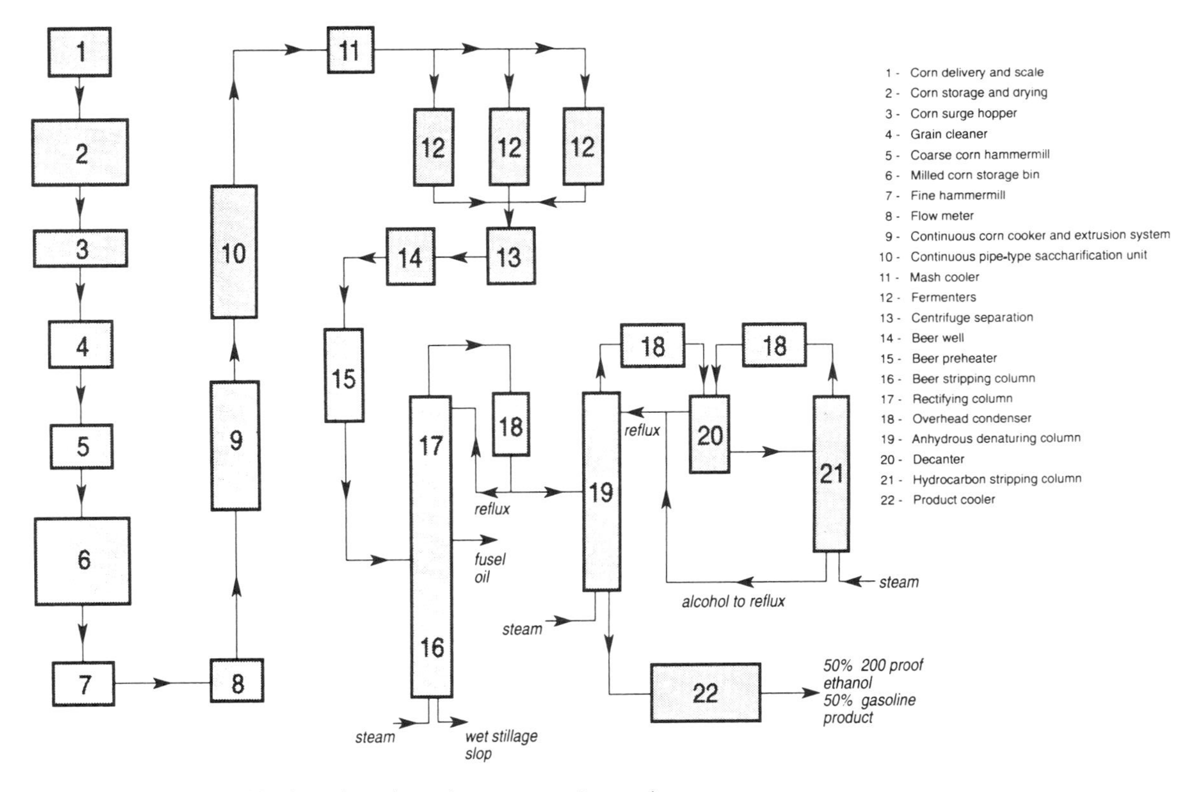

Figure 5.8 Simplified grain ethanol process schematic.

attrition mill, where material is reduced to 20-mesh screening. Extruder cooking greatly reduces the required amount of process heat over conventional steam-jacketed cooking tanks. Heat generated by friction in addition to steam injected directly in the corn extrusion process provides heat needed for cooking.

Material from the corn cooker extruder facility is next fed to a presaccharification mixer. At this stage, water, enzymes, thin stillage, and pH controls (such as sodium hydroxide for low water pH or concentrated sulfuric acid for high water pH) are added to the tank and continuously mixed. Residence time in the premixing tank is approximately 2 min. Sufficient liquid input to the premixing process produces approximately 30 gal of mash per bushel of grain input. Cooling water is provided to ensure that the mash mixture is at 66°C (150°F) at discharge. The mash consists of approximately 33% solids.

The mash is pumped from the mixing tank by means of a mash transfer pump through a continuous-pipe saccharification process. In this process, mash starch mixture is converted into fermentable sugars at 60°C (140°F). Residence time in the unit is approximately 7 min. After the saccharification process, mash passes through a mash cooler, where it is cooled to 27°C (80°F) by water.

Saccharificated corn mash is next transferred to alcohol fermentation tanks. Mash is pumped into a tank while fermentation yeast is concurrently added. Multiple tank sections allow continuous feed to be maintained so that one tank is being cleaned and filled, one tank is being discharged, and the remaining fermenters are kept at 27°C (80°F). Fermentation converts sugar produced by cooking and saccharification into a alcohol-water wet mash mixture, termed *beer*. The beer, a 6–12% ethanol + water mixture, can be commercially produced using batch or continuous processes at room temperature over a 36–48-hr period. Techniques using advanced enzymes, which may accelerate this step, are still in developmental states.

Each tank has a provision for venting carbon dioxide generated during fermentation. The CO_2 is vented to a scrubber, where alcohol is stripped from the gases and redirected to the beer well. After fermentation, the beer is pumped from the tank to a solid separation system.

After leaving the fermenter, water, ethanol, and solids pass through a centrifuge, which separates heavy solids from the alcohol-water stream for removal, treatment, and recovery of *distiller dried grains* with solids (DDGS) for separate marketing. The water-alcohol mixture, which is approximately 10% alcohol, is pumped to the beer well.

Alcohol recovery and stillage separation of a dilute alcohol-water-solids beer mixture occur in the distillation and drying section of the plant. Beer is pumped from a beer well to distillation columns through a beer preheater, where it is heated from 33°C (90°F) to 93°C (199°F). Saturated beer having an alcohol concentration of 6–12% enters the distillation column for product separation.

Steam distillation of the alcohol-water mixture will produce an azeotropic mixture of approximately 96% ethanol. Further drying of the product produces a 200-proof anhydrous ethanol product, which can be denatured and blended with gasoline to produce a fuel-grade blend of 10% alcohol–90% gasoline, commercially termed *gasohol.*

Alcohol is stripped from the water and solids by the rectifying column. The column produces an overhead azeotropic mixture having an 95.7% alcohol content at 79°C (173°F). The overhead vapor passes through a rectifying condenser. A fraction of the condensed product is sent to a denaturing and drying column for further drying and handling.

Higher molecular weight alcohols, termed fusel oils, are produced in the fermentation stage. These components must be removed from the upper section of the rectifying column by side extraction, and then cooled, washed, and stored.

Stillage is removed from the beer in the beer stripping column below the rectifying section. The column produces a bottom slop having approximately 8% solids and less than 0.05% ethanol. The wet stillage leaves the column at approximately 100°C (212°F). This material is pumped to the distillers dried grain recovery section of the plant.

The azeotropic alcohol-water product is pumped to an anhydrous denaturing column. The column produces an anhydrous ethanol–unleaded gasoline blend, which is pumped to a tank and cooled for storage. Steam supplied at the base produces a tertiary azeotrope of hydrocarbon alcohol and water. After condensing, the overhead is fed to a decanter, where material separates into a top hydrocarbon-rich layer and a bottom alcohol-hydrocarbon-water mixture. The upper section is pumped as reflux back into the drying column. The bottom of the decanter is pumped to the hydrocarbon stripping column, where alcohol and hydrocarbon are separated.

A major by-product of the main plant will be a high-protein livestock feed of dark distillers dried grains (DDDG). Nonsoluable grain residuals are separated from dissolved solids, i.e., sugars, in the protein by a product drying system. These grain residuals are further dewatered to produce a DDDG having at least 30–50% solids. Energy for drying these grains is expensive, and stack gases from the steam generation system could be used to further dry the product. Liquid waste from this process, after partial recirculation to the plant processing, would be passed to a waste stream discharging from the plant to a sewer system.

PROBLEMS

5.1 Consider the stoichiometric combustion of the paraffin fuel C_xH_{2x+2} and air. Find an expression in terms of x for (a) the molar *FA* ratio, kg mole fuel/kg mole air; (b) the *AF* ratio, kg air/kg fuel; (c) the air-fuel mixture molecular weight, kg/kg mole; and (d) the ideal product mole fractions.

5.2 NASA has developed a thermodynamic model for gasoline. The analysis approximates gasoline as a mixture of 25% benzene, 25% octane, and 50% dodecane, by weight. For this hypothetical fuel, calculate (a) the molar $\overline{FA}$ ratio, lb mole fuel/lb mole air; (b) the *AF* ratio, lbm air/lbm fuel; (c) the fuel molecular weight, lbm/lb mole; and (d) the fuel higher heating value, Btu/lbm.

5.3 Gasohol, a potential unleaded gasoline applications substitute fuel, can be represented as a liquid mixture of 90% octane and 10% ethanol by weight. For this fuel, determine (a) the stoichiometric *AF* ratio, lbm air/lbm fuel; (b) the fuel molecular weight, lbm/lb mole; (c) the fuel higher heating value, Btu/lbm fuel; and (d) the gasohol-to-*n*-octane mass ratio necessary ideally to produce equal *STP* heat releases burning in equal amounts of air.

5.4 Iso-octane is used as a standard fuel in rating spark-ignition engine fuels. For iso-octane, calculate (a) the higher heating value for this liquid fuel, Btu/lbm; (b) the lower heating value, Btu/lbm; (c) the higher heating value, Btu/gal; and (d) lower heating value, Btu/gal.

5.5 Dodecane is used as a standard fuel in rating compression-ignition engine fuels. For dodecane, calculate (a) the higher heating value of this fuel, kJ/kg; (b) the higher heating value, kJ/m^3; (c) the lower heating value, kJ/kg; and (d) the lower heating value, kJ/m^3.

5.6 The *API* gravity of a distillate fuel oil at 78°F is 64.5. Determine (a) the *API* gravity at 60°F; (b) the specific gravity at 60°F; (c) the density at 60°F, lbm/ft^3; and (d) the density at 78°F, lbm/gal.

5.7 A gasoline blend has a specific gravity of 0.707 at 15°C. Calculate (a) the specific gravity, *API*; (b) the fuel density, kg/m^3; (c) the fuel lower heating value, kJ/kg; and (d) the fuel higher heating value, kJ/m^3.

5.8 Gasoline, having 85 wt % carbon and 15 wt % hydrogen, has a density of 6 lbm/ft^3 and a higher heating value of 20,700 Btu/lbm. For ideal complete combustion and $\Phi = 0.8$, find (a) the ratio of dry moles of product to moles of reactant; (b) the *AF* ratio, lbm air/lbm fuel; (c) the ratio of water formed to fuel burned, gal H_2O/gal fuel; and (d) the approximate overall chemical formula of the fuel, C_xH_y.

5.9 A proposal has been made to replace fuel oil by combustion of a liquid methanol and pulverized coal mixture. For a fuel having a gravimetric analysis of 50% liquid and 50% solid (represented by pure carbon), find (a) the stoichiometric equation; (b) the *AF* ratio, kg air/kg fuel; (c) the fuel molecular weight, kg/kg mole; (d) the fuel lower heating value, kJ/kg mole fuel; and (e) the fuel higher heating value, kJ/kg fuel.

5.10 A pulverized anthracite coal source is to be used to produce a synthetic gasoline, $C_{8.5}H_{18.5}$, by hydrogenation. The first step required in the conversion process is to remove the sulfur, oxygen, nitrogen, and ash to produce a pure stream of carbon and hydrogen. The carbon-hydrogen stream is then further reacted with additional hydrogen to upgrade the material to the specified synthetic fuel conditions. For an ultimate coal analysis of 90% carbon, 4% hydrogen, 2% oxygen, 1% nitrogen, and 3% ash, determine (a) the refined coal carbon-to-hydrogen ratio, kg/kg; (b) the refined coal carbon and hydrogen mole fractions, %; (c) the refined coal ideal hydrogenation reaction; (d) the mass of synthetic fuel to mass of raw coal, kg fuel/kg coal; and (e) the mass of required hydrogen to mass of raw coal, kg H_2/kg coal.

5.11 The gravimetric analysis of a liquid marine boiler fuel on an ash- and moisture-free basis is given as:

Carbon	86.63%	Oxygen	0.19%	Sulfur	1.63%
Hydrogen	11.27%	Nitrogen	0.28%		

Calculate (a) the stoichiometric equation for ideal complete combustion of the fuel; (b) the ratio of weight of water formed to weight of fuel burned, lbm H_2O/lbm fuel; and (c) the higher heating value of the fuel, Btu/lbm fuel.

5.12 A fuel oil has the following gravimetric ash- and moisture-free analysis:

Carbon	87.0%	Oxygen	0.7%	Sulfur	0.2%
Hydrogen	12.0%	Nitrogen	0.1%		

The dry exhaust gas volumetric analysis is found to be:

CO_2	12%	O_2	8%
N_2	78%	CO	2%

For constant-pressure combustion at 1-atm pressure, calculate (a) the mass of water formed per mass of fuel burned, kg H_2O/kg fuel; (b) the dew point temperature of the stack gases, °C; (c) the mass of CO_2 produced per mass of fuel burned, kg CO_2/kg fuel; and (d) the percent of excess air required for combustion, %.

5.13 A boiler burns 50 lbm/hr of *n*-decane in 130% theoretical air. If both fuel and air are supplied at 60°F, find (a) the required fuel flow rate, gal fuel/hr; (b) the required air flow rate, ft^3 air/min; (c) the required stack gas condensation rate for removal of water, lbm H_2O/hr; and (d) the dry exhaust gas flow rate for exhaust gases at 300°F, ft^3/hr.

5.14 An iso-pentane and 135% theoretical air mixture ignites and burns at constant pressure. Both liquid fuel and air are supplied to the unit at 27°C and 103 kPa. Exhaust gases leave the unit at 327°C. Calculate (a) the energy released by ideal combustion, kJ/kg mole fuel; (b) the exhaust dew point temperature, °C; (c) are specific heat ratio of dry exhaust gases, kJ/kg C; and (d) the combustion efficiency for a product temperature of 127°C, %.

5.15 An internal combustion engine runs on liquid octane and 150% theoretical air. At steady-state operation, the fuel-air mixture enters the engine at 25°C and 101 kPa. The heat loss from the engine combustion is equal in magnitude to 20% of the work output. Assuming complete combustion of the mixture and the exhaust = 600K, find (a) the reaction equivalence ratio; (b) the net work, kJ/kg mole fuel; (c) the engine thermal efficiency, %; and (d) the fuel consumption required to produce 100 kW, kg/sec.

5.16 An 8-cylinder diesel engine burns #2 fuel oil in 200% theoretical air combustion. The fuel, *API*<60°F> = 32, has an ultimate analysis of 86.4% carbon, 12.7% hydrogen, 0.4% sulfur, 0.3% oxygen, and 0.2% nitrogen. The engine is rated at a maximum brake power output of 5220 hp and a brake specific fuel consumption of 0.284 lbm/hp hr. For these conditions, calculate (a) the *AF* ratio, lbm air/lbm fuel; (b) the fuel density, lbm/ft^3; (c) the fuel consumption, gal/hr; and (d) the engine thermal efficiency, %

5.17 A synthetic colloidal fuel is to be made using pulverized coal and liquid methanol. The fuel contains 80% methanol and 20% pulverized coal on a mass basis. Assume that the liquid slurry can be represented as $[\alpha Cs + \beta CH_3OH]$, where α and β are mole fractions of carbon and methanol, respectively. For this fuel, determine (a) the values of α and β; (b) the ideal stoichiometric combustion reaction; (c) the *FA* ratio, lbm fuel/lbm air; and (d) constant-pressure combustion efficiency for stoichiometric conditions with reactants @ 77°F and products at 1080°R, %.

5.18 Liquid butane is being used as a fuel for space heating. The required energy from combustion is 50 kW. Design specifications require that the exhaust gas temperature not exceed 400K. The dew point temperature of the exhaust is 41.5°C, while both fuel and air are supplied at 1-atm pressure and 25°C. Calculate (a) the percent of excess air,

%; (b) the ideal minimum fan power required for the air supply, kW; (c) the fuel rate of consumption, kg fuel/hr; and (d) the maximum volumetric flow rate of dry exhaust gases, m^3/hr.

5.19 A portable kerosene burner is being used for the purpose of providing space heating. The heater delivers 15,200 Btu/hr of thermal energy and uses kerosine having a specific gravity @ 60°F of 0.8. For the unit, determine (a) the higher heating value of the kerosine, Btu/lbm; (b) the density of the fuel, lbm/gal; and (c) the steady burn period for the space heater if the unit has an 85% combustion efficiency and a 2-gal fuel tank, hr.

5.20 An oil-fired boiler has an 80% boiler efficiency. Cold water enters the tank at 65°F and leaves at 165°F. The heater, when running at steady-state and steady-flow conditions, produces 200 gal H_2O/hr. The burner is fired using liquid benzene and 115% theoretical air. Fuel supply is at 77°F, while the air enters the burner at 100°F. Find (a) the exhaust flue gas temperature for complete combustion, °F; (b) the fuel consumption, gal fuel/hr; (c) the minimum fan power for the air supply, hp; and (d) the combustion efficiency, %.

5.21 A steam generator in a turboelectric plant uses fuel oil and produces the following performance results:

Fuel analysis		*Stack analysis*	
$C_{12}H_{26}$		CO_2	12.01%
184,000	lbm/hr	O_2	3.71%
T_{inlet}	80°F (fuel and air)	T_{outlet}	450°F

feedwater @	570°F
sat steam @	2400 psia
steam flow	3,500,000 lbm/hr

For the unit, estimate (a) the amount of excess air used, %; and (b) the thermal efficiency of this steam generator, %.

6

Gaseous Fuels

6.1 INTRODUCTION

Several important solid and liquid fuels, along with their general thermochemical properties and combustion characteristics, were treated separately in Chapters 4 and 5. In this chapter, the remaining fuel category by phase, i.e., gaseous fuels, will be discussed. Obviously, all gases can be liquefied since every pure compound, depending on temperature and pressure, can exist as a gas, a liquid, and/or a vapor. Many gases, in fact, can be condensed simply by compression. Some, such as liquefied natural gas (LNG) and liquid hydrogen, require a supercold, or *cryogenic,* state in order to exist and will not remain in this condition without extreme refrigeration and insulation. For the sake of this discussion, therefore, a fuel will be considered to be a gas if it is noncondensable over *normal* temperatures and pressures. Examples of gaseous fuels include hydrogen gas, methane, and natural and synthetic gas (SNG). Vapors such as propane, butane, and liquefied petroleum gas (LPG) are fuels that are condensable over normal temperatures and pressures. Organizations such as the American Gas Association (AGA) are involved in many aspects of fuel gas technology, including supporting research and development within the natural gas industry, producing synthetic gas via coal gasification, and developing biomass-generated methane systems. Gaseous fuel science will continue to play an important role in combustion engineering in part because of the clean-burning nature of these energy resources, the fact that most fuels actually burn in the gas phase, and insights into the complex nature of general combustion processes being provided by studies of simple gas-phase molecular reactions.

EXAMPLE 6.1 The specific gravity (SG) of gaseous fuels is expressed as the density, or specific weight, of the fuel at 15.56°C and 1 atm to that of air at 15.56°C and 1 atm. For these conditions, determine (a) the density of methane, CH_4, kg/m^3; (b) the density of propane, C_3H_8, kg/m^3; (c) the specific gravity of methane; and (d) the specific gravity of propane.

Solution:

1. Density:

$$P = \rho RT = \rho\left(\frac{\overline{R}}{MW}\right)T, \qquad \rho = \left(\frac{MW}{\overline{R}}\right)\frac{P}{T}$$

CH_4:

$$\text{a. } \rho = \frac{(16\text{ kg/kg mole})(101\text{ kN/m}^3)}{(8314\text{ kN m/kg mole K})(288.56\text{K})} = 6.736 \times 10^{-4}\text{ kg/m}^3$$

C_3H_8:

$$\text{b. } \rho = \frac{(44)(101)}{(8314)(288.56)} = 1.852 \times 10^{-3}\text{ kg/m}^3$$

2. Specific gravity of gases:

$$SG <15.56°\text{C}> = \frac{\rho_{\text{gas}}<15.56°\text{C, 1 atm}>}{\rho_{\text{air}}<15.56°\text{C, 1 atm}>} = \frac{MW_{\text{gas}}}{MW_{\text{air}}}$$

$$\rho_{\text{air}} = \frac{(28.97)(101)}{(8314)(288.56)} = 1.22 \times 10^{-3}\text{ kg/m}^3$$

$$\text{c. } SG_{CH_4} = \frac{6.736 \times 10^{-4}}{1.220 \times 10^{-3}} = \frac{16}{28.97} = 0.552$$

$$\text{d. } SG_{C_3H_8} = \frac{1.852 \times 10^{-3}}{1.220 \times 10^{-3}} = \frac{44}{28.97} = 1.519$$

Comments Since the specific gravity of methane is less than 1, a methane leak will fill an entire room, while propane, with a specific gravity greater than 1, tends to sink to the floor. This effect causes fire hazards associated with methane leaks to be different from those for propane. For methane, a spark anywhere could ignite an entire room whereas, for propane, the ignition would have to occur near the floor.

6.2 GASEOUS FUEL PROPERTIES

Recall from Chapters 1 and 5 that the specific gravity (SG) of gases and liquids is equal to the ratio of density for a particular fluid of interest to that of a reference compound. For liquids, the reference is water; for gases, however, the reference is air. Liquid density is only a function of temperature but, for a gas, density is a function of both temperature and pressure and, therefore, the specific gravity of a gas is given as

$$SG_{gas} = \frac{\rho_{fuel} < 1 \text{ atm } 15.56°\text{C } (60°\text{F}) >}{\rho_{air} < 1 \text{ atm } 15.56°\text{C } (60°\text{F}) >} \tag{6.1}$$

If ideal-gas behavior is assumed, the molar density $\bar{\rho}$ of fuel gases is then equal to

$$\bar{\rho} = \frac{P}{\bar{R}T} \frac{\text{kg mole}}{\text{m}^3} \left(\frac{\text{lb mole}}{\text{ft}^3} \right) \tag{6.2}$$

and the specific gravity of these fuels is, therefore, equal to

$$SG_{gas} = \left[\left(\frac{MW}{\bar{R}} \right) \left(\frac{P}{T} \right) \right]_{gas} \left[\left(\frac{MW}{\bar{R}} \right) \left(\frac{P}{T} \right) \right]_{air} = \frac{MW_{gas}}{MW_{air}} \tag{6.3}$$

Equation (6.3) shows that a gaseous fuel having a specific gravity of less than 1.0 is lighter than air whereas a fuel having a specific gravity greater than 1.0 is heavier than air. This can be an important factor in gas leakage and safety requirements for storing and handling gaseous fuels. Table 6.1 lists specific gravities and critical constants for several gaseous fuels.

Solid and liquid fuel constant-volume bomb calorimetry, described in Chapter 4, is not used with gaseous fuels. Instead, heating values for gaseous fuels are experimentally determined using a constant-pressure flow calorimeter shown in Figure 6.1. Gas and air at room temperature are supplied to the calorimeter unit, and energy released by burning gases is absorbed by cooling water flowing through a water jacket surrounding the burner. Regulated cooling water mass flow rate ensures that products of combustion exhaust from the calorimeter at room temperature. Precise measurement of the cooling water temperature rise and water mass flow rate enables one to calculate the energy release rate of the fuel, i.e., kJ/sec (Btu/min). Gaseous fuel volumetric flow rates are obtained by use of a wet test meter. Flow calorimeters can experimentally measure heating values in a 3.7–120 MJ/m^3 (100–3200 Btu/ft^3) range. An inability to maintain steady-state and steady-flow conditions, caused, for example, by fluctuations in water and/or gas source supply conditions as well as gas leaks within the calorimeter, can introduce significant experimental error and, hence, can predict erroneous heating values for a gaseous fuel.

Table 6.1 Gaseous Fuel Specific Gravities

		Critical constants		
			T	
	SG	*P*, atm	°C	(°F)
Paraffins				
Methane	0.554	45.8	−81.9	−116
Ethane	1.049	48.2	32.5	90
Propane	1.562	42.0	97.0	206
Butane (iso)	2.066	36.0	135.3	275
Butane (N)	2.066	37.5	152.6	306
Pentane (nee) 2.2-dimethyl propane	2.487	31.6	160.9	321
Pentane (iso) 2-methylbutane	2.487	32.9	188.1	370
Pentane (N)	2.487	33.3	197.0	386
Hexane (nee) 2.2-dimethylbutane	2.973	30.7	216.4	421
Hexane 2.3-dimethylbutane	2.973	30.9	227.6	441
Hexane (iso) 2-methylpentane	2.973	29.9	225.3	437
Hexane 3-methylpentane	2.973	30.8	231.4	448
Hexane (N)	2.973	−29.9	234.8	454
Heptane 2.2.3-dimethylbutane	3.459	−29.7	258.7	497
Heptane (iso) 2-methylhexane	3.459	27.2	258.1	496
Heptane (N)	3.459	27.0	267.6	513
Octane (iso) 2.2.4-trimethylpentane[a]	3.944	25.5	271.4	520
Octane 2.5-dimethylhexane	3.944	25.0	279.2	534
Octane (N)	3.944	24.6	296.4	565
Nonane (N)	4.428	22.5	322.0	611
Decane (N)	4.915	20.8	345.9	654
Naphthenes				
Cyclopropane	1.451	—	—	—
Cyclobutane	1.938	—	—	—
Cyclopentane	2.422	—	—	—
Cyclohexane	2.905	40.4	281.4	538

Table 6.1 Continued

		Critical constants		
			T	
	SG	*P*, atm	°C	(°F)
Olefins				
Ethylene-ethene	0.974	50.5	259.2	498
Propylene-propane	1.451	45.4	92.0	197
Butylene (iso) 2-methylpropane	1.934	39.5	145.3	293
Butylene (a) butene 1	1.934	39.7	147.0	296
Butylene (b) butene 2	2.004	40.8	155.3	311
Amylene (N) pentene 1	2.420	39.9	201.4	394
Diolefins				
Butadiene 1.3	1.869	42.7	152.6	306
Butadiene 1.2	1.869	—	—	—
Acetylene				
Acetylene-ethyne	0.911	61.6	36.4	97
Aromatics				
Benzene	2.692	48.6	289.8	553
Toluene-methylbenzene	3.176	40.1	321.4	610
Miscellaneous gases				
Air	1.000	37.2	−140.0	−221
Ammonia	0.596	111.5	132.6	270
Carbon monoxide	0.967	35.0	−138.6	−218
Carbon dioxide	1.528	72.9	31.4	88
Chlorine	2.449	76.0	144.2	291
Hydrogen	0.0696	12.8	−239.7	−400
Hydrogen sulfide	1.190	88.8	100.9	213
Nitrogen	0.972	33.5	−146.9	−233
Oxygen	1.105	49.7	−118.6	−182
Sulfur dioxide	2.44	77.8	157.6	315

[a]Refers to isooctane, used as a standard in fuel testing. Organic chemists apply the same name to 2-methylheptane.

Source: Flow Meter Engineering Handbook, 3rd Edition, Philadelphia, PA (1961).

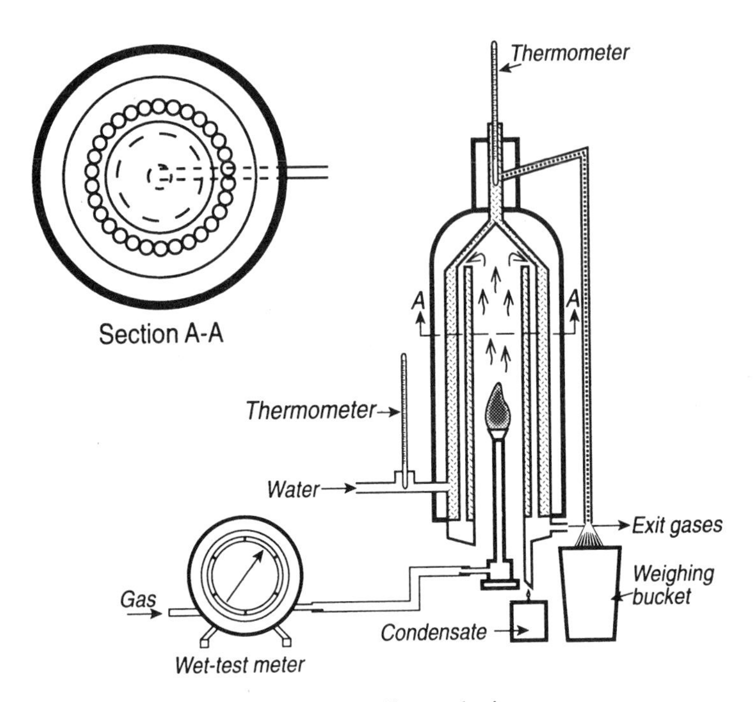

Figure 6.1 Constant-pressure flow calorimeter.

The molar heat of combustion of methane, a major gaseous fuel constituent, is found in Appendix B to equal

$$HHV_{CH_4} = (212{,}800 \text{ cal/g mole})(4.187) = 890{,}994 \text{ kJ/kg mole} \quad (6.4a)$$

or

$$= (212{,}800 \text{ cal/g mole})(1.8001) = 383{,}040 \text{ Btu/lb mole} \quad (6.4b)$$

The molar density of methane at *STP* can be obtained from Equation (6.2) and is equal to

$$\bar{\rho}_{CH_4} = \frac{(101{,}000 \text{ N/m}^2)}{(8314 \text{ kN m/kg mole K})(298\text{K})} = 0.0408 \text{ kg mole/m}^3 \quad (6.5a)$$

or

$$= \frac{(14.7 \text{ lbf/in}^2)(144 \text{ in}^2/\text{ft}^2)}{(1545 \text{ ft lbf/lb mole R})(537\ °\text{R})} = 0.00255 \text{ lb mole/ft}^3 \quad (6.5b)$$

By combining Equations (6.4) and (6.5), the volumetric heating value for methane is given as

$$HHV_{CH_4} = (890{,}994 \text{ kJ/kg mole})(0.0408 \text{ kg mole/m}^3) = 36{,}350 \text{ kJ/m}^3 \quad (6.6a)$$

$$= (383{,}040 \text{ Btu/lb mole})(0.00255 \text{ lb mole/ft}^3) = 977 \text{ Btu/ft}^3 \quad (6.6b)$$

The heating value of gaseous fuels is most often specified on a per-unit volume basis, whereas solid and liquid fuels are usually expressed in terms of a unit mass of fuel. It is essential that energy for different fuels be compared on a consistent basis, i.e., per unit mole, unit volume, or unit mass. To illustrate, consider the molar heating values of methane and hydrogen:

$$\overline{HHV}_{CH_4} = 890{,}994 \text{ kJ/kg mole} \quad (6.7a)$$

$$(383{,}040 \text{ Btu/lb mole})$$

and

$$\overline{HHV}_{H_2} = 286{,}043 \text{ kJ/kg mole}$$

$$(122{,}977 \text{ Btu/lb mole}) \quad (6.7b)$$

Equations (6.7a) and (6.7b) clearly show that, on a molar basis, methane has a greater heating value than hydrogen.

The heating value for these two fuels on a volumetric basis is equal to

$$HHV = \overline{HHV}\,\overline{\rho} \quad (6.8)$$

and

$$HHV_{CH_4} = (890{,}994)(0.0408) = 36{,}350 \text{ kJ/m}^3 \ (977 \text{ Btu/ft}^3) \quad (6.8a)$$

$$HHV_{H_2} = (286{,}043)(0.0408) = 11{,}670 \text{ kJ/m}^3 \ (314 \text{ Btu/ft}^3) \quad (6.8b)$$

Equations (6.8a) and (6.8b) show that, on a unit volume basis, methane also has a greater heating value than hydrogen.

Finally, the heating value for these two fuels on a mass basis is equal to

$$HHV = \frac{\overline{HHV}}{MW} \quad (6.9)$$

with

$$HHV_{CH_4} = 890{,}994/16 = 55{,}687 \text{ kJ/kg} \ (23{,}940 \text{ Btu/lbm}) \quad (6.9a)$$

and

$$HHV_{H_2} = 286{,}043/2 = 143{,}022 \text{ kJ/kg} \ (61{,}489 \text{ Btu/lbm}) \quad (6.9b)$$

Equations (6.8a) and (6.8b) indicate that, on a unit mass basis, however, hydrogen has a greater heating value than methane.

Propulsion heat engine utilization usually involves vehicle weight and/or volume design restrictions. In these instances, the fuel-engine energy density requirements of such machinery may dictate the use of fuels having high heats of combustion on both a unit mass and volume basis. This suggests that gaseous fuels may not be best suited to mobility power requirements but, instead, are more compatible with heating and stationary power needs. In addition, when the combustion characteristics of gaseous fuels are matched with a particular gas burner, the fuel-engine energy interface requirements cannot be based on selecting fuels having an equal heating value or specific gravity alone. In fact, in order to pass a constant energy rate through a given burner orifice, gaseous fuels with equal operating gas pressure levels and pressure drop across the orifice must have an equal *Wobble number,* W_0, where

$$W_0 = \frac{\text{higher heating value of gaseous fuel}}{\sqrt{SG}} \tag{6.10}$$

EXAMPLE 6.2 Hydrogen has been suggested to be a potential long-term replacement for natural gas. Considering methane to represent natural gas, calculate the heating value per unit volume, kJ/m, specific gravity, and Wobble number, kJ/m^3 for (a) 100% CH_4, (b) 75% CH_4–25% H_2, (c) 50% CH_4–50% H_2, (d) 25% CH_4–75% H_2, and (e) 100% H_2 on a volumetric basis.

Solution

1. Molar density for $P = \bar{\rho}\bar{R}T$:

$$\bar{\rho} = 101{,}000 \text{ N/m}^2/(8314 \text{ kN m/kg mole K})(298\text{K})$$
$$= 0.04077 \text{ kg mole/m}^3$$

2. Heating value for CH_4-H_2 mixture, using Appendix B.1:

$$HHV_{CH_4} = 212{,}800 \text{ cal/g mole} \qquad HHV_{H_2} = 68{,}317 \text{ cal/g mole}$$

and

$$HHV_{CH_4} = (212{,}800)(4.187 \text{ kJ/kg mole})(0.04077 \text{ kg mole/m}^3)$$
$$= 36{,}320 \text{ kJ/m}^3$$
$$HHV_{H_2} = (68{,}317)(4.187)(0.4077) = 11{,}660 \text{ kJ/m}^3$$

where

$$HHV_{mixt} = \bar{x}_{CH_4} HHV_{CH_4} + (1 - \bar{x}_{CH_4})\, HHV_{H_2}$$

a.

$\bar{x}_{CH_4}$	$\bar{x}_{H_2}$	HHV <kJ/m³>
100	0	36,320
0.75	0.25	30,155
0.50	0.50	23,990
0.25	0.75	17,825
0	100	11,660

3. Specific gravity for CH_4-H_2 mixture:

$$SG = [MW_{mixt}/MW_{air}]$$

where

$$MW_{mixt} = \bar{x}_{CH_4} MW_{CH_4} + (1 - \bar{x}_{CH_4}) MW_{H_2}$$

b.

$\bar{x}_{CH_4}$	$\bar{x}_{H_2}$	MW	SG
100	0	16.0	0.552
0.75	0.25	12.5	0.431
0.50	0.50	9.0	0.311
0.25	0.75	5.5	0.190
0	100	2.0	0.069

4. Wobble number for CH_4-H_2 mixture:

$$W_0 = HHV/\sqrt{SG}$$

c.

$\bar{x}_{CH_4}$	$\bar{x}_{H_2}$	W_0
100	0	48,890
0.75	0.25	45,920
0.50	0.50	43,050
0.25	0.75	40,920
0	100	44,390

Comments A review of these results shows that the energy density per unit volume decreases as hydrogen is substituted for methane to almost one-third the original value; the specific gravity also drops as hydrogen substitution increases. However, the Wobble number remains fairly constant for the different CH_4-H_2 mole fractions.

Gaseous fuels had a greater role as a stationary power fuel because of environmental benefits associated with their clean burning until the energy crisis placed an economic premium on fuel alternatives. The concentrations of CO, CO_2, and O_2 in the exhaust from natural gas and even oil-fired boilers and furnaces can be measured by chemical absorption devices such as the Orsat analyzer shown in Figure 6.2. Exhaust, or *flue,* gases are monitored in many boiler applications, and major exhaust gas composition measurements via an Orsat analysis can be used, in conjunction with the boiler energy balance presented in Chapter 4, to operate an efficient burner stoichiometry. A discussion of more complex and sensitive methods of analysis, such as gas chromatography, ultraviolet infrared emissions spectroscopy, ultraviolet absorption spectroscopy, and mass spectrometry, which are used to determine the presence of minute amounts of unburned hydrocarbons and nitric oxides, can be found in Chapter 9.

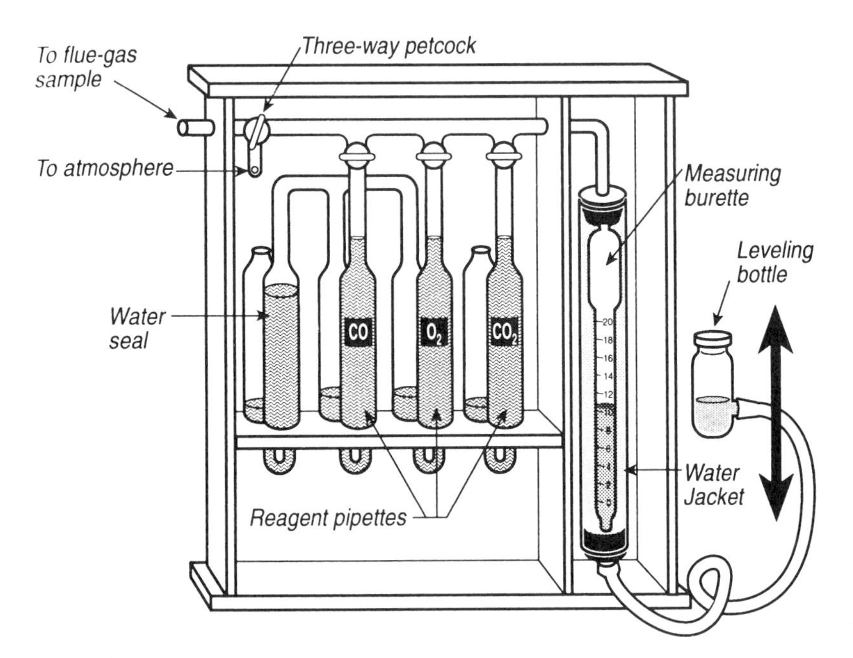

Figure 6.2 Orsat flue gas analyzer.

A 100-ml flue gas sample can be drawn into an Orsat apparatus for analysis by lowering a displacing water bottle, which changes the hydrostatic head in the sampling loop. During fill, all the reagent bottle valves, as well as the atmospheric vent line valves, are closed. After the 100-ml volume is filled with exhaust, the sample line is closed, and the valve to the CO_2 absorbent is then opened. The water-leveling bottle is raised until 100 ml of gas are passed into the CO_2 absorbent, a 20% aqueous solution of potassium hydride. When the water-leveling bottle is returned to its original position, the reduction in the original 100-ml volume can be measured and is equal to the volume of CO_2 in the original gas mixture sample. This process is repeated several times to ensure complete CO_2 removal. Next, oxygen is removed by passing the gas sample back and forth several times between the 100-ml sample volume and the second reagent bottle, which contains an aqueous alkaline solution of pyrogallic acid. The percentage of oxygen by volume in the original gas sample is determined by noting an additional reduction of the remaining gas contained in the 100-ml sample volume. The last reagent pipette contains an aqueous solution of cuprous chloride, which absorbs CO and allows the volume percentage of CO in the sample to be determined as well. This order of absorption is critical to ensure reliable results. The Orsat analysis yields a *dry* product analysis, as shown in Example 6.3, and assumes that the remaining gas is N_2.

EXAMPLE 6.3 An exhaust gas sample consists of the following numbers of moles of gaseous species:

Moles $CO_2 = N_1$

Moles $CO = N_2$

Moles $H_2O = N_3$

Moles $O_2 = N_4$

Moles $N_2 = N_5$

The sample is analyzed using an Orsat analyzer. Show that the actual volumetric analysis for CO_2 (using an Orsat apparatus) equals the sample volume of CO_2, based on a dry analysis of the products.

Solution

1. Initial total gas sample volume:

$$P_{tot}V_{tot} = N_{tot}\overline{R}T$$

where

$$P_{tot} = 1 \text{ atm} = \text{constant during the analysis}$$
$$T = \text{constant during analysis}$$
$$V_{tot} = 100 \text{ units}$$

$$N_{tot} = N_1 + N_2 + N_3 + N_4 + N_5$$
$$P_{tot}V_{tot} = (N_1 + N_2 + N_3 + N_4 + N_5)\bar{R}T$$

2. Since liquid and water vapor are in direct contact during the entire analysis, the partial pressure of the water vapor remains constant in the apparatus, or

$$P_{H_2O} = P_{sat}<T> = \text{const}$$

thus,

$$\bar{x}_{H_2O} = \frac{P_{H_2O}}{P_{tot}} = \frac{N_3}{N_1 + N_2 + N_3 + N_4 + N_5} = \frac{N_3'}{N_2 + N_3' + N_4 + N_5}$$

NOTE: N_{tot} is changing because of the absorption of CO_2 and, thus, the moles of water vapor, N_{H_2O}, also have to change.

3. Total gas sample after removal of CO_2:

$$P_{tot}V_1 = (N_2 + N_3' + N_4 + N_5)\bar{R}T$$

where

$$N_3' = \text{moles of } H_2O \text{ after } CO_2 \text{ analysis} \neq N_3$$

4. Volume percent associated with CO_2 removal:

$$\% \ CO_2 = \left[\frac{V_{tot} - V_1}{V_{tot}}\right] \times 100$$

$$V_{tot} = \frac{[N_1 + N_2 + N_3 + N_4 + N_5]\bar{R}T}{P_{tot}}$$

and

$$V_1 = \frac{[N_2 + N_3' + N_4 + N_5]\bar{R}T}{P_{tot}}$$

where

$$\left[\frac{V_{tot} - V_1}{V_{tot}}\right] = \frac{(N_1 + N_2 + N_3 + N_4 + N_5) - (N_2 + N_3' + N_4 + N_5)}{(N_1 + N_2 + N_3 + N_4 + N_5)}$$

$$\left[\frac{V_{tot} - V_1}{V_{tot}}\right] = \frac{N_1 + N_3 - N_3'}{(N_1 + N_2 + N_3 + N_4 + N_5)}$$

$$= \frac{N_1 + N_3 - [(N_2 + N_4 + N_5)/(N_1 + N_2 + N_4 + N_5)]N_3}{(N_1 + N_2 + N_3 + N_4 + N_5)}$$

$$= \frac{N_1(N_1 + N_2 + N_3 + N_4 + N_5)/(N_1 + N_2 + N_4 + N_5)}{(N_1 + N_2 + N_3 + N_4 + N_5)}$$

$$= \frac{N_1}{N_1 + N_2 + N_4 + N_5} = \bar{x}_{CO_2}\Big)_{\text{dry basis}}$$

6.3 NATURAL GAS

The world's most readily available and abundant gaseous fuel resources are found in *natural gas* reserves. Gaseous fuels have been used for centuries in China and for over 100 years in both the United States and Europe. In the United States, when natural gas was originally discovered at oil wells, it was burned, or *flared* off, as a useless by-product of oil production. Today, natural gas is a major industry that transports fuel throughout the United States by a complex interstate pipeline network. Natural gas was formed by *anaerobic,* or bacterial-assisted, decomposition of organic matter under heat and pressure and, therefore, like coal and crude oil, is a variable-composition hydrocarbon fuel. Table 6.2 lists properties of certain natural and synthetic gas resources. Synthetic natural gas (SNG) production and anaerobic digester technology will be discussed in Sections 6.4 and 6.5.

Natural gas consists chiefly of methane, ranging anywhere from 75% to 99% by volume, with varying concentrations of low molecular weight hydrocarbons, CO, CO_2, He, N_2, and/or H_2O. Conventional gas well drilling has proved successful in or near oil fields. New and additional unconventional drilling methods are finding reserves in deep wells and coal beds, as well as in shale and tar sands. Natural gas is practically colorless and odorless and, for safety reasons, is "soured" with the familiar rotten egg odor by adding hydrogen sulfide, H_2S. The American Gas Association classifies natural gas as sweet or sour gas and, additionally, as being associated or non associated gas. Associated, or *wet,* gas is either dissolved in crude oil reserves or confined in pressurized gas caps located on the top of oil ponds. Wet gas has appreciable concentrations of ethene, butene, propane, propylene, and butylenes. Nonassociated, or *dry,* gas can be found in gas pockets trapped under high pressure that have migrated from oil ponds or are the results of an early coalization gasification stage. Natural gas, like coal and oil, has regional characteristics. For example, western U.S. natural gas fields generally

Table 6.2 Gaseous Fuel Characteristics

Type	CO_2	O_2	N_2	CO	H_2	CH_4	C_2H_6	C_3H_8	C_4H_{10}	Illuminants and Others	*SG*	Heating Value‖ kJ/m³ Gross	Heating Value‖ kJ/m³ Net	Heating Value‖ Btu/ft³ Gross	Heating Value‖ Btu/ft³ Net
Natural	....	....	5.0	....	....	90.0	5.0	....	.....	.....	0.60	37,300	33,680	1,002	904
Natural	....	....	0.8	....	....	83.4	15.8	....	.....	.....	0.61	42,060	38,030	1,129	1,021
Natural	6.5	....	....	....	....	77.5	16.0	....	.....	.....	0.70	39,970	36,170	1,073	971
Natural	0.8	....	8.4	....	....	84.1	6.7	....	.....	.....	0.63	36,280	32,740	974	879
Natural	....	....	....	....	....	36.7	14.5	23.5	14.9	10.4*	1.29	79,460	72,980	2,133	1,959
Propane	....	....	....	....	....	....	2.2	97.3	0.5	.....	1.55	95,290	87,840	2,558	2,358
Propane	....	....	....	....	....	....	2.0	72.9	0.8	24.3#	1.77	93,280	86,280	2,504	2,316
Butane	....	....	....	....	....	....	....	6.0	94.0	.....	2.04	119,580	110,300	3,210	2,961
Butane	....	....	....	....	....	....	....	5.0	66.7	28.3§	2.00	118,610	109,300	3,184	2,935
Refinery oil	....	0.2	0.6	1.2	6.1	4.4	72.5	....	.....	.....	1.00	61,470	56,770	1,650	1,524
Refinery oil	0.2	0.2	0.5	1.2	13.1	23.3	21.9	....	.....	39.6	0.89	54,950	50,330	1,475	1,351
Oil gas	1.2	0.5	2.4	7.7	54.2	30.1	....	....	.....	3.9	0.37	21,230	19,000	570	510
Coal gas	2.4	0.8	11.3	7.4	48.0	27.1	....	....	.....	3.0	0.47	20,190	18,100	542	486
Coal gas	1.7	0.8	8.1	7.3	49.5	29.2	....	....	.....	3.4	0.47	22,310	20,120	599	540
Coal gas	2.1	0.4	4.4	13.5	51.9	24.3	....	....	.....	3.4	0.42	19,370	17,360	520	466
Coke oven	2.2	0.8	8.1	6.3	46.5	32.1	....	....	.....	4.0	0.44	21,200	18,960	569	509
Producer	8.0	0.1	50.0	23.2	17.7	1.0	....	....	.....	.....	0.86	5,330	4,950	143	133
Producer	4.5	0.6	50.9	27.0	14.0	3.0	....	....	.....	.....	0.86	6,070	5,700	163	153
Blast furnace	11.5	....	60.0	27.5	1.0	....	....	....	.....	.....	1.02	3,430	3,430	92	92
Blue gas (water gas)	5.4	0.7	8.3	37.0	47.3	1.3	....	....	.....	.....	0.57	10,690	9,760	287	262
Blue gas (water gas)	5.5	0.9	27.6	28.2	32.5	4.6	....	....	.....	0.7	0.70	9,690	8,900	260	239
Carbureted water	3.6	0.4	5.0	21.9	49.6	10.6	2.5	....	.....	6.1	0.54	19,970	17,170	536	461
Carbureted water	6.0	0.9	12.4	26.8	32.2	13.5	....	....	.....	8.2	0.66	19,740	16,800	530	451
Carbureted water	0.7	0.3	5.8	11.7	28.0	36.1	....	....	.....	17.4	0.63	31,290	28,680	840	770
Sewage	22.0	....	6.0	....	2.0	68.0	....	....	.....	.....	0.79	25,700	23,130	690	621

Source: American Gas Association (1948).

*C_5H_{12} #C_3H_6 §C_4H_8 ‖ At 60 and 30 in. Hg.

contain substantial amounts of CO_2, midwestern reserves have higher N_2 concentrations and some He, and eastern gas is high in paraffins. European gas reserves are basically high in CO_2, H_2, and olefin hydrocarbons.

Liquid petroleum gas, or LPG, consists of condensable hydrocarbon vapors recovered by expansion of wet gas reserves. By compressing the condensable fractions, liquefied fuel vapors, such as commercial propane and butane, can be stored and transported at ambient temperatures as a liquid. Liquefied natural gas, LNG, is the condensed state of dry natural gas but requires a cryogenic refrigeration for storage and handling at −102°C (−260°F). At present, efficient transportation of large Middle Eastern natural gas to the United States, Europe, and Asia by sea requires specially designed LNG tankers.

EXAMPLE 6.4 A natural gas has a volumetric analysis of 95% CH_4, 3% C_2H_6, and 2% CO_2. For conditions of 14.7 psia and 77°F, calculate (a) the higher heating value of the fuel, Btu/ft^3 of gas; and (b) the lower heating value of the fuel, Btu/ft^3 of gas.

Solution:

1. Stoichiometric equation:

$$[0.95\ CH_4 + 0.03\ C_2H_6 + 0.02\ CO_2] + a[O_2 + 3.76\ N_2]$$
$$\rightarrow bCO_2 + cH_2O + dN_2$$

Carbon atom balance:

$$0.02 + 0.95 + 0.06 = b = 1.03$$

Hydrogen atom balance:

$$(4)(0.95) + (6)(0.03) = 2c \qquad c = 1.99$$

Oxygen atom balance:

$$(2)(0.02) + 2a = (2)(1.03) + 1.99 \qquad a = 2.005$$

Nitrogen atom balance:

$$d = 3.76a = (3.76)(2.005) = 7.539$$

2. Energy balance:

$$Q = \sum_{i=\text{prod}} N_i[\bar{h}_f^0 + \Delta\bar{h}]_i - \sum_{j=\text{react}} N_j[\bar{h}_f^0 + \Delta\bar{h}]_j$$

or

$$
\begin{aligned}
Q = {} & 1.03[\bar{h}_f^0 + \Delta\bar{h}]_{CO_2} + 1.99[\bar{h}_f^0 + \Delta\bar{h}]_{H_2O_l} \\
& + (3.76)(2.005)\,[\bar{h}_f^0 + \Delta\bar{h}]_{N_2} - 0.95[\bar{h}_f^0 + \Delta\bar{h}]_{CH_4} \\
& - 0.03[\bar{h}_f^0 + \overline{\Delta} h]_{C_2H_6} - 0.02[\bar{h}_f^0 + \Delta\bar{h}]_{CO_2} \\
& - 2.005\,[\bar{h}_f^0 + \Delta\bar{h}]_{O_2} - (2.005)(3.76)\,[\bar{h}_f^0 + \Delta\bar{h}]_{N_2}
\end{aligned}
$$

Recall that the higher heating value assumes water in the products is a liquid.

From Appendix B,

$$
\begin{aligned}
HHV = {} & (1.03)(-94{,}054) + 1.99\,(-68{,}317) - 0.95(-17{,}889) \\
& - 0.03(-20{,}236) - 0.02(-94{,}054) \\
= {} & -213{,}340 = 213{,}340 \text{ cal/g mole} \\
= {} & (213{,}340 \text{ cal/g mole})\left(1.8001 \frac{\text{Btu/lb mole}}{\text{cal/g mole}}\right) \\
= {} & 384{,}033 \text{ Btu/lb mole}
\end{aligned}
$$

3. Density, *fuel:*

$$
\bar{\rho} = \frac{P}{\bar{R}T} = \frac{(14.7 \text{ lbf/in}^2)(144 \text{ in}^2/\text{ft}^2)}{(1545 \text{ ft lbf/lb mole R})(537°\text{R})}
$$
$$
= 0.00255 \text{ lb mole/ft}^3
$$

4. Higher heating value, *water as liquid in product:*

$$HHV = (384{,}033 \text{ Btu/lbmole})(0.00255 \text{ lb mole/ft}^3)$$

a. $$HHV = 980 \text{ Btu/ft}^3$$

5. Lower heating value, *water as vapor in product:*

$$
\begin{aligned}
LHV &= HHV - 1.99\, h_{fg} <68°\text{F}> \\
LHV &= +384{,}033 \text{ Btu/lb mole fuel} \\
&\quad - \left(1.99 \frac{\text{lb mole } H_2O}{\text{lb mole fuel}}\right)\left(1054 \frac{\text{Btu}}{\text{lbm } H_2O}\right)\left(18 \frac{\text{lbm}}{\text{lb mole } H_2O}\right) \\
&= 346{,}279 \text{ Btu/lb mole}
\end{aligned}
$$

b. $$LHV = (346{,}280)(0.00255) = 880 \text{ Btu/ft}^3$$

6.4 COAL-DERIVED GASEOUS FUELS

Synthetic, or manufactured, gaseous fuels have been generated using coal resources for more than 100 years. *Coke* and *coke gas,* by-products of coal used in the iron industry, were produced as early as the eighteenth century. *Towngas,* a commercial and residential grade "low-Btu" fuel, was used during the late nineteenth and early twentieth centuries until replaced by electric power as well as by the oil and natural gas industries. Most coal-derived fuel gas technologies fall into one of three general categories: coal pyrolysis, coal gasification, or coal catalytic synthesis. Today, many factors indicate a need for redeveloping a coal-derived syngas industry. These factors include:

Large U. S. coal reserves ill suited for direct combustion
Environmental pressure for greater coal combustion pollution abatement
Sulfur removal as H_2S during gasification versus sulfur generation as SO_2 during combustion
Market for clean-burning, coal-based fuel gas in stationary power plants
Synthetic gaseous fuel transportable via pipelines

Coal is a poor feedstock, however, for making a commercially viable gaseous fuel, in part because of the following properties:

Natural occurrence as a solid
Variable and nonuniform nature
Poor conversion energetics and economics
Lower overall gasification energy efficiencies as compared to direct combustion efficiencies

Carbonization, or pyrolysis, is a destructive thermal distillation process in which volatile combustible fractions contained in raw coal or coke (such as hydrogen, methane, ethyene, and carbon monoxide) are driven off. Yields are relatively low with 70 wt % of the original coal remaining as a solid residue after devolatilization. To release these gases, coal or coke is placed in a closed vessel, or *retort,* and heated by external coal combustion to a temperature range of 530–1000°C (985–1830°F). Coal and coke gas have properties that are a function of the particular coal supply, actual pyrolysis temperature, specific type of retort being used, and total residence time of reactants within the vessel. The energy volume of these gases covers the range 18,630–24,220 kJ/m^3 (500–650 Btu/ft^3).

An alternate means of producing syngas is by partial combustion and thermal cracking of an incandescent solid fuel bed of coal, coke, peat, or even wood with air. *Producer gas,* for example, is generated by fuel-lean combustion, which occurs in a reacting bed of coal or coke. The initial

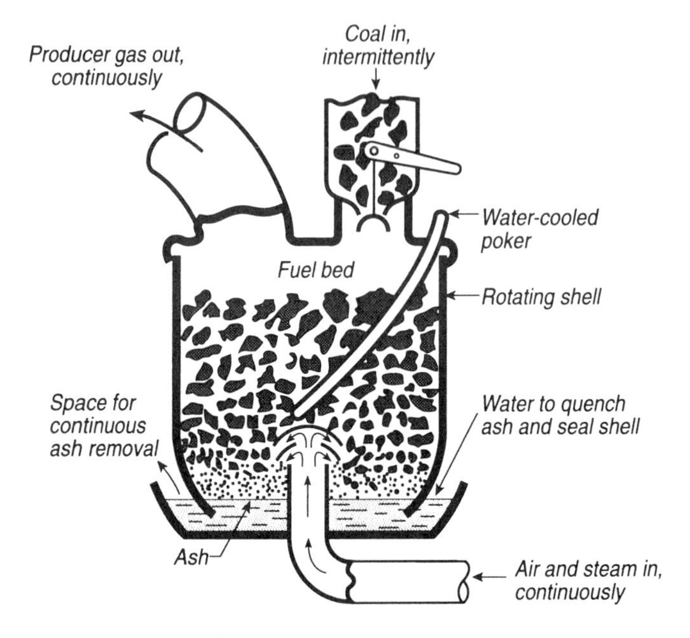

Figure 6.3 Producer gas generator.

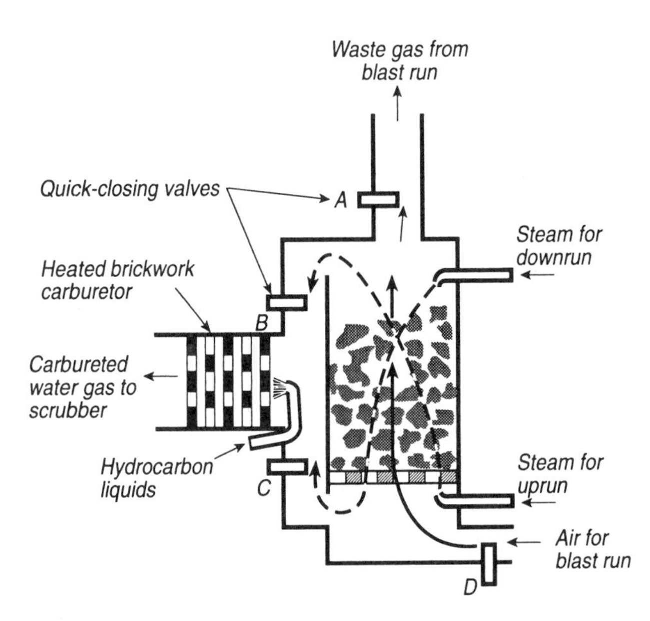

Figure 6.4 Carbureted water gas generator.

stages of gasification result from partial combustion of the bottom coal to yield carbon dioxide and heat (see Fig. 6.3). Thermodynamically, this process can be represented by the reaction

$$C_S + O_2 \rightarrow CO_2 + \text{heat} \tag{6.11}$$

Exothermic reactions within the remaining coal bed with carbon dioxide and heat then produce carbon monoxide or

$$\text{Heat} + C_S + CO_2 \rightarrow 2CO \tag{6.12}$$

Overall, then, the gasification of coal can be expressed as

$$2C_S + O_2 \rightarrow 2CO \tag{6.13}$$

Producer gas has a low energy volume of 5220–6700 kJ/m^3 (140–180 Btu/ft^3) because of a high percentage of nitrogen present in the gas, approximately 50 vol %.

The low yields of gas resulting from coal pyrolysis and the relatively low energy content of producer gas, approximately 10–20% that of natural gas, limit their use to special applications at local production sites. One means of upgrading producer gas, for example, would be to reduce the mole fraction of nitrogen in product gas by using oxygen instead of air. Steam-cracking coal at elevated temperature and pressure generates *carbureted water gas,* a synthesis gas that also has a lower nitrogen content and, therefore, greater energy volume than producer gas. Carbureted water gas production is initiated by blowing pyrolysis air through coal for 1–2 min, followed in turn by a 2–4-min blast of steam through the incandescent bed to hydrogasify the coal. This entire process is repeated continuously to produce a steady stream of gas having an energy volume of approximately 10,060 kJ/m^3 (270 Btu/ft^3) (see Fig. 6.4). Steam cracking of coal can be represented thermodynamically by the equilibrium reaction

$$C_S + H_2O \overset{T,P}{\rightleftarrows} CO + H_2 \tag{6.14}$$

Oil gas or fuel oil can be sprayed into carbureted water gas during processing to further raise the heating value to 18,630–22,350 kJ/m^3 (500–600 Btu/ft^3). Carbureted water gas is often referred to as *blue gas* since carbon monoxide in this gaseous fuel burns with a characteristic short blue flame.

Russian engineering in the 1930s pioneered underground hydrogasification of coal by injecting steam directly into coal beds. By 1938, this technique was supplying coal gas on a commercial basis within the Soviet Union. This technology has been considered by the U.S. coal industry as a potential method for utilizing the vast reserves of relatively low-grade, environmentally unburnable western coal. However, because of the large

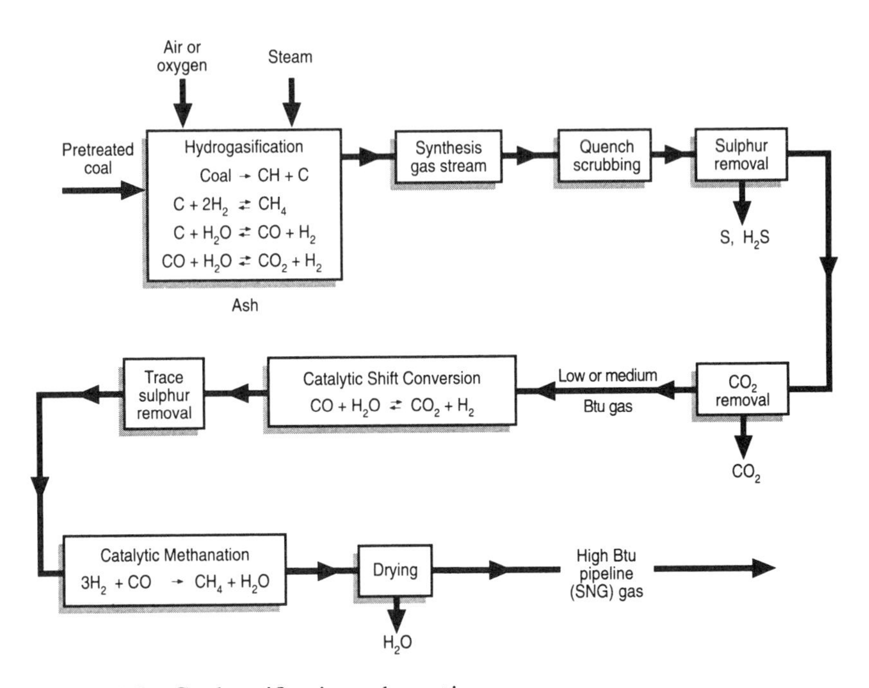

Figure 6.5 Coal gasification schematic.

amount of water required, the limited availability of usable water resources in the region, and the world's growing critical water crisis, this is a questionable technology for gasifying U.S. coal.

Producer gas, a "low-Btu" coal gas made with nineteenth-century coal conversion technology, cannot be economically transported and, therefore, is not a viable substitute for natural gas. A full-scale commercial industry that could supply a coal-derived *synthetic natural gas,* SNG, would require large water and energy inputs, along with development of more complex and yet unproven technologies. Environmental concerns over burning high-sulfur coal may spur development of this alternate coal utilization even though SNG will be more expensive and less efficient than direct combustion of coal. In addition, future depletion of both proven and projected natural gas reserves would impact existing gas supplies and pipeline distribution.

EXAMPLE 6.5 One mole of CO and 220% theoretical oxygen, both at 25°C, undergo a steady flow reaction at 1-atm pressure. Neglecting dissociation of O_2, determine (a) the final equilibrium composition, and (b) the final temperature if the process is adiabatic.

Solution:

1. Stoichiometric reaction:

$$CO + \tfrac{1}{2}O_2 \rightarrow CO_2$$

2. Actual reaction:

$$CO + (2.2)(0.5)O_2 \rightarrow aO_2 + bCO + cCO_2$$

Carbon atom balance:

$$1 = b + c$$

Oxygen atom balance:

$$3.2 = 2a + b + 2c$$

$$3.2 = 2a - b + 2$$

$$1.2 = 2a - b$$

3. Energy balance for $P = C$ (adiabatic):

$$dE = \delta Q - \delta W + \Sigma N_i \bar{e}_i - \Sigma N_j \bar{e}_j$$

$$\Sigma N_i \bar{e}_i = \Sigma N_j \bar{e}_j$$

$$[\bar{h}_f^0 + \Delta\bar{h}]_{CO} + 1.1[\bar{h}_f^0 + \Delta\bar{h}]_{O_2}$$

$$= a[\bar{h}_f^0 + \Delta\bar{h}]_{O_2} + b[\bar{h}_f^0 + \Delta\bar{h}]_{CO} + c[\bar{h}_f^0 + \Delta\bar{h}]_{CO_2}$$

4. Equilibrium constant:

$$CO + \tfrac{1}{2}O_2 \rightleftarrows CO_2$$

$$\log K_P = \log K_P)_{CO_2} - \tfrac{1}{2}\log K_P)_{O_2} - \log K_P)_{CO}$$

and

$$K_P = \frac{[P/P_0]_{CO_2}}{[P/P_0]_{O_2}^{1/2}[P/P_0]_{CO}}$$

$$= \left[\frac{P_{CO_2}}{P_{O_2}^{1/2}\, P_{CO}}\right] \times P_0^{-1/2}$$

Now,

$$P_0 \equiv 1 \text{ atm} = P_{tot}$$

and

$$P_i = \bar{x}_i\, P_{tot}$$

$$\bar{x}_{O_2} = \frac{a}{a + b + c}$$

$$\bar{x}_{CO} = \frac{b}{a + b + c}$$

$$\bar{x}_{CO_2} = \frac{c}{a + b + c}$$

or

$$K_P = \frac{(c)(a + b + c)^{1/2}}{(a)^{1/2}(b)}$$

5. From item 2, then, one can eliminate b and c, obtaining

$$a = a$$
$$b = 2a - 1.2$$
$$c = 2.2 - 2a$$

and

$$K_P = \frac{(2.2 - 2a)(a + 1)^{1/2}}{a^{1/2}(2a - 1.2)}$$

6. Since the final state, i.e., T_2, a, b, and c, is unknown, an iterative technique is required for solution.

7. Initial guess for $T_2 = 2800\text{K}$:

Energy balance using Appendix B data:

$$-26{,}416 = a(+21{,}545) + (2a - 1.2)(-26{,}416 + 20{,}582) + (2.2 - 2a)(-94{,}051 + 33{,}567)$$

or

$$a = 0.762$$

Equilibrium constant:

$$K_P = \frac{[(2.2) - (2)(0.762)][1.762]^{1/2}}{[0.762]^{1/2}[(2)(0.762) - 1.2]} = 3.173$$

From Appendix B,

$$\log K_P = 7.388 - 6.563 = 0.825$$
$$K_P = 6.683$$

8. Second estimation for $T_2 = 3000\text{K}$:

$$-26{,}416 = a(23{,}446) + (2a - 1.2)(-26{,}416 + 22{,}357) + (2.2 - 2a)(-94{,}051 + 36{,}535)$$
$$a = 0.731$$

Equilibrium constant:

$$K_P = \frac{[2.2 - (2)(0.731)][1.731]^{1/2}}{[0.731]^{1/2}[(2)(0.731) - 1.2]} = 4.335$$
$$\log K_P = 6.892 - 6.407 - 0.485$$
$$K_P = 3.055$$

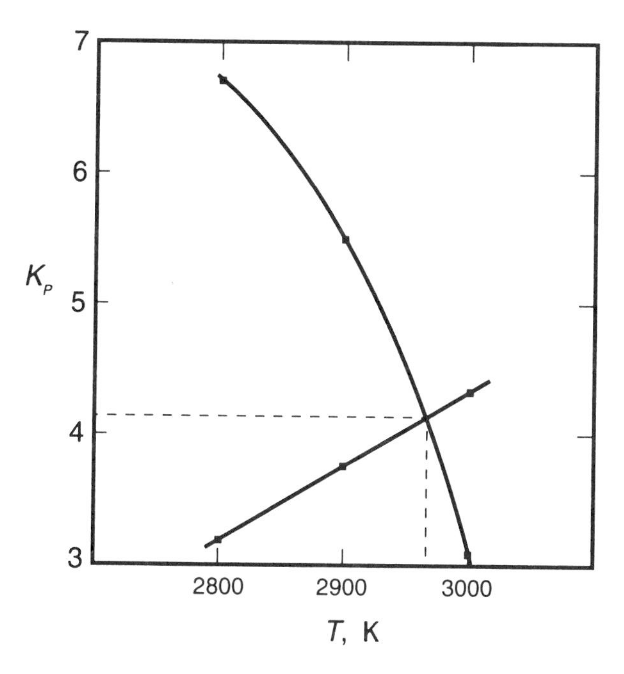

9. Third estimation for $T_3 = 2900\text{K}$:

$$-26{,}416 = a(22{,}493) + (2a - 1.2)(-26{,}416 + 21{,}469) + (2.2 - 2a)(-94{,}051 + 35{,}049)$$
$$a = 0.746$$

Equilibrium constant:

$$K_P = \frac{[2.2 - (2)(0.746)][1.746]^{1/2}}{[0.746]^{1/2}[(2)(0.746) - 1.2]} = 3.709$$

$$\log K_P = 7.132 - 6.483 = 0.649$$
$$K_P = 4.46$$

10. From the graph, $T_2 \sim 2970K$:

$a \sim 0.740$

$b \sim 0.280$

$c \sim 0.720$

or

$$CO + 2.2\ O_2 \rightarrow 0.740\ O_2 + 0.280\ CO + 0.72\ CO_2$$

There are no known direct coal-methanization conversion processes. The overall equilibrium reaction

$$C_S + 2H_2O_g \rightleftarrows CH_4 + CO_2 \tag{6.15}$$

suggests favorable thermodynamics in that the heat of reaction is approximately zero, but no catalyst has been found that will allow this overall process to occur. Bioconversion of cellulose and other organic materials, covered in the next section, can generate methane directly.

Current and conversion technologies require pretreatment gasification and methanization in order to produce a pipeline-quality SNG, (Fig. 6.5). Development and application of various designs will differ in their coal selection as well as the means they use to introduce coal and either air or oxygen into the gasifier. Pretreatment of coals that cake requires mild oxidation to prevent caking during gasification. Gasification begins with heating and drying of the coal. Devolatilization or distillation will drive off evolved gases, with the initial heating raising the coal to near its softening temperature. Chemical conversion reacts coal with oxygen, or air, and steam. Recall that this chemical conversion will yield a low-Btu product if air is used, while a low-nitrogen or medium-Btu gas will be produced with the use of oxygen or water. The degree of chemical conversion is related to equilibrium shifts in the products CO, CH_4, and H_2 or

$$2C + O_2 \rightleftarrows 2CO_2 \tag{6-16}$$
$$C + H_2O \rightleftarrows CO + H_2 \tag{6-17}$$
$$C + CO_2 \rightleftarrows 2CO \tag{6-18}$$

Several coal gasifiers are being developed and include fixed or moving beds, entrained flow, fluidized beds, and molten salt systems. The Lurgi fixed-bed gasification technique feeds coal at the top of a gasifier, with steam or air supplied at the bottom, allowing gasification to occur as the

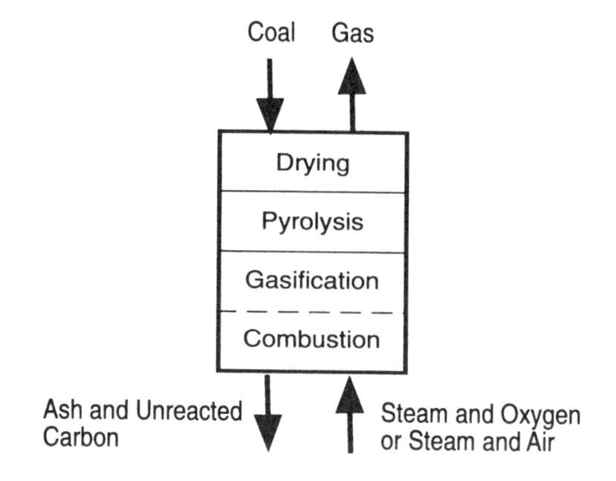

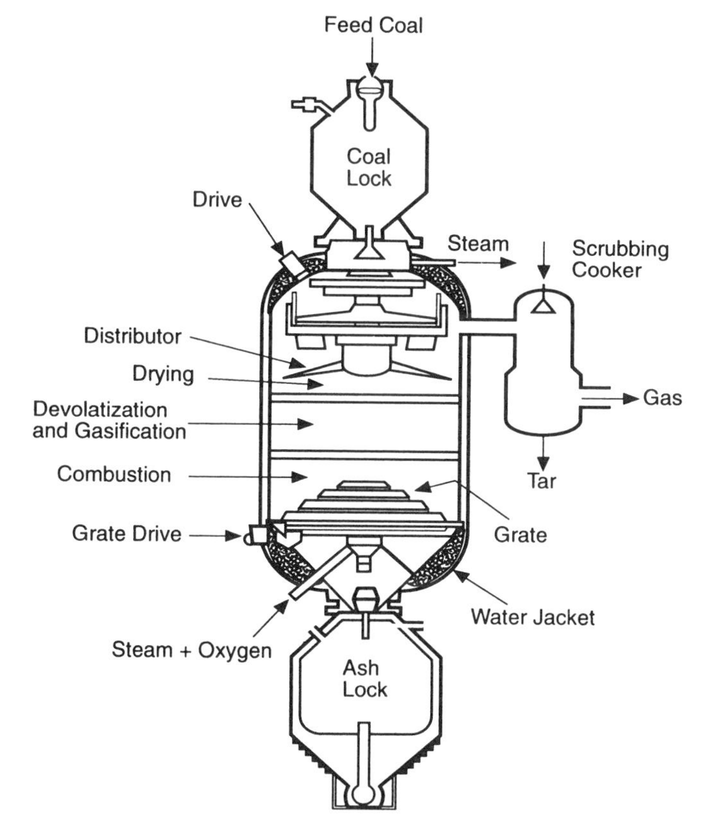

Figure 6.6 Lurgi coal gasification.

coal passes down through the gasifier; see Figure 6.6. The Kopper, an entrained flow system, feeds a mixture of pulverized coal and air/steam or oxygen/steam to the gasifier. Gasifiers can be further categorized as either slagging or dry, depending on whether they are operated above or below ash fusion temperature. Coal gasification chemically reduces sulfur to hydrogen sulfide, H_2S, during the pyrolysis stage. Extraction of H_2S yields a sulfur-free synthesis gas.

Catalytic methanization of synthesis gas is required to produce an SNG of approximately 95–98% methane and occurs via the reaction

$$CO + 3H_2 \rightleftarrows CH_4 + H_2O \tag{6-19}$$

The major thrust of current coal gasification programs is to optimize methane production by conducting hydrogasification at much higher pressures and temperatures than those used in traditional coal gas production. Future technological breakthroughs are hampered by the very complex, nonuniform, and variable composition of U.S. coal resources. Development of coal-based SNG having pipeline quality will be strongly influenced by the going price of energy alternatives and the ready availability of alternate gas resources.

6.5 BIOMASS AND SYNTHETIC NATURAL GAS

Resources other than coal or crude oil can be used as feedstock for generating a synthetic gaseous fuel. This synthetic natural gas, or *biogas,* can be produced from a variety of organic materials, including vegetation, animal and plant residue, and municipal solid wastes (MSW). A renewable methane fuel resource can be derived from solar energy stored in naturally occurring organic materials, as well as from certain wasteful by-products of modern industrialized civilizations by microbiological conversion processes, direct material pyrolysis, or thermochemical technologies.

Microbiological conversion, based on the metabolic process of certain bacteria, can consume cellulose found in vegetation or manures produced by livestock and yield methane, carbon dioxide, and undecomposable or digested sludge as their by-products. *Marsh gas* is a natural methane produced from decayed organic matter submerged in stagnant water and is itself the source of the ghostly will-o'-the wisp observed in swampy areas. Bacterial required for natural or commercial microbiological methane conversion processes need oxygen to reproduce. *Anaerobic* bacteria are those that thrive on chemically combined oxygen, while *aerobic* bacteria can exist only on dissolved oxygen and are therefore referred to as being oxygen-free (Fig. 6.7).

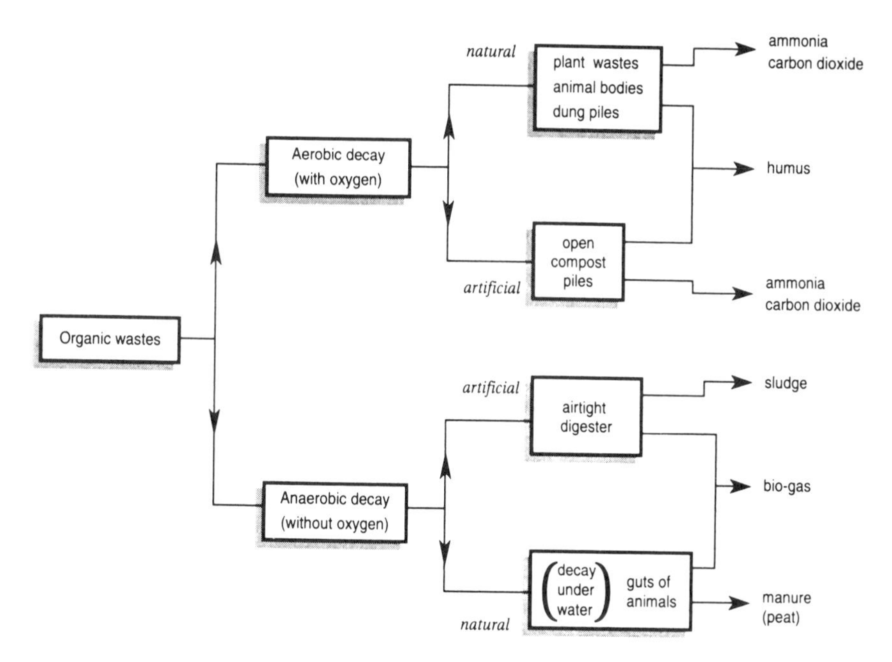

Figure 6.7 Aerobic and anaerobic biogas production.

Moist biomass feedstock is first converted by acid-forming, or aerobic, bacteria to simple organic compounds, chiefly organic acids and carbon dioxide. These intermediate products are then fermented by anaerobic bacteria, yielding methane and additional carbon dioxide as by-products. This *anaerobic digester* gas product is approximately 50–67% methane, 50–33% CO_2 by volume with approximately 10–40 ppm hydrogen sulfide.

Specially designed digestion tanks, or *digesters,* are used commercially to accomplish this anaerobic biomass conversion process. In order to sustain a conversion process within an artificial digester environment, bacteria health and growth require precise monitoring and control of the following:

Proper balance between acid-forming and methane-forming bacteria
Proper solids-to-liquid ratio (7–9% solids):

sewage (5% solids) *animal manure* (18% solids) *vegetable wastes* (30–40% solids)	raw stock percentages

Proper pH between 7.5–8.5 or near neutral (7)
Proper phosphorus, nitrogen, and other nutrients levels
Proper digester temperature of 20–60°C (68–140°F)

The anaerobic mesophilic bacteria, which can exist between 20–40°C (68–104°F), require 20 days to digest, while the aerobic thermophilic bacteria can exist in a 49–60°C (120–140°F) environment and require only 7–8 days for conversion.

Methane digester history can be traced back to ancient China, where methane was produced in sewage lagoons and to nineteenth-century England, where sewer gas was used to supply *lamp gas* for street illumination in London. Small-scale digester technology applications can be found throughout the world today, including areas of India, France, Germany, and the United States.

Large land acreage suitable for farming would be required to grow sufficient photosynthetic fuel crops necessary to sustain a viable biomass gas industry. Such an industry, which must also have water for the biomass conversion process, would therefore compete directly with the current utilization of fertile land and water resources for producing food. Regions of humid tropics and temperate zones where large parcels of uncultivated lands and low per capita energy consumption exist hold greater promise for biomass fuel production and use.

Aerobic destruction, or *composting,* converts organic wastes into a stable humuslike product, the chief value of which is its use as a soil conditioner and fertilizer. Composting will not generate a gaseous fuel but can stabilize and neutralize organic wastes prior to their disposal. Suitable geological conditions at landfill sites where untreated organic and municipal wastes have been buried, however, can promote anaerobic methane generation. Methane in old sanitary landfills is often vented and/or flared off to minimize the explosive risks of gas pockets building up within disposal sites. At selected dump sites, unnatural or landfill gas can be drilled, piped, and treated to remove carbon dioxide, moisture, hydrogen sulfide, and other contaminants and can yield a pipeline-quality gas.

Several biomass gasification techniques for producing a commercial fuel gas are in various stages of research and development today. Biomass gasification is difficult as a result of, among many factors, the nonhomogeneous and fibrous complexity of these organic materials. As such, proposed conversion technologies differ in their design approach to selecting operating pressure, temperature, biomass preparation, and time needed for the conversion reaction. Endothermic destructive distillation, or pyrolysis, of organic materials at 430–450°C (805–841°F) can yield a gas having an energy density near that of a coal-derived medium-Btu gas. Municipal solid waste, for example, can be used to produce a low [3.73–5.59 MJ/m^3 (100–150 Btu/ft^3)] to medium [11.2–14.9 MJ/m^3 (300–400 Btu/ft^3)] gas by pyrolysis. Thermal decomposition of these organic wastes in substoichiometric or oxygen-deficient atmospheres will shift from gas product mixtures rich in methane–carbon dioxide to mixtures rich in hydrogen–carbon monoxide as pyrolysis temperature increases. In addition to *pyrogas,* small amounts of oil, charred metals, solid materials containing glassy aggregate, carbon char, and considerable bottom and fly ash will result from municipal waste pyrolysis. Gasification techniques differ in degree of carbon gasification and amount of air utilized per unit of organic mass converted. Catalytic hydrogenation uses a hydrogen-rich environment, which is heated to temperatures in excess of 300°C (908°F) for biomass gasification. Newer gasification systems utilizing fixed-bed and/or fluidized-bed reactors, similar to the solid fuel combustion systems described earlier in this text, are currently being evaluated and pursued.

In the previous sections, several alternative and potential substitutes for natural gas were considered. The viability and development of any and all of these synthetic gaseous fuel alternatives will depend on factors including:

Successful economics of scale in their development
Ability to compete successfully against other alternatives
Environmental impact of their production

Alternatives are apparently available to substitute for any major worldwide natural gas shortage in the near-term future. The long-term gaseous alternative, hydrogen, will be discussed in the following section.

6.6 HYDROGEN

Environmentally, diatomic hydrogen is the cleanest-burning fuel since its combustion produces only water and oxides of nitrogen.

$$H_2 + a[O_2 + 3.76N_2] \rightarrow bH_2O + cNO_x + dH_2 + eO_2 + fN_2 \qquad (6.20)$$

Hydrogen, therefore, holds promise as a long-term fuel option because, when burned in air, it also generates no unburned hydrocarbons or carbon dioxide, a compound that contributes significantly to the greenhouse effect of the earth's upper atmosphere.

Certain thermochemical properties give diatomic hydrogen gas many of its unique fuel characteristics. The specific gravity of gaseous hydrogen, for example, is approximately 0.07, meaning that it is less dense than methane, is lighter than air, and requires a greater storage volume than natural gas.

Hydrogen burns differently than natural gas, partly because of its lower ignition energy, invisible but hotter flame, and broader explosive fuel-air mixture limits. Furthermore, the heat of combustion for gaseous hydrogen on a volumetric basis of 12.0 MJ/m^3 (320 Btu/ft^3) is approximately one-third that for natural gas, whereas the heat of combustion on a mass basis is 2.75 times that of most hydrocarbon fuels. Hydrogen's high diffusivity also means that it has a greater leak rate than natural gas, is able to embrittle metals, and can be "absorbed" by certain solid materials termed *hydrides*. The saturation state for hydrogen at 1 atm of −253°C (−424°F) requires that liquid storage must be at *cryogenic* conditions, a state that cannot be achieved by simply compressing and cooling hydrogen gas.

Unfortunately, hydrogen gas does not occur naturally and, hence, must be generated, which requires both a material and an energy input. Hydrogen, however, is an abundant element contained in many different substances including solids like coal and shale, liquids such as crude oil and water, and gases like ammonia and methane. In the short term, hydrogen can be produced by steam-cracking coal via the *carbureted water–gas* reaction or

$$C_s + H_2O \overset{T,P}{\rightleftarrows} CO + H_2 \tag{6.21}$$

and by steam-cracking natural gas via the *water–gas* reaction

$$CH_4 + 2H_2O \overset{T,P}{\rightleftarrows} CO_2 + 4H_2 \tag{6.22}$$

Note that many of these resources are also suitable feedstocks for producing alternate liquid and gaseous synfuels.

In the long term, hydrogen can be generated by the electrolysis of water.

$$2H_2O \overset{T,P}{\rightleftarrows} H_2 + O_2 \tag{6.23}$$

Any use of electrical energy to make hydrogen via water electrolysis competes with the direct utility use of electricity. Hydrogen production and electric power generation could be combined during off-peak hours of electrical power demand, though, as a means of load leveling and energy storage at utility power plants. An unusual concept, shown in Figure 6.8, uses heat generated from fissionable radioactive waste to produce steam which, in turn, generates electricity required for water electrolysis. Eventually, a solar electrical energy/hydrogen economy would be independent of any petrochemical, coal, or even nuclear energy systems.

Several positive reasons for developing a viable hydrogen fuel industry can be identified, including the obvious fact that it is the long-term renew-

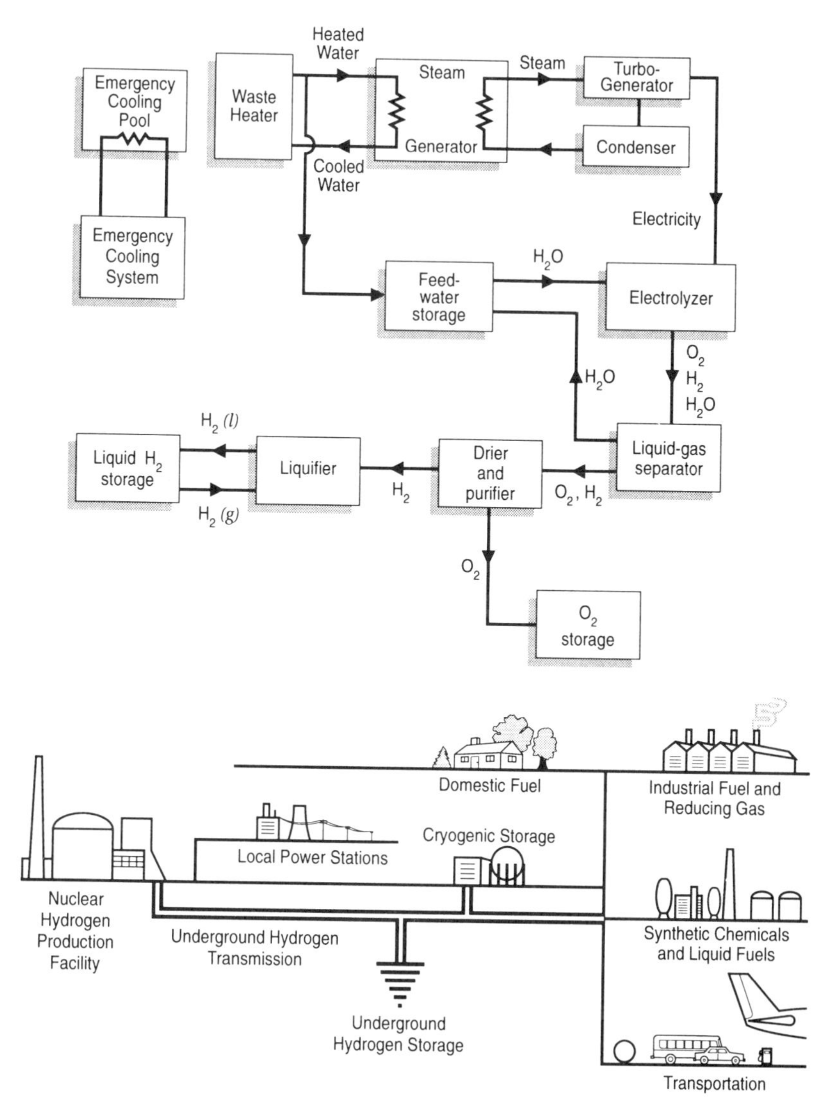

Figure 6.8 The hydrogen economy.

able fuel of the future. Hydrogen, like electricity, is an energy carrier that can be transported from production to particular sites by pipeline for specific uses. Hydrogen is a clean-burning fuel that would become a viable alternative if the use of coal-derived fuels is environmentally restricted. However, when discussing the environmental benefits of hydrogen as a fuel, one must also consider the pollution associated with generating hydrogen, for example, from coal. Hydrogen is also an essential ingredient required to upgrade many of the liquid syncrudes mentioned in Chapter 5.

Some of the more prominent negative aspects of a hydrogen fuel technology are economic, including the costs of production, transportation, storage, safety, and public acceptance. The fateful accident at Lakehurst, New Jersey, in 1937, when the hydrogen-filled German zeppelin *Hindenburg* caught fire and burned, has created public rejection of hydrogen, termed the *Hindenberg syndrome.* Many now believe, however, that the explosion and accident were associated with diesel fuel and not hydrogen. In fact, when a hydrogen leak occurs, it tends to escape rapidly, to rise vertically into the atmosphere and, if it catches fire, to provide little or no radiative heating because of the absence of carbon. Transportation uses of hydrogen are severely limited by its need for compact storage either as a cryogenic liquid in anhydrides like iron titanium or in compounds like methanol. Finally, the world's entire energy infrastructure would have to be changed drastically to utilize hydrogen effectively as a major fuel.

Hydrogen has been suggested by some as a possible replacement for natural gas, but there now appear to be significant new natural gas resources, as well as additional syngas resources, that will meet future gas needs. Hydrogen has been used in spark-ignition engines but, with no fuel storage and distribution system and poor public acceptance, this option has limited short-range potential. In certain high–pollution density metropolitan areas, however, there may be an environmental motivation for developing and using some hydrogen-fueled vehicles. Another use of hydrogen might be in advanced H_2-O_2 fuel cell technology for use in electric vehicles. Hydrogen in liquid form does appear to be a leading contender for future commercial aviation application, a field in which weight is a critical factor. Hydrogen is seen in the short term as not being a major fuel option, but it will play a role in the future. Fuel requirements in the coming decades will not be met by utilizing any one resource such as petroleum, coal, synfuels, or hydrogen. In fact, future energy needs will be based on a mix of all the fuels discussed in the last three chapters. The combustion engineer will, therefore, require a broad overview of fuel science in order to understand and to utilize most effectively the appropriate fuel option for each application.

6.7 LIQUID AND GASEOUS FUEL BURNERS

Solid fuels have been replaced today in most heat and/or power applications by liquid and gaseous hydrocarbon fuels. This is due, in part, to the thermochemical properties and portable nature of fluids, which allow such materials to be easily stored, transported, handled, and prepared for combustion. Note that liquid and gaseous fuel delivery components will be simpler than those systems described in Chapter 4 for use with solid fuels. In addition, the required reaction volume for burning fuel oil or gas is considerably less than that needed for most solid fuel reactions. Commercial use of liquid and/or gaseous fuels for heat transfer applications, such as space or industrial process heating, and/or steam generation, requires burners that will sustain atmospheric combustion within a particular furnace or oven.

Specific burner–fuel interactions are strongly influenced by properties of the particular fuel being fired. For example, the basic elements needed to properly and efficiently burn fuel oils using a burner are:

- Proper and safe storage of the fuel
- Settling equipment as well as fuel filters and strainers to ensure constant fuel quality
- Use of fuel heaters when necessary, such as when burning heavy fuel oils
- Suitable fuel delivery pumps to ensure consistent and sufficient fuel flow
- Burner elements that supply fuel to the furnace by
  - Injecting atomized fuel oil droplets
  - Vaporizing fuel and preheating fuel vapor
  - Mixing with sufficient air for combustion
  - Igniting the fuel–air mixture
- Suitable furnace volume for complete combustion

An essential requirement of any oil-burning system is to ensure that fuel is supplied at a proper temperature, pressure, and viscosity to enable the burner to mix suitably atomized fuel into air for combustion. Oil burner designs basically consist of two parts; a burner barrel, or *atomizer,* which allows a stream of small droplets of fuel to be supplied to the combustion chamber, and a *register,* which supplies as well as regulates combustion air to the furnace. Some burners use mechanical techniques, such as a high pressure fuel supply, to produce atomized fuel oil. Other designs use fluid techniques, such as high pressure fuel and air or fuel and steam injection to generate atomized fuel droplets. Several types of oil burners have found use and include *swirl atomizers, rotary cup burners,* and *high pressure jet atomizers.*

Gaseous fuels do not require charge preparation since they already exist naturally in a reactive state. The main function of the gas burner is therefore to introduce the fuel at the proper *AF* ratio. Gas burners are either low or high pressure configurations. Low pressure units are most frequently used with natural gas systems that are generally found in domestic applications, such as cooking and heating. These burners, often located near the floor level of the unit, use a multi-jet arrangement to discharge fuel gas vertically into the furnace. Combustion air flows around the burner orifaces and fuel–air ignition and combustion then occur above the burner. High pressure gas burner designs were incorporated in certain steam generation plants by the 1920s to 1930s in regions where natural gas was available.

An important operational parameter for a particular burner is its *turndown ratio,* i.e., the ratio of maximum fuel input rate to minimum fuel input. The upper range of this parameter is limited by incomplete combustion, blowoff, or inability to supply sufficient primary air. The lower limit is governed by the extinction of the flame or flashback tendency of the reaction zone. Turndown ranges in excess of 5 to 1 are usually difficult to sustain.

PROBLEMS

6.1 A homogeneous gaseous mixture of iso-octane and air has a specific gas constant of 51.255 ft lbf/lbm °R. For the mixture, find (a) the % theoretical air; (b) the *STP* mixture density, kg/m^3; (c) the equivalence ratio; and (d) the reaction mass *FA* ratio, lbm fuel/lbm air.

6.2 A homogeneous mixture of methane and air at *STP* has a specific heat ratio of 1.35. For this mixture, calculate (a) the molar constant pressure specific heat, kJ/kg mole K; (b) the reactant mole fractions, %; (c) the mixture density, kg/m^3; and (d) the mixture specific gravity.

6.3 A natural gas supply is composed of 20% CH_4, 40% C_2H_6, and 40% C_3H_8, where all percentages are by volume. The Orsat analysis of dry combustion products yields 10.6% CO_2, 3.0% O_2, and 1.0% CO. Determine (a) the gravimetric fuel analysis, %; (b) the required theoretical *AF* ratio, kg air/kg fuel; (c) the reaction excess air, %; and (d) the mass of dry exhaust gases to mass of fuel fired, kg gas/kg fuel.

6.4 The volumetric analysis of a natural gas supply is 22.6% C_2H_6 and 77.4% CH_4. Find (a) the mass stoichiometric *FA* ratio, lbm fuel/lbm air; (b) the mass of CO_2 and H_2O formed per mass of fuel, lbm gas/lbm fuel; (c) the gravimetric percentages of C and H_2 in the fuel, %; (d) the dew point temperature for ideal stoichiometric combustion,

°F; and (e) the specific gravity of the dry exhaust gases for this natural gas.

6.5 A gaseous fuel having a volumetric analysis of 65% CH_4, 25% C_2H_6, 5% CO, and 5% N_2 is burned with 30% excess air. Find (a) the ideal mass of air supplied per unit mass of fuel, kg air/kg fuel; (b) the volumetric flow rate of *STP* air to that of gaseous fuel at *STP* conditions; (c) the reaction equivalence ratio; and (d) the moles of CO_2 produced per mole of fuel.

6.6 A natural gas supply is to be augmented using a mixture of methane and propane. The supplier indicates that the specific gravity of the gas at *STP* conditions is 1.0. For this fuel, determine (a) the mole fractions of methane and propane, %; (b) the fuel density at *STP*, kg/m^3; (c) the fuel higher heating value at *STP*, kJ/kg; and (d) the ratio of the answer to part c to the higher heating value of methane.

6.7 Propane, an LPG fuel resource, is supplied to a constant-pressure atmospheric burner at 77°F. For a combustion efficiency of 87%, calculate (a) the ideal burner exhaust gas temperature, °F; (b) the excess air, %; (c) the dry exhaust gas mole fractions, %; and (d) the dew point stack temperature, °F.

6.8 A synthetic natural gas, or SNG, is to be generated using methane and propane. The lower heating value of the fuel at *STP* has a value of 37.25 MJ/m^3. Calculate (a) the mixture molar density, kg mole/m^3; (b) the fuel component mole fractions, %; (c) the fuel density at *STP*, kg/m^3; and (d) the specific gravity of the fuel.

6.9 Gas burners can be fired with different gaseous fuels yet produce an equivalent performance as long as the gases have equal Wobble numbers. Consider a gaseous fuel mixture of 85% methane–15% butane at *STP*, by volume. For stoichiometric combustion, find (a) the molar $\overline{FA}$ ratio, m^3 fuel/m^3 air; (b) the fuel specific gravity; (c) the fuel density, kg/m^3; (d) the fuel *HHV*, MJ/m^3; and (e) the fuel Wobble number, MJ/m^3.

6.10 A cigarette lighter uses butane combustion in 200% theoretical air. Both fuel and air are at 25°C, and the complete combustion products are at 127°C. The gas-phase combustion occurs at 101 kPa. Determine (a) the molar $\overline{AF}$, kg mole air/kg mole fuel; (b) the equivalence ratio; (c) the reactant mass fractions, %; (d) the dew point temperature, °C; and (e) the heat released during combustion, kJ/kg fuel for the butane combustion.

6.11 A flow calorimeter is used to measure the higher heating value of a natural gas. The gas is delivered to the calorimeter at 14.7 psi and 77°F. The water supplied to the calorimeter has a mass flow rate of 1.3 lbm/min, with a corresponding temperature rise of 8.3°F. The volume flow rate of the gas, measured using a wet test meter, is 0.01 ft^3/min. Find (a) the heat absorbed by the water jacket, Btu/min; and (b) the higher heating value of the fuel, Btu/ft^3.

6.12 A gas furnace burns 2.5 ft^3/h of a gaseous fuel in 25% excess air. The fuel volumetric analysis yields 90% CH_4, 7% C_2H_6, and 3% C_3H_8. Both fuel and air enter the atmospheric burner at 77°F, while flue gases exit at a temperature of 1880°F. Assuming ideal combustion, calculate (a) the dry flue gas analysis, %; (b) the product volumetric flow rate, ft^3/hr; (c) heat release, Btu/hr; and (d) the furnace efficiency, %.

6.13 A portable furnace burns a propane-air mixture having a 0.8 equivalence ratio. The fuel-air supply to the burner is at 25°C and 101 kPa, with the gases leaving the stack at 127°C. Assuming ideal complete combustion, find (a) the ideal flue gas molecular weight, kg/kg mole; (b) the exhaust dew point temperature, °C; (c) the mass of CO_2 produced per mass of fuel supplied, kg gas/kg fuel; and (d) the furnace efficiency, %.

6.14 Methane and air are supplied to a constant-pressure burner at 108 kPa and 25°C. The dry products of combustion mole fractions are 6.95% CO_2, 3.02% O_2, and 90.00% N_2. The dew point temperature of the products is 100°C. Determine (a) the molar $\overline{AF}$ ratio, kg mole air/kg mole fuel; (b) the exhaust gas molecular weight, kg/kg mole; (c) the combustion heat transfer if the exhaust gases are at 100°C; and (d) the combustion efficiency, %.

6.15 Acetylene is burned using a constant-pressure water-cooled burner. Fuel is supplied at 200 kPa and 15°C, while air supply enters the burner at 200 kPa and 20°C. The products leave the unit at 150°C. Water supplied at a flow rate of 0.25 kg/sec enters the jacket at 20°C and leaves at 120°C. Calculate (a) the heat transfer rate to cooling water, W; (b) the total mass flow rate of fuel and air, kg/sec; (c) the volumetric flow rate of dry exhaust gases, m^3/sec; and (d) the rate of condensation of water in the exhaust, kg/sec.

6.16 A hot water heater receives feed water at 60°F and produces hot water at 120°F. The gas-fired unit produces 30-gal water/h and burns methane at *STP* reactant conditions. The burner design requires 130% theoretical air combustion. Calculate (a) the required heat transfer rate to the water, Btu/hr; (b) the gas flow rate, ft^3/min; (c) the dew point temperature, °F; and (d) the boiler efficiency, %.

6.17 Methane is burned initially at 1 atm and 25°C in 20% excess air. The adiabatic flame temperature for the constant-pressure combustion process is 2500K. For complete combustion, calculate (a) the ideal reaction equivalence ratio; (b) the frozen product mole fractions, %;

and (c) the air preheat temperature needed to produce these results, K.

6.18 A mixture of 1 part by volume of ethylene to 50 parts by volume of air is ignited in a closed rigid vessel. The initial pressure and temperature of the charge mixture is 10 atm and 300K. Calculate (a) the mixture equivalence ratio; (b) the maximum adiabatic temperature, K; and (c) the maximum adiabatic pressure, atm.

6.19 A furnace designed to burn coal gas in 60% excess air is to be converted to a biomass methane gas. Design requirements are to provide the same energy transfer, using the same volumetric supply rate of air. Fuel and air are supplied at *STP.* Composition of the two gaseous fuels are:

Coal gas	Biogas
30.0% CH_4	90.0% CH_4
3.6% C_2H_4	2.0% CO_2
8.0% CO	8.0% N_2
52.0% H_2	
0.4% O_2	
2.0% CO_2	
4.0% N_2	

For these conditions, determine (a) the maximum energy release rate for 100 m^3/sec of air, kW; (b) the excess air requirements for the converted burner, %; (c) the ratio of volumetric fuel flow rates; and (d) the Wobble number for the two fuels, mJ/m^3.

6.20 A carbureted water gas has the following composition:

16.0% C_2H_4	32.3% H_2	2.9% CO_2
19.9% CH_4	26.1% CO	2.8% N_2

Flue gas analysis of the combustion process yields:

11.83% CO_2	4.53% O_2
83.24% N_2	0.4% CO

For combustion at *STP,* compute (a) the volumetric flow rate of air to the volumetric flow rate of fuel, ft^3 air/ft^3 fuel; (b) the volumetric flow rate of flue gases at 620°F to flow rate of fuel, ft^3 gas/ft^3 fuel; (c) the equivalence ratio for the reaction; and (d) the ideal combustion efficiency for the reaction with the stack gases at 620°F.

6.21 A hot water heater receives feed water at 60°F and produces hot water at 120°F. The gas-fired unit burns 0.73 lbm/hr of methane, CH_4, and produces 30 gal/hr of hot water. If the heater is supplied with fuel and air at *STP*, calculate (a) the required heat transfer to the water, Btu/hr; (b) the boiler efficiency, %; (c) the gas consumption, ft^3/min; (d) the ideal dry flue gas analysis for 130% theoretical air combustion; and (e) the product dew point temperature, °F.

6.22 The preliminary design for a low-pollution steam power plant is required to satisfy the following design specifications: 1×10^6-kW net power plant output; 40% steam cycle thermal efficiency; CH_4 fuel source; 15% excess air for combustion; 1 atm and 25°C burner air supply; 1 atm and 227°C flue gas conditions. For complete ideal combustion, determine (a) the ideal Orsat analysis, %; (b) the dew point temperature, °C; (c) the fuel mass flow rate for a stack temperature that is 10°C greater than the dew point temperature, kg fuel/hr; and (d) the combustion efficiency, %.

7

Fluid Mechanics

7.1 INTRODUCTION

In the first three chapters, basic thermochemical principles were developed and used to describe the ideal energetics of various combustion processes. Using both frozen and equilibrium chemical reactions, several important topics were introduced, including the conservation concept for atomic species in reactive mixtures, standard-state heats of formation for reactive species, and the ideal heat release rates for many important solid, liquid, or gaseous fuels. Many facets of combustion cannot be inferred from a thermodynamic analysis of chemical energy conversion alone. For example, most high-temperature chemical reactions in combustion involve fluid motion and, thus, the necessity to understand basic fluid mechanics is obvious. In the following chapter, conservation relationships for mass, momentum, and energy will be written for one-dimensional reactive fluid flow in which the particular energy released by exothermic reactions will influence the fluid properties of the flow. An extensive investigation of these three conservation equations is the subject of gas dynamics, which is a separate engineering area of study in and of itself. Analysis of one-dimensional chemically reactive gas dynamic relations developed in this chapter will indicate that two distinctly different combustion wave propagation phenomena theoretically can occur. The characteristics of these two classic results will be treated in detail. An understanding of this ideal case will provide

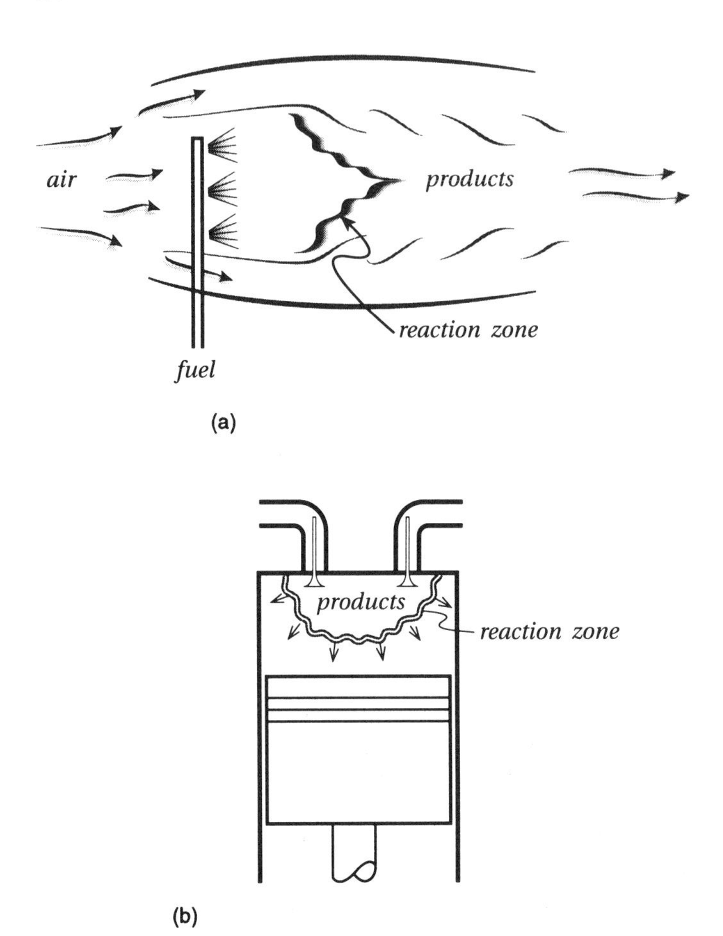

Figure 7.1 Combustion zone fluid motion: (a) continuous burner flame zone; (b) internal combustion engine flame zone.

some insight into the combustion processes that occur in internal combustion engines as well as in gas turbine combustors.

7.2 BASIC CONSERVATION EQUATIONS

The concept of wave propagation is common to a variety of engineering problems and is useful in dealing with light physics, sound wave propagation, and shock wave formation. Analysis of spatial and time characteristics of an actual combustion wave propagating through a reactive medium can be quite complex; that is to say, motion of a moving reaction zone, in general, will be an unsteady three-dimensional translation through a nonhomogeneous medium. One important case is that in which the reaction zone, or flame front, passes through a homogeneous premixed fuel-air mixture. A reaction wave passing through such a medium will separate an unburned zone ahead of the wave front from a burned product zone behind. The actual region of separation is quite small, on the order of several millimeters and, to a good approximation, can be thought of as a line of discontinuity separating unburned and burned gases (Figure 7.1).

The physics associated with one-dimensional wave propagation can be formulated using either differential or integral calculus. Furthermore, differential or integral form of the conservation equations can be expressed using particle, i.e., *Lagrangian,* or field, i.e., *Eulerian,* mathematics. Recall that, in Lagrangian analysis, the characteristics of a fixed mass, frequently referred to as a control mass or closed system, are described for specific material undergoing a change of state as well as location. In Eulerian analysis, characteristics of a fluid flux undergoing a change of state and position are described in terms of a specified volume in space, called a control volume, or open system.

Consider the nonaccelerating one-dimensional planar reaction zone shown in Figure 7.2. In Figure 7.2a, the actual process is seen to be an unsteady wave propagation moving from left to right as time proceeds. In Figure 7.2b, a simple coordinate transformation changes the problem to that of a stationary wave front through which a fluid field passes.

The following assumptions will be made in the analysis of the classic one-dimensional wave propagation problem:

1. Differential formulation
2. One-dimensional stationary wave front
3. Steady-state and steady-flow conditions
4. Inviscid flow
5. Irreversible process
6. Stationary discontinuity at wave front
7. Nonreactive homogeneous ideal-gas mixture on either side of wave

From physics, for a fixed total mass M, the conservation of mass can be stated as

$$M = \text{const} \qquad \text{kg (lbm)} \tag{7.1}$$

or, for the control mass CM,

$$\left.\frac{DM}{Dt}\right)_{\text{CM}} = 0 \qquad \frac{\text{kg}}{\text{sec}}\left(\frac{\text{lbm}}{\text{sec}}\right) \tag{7.2}$$

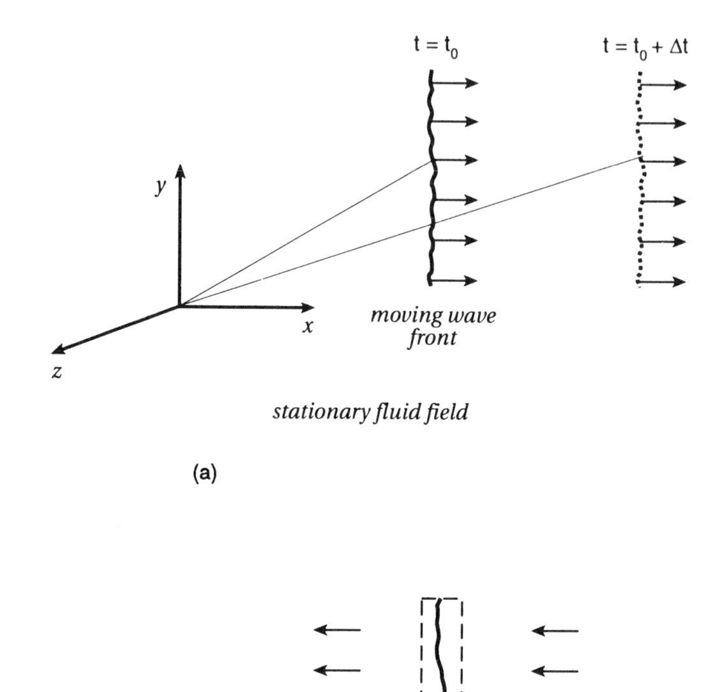

Figure 7.2 One-dimensional nonaccelerating wave front: (a) propagating wave description; (b) stationary wave description.

Written in terms of a Field control volume, *CV*, approach, the above statement can also be expressed as

$$\left.\frac{DM}{Dt}\right)_{\text{CM}} = \left.\frac{\partial M}{\partial t}\right)_{\text{CV}} + \left.\Sigma \dot{m}_i\right)_{\text{CV}} = 0 \tag{7.2a}$$

where

$\left.\frac{\partial M}{\partial t}\right)_{\text{CV}}$ = local change of mass in *CV* with time

$\left.\Sigma \dot{m}_i\right)_{\text{CV}}$ = net mass efflux out of(+) into(−) *CV*

For steady-state, steady-flow, Equation (7.2a) reduces to

$$\left.\cancel{\frac{\partial M}{\partial t}}\right)_{\text{CV}} + \left.\Sigma \dot{m}_i\right)_{\text{CV}} = 0 \tag{7.2b}$$

which, for the one-dimensional case with uniform properties in regions 1 and 2, becomes

$$\dot{m}_1 = \rho_1 A_1 V_1 = \rho_2 A_2 V_2 = \dot{m}_2 \qquad \text{kg/sec (lbm/sec)} \tag{7.3}$$

Defining a mass flow rate per unit area, G, one obtains

$$G_1 = \rho_1 V_1 = \rho_2 V_2 = G_2 \qquad \text{kg/sec m}^2 \text{ (lbm/sec ft}^2\text{)} \tag{7.4}$$

Again, from physics, conservation of momentum for a fixed total mass M is given by

$$\Sigma \vec{F} = \left. M\vec{a}\right)_{\text{CM}} = \left.\frac{D}{Dt}\{M\vec{V}\}\right)_{\text{CM}} = \left. M\frac{D\vec{V}}{Dt}\right)_{\text{CM}} + \left.\vec{V}\frac{DM}{Dt}\right)_{\text{CM}} = \left. M\frac{D\vec{V}}{Dt}\right)_{\text{CM}} \tag{7.5}$$

or expressed using a Field approach for forces in the x direction Equation (7.5) becomes

$$\Sigma F_x = \left.\frac{DMV_x}{Dt}\right)_{\text{CM}} = \left.\frac{\partial MV_x}{\partial t}\right)_{\text{CV}} + \left.\Sigma_i(\dot{m}_x V_x)_i\right)_{\text{CV}} \qquad \text{N (lbf)} \tag{7.5a}$$

where

$\left.\frac{\partial MV_x}{\partial t}\right)_{\text{CV}}$ = local change in *CV* of x − momentum with time

$\left.\Sigma \dot{m}_x V_x\right)_{\text{CV}}$ = net x − momentum flux out of (+) into(−) *CV*

For steady-state, steady-flow conditions,

$$\Sigma F_x = \left.\cancel{\frac{\partial MV_x}{\partial t}}\right)_{\text{CV}} + \left.\Sigma_i(\dot{m}_x V_x)_i\right)_{\text{CV}} \tag{7.5b}$$

and, for inviscid flow, the conservation of momentum, assuming only normal pressure forces, equation (7.5b) becomes

$$g_0(P_1A - P_2A) = -(\rho_1 AV_1)\, V_1 + (\rho_2 AV_2)\, V_2 \tag{7.6a}$$

$$g_0 P_1 + \rho_1 V_1^2 = g_0 P_2 + \rho_2 V_2^2 \tag{7.6b}$$

or

$$P_1 + \frac{G_1^2}{g_0 \rho_1} = P_2 + \frac{G_2^2}{g_0 \rho_2} \qquad \frac{\text{N}}{\text{m}^2}\left(\frac{\text{lbf}}{\text{ft}^2}\right) \tag{7.7}$$

Note the use of g_0 to express mass and weight in consistent dimensions and units.

EXAMPLE 7.1 The speed of sound c through an undisturbed ideal-gas medium can be determined using the one-dimensional formulation for compressible wave propagation. Sound can be modeled as a small-amplitude isentropic wave phenomenon. For the condition shown below, write (a) the appropriate continuity equation, (b) the momentum equation, (c) the differential equation for sonic velocity using parts (a) and (b); and (d) show that an ideal gas $c^2 = \gamma RTg_0$.

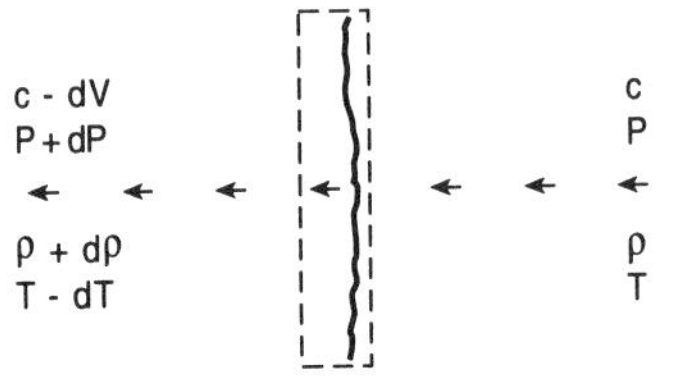

c - dV
P + dP
ρ + dρ
T - dT

c
P
ρ
T

Stationary Wave

Solution:

1. Continuity:

$$G = \rho_1 V_1 = \rho_2 V_2$$

$$\rho c = (\rho + d\rho)(c - dV)$$

$$\rho c = \rho c + c\, d\rho - \rho\, dV - \cancel{d\rho\, dV}$$

or

a. $\quad c\, d\rho = +\, \rho\, dV$

2. Momentum:

$$g_0(P_2 - P_1) = \rho_1 V_1^2 - \rho_2 V_2^2$$

$$g_0[(P + dP) - P] = \rho c^2 - (\rho + d\rho)(c - dV)^2$$

$$g_0\ dP = \rho c^2 - (\rho + d\rho)(c^2 - 2c\ dV + \cancel{dV^2})$$

$$g_0\ dP = \rho c^2 - \rho c^2 + 2\rho c\ dV - c^2\ d\rho + 2c\,\cancel{d\rho\ dV}$$

b. $$g_0\ dP = 2\rho c\ dV - c^2\ d\rho$$

3. Combining items 1 and 2

$$g_0\ dP = 2c(c\ d\rho) - c^2\ d\rho = c^2\ d\rho$$

$$c^2 = g_0 \frac{dP}{d\rho}$$

and for small-amplitude sound waves where entropy is conserved

c. $$c^2 = g_0 \left.\frac{\partial P}{\partial \rho}\right)_{s=c}$$

4. Sonic velocity in an ideal gas:

State	Process
$P = \rho RT$	$P\rho^{-\gamma} = \text{const} = c_1$

then, expressing $P = P<\rho>$ using the process relationship,

$$c^2 = g_0 \frac{\partial}{\partial \rho}[c_1 \rho^\gamma]_{s=c} = g_0 \left[\ \gamma c_1 \rho^{\gamma - 1}\right]$$

$$c^2 = g_0 \gamma \left[\ P\rho^{-\gamma}\right] \rho^{\gamma - 1}$$

substituting from the ideal-gas law, the equation becomes

$$= \gamma g_0 P/\rho$$

$$c^2 = \gamma g_0 RT$$

or

$$c = [\gamma RT g_0]^{1/2}$$

In addition to the conservation of mass and momentum, the conservation of energy is needed before a discussion of the reactive wave propagation process can be considered in detail. For a fixed total mass M, the conservation of energy is equal to

$$\left.\frac{DE}{Dt}\right)_{\text{CM}} = \dot{Q} - \dot{W} \qquad \text{kJ (Btu)} \tag{7.8}$$

and, in Field *CV* terminology,

$$\left.\frac{DE}{Dt}\right)_{\text{CM}} = \left.\frac{\partial E}{\partial t}\right)_{\text{CV}} + \left.\Sigma_i \dot{m}_i e_i\right)_{\text{CV}} = \dot{Q} - \dot{W} \qquad \text{kW}\left(\frac{\text{Btu}}{\text{hr}}\right) \tag{7.9}$$

and

$$\left.\frac{\partial E}{\partial t}\right)_{\text{CV}} = \text{local change in } CV \text{ energy with time}$$

$$\left.\Sigma \dot{m}_i e_i\right)_{\text{CV}} = \text{net energy flux out of}(+)\text{ into}(-)\ CV$$

Again, for steady-state, steady-flow conditions,

$$\left.\frac{\partial E}{\partial t}\right)_{\text{CV}} + \dot{m}_2 e_2 - \dot{m}_1 e_1 = \dot{Q} - \dot{W} \tag{7.10a}$$

or

$$\dot{Q} - \dot{W} = \dot{m}_2[u_2 + \frac{P_2}{\rho_2} + \frac{V_2^2}{2g_0} + \frac{g}{g_0}Z_2] - \dot{m}_1[u_1 + \frac{P_1}{\rho_1} + \frac{V_1^2}{2g_0} + \frac{g}{g_0}Z_1] \tag{7.10b}$$

Assuming one-dimensional flow and applying conservation of mass equation (7.3), Equation (7.10b) becomes

$$u_1 + \frac{P_1}{\rho_1} + \frac{V_1^2}{2g_0} + {}_1q_2 = u_2 + \frac{P_2}{\rho_2} + \frac{V_2^2}{2g_0} + {}_1w_2 \tag{7.11}$$

or

$$h_1 + \frac{V_1^2}{2g_0} + {}_1q_2 = h_2 + \frac{V_2^2}{2g_0} + {}_1w_2 \qquad \frac{\text{kJ}}{\text{kg}}\left(\frac{\text{Btu}}{\text{lbm}}\right) \tag{7.12}$$

Equation (7.12) relates changes in flow properties to the heat added, i.e., energy released by combustion. Table 7.1 summarizes the conservation equations for the one-dimensional wave propagation process.

EXAMPLE 7.2 The basic conservation relationships developed in this chapter can be used to predict the local static and corresponding stagnation properties for one-dimensional homogeneous fluid flow. Recall that the local stagnation state is the condition achieved if the fluid at that point is isentropically brought to rest. For these conditions, determine an expression for the ratio of the stagnation to local values for (a) T_0/T, (b) ρ_0/ρ, (c) P_0/P. For sonic conditions in a flow, the above expressions reduce to the critical ratios. Determine the critical values for parts a and c in terms of sonic parameters $T*$, $P*$, $\rho*$.

Solution:

1. Energy equation:

$$h_1 + \frac{V_1^2}{2g_0} + \cancel{{}_1q_2} = h_2 + \frac{V_2^2}{2h_0} + \cancel{{}_1w_2}$$

Let state 1 ≡ the stagnation state and state 2 = the arbitrary state.

$$h_0 = h + \frac{V^2}{2g_0}$$

Assume ideal-gas mixture with constant specific heats.

$$T_0 - T = \frac{V^2}{2g_0C_p}$$

where

$$C_p - C_v = R \qquad C_p/C_v = \gamma$$

and

$$C_p = \frac{R\gamma}{\gamma - 1}$$

$$T_0 - T = \frac{V^2(\gamma - 1)}{2g_0R\gamma} = \frac{V^2(\gamma - 1)T}{2g_0R\gamma T}$$

but

$$c = \sqrt{g_0R\gamma T}$$

$$T_0 - T = \frac{V^2(\gamma - 1)T}{2c^2}$$

Mach number, $N_m \equiv V/c$,

or

$$T_0 - T = N_m^2\left(\frac{\gamma - 1}{2}\right)T$$

$$\frac{T_0}{T} - 1 = N_m^2\left(\frac{\gamma - 1}{2}\right)$$

a. $$\frac{T_0}{T} = 1 + \left(\frac{\gamma - 1}{2}\right)N_m^2$$

2. Isentropic process, ideal-gas mixture:

$$\frac{P}{\rho^{\gamma}} = \text{const}$$

or

$$\frac{P}{\rho^{\gamma}} = \frac{P_0}{\rho_0^{\gamma}}$$

and

$$\frac{P}{\rho} = RT$$

$$\frac{P_0}{P} = \left(\frac{\rho_0}{\rho}\right)^{\gamma} = \left(\frac{RT}{P}\right)^{\gamma}\left(\frac{P_0}{RT_0}\right)^{\gamma}$$

$$\frac{P_0}{P} = \left(\frac{T}{T_0}\right)^{\gamma}\left(\frac{P_0}{P}\right)^{\gamma}$$

$$\left(\frac{P_0}{P}\right)^{1-\gamma} = \left(\frac{T}{T_0}\right)^{\gamma}$$

$$\left(\frac{P_0}{P}\right) = \left(\frac{T}{T_0}\right)^{\gamma/(1-\gamma)} = \left(\frac{T_0}{T}\right)^{\gamma/(\gamma-1)}$$

b. $$\left(\frac{P_0}{P}\right) = \left[1 + \left(\frac{\gamma - 1}{2}\right) N_m^2\right]^{\gamma/(\gamma-1)}$$

$$\frac{\rho_0}{\rho} = \left(\frac{P_0}{P}\right)^{1/\gamma} = \left(\frac{T_0}{T}\right)^{1/(\gamma-1)}$$

c. $$\frac{\rho_0}{\rho} = \left[1 + \left(\frac{\gamma - 1}{2}\right) N_m^2\right]^{1/(\gamma-1)}$$

3. Critical ratios, sonic conditions:

Let

$$V = c$$
$$P = P*$$
$$T = T*$$
$$\rho = \rho*$$

i.e., $N_m = 1.0$.

$$\frac{T_0}{T*} = \left[1 + \left(\frac{\gamma - 1}{2}\right)\right] = \frac{\gamma + 1}{2}$$

or

d. $$\frac{T*}{T_0} = \left(\frac{2}{\gamma + 1}\right)$$

e. $$\frac{P*}{P_0} = \left(\frac{2}{\gamma + 1}\right)^{\gamma/(\gamma-1)}$$

f. $$\frac{\rho*}{\rho_0} = \left(\frac{2}{\gamma + 1}\right)^{1/(\gamma-1)}$$

Comments One-dimensional isentropic flow for an ideal gas is seen to be a function of γ and Mach number. These dimensionless equations are tabulated as a function of N_m in Gas Tables for convenience in numerical analysis.

Table 7.1 1D Diabatic Flow Conservation Equations

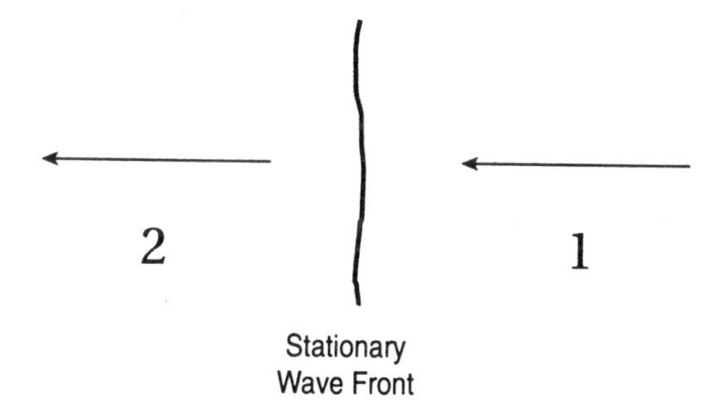

Conservation of mass:

$$G_1 = \rho_1 V_1 = \rho_2 V_2 = G_2$$

Conservation of momentum:

$$P_1 + \frac{G_1^2}{g_0 \rho_1} = P_1 + \frac{\rho_1 V_1^2}{g_0} = P_2 + \frac{\rho_2 V_2^2}{g_0} = P_2 + \frac{G_2^2}{g_0 \rho_2}$$

Conservation of energy:

$$h_1 + \frac{G_1^2}{2g_0 \rho_1} + {}_1q_2 = h_1 + \frac{V_1^2}{2g_0} + {}_1q_2 = h_2 + \frac{V_2^2}{2g_0} + {}_1w_2 = h_2 + \frac{G_2^2}{2g_0 \rho_2} + {}_1w_2$$

7.3 THE RAYLEIGH LINE

As stated above, continuity, momentum, and energy equations are the primary gas dynamic relationships that can be used to further investigate such subjects as subsonic and supersonic compressible flow, converging-diverging nozzle performance, and normal shock wave propagation. In this chapter, discussion will be limited to the problem in which heat release (combustion) and compressible flow (wave propagation) are coupled. Furthermore, the treatment will be restricted to conditions for which the following assumptions are made:

1. One-dimensional discontinuous compressible flow
2. Irreversible but invisid flow
3. Steady stationary combustion zone (wave front)
4. Uniform properties upstream and downstream from reaction
5. Ideal gas with constant-mixture molecular weight

Two terms come into use when the general subject of reactive flow is considered: *adiabatic* and *diabatic* flow. In *adiabatic* flow, there is no heat transfer between the fluid and surroundings while, in *diabatic* flow, heat transfer is allowed to occur. A combustion wave is an irreversible wave phenomenon with local heat release and thus can be thought of as a particular type of diabatic flow.

In one-dimensional frictionless flow with heat transfer, a relationship can be written that combines the continuity and momentum equations. The resulting expression is termed the *Rayleigh line,* and its solution is often displayed on either P–v or T–s coordinates. The continuity equation and momentum relationships were developed earlier and are stated again as

Continuity:

$$\rho_1 V_1 = \rho_2 V_2 = G \qquad \frac{\text{lbm}}{\text{ft}^2\ \text{sec}} \left(\frac{\text{kg}}{\text{m}^2\ \text{sec}} \right) \tag{7.4}$$

Momentum:

$$P_1 + \frac{G_1^2}{g_0 \rho_1} = P_2 + \frac{G_2^2}{g_0 \rho_2} \qquad \frac{\text{N}}{\text{m}^2} \left(\frac{\text{lbf}}{\text{ft}^2} \right) \tag{7.7}$$

Combining Equations (7.4) and (7.7) yields

$$P_1 - P_2 = \frac{1}{g_0} \left(\frac{G_2^2}{\rho_2} - \frac{G_1^2}{\rho_1} \right) \tag{7.13a}$$

$$P_1 - P_2 = \frac{G^2}{g_0} \left(\frac{1}{\rho_2} - \frac{1}{\rho_1} \right) \tag{7.13b}$$

or

$$\frac{P_1 - P_2}{(1/\rho_2 - 1/\rho_1)} = \frac{G^2}{g_0} \tag{7.13c}$$

$$\frac{P_1 - P_2}{1/\rho_1 - 1/\rho_2} = \frac{P_1 - P_2}{v_1 - v_2} = -\frac{G^2}{g_0} \tag{7.14}$$

Equation (7.14) is called the *Rayleigh line*. For a fixed rate of flow, as is the case for a one-dimensional, steady-state, steady-flow problem, an inspection of Equation (7.14) indicates that

$$\frac{\Delta P}{\Delta v} = \frac{\Delta P}{\Delta(1/\rho)} = -\text{ const}$$

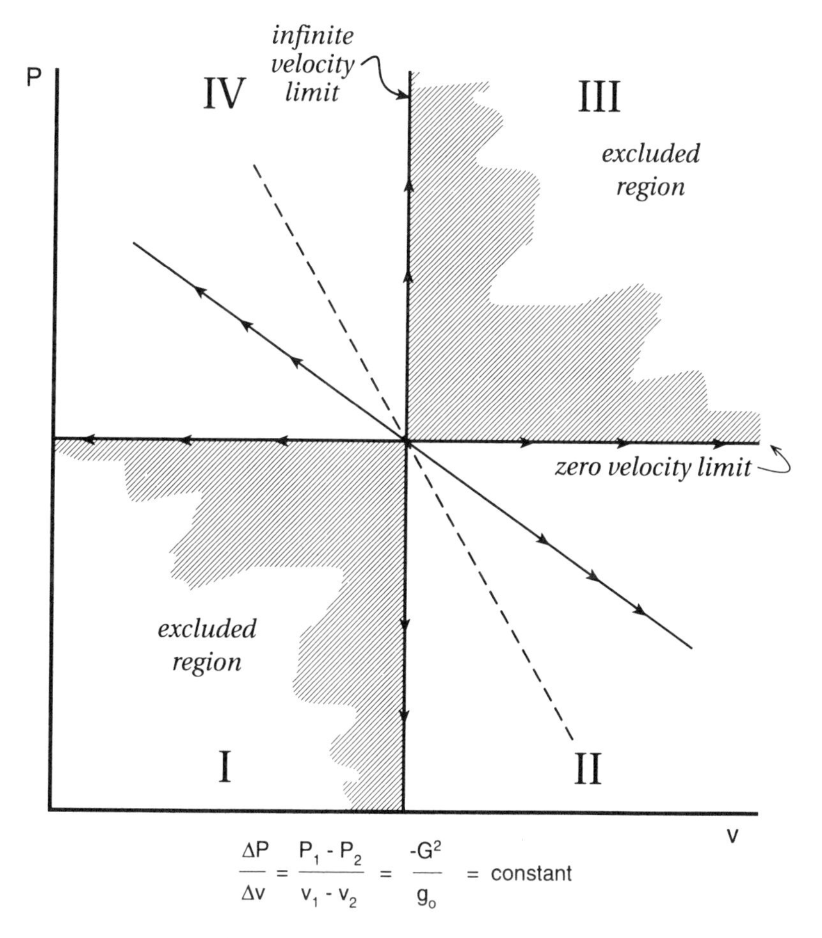

Figure 7.3 Rayleigh flow P–v diagram.

The Rayleigh curve is a straight line relation of negative slope on the P–v plane, see Figure 7.3. The steeper the slope, the greater the flow velocity. Now in the limit, for zero slope, the velocity approaches zero while, at an infinite slope, velocity would approach an infinite value. A further inspection of Equation (7.14) reveals that two quadrants, regions I and II in Figure 7.3, are excluded regions. In other words, all points of a given flow must lie on a common Rayleigh line and can pass only through regions II and IV.

EXAMPLE 7.3 Air at a pressure of 14.7 psia and temperature of 530°R is flowing with a velocity of 800 ft/sec. Treating the flow as a one-dimensional, evaluate (a) the density of the air, ρ, lbm/ft^3; (b) the specific volume v, ft^3/lbm; (c) the slope of the Rayleigh line on a P–v coordinate; and (d) the conditions for Rayleigh flow for a specific volume of 5 ft^3/lbm.

Solution:

1. Continuity:

$$G = \rho_1 V_1 = \frac{P_1 V_1}{RT_1}$$

2. Density:

$$\rho_1 = \frac{(14.7 \text{ lbf/in}^2)(144 \text{ in}^2/\text{ft}^2)}{(53.34 \text{ ft lbf/lbm °R})(530\text{°R})}$$
$$= 0.07488 \text{ lbm/ft}^3$$
$$v_1 = 13.355 \text{ ft}^3/\text{lbm}$$

3. Mass flow:

$$G = \rho_1 V_1 = (0.07488 \text{ lbm/ft}^3)(800 \text{ ft/sec})$$
$$= 59.9 = 60 \text{ lbm/sec ft}^2$$

4. P–v Rayleigh line, *slope:*

$$\frac{\Delta P}{\Delta v} = \frac{-G^2}{g_0} = \frac{(60 \text{ lbm/sec ft}^2)^2}{(32.2 \text{ ft lbm/lbf sec}^2)}$$

$$= -111.8 \frac{\text{lbm}^2/\text{sec}^2 \text{ ft}^4}{\text{ft lbm/lbf sec}^2}$$

$$= -111.8 \left(\frac{\text{lbm}}{\text{ft}^4}\right)\left(\frac{\text{lbf}}{\text{ft}}\right)$$

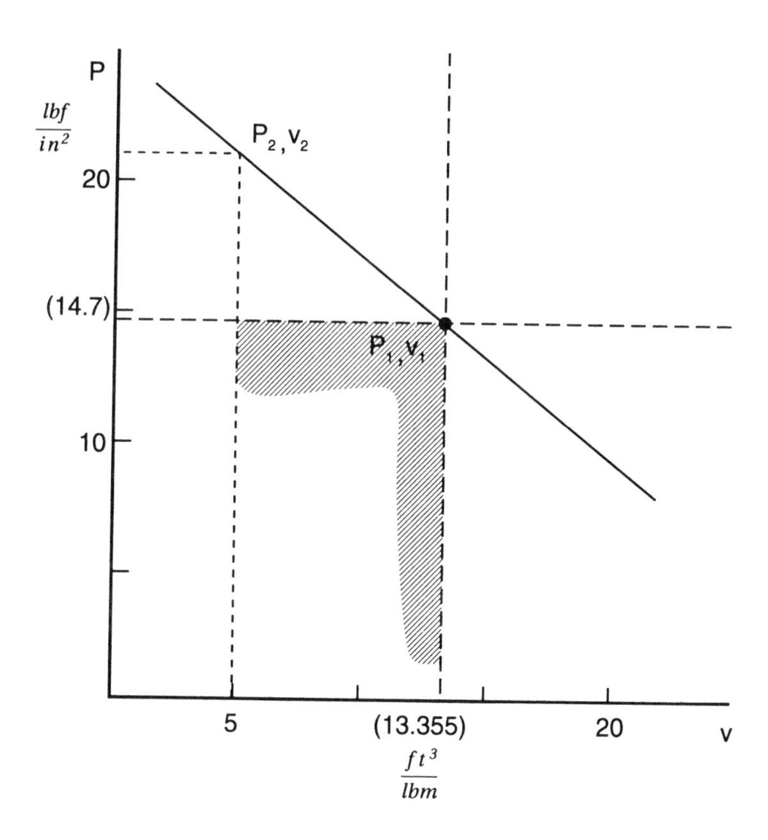

$$\frac{\Delta P}{\Delta v} = -\frac{111.8\ (\text{lbm/ft}^3)(\text{lbf/ft}^2)}{144\ \text{in}^2/\text{ft}^2}$$

$$= -0.78\ \frac{\text{psi}}{\text{ft}^3/\text{lbm}}$$

at $v_2 = 5\ \text{ft}^3/\text{lbm}$

$$V_2 = \frac{60\ \text{lbm/sec ft}^2}{(1/5)\ \text{lbm/ft}^3} = 300\ \text{ft/sec}$$

$$P_2 = 14.7 + 0.78\ (13.355 - 5) = 21.2\ \text{psia}$$

7.4 THE RANKINE–HUGONIOT CURVE

In Section 7.3, the continuity and momentum expressions developed for one-dimensional compressible flow were combined to form the Rayleigh line. The combination of the continuity, momentum, and energy relationships will produce an additional equation, termed the Rankine–Hugoniot curve, which is of use when modeling a combustion zone within a diabatic flow. Recall that continuity is equal to

$$G = \rho V \tag{7.4}$$

and

$$\frac{1}{\rho_1} = \frac{V_1}{G_1} \tag{7.15a}$$

$$\frac{1}{\rho_2} = \frac{V_2}{G_2} \tag{7.15b}$$

The Rayleigh line, a combination of continuity and momentum, Equation (7.14), was equal to

$$\frac{P_2 - P_1}{(1/\rho_2 - 1/\rho_1)} = \frac{-G^2}{g_0} \tag{7.14}$$

Conservation of energy for a flow with no work transfer is given as

$$q + h_1 + \frac{V_1^2}{2g_0} = h_2 + \frac{V_2^2}{2g_0} \tag{7.12}$$

$$q = (h_2 - h_1) + \frac{1}{2g_0}(V_2^2 - V_1^2)$$

Assuming an ideal gas with constant specific heats, the above becomes

$$q = C_p(T_2 - T_1) + \frac{1}{2g_0}(V_2^2 - V_1^2) \tag{7.16}$$

$$= \left(\frac{R\gamma}{\gamma - 1}\right)(T_2 - T_1) + \frac{1}{2g_0}(V_2^2 - V_1^2) \tag{7.17}$$

and, for an ideal gas where $P = \rho RT$,

$$q = \left(\frac{\gamma}{\gamma - 1}\right)\left(\frac{P_2}{\rho_2} - \frac{P_1}{\rho_1}\right) + \frac{1}{2g_0}[V_2^2 - V_1^2] \tag{7.18}$$

substituting the definition $G = \rho V$,

$$q = \left(\frac{\gamma}{\gamma - 1}\right)\left(\frac{P_2}{\rho_2} - \frac{P_1}{\rho_1}\right) + \frac{1}{2g_0}\left(\frac{G_2^2}{\rho_2^2} - \frac{G_1^2}{\rho_1^2}\right) \tag{7.19}$$

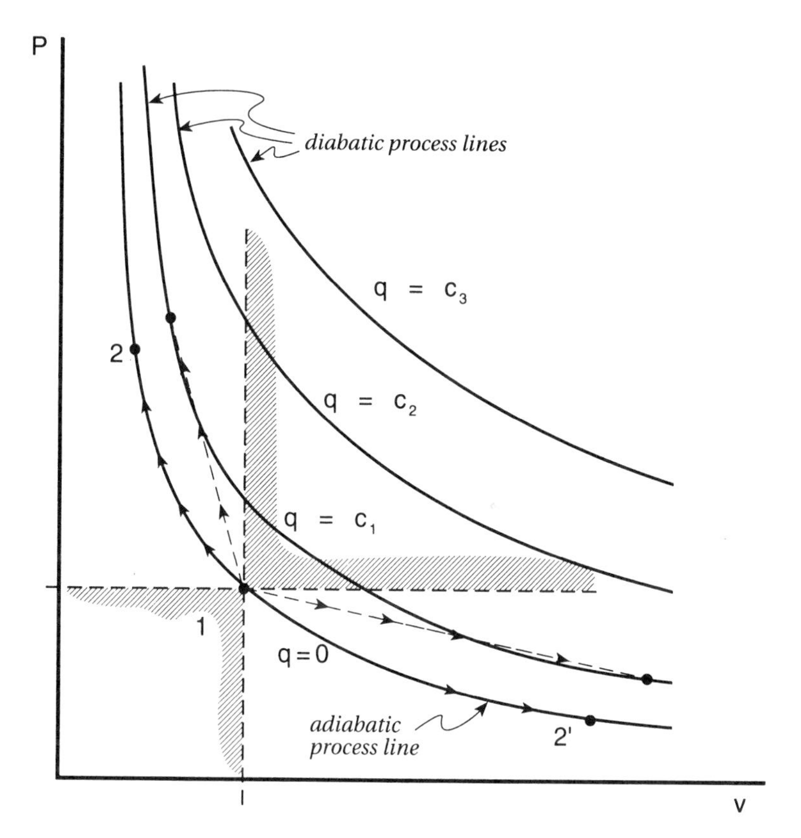

Figure 7.4 Rankine–Hugoniot curve.

$$= \left(\frac{\gamma}{\gamma - 1}\right)\left(\frac{P_2}{\rho_2} - \frac{P_1}{\rho_1}\right) + \frac{1}{2g_0}\left(\frac{1}{\rho_2^2} - \frac{1}{\rho_1^2}\right) G^2$$

$$= \left(\frac{\gamma}{\gamma - 1}\right)\left(\frac{P_2}{\rho_2} - \frac{P_1}{\rho_1}\right) + \frac{1}{2g_0}\left[\left(\frac{1}{\rho_2} + \frac{1}{\rho_1}\right)\left(\frac{1}{\rho_2} - \frac{1}{\rho_1}\right)\right] G^2$$

From Equation (7.14), substitution for G^2/g_0 above yields

$$q = \left(\frac{\gamma}{\gamma - 1}\right)\left(\frac{P_2}{\rho_2} - \frac{P_1}{\rho_1}\right) + \frac{1}{2}\left[\left(\frac{1}{\rho_2} + \frac{1}{\rho_1}\right)\left(\frac{1}{\rho_2} - \frac{1}{\rho_1}\right)\right]\left[\left(\frac{-(P_2 - P_1)}{1/\rho_2 - 1/\rho_1}\right)\right] \quad (7.20)$$

$$q = \left(\frac{\gamma}{\gamma - 1}\right)\left(\frac{P_2}{\rho_2} - \frac{P_1}{\rho_1}\right) - \frac{1}{2}\left(\frac{1}{\rho_2} + \frac{1}{\rho_1}\right)(P_2 - P_2) \quad (7.20a)$$

or

$$q = (h_2 - h_1) - \frac{1}{2}(P_2 - P_1)(v_2 + v_1) \tag{7.20b}$$

Equation (7.20) is called the Rankine–Hugoniot curve and shows that the fluid flow parameters P and ρ are functions of the local heat addition to the reaction zone. The relation is often shown on P–v coordinates, see Figure 7.4. On this diagram, each curve corresponds to a given heat release, i.e., q = const. An adiabatic flow will lie on the curve for $q = 0$. A diabatic flow for an exothermic reaction has to move from an initial point on the $q = 0$ curve, passing through the state P_1 and v_1 and terminating on a particular Hugoniot curve where q = const. Two separate conditions along the Hugoniot curve will satisfy the diabatic flow case for P_2 and v_2. The two solutions are labeled states 2 and 2′; see Figure 7.4.

For conditions at state 2, the curve indicates that the pressure behind the reaction zone, P_2, will be greater than the pressure ahead of the wave front, P_1. Also, the temperature of the burned gases, T_2, will be greater than the initial reactant temperature T_1. The specific volume of the products of combustion, v_2, will be less than the reactants v_1.

For the state defined by solution 2′, analysis shows that the pressure behind the wave front, P_2', will be less than that ahead of the reaction zone, P_1. The temperature of the combustion products, T_2', will be greater than the temperature of the initial reactants, T_1. The burned gases will have a specific volume, v_2', that will be larger than that of the original unburned mixture.

7.5 THE CHAPMAN–JOUQUET POINTS

The condition of the burned gases behind a stationary combustion zone is defined by a solution to both the Rayleigh and Hugoniot curves, see Figure 7.5. An inspection of Figure 7.5 reveals that a given Rayleigh line will intersect a particular Hugoniot curve on either the upper or lower branch. Solutions defined by points of intersection along the upper Hugoniot branch are called *detonation waves,* while lower branch solutions are termed *deflagration waves.* Recall from Section 7.3 that Rayleigh flow cannot pass into regions I and II. Conditions at 2 or 2′ must therefore lie on

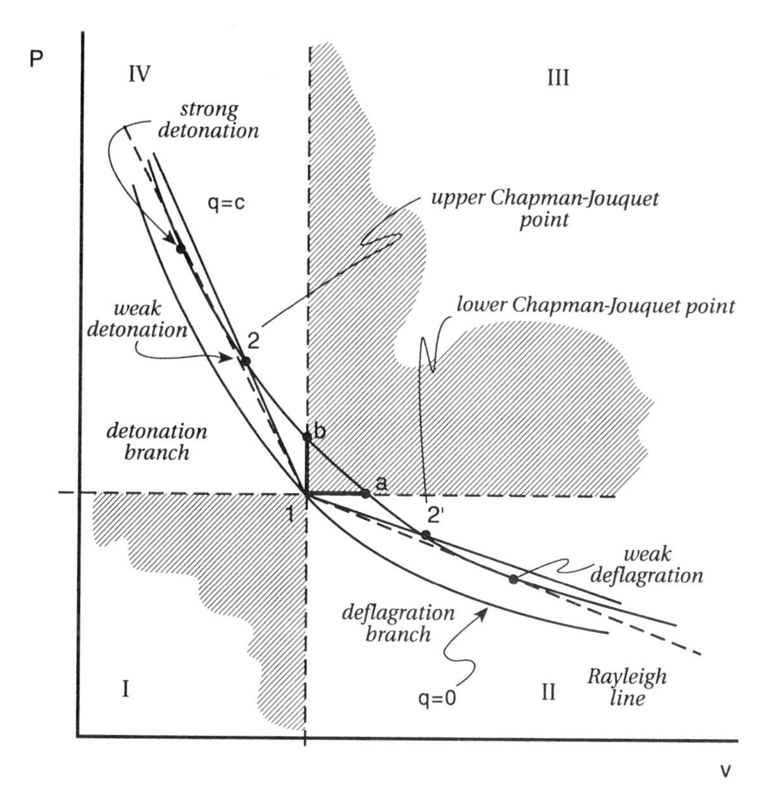

Figure 7.5 Rayleigh–Hugoniot diabatic flow analysis.

a Rayleigh line passing through the initial state P_1, v_1 and intersecting a particular Hugoniot curve in regions II or IV for a specific q.

Several characteristics of the upper, or detonation, branch can be inferred from an inspection of Figure 7.5. Recall that the burned-gas temperature and pressure T_2 and P_2 must be greater than the initial values T_1 and P_1. In addition, the specific volume at state 2 will be compressed from its original value. A Rayleigh line having the smallest slope yet still intersecting a particular Hugoniot curve is seen to be a line of tangency. This unique solution is defined as the upper Chapman–Jouquet point and can be seen to be the solution with the minimum velocity at 2. Example 7.4 shows that the velocity of the gases at the Chapman–Jouquet points *relative* to the reactive wave front is at sonic speeds. Additional solutions obtained from intersections above the Chapman–Jouquet point are supersonic at state 2, with points at pressures greater than the Chapman–Jouquet point classified as strong detonations and those below called weak detonations.

EXAMPLE 7.4 The velocity V_2 of the burned gases at the Chapman–Jouquet points must satisfy both the Rayleigh line and the Hugoniot curve. For this condition, show that the velocity V_2 is equal to the sonic velocity c_2.

Solution:

1. Rayleigh line:

$$\frac{dP_2}{dv_2} = \frac{P_2 - P_1}{v_2 - v_1} = \frac{-V_2^2}{g_0 v_2^2}$$

2. Rankine–Hugoniot relationship:

$$q = (h_2 - h_1) - \tfrac{1}{2}(P_2 - P_1)(v_2 + v_1)$$

differentiating for fixed initial conditions at 1,

$$dq = 0 = d\{(h_2 - h_1) - \tfrac{1}{2}(P_2 - P_1)(v_2 + v_1)\}$$
$$dh_2 = \tfrac{1}{2}(v_2 + v_1)dP_2 + \tfrac{1}{2}(P_2 - P_1)dv_2$$

3. Using the Rayleigh line,

$$dv_2 = \frac{(v_2 - v_1)}{(P_2 - P_1)} dP_2$$

and substituting into the Hugoniot curve,

$$dh_2 = \tfrac{1}{2}(v_2 + v_1)dP_2 + \tfrac{1}{2}(P_2 - P_1)\frac{(v_2 - v_1)}{(P_2 - P_1)} dP_2$$

or

$$dh_2 = v_2\, dP_2$$

4. Assuming an ideal gas with constant specific heats,

$$dh_2 = C_P\, dT_2 = \frac{\gamma R}{(\gamma - 1)} dT_2$$

and

$$P_2 v_2 = RT_2$$
$$d(P_2 v_2) = R\, dT_2$$

5. Combining parts 3 and 4,

$$\gamma R \, dT_2 = (\gamma - 1) \, v_2 \, dP_2$$
$$\gamma[P_2 \, dv_2 + v_2 \, dP_2] = (\gamma - 1) \, v_2 \, dP_2$$
$$\gamma P_2 \, dv_2 = -v_2 \, dP_2$$

and

$$\frac{dP_2}{dv_2} = -\gamma \frac{P_2}{v_2}$$

6. From parts 1 and 5,

$$\frac{dP_2}{dv_2} = -\gamma \frac{P_2}{v_2} = -\frac{V_2^2}{g_0 v_2^2}$$

$$V_2^2 = \gamma P_2 v_2 g_0 = \gamma R T_2 g_0$$

$$V_2 = c_2$$

Deflagration wave characteristics along the lower Hugoniot branch can also be deduced from consideration of Figure 7.5. Again, burned-gas temperature T_2' will be greater than the initial gas temperature T_1; but the final pressure P_2' is less than the initial pressure P_1. The gas expands from the original specific volume v_1 to its final value v_2'. An inspection of Figure 7.5 reveals that the lower Chapman–Jouquet point is defined by the tangent Rayleigh line of smallest slope. Since the velocity at the Chapman–Jouquet points is sonic (see Example 7.4), all other deflagrations obtained from the intersections of Rayleigh and Hugoniot curves will be subsonic. Points of intersection for P_2' greater than the Chapman–Jouquet pressure are referred to as weak deflagrations, whereas values of P_2' less than the Chapman–Jouquet value are called strong deflagrations.

7.6 CALCULATION OF CHAPMAN-JOUQUET NORMAL DETONATION PARAMETERS

In Section 7.5, it was shown that the equations for mass, momentum, and energy for one-dimensional, steady compressible flow were satisfied by two different reactive wave propagations, i.e., detonations and deflagrations.

Chapters 2 and 3 developed methods for predicting the energy released by various chemical energy conversion processes. By combining the material in Chapters 2 and 3 with results in Section 7.5, a simplified method for estimating upper Chapman–Jouquet properties can be obtained. The following additional assumptions will be made:

1. Nonreactive ideal-gas mixtures at states 1 and 2.
2. Conditions at state 1 are those for a homogeneous mixture of fuel and air.
3. Conditions at state 2 are based on mixture properties for the products of combustion.
4. Diabatic heat transfer is approximated by the change in enthalpy of formation between the reactants and products.
5. $P_2 >> P_1$.

The conservation of energy equation (7.12) can be used to obtain an expression for burned-gas temperature T_2 as

$$h_2 - h_1 + \frac{V_2^2}{2g_0} - \frac{V_1^2}{2g_0} = q \tag{7.12}$$

$$C_{P_2}T_2 = C_{P_1}T_1 + q + \frac{V_1^2}{2g_0} - \frac{c_2^2}{2g_0} \tag{7.21a}$$

or

$$C_{P_2}T_2 = C_{P_1}T_1 + q + \frac{V_1^2}{2g_0} - \frac{\gamma_2 R_2 T_2}{2} \tag{7.21b}$$

Using the conservation of momentum equation (7.6b), V_1 is given by

$$\rho_1 V_1^2 = g_0(P_2 - P_1) + \rho_2 V_2^2 \tag{7.6b}$$

$$\rho_1 V_1^2 = g_0 P_2 \left[1 - \frac{P_1}{P_2} \right] + \rho_2 V_2^2$$

and, assuming $P_2 >> P_1$, the expression for the initial velocity becomes

$$V_1^2 = g_0 \frac{P_2}{\rho_1} + \frac{\rho_2}{\rho_1} V_2^2 \tag{7.22}$$

Substituting Equation (7.22) into Equation (7.21) yields an expression for the burned-gas temperature T_2 solely in terms of thermodynamic properties at states 1 and 2 as

$$C_{P_2}T_2 = C_{P_1}T_1 + q - \frac{\gamma_2 R_2 T_2}{2} + \frac{1}{2g_0}\left[g_0 \frac{P_2}{\rho_1} + \frac{\rho_2}{\rho_1} V_2^2 \right]$$

$$C_{P_2}T_2 = C_{P_1}T_1 + q - \frac{\gamma_2 R_2 T_2}{2} + \frac{1}{2}[R_2T_2 + \gamma_2 R_2 T_2]\frac{\rho_2}{\rho_1} \tag{7.23}$$

Now, again using the momentum equation, one can write for $P_1 << P_2$

$$g_0 P_2 \left[1 - \frac{P_1}{P_2} \right] = \rho_1 V_1^2 - \rho_2 V_2^2$$

$$\frac{\rho_1 V_1^2}{\rho_2 V_2^2} = \frac{g_0 P_2}{\rho_2 V_2^2} + 1 \tag{7.24}$$

where utilizing the conservation of mass, Equation (7.24) then becomes

$$\left(\frac{\rho_1}{\rho_2}\right)\left(\frac{\rho_2}{\rho_1}\right)^2 \left(\frac{V_2^2}{V_2^2}\right) = \frac{g_0 P_2}{\rho_2 V_2^2} + 1$$

$$\frac{\rho_2}{\rho_1} = \left[\frac{g_0 P_2}{\rho_2 V_2^2} + 1 \right]$$

Since the velocity of the upper Chapman–Jouquet point is sonic, the above expression reduces to

$$\frac{\rho_2}{\rho_1} = \left[\frac{1}{\gamma_2} + 1 \right] = \frac{1 + \gamma_2}{\gamma_2} \tag{7.25}$$

Substituting Equation (7.25) into Equation (7.22) yields

$$C_{P_2}T_2 = C_{P_1}T_1 + q - \frac{\gamma_2 R_2 T_2}{2} + \frac{R_2 T_2 \gamma_2}{2}\left[\frac{1 + \gamma_2}{\gamma_2} \right]^2$$

$$T_2 = \left(\frac{C_{P_1}}{C_{P_2}}\right) T_1 + \frac{q}{C_{P_2}} - \frac{\gamma_2 R_2 T_2}{2C_{P_2}} + \frac{\gamma_2 R_2 T_2}{2C_{P_2}}\left[\frac{1 + \gamma_2}{\gamma_2} \right]^2$$

Recall, for an ideal gas, the following relationships:

$$C_P - C_V = R \qquad \gamma = C_P/C_V$$

or

$$C_P = \frac{R\gamma}{\gamma - 1}$$

and

$$T_2 = \left(\frac{C_{P_1}}{C_{P_2}}\right) T_1 + \frac{q}{C_{P_2}} - \frac{\gamma_2 R_2 T_2(\gamma_2 - 1)}{2R_2\gamma_2}$$

$$+ \frac{(\gamma_2 R_2 T_2)(\gamma_2 - 1)}{2R_2\gamma_2} \left[\frac{1 + \gamma_2}{\gamma_2}\right]^2$$

$$T_2 = \left(\frac{C_{P_1}}{C_{P_2}}\right) T_1 + \frac{q}{C_{P_2}} + \left[\frac{\gamma_2 - 1}{2}\right] T_2 \left[\frac{1 + 2\gamma_2}{\gamma_2^2}\right]$$

$$\left[1 - \left(\frac{\gamma_2 - 1}{2}\right)\left(\frac{1 + 2\gamma_2}{\gamma_2^2}\right)\right] T_2 = \left(\frac{C_{P_1}}{C_{P_2}}\right) T_1 + \frac{q}{C_{P_2}}$$

$$T_2 = \frac{2\gamma_2^2}{\gamma_2 + 1}\left[\left(\frac{C_{P_1}}{C_{P_2}}\right) T_1 + \frac{q}{C_{P_2}}\right] \quad \text{K (°R)} \qquad (7.26)$$

Equation (7.26) can be used to estimate the detonation temperature T_2, using only thermodynamic properties at states 1 and 2. Example 7.5 illustrates the solution technique with a specific problem.

EXAMPLE 7.5 The reaction of a combustible mixture can, under certain conditions in closed vessels, be observed to produce a detonation wave. Consider a stoichiometric mixture of methane and air initially at *STP*, and calculate (a) the detonation temperature, K; (b) the detonation pressure, atm; and (c) the detonation velocity relative to the flame front, m/sec. Assume complete combustion in your calculations.

Solution:

1. Stoichiometric reaction:

$$CH_4 + 2[O_2 + 3.76\ N_2] \rightarrow CO_2 + 2H_2O + (2)(3.76)N_2$$

2. Initial state for $T_1 = 298\text{K}$, $P_1 = 1$ atm:

$$\bar{x}_{CH_4} = \frac{1}{1 + (2)(4.76)} = 0.095$$

$$\bar{x}_{O_2} = \frac{2}{10.520} = 0.190$$

$$\bar{x}_{N_2} = 1.000 - 0.095 - 0.190 = 0.715$$

$$MW_1 = (0.095)(16) + (0.190)(32) + (.715)(28)$$

$$= 27.62 \text{ g/g mole}$$

From Appendix B,

$$\overline{C}_{P_1} = (0.095)(8.518) + (0.190)(7.020) + (0.715)(6.961)$$
$$= 7.120 \text{ cal/g mole K}$$

$$\gamma_1 = \frac{\overline{C}_{P_1}}{\overline{C}_{P_1} - \overline{R}} = \frac{7.120}{7.120 - 1.987} = 1.387$$

3. The heat addition can be approximated as the difference in enthalpies of formation of the initial and final mixtures.

$$|{}_1Q_2| \cong |\sum_2 N_i \bar{h}^0_{f_i}| - |\sum_1 N_j \bar{h}^0_{f_j}|$$
$$= (1)(94{,}054) + 2(57{,}798) + (2)(3.76)(0) - (1)(17{,}895)$$
$$= 191{,}755 \text{ cal}$$

or

$${}_1\bar{q}_2 = \frac{191{,}755 \text{ cal}}{3 + (2)(3.76)} = 18{,}228 \text{ cal/g mole product}$$

4. Energy balance on a molar basis:

$$T_2 = \frac{2\gamma_2^2}{\gamma_2 + 1}\left\{\left[\frac{\overline{C}_P}{MW}\right]_1 \left[\frac{MW}{\overline{C}_P}\right]_2 T_1 + \frac{{}_1\bar{q}_2}{\overline{C}_{P_2}}\right\}$$

an iterative technique is necessary since the conditions at state 2 are unknown.

5. Final state, assuming $T_2 = 3200\text{K}$:

$$\bar{x}_{CO_2} = \frac{1}{3 + (2)(3.76)} = 0.095$$

$$\bar{x}_{H_2O} = \frac{2}{10.520} = 0.190$$

$$\bar{x}_{N_2} = 1.0 - 0.095 - 0.190 = 0.715$$

$$MW_2 = (0.095)(44) + (0.190)(18) + (0.715)(28)$$
$$= 27.62 \text{ g/g mole}$$

$$\overline{C}_{P_2} = (0.095)(14.930) + (0.190)(13.441) + (0.715)(8.886)$$
$$= 10.326 \text{ cal/g mole K}$$

$$\gamma_2 = \frac{10.326}{10.326 - 1.987} = 1.238$$

$$T_2 = \frac{(2)(1.238)^2}{(2.238)} \left\{ \left[\frac{7.120}{27.62} \right] \left[\frac{27.62}{10.326} \right] (298) + \frac{18{,}228}{10.326} \right\}$$

$T_2 = 2838.5$ (too high a guess)

6. Final state, assuming that $T_2 = 2800\text{K}$:

$$\overline{C}_{P_2} = (0.095)(14.807) + (0.190)(13.146) + (0.715)(8.820)$$
$$= 10.211$$

$$\gamma_2 = \frac{10.211}{10.211 - 1.987} = 1.242$$

$$T_2 = \frac{(2)(1.242)^2}{(2.242)} \left\{ \left[\frac{7.120}{27.62} \right] \left[\frac{27.62}{10.211} \right] (298) + \frac{18{,}228}{10.211} \right\}$$

$$T_2 = 2742.4\text{K}$$

7. Assume that $T_2 = 2700\text{K}$.

$$\overline{C}_{P_2} = (0.095)(14.771) + (0.190)(13.059) + (0.715)(8.800)$$
$$= 10.176$$

$$\gamma_2 = \frac{10.176}{10.176 - 1.987} = 1.243$$

$$T_2 = \frac{(2)(1.243)^2}{(2.243)} \left\{ \left[\frac{7.120}{27.62} \right] \left[\frac{27.62}{10.176} \right] (298) + \frac{18{,}228}{10.176} \right\}$$
$$= 2755\text{K}$$

8. Assume that $T_2 = 2750\text{K}$.

$$\overline{C}_{P_2} = (0.095)(14.789) + (0.190)(13.103) + (0.715)(8.810)$$
$$= 10.194$$

$$\gamma_2 = \frac{10.194}{10.194 - 1.987} = 1.242$$

$$T_2 = \frac{(2)(1.242)^2}{(2.242)} \left\{ \left[\frac{7.120}{27.62} \right] \left[\frac{27.62}{10.194} \right] (298) + \frac{18{,}228}{10.194} \right\}$$
$$= 2747\text{K}$$

a. $T \cong 2750\text{K}$

9. Relative velocity at state 2:

$$V_2 = c_2 = \sqrt{\gamma_2 R_2 T_2 g_0}$$

$$= \left[\frac{(1.242)(8314 \text{ N m/kg mole sec}^2 \text{ K})(2750 \text{ K})(1.0 \text{ kg m/N sec}^2)}{(27.62 \text{ kg/kg mole})} \right]^{1/2}$$

$$V_2 = 1{,}014 \text{ m/sec}$$

10. Conditions at state 2:

$$\frac{\rho_2}{\rho_1} = \frac{\gamma_2 + 1}{\gamma_2} = \frac{1.242 + 1}{1.242} = 1.805$$

$$\frac{P_2}{P_1} = \frac{\rho_2 R_2 T_2}{\rho_1 R_1 T_1} = \frac{\rho_2 MW_1 T_2}{\rho_1 MW_2 T_1}$$

$$= (1.805)\left(\frac{27.62}{27.62}\right)\left(\frac{2750}{298}\right)$$

$$= 16.66$$

and

b. $P_2 = 16.66\ P_1 = 16.66$ atm

11. Detonation velocity, V_1

c. $V_1 = (\rho_2/\rho_1)V_2 = (1.805)(1014 \text{ m/sec}) = 1830 \text{ m/sec}$

7.7 DETONATION THEORY AND EXPERIMENTAL EVIDENCES

Experimental measurements have verified the existence of both detonation and deflagration wave propagations. The potential of such phenomena will influence the design, fabrication, and operation of a variety of internal and external combustion devices.

Formation of detonation waves has been experimentally observed in various closed-vessel geometries and has involved a variety of fuel-air mixtures, including coal-dust mine explosions, natural gas leak explosions in buildings, and combustion knock in internal combustion engines. The one-dimensional Chapman–Jouquet detonation model, although somewhat an oversimplification of the actual process, does give results that are in reasonable agreement with engineering measurements. Table 7.2 lists the order of magnitude of burned to unburned properties for a normal detonation wave, while Table 7.3 gives detonation parameters for several premixed fuel-air mixtures.

Strong detonation can occur in rare instances but is usually produced when a very strong shock wave passes through a homogeneous reactive mixture. Weak detonations also are very seldom found to occur but can be generated experimentally using special gas mixtures.

EXAMPLE 7.6 Dissociation of the combustion products can become significant for temperatures above 1250K. This phenomenon can affect the detonation conditions calculated for different end gas conditions. Consider a stoichiometric H_2-O_2 mixture initially at *STP* and calculate the detonation temperature, K, assuming complete combustion. Repeat the calculations, assuming that the products consist of H_2, O_2, and H_2O.

Solution:

1. Stoichiometric equation:

$$H_2 + \tfrac{1}{2}O_2 \rightarrow H_2O$$

2. Initial state for $T_1 = 298\text{K}$:

$$\bar{x}_{H_2} = \frac{1.0}{1.5} = 0.667 \qquad \bar{x}_{O_2} = 0.333$$

$$MW_1 = (0.667)(2) + (0.333)(32) = 12.000 \text{ g/g mole}$$

From Appendix B,

$$\overline{C}_{P_1} = (0.667)(6.892) + (0.333)(7.020) = 6.935 \text{ cal/g mole K}$$

$$\gamma_1 = \frac{\overline{C}_{P_1}}{\overline{C}_{P_1} - \overline{R}} = \frac{6.935}{6.935 - 1.987} = 1.402$$

3. Heat addition can be approximated as the difference in enthalpies of formation of the products and reactants.

$$|Q| \cong \sum_2 N_i \bar{h}^0_{f_i} - \sum_1 N_j \bar{h}^0_{f_j}$$

$$|Q| = (1 \text{ g mole})(57{,}798 \text{ cal/g mole}) - (0.5)(0) - (1)(0)$$
$$= 57{,}798$$

and

$$|\bar{q}| = \frac{57{,}798}{1.0} = 57{,}798 \text{ cal/g mole product}$$

4. Energy balance:

$$T_2 = \frac{2\gamma_2^2}{\gamma_2 + 1}\left\{\left[\frac{\overline{C}_P}{MW}\right]_1 \left[\frac{MW}{\overline{C}_P}\right]_2 T_1 + \frac{\bar{q}}{\overline{C}_{P_2}}\right\}$$

Since the condition at item 2 is unknown an iterative technique is necessary.

5. Final state:

$$\bar{x}_{H_2O} = 1.0$$
$$MW_2 = 18.0$$
$$\overline{C}_{P_2} = \overline{C}_P <T_2>_{H_2O}$$
$$\gamma_2 = \frac{\overline{C}_{P_2}}{\overline{C}_{P_2} - \overline{R}}$$

$(T_2)_{\text{assumed}}$	$\overline{C}_{P_2}$	γ_2	$(T_2)_{\text{calculated}}$
4500	14.030	1.165	5442
5000	14.174	1.163	5373
5300	14.254	1.162	5336

NOTE: $T_2 \sim 5300$K.

6. Product dissociation:

$$H_2 + ½O_2 \rightarrow aH_2 + bO_2 + cH_2O$$

Hydrogen atom balance:

$$a + c = 1$$

Oxygen atom balance:

$$2b + c = 1$$

or

$$b = \frac{a}{2} \qquad c = 1 - a$$

7. Equilibrium constant

$$H_2 + ½O_2 \rightleftarrows H_2O$$

$$K_p = \frac{[P/P_0]_{H_2O}}{[P/P_0]_{O_2}^{½}[P/P_0]_{H_2}}$$

where $P_0 \equiv 1$ atm

$$K_p = \frac{P_{H_2O}}{P_{O_2}^{½} P_{H_2}}$$
$$= \frac{\bar{x}_{H_2O}}{\bar{x}_{O_2}^{½} \bar{x}_{H_2}} (P_{\text{tot}})^{-½}$$

$$= \left[\frac{(1-a)}{\left(\frac{a}{2}\right)^{1/2}(a)} \right] \left[a + \frac{a}{2} + 1 - a\right]^{1/2} P_{tot}^{-1/2}$$

$$K_p = \frac{(1-a)\ \left(\frac{a}{2}+1\right)^{1/2}}{\left(\frac{a}{2}\right)^{1/2}(a)} (P_{tot})^{-1/2}$$

assume that $T_2 = 4000\text{K}$ and $P_2 = 18$ atm.

From Appendix B

$$\log K_p = 0.238$$
$$K_p = 1.7298$$

or

$$(1.7298)(18)^{1/2} = \frac{(1-a)\left(\frac{a}{2}+1\right)^{1/2}}{(a/2)^{1/2}(a)}$$

using numerical techniques yields:

$$a = 0.280$$
$$b = 0.140$$
$$c = 0.720$$

8. Final state for $T_2 = 4000\text{K}$:

$$\bar{x}_{H_2} = \frac{a}{a+b+c} = \frac{0.280}{0.280 + 0.140 + 0.720} = 0.2456$$

$$\bar{x}_{O_2} = \frac{b}{a+b+c} = \frac{0.140}{1.140} = 0.1228$$

$$\bar{x}_{H_2O} = \frac{c}{a+b+c} = \frac{0.720}{1.140} = 0.6316$$

$$MW_2 = (0.2456)(2) + (0.1228)(32) + (0.6316)(18) = 15.79$$
$$\overline{C}_{P_2} = (0.2456)(9.342) + (0.1228)(9.932) + (0.6316)(13.850)$$
$$= 12.262$$

$$\gamma_2 = \frac{12.262}{12.262 - 1.987} = 1.193$$

$$|Q| = \Sigma N_i \bar{h}^0_{f_i} - \Sigma N_j \bar{h}^0_{f_j}$$
$$= (0.280)(0) + (0.140)(0) + (0.720)(57.798)$$
$$- (1.0)(0) - (0.5)(0) = 41{,}615 \text{ cal}$$

and

$$\bar{q} = \frac{41{,}615}{1.140} = 36{,}504 \text{ cal/g mole product}$$

$$T_2 = \frac{(2)(1.193)^2}{(2.193)} \left\{ \frac{(6.935)(15.79)}{(12.0)(12.262)} (298) + \frac{36{,}504}{12.262} \right\}$$

$$= 4152 \quad \text{or}$$

$$T_2 \simeq 4100\text{K}$$

Table 7.2 Orders of Magnitude for Detonation/Deflagration Parameters

Parameter	Detonation	Deflagration
u_2/c_1	5–10	0.0001–0.03
u_2/u	0.4–0.7	4–16
P_2/P_1	13–55	0.98–0.976
ρ_2/ρ_1	1.4–2.6	0.06–0.25
T_2/T_1	8–21	4–16

Source: Friedman, R, American Rocket Soc. J., Vol. 23, p. 349 (1953)

Table 7.3 Stoichiometric Detonation Velocities for Gaseous Mixtures

Mixture	P_2 atm	T_2 K	u_1, m/sec Calculated	u_1, m/sec Measured
$(2H_2 + O_2)$	18.05	3583	2806	2819
$(2H_2 + O_2) + 5O_2$	14.13	2620	1732	1700
$(2H_2 + O_2) + 5N_2$	14.39	2685	1850	1822
$(2H_2 + O_2) + 5H_2$	15.97	2975	3627	3527
$(2H_2 + O_2) + 5He$	16.32	3097	3617	3160
$(2H_2 + O_2) + 5Ar$	16.32	3097	1762	1700
$CO + O_2$	18.6	3500	—	(1790)
$C_2H_2 + O_2$	44	4200	—	(2400)
C_2H_2 + air	19	3100	—	(1900)
CH_4 + air	17.2	2736	—	(1800)
C_3H_8 + air	18.3	2823	—	(1800)

NOTE: P_1 = 1 atm; T_1 = 291K.

Source: Lewis, B and Von Elbe, G., Combustion, Flames, and Explosions in Gases, 2nd Edition, Academic Press, NY (1961)

The precise structure of a normal detonation wave has been molded by Zeldovich (1950), van Neumann (1942), and Doring (1943); see Figure 7.6. A ZND detonation wave is produced by a moving shock wave traveling at

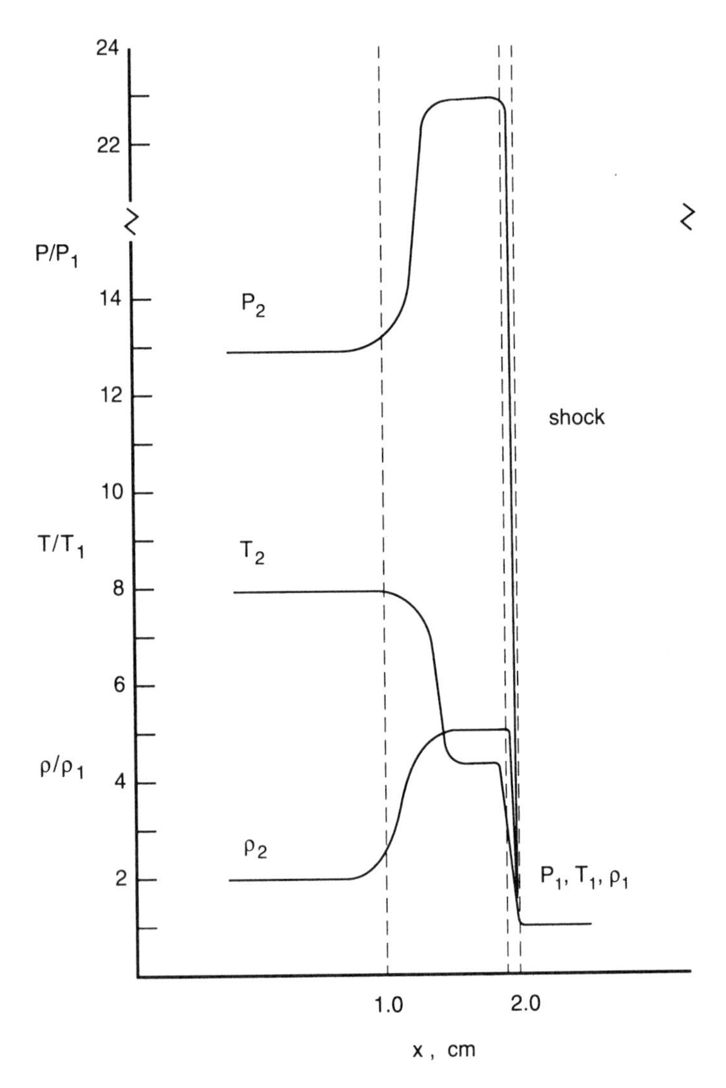

Figure 7.6 ZND detonation wave structure.

Table 7.4 Detonation Limits for Gaseous Mixtures

Mixture	% Fuel in mixture	
	Lean	Rich
$H_2 - O_2$	15	90
H_2 – air	18	59
$CO - O_2$	38	90
$NH_3 - O_2$	25	75
$C_3H_8 - O_2$	3	37

Source: Lewis, B. and Von Elbe, G., Combustion, Flames, and Explosions in Gases, 2nd Edition, Academic Press, NY (1961)

the detonation velocity. Chemical reaction is initiated behind the shock front in the compression-heated mixture. Temperature continues to rise behind the moving wave front as a result of chemical reactions. Behind the short transition region, the properties P, T, and ρ relax to conditions of chemical equilibrium, as predicted by the Chapman–Jouquet analysis.

Heat released by combustion is strongly influenced by the mixture air-fuel ratio. Detonation wave parameters should thus be influenced by the stoichiometry of the fuel-air mixture. Detonation limits are experimentally obtained using closed tubes with ignition of the combustible mixture at one end. Table 7.4 gives the approximate detonation limits for several reactive mixtures.

Propagation of a low-pressure subsonic reaction zone is common knowledge. Deflagrations, called flames, are observed in acetylene torch and spark-ignition internal combustion engines, as well as in gas turbine combustion cans. Results predicted using the one-dimensional deflagration model are not in good agreement with measured flame properties since the simplified model developed in this chapter is inadequate and this case will be considered again in Chapter 8.

PROBLEMS

7.1 An ideal-gas mixture consists of methane, CH_4, and 120% theoretical air at *STP*. For these conditions, calculate (a) constant-pressure mixture molar specific heat, kJ/K kgmole; (b) constant-volume mixture molar specific heat, kJ/kg mole K; (c) mixture specific heat ratio at *STP*, and (d) sonic velocity at *STP*, m/sec.

7.2 A mixture of hydrogen and air at a temperature of 1000K has a corresponding sonic velocity of 1200 m/sec. Determine (a) equivalence ratio for the mixture; (b) mixture molecular weight, kg/kg mole; (c) mixture specific heat ratio; and (d) mixture molar specific heats, kJ/kg mole K.

7.3 Consider a one-dimensional steady-state and steady-sonic flow. Show that the speed of sound for these conditions can be related to the stagnation temperature by the relationship

$$c = \left[\frac{2\gamma R T_0 g_0}{\gamma + 1} \right]^{1/2}$$

7.4 A mixture of methane and air at *STP* has a specific heat ratio of 1.35. For these conditions, find (a) constant-pressure mixture molar specific, Btu/lb mole R; (b) reactant mixture mass fraction, (c) mixture sonic velocity at *STP*, ft/sec; and (d) mixture mass flux lbm/sec ft^2.

7.5 Show that, for Rayleigh flow,

$$\frac{V_2}{V_1} = \frac{(1 + \gamma N_{m1}^2)(N_{m2}^2)}{(1 + \gamma N_{m2}^2)(N_{m1}^2)}$$

7.6 The Rankine–Hugoniot relationship can be expressed in terms of the local stagnation conditions for the gas at states 1 and 2. Show that, in this case, the Hugoniot curve is expressed as

$$\frac{q}{C_p T_1} = \left[1 + \left(\frac{\gamma - 1}{2} \right) N_{m_1}^2 \right] \left[\frac{T_{02}}{T_{01}} \right]$$

7.7 Show that the ratio of the density of burned gases, ρ_2, to that of the unburned gases, ρ_1, is given as

$$\frac{\rho_2}{\rho_1} = \frac{\gamma_2 + 1}{\gamma_1}$$

7.8 The solution of the one-dimensional shock wave problem is given by the intersection of the Rayleigh line and the Rankine–Hugoniot curve for $q = 0$. Show that (a) stagnation states for points 1 and 2 are equal for the adiabatic case, i.e., $T_{0_1} = T_{0_2}$; (b) the temperature ratio across the shock wave is

$$\frac{T_2}{T_1} = \frac{1 + \left[\frac{(\gamma - 1)}{2} \right] \left(N_{m_1}^2 \right)}{1 + \left[\frac{(\gamma - 1)}{2} \right] \left(N_{m_1}^2 \right)}$$

and (c) the velocity ratio is given by

$$\frac{V_2}{V_1} = \frac{N_{m_2}}{N_{m_1}} \left[\frac{1 + \left[\frac{(\gamma - 1)}{2} \right] \left(N_{m_1}^2 \right)}{1 + \left[\frac{(\gamma - 1)}{2} \right] \left(N_{m_1}^2 \right)} \right]^{1/2}$$

7.9 The absolute velocity of the burned gases behind a detonation wave with respect to a stationary observer is written as

$$V_{prod} = V_1 - V_2$$

(a) Show that this velocity is given by the relation

$$V_{prod} = [\,1 - (\rho_1/\rho_2)\,]\,V_1$$

(b) For a detonation, show that the above expression indicates that gas products travel behind the moving wave front but at a lower speed. (c) Show that, for a deflagration, the particular speed of the products travels in the opposite direction from the wave front. (d) Show that the absolute burned gas speed is also equal to

$$V_{prod} = [(v_1 - v_2)(P_2 - P_1)]^{1/2}$$

7.10 Show that, for an ideal-gas mixture, the Hugoniot relationship can be expressed as

$$q = u_2 - u_1 - (1/2)(P_2 - P_1)(v_1 - v_2)$$

7.11 The Rankine–Hugoniot curve for a fixed heat release is given as

$$q = h_2 - h_1 - (1/2)(P_2 - P_1)(v_2 + v_1)$$

Show that, for adiabatic flow, $P_1 v_1^{\gamma} = P_2 v_2^{\gamma} =$ const.

7.12 A combustible mixture is initially at rest in a duct of constant cross-sectional area at a temperature of 500°R. A Chapman–Jouquet detonation wave propagates through the mixture, with the heat liberated by reaction equal to 960 Btu/lbm. Calculate the propagation Mach number of the detonation wave.

7.13 A Chapman–Jouquet deflagration is propagated through a combustible gaseous mixture in a duct of constant cross-sectional area. The heat release is equal to 480 Btu/lbm. The Mach number and flow velocity relative to the walls are 0.8 and 800 ft/sec in the unburned gas. Assuming that γ is 7/5 for both burned and unburned gases, estimate (a) the velocity of the flame relative to the walls, ft/sec; and (b) the velocity of the burned gas relative to the walls, ft/sec.

7.14 A mixture of H_2, O_2, and N_2 initially at rest are at 1 atm and 25°C. The heat release for the mixture is given by the reaction

$$2H_2 + O_2 + 5N_2 \rightarrow 2H_2O + 5N_2$$

Determine (a) the detonation temperature, K; (b) the detonation pressure, atm; and (c) the detonation velocity relative to the flame, m/sec.

7.15 A mixture of H_2 and O_2 at *STP* has a constant-volume molar specific heat of 5.013 cal/K g mole. For these conditions, calculate (a) the reactant mixture mole fractions; (b) the mixture sonic velocity at *STP*, m/sec; (c) the mixture mass flux G at *STP* kg/sec m^2; (d) the detonation temperature, K; and (e) the detonation velocity, m/sec.

7.16 Ignition of homogeneous coal dust–air mixtures confined within an enclosed space, such as mines or coal-fired steam plants, can produce explosive reactions. Consider a carbon dust–air mixture having an 0.8 equivalence ratio initially at 101 kPa and 25°C. For these conditions, calculate (a) the ideal detonation temperature, K; (b) the ideal detonation pressure, kPa; (c) the burned gas density, kg/m^3; (d) the detonation wave velocity, m/sec; and (e) the burned gas velocity, m/sec.

7.17 A detonation occurs in a stoichiometric mixture of propane and air initially at 1 atm and 25°C. The steady detonation wave speed relative to a fixed observer is 1825 m/sec, and the product gas temperature is 2800K. Assuming ideal complete combustion, calculate (a) product specific heat ratio; (b) product velocity relative to the detonation wave, m/sec; (c) the product mass flux, kg/sec m^2; and (d) detonation pressure, atm.

7.18 Gasoline can audibly knock, or detonate when burned in many spark-ignition engines. Consider that the reactants at TDC are initially at 260 psia and 1080°R. Use one-dimensional detonation theory, and assume that gasoline can be represented by normal octane C_8H_{18}. For ideal complete combustion of a stoichiometric mixture, estimate (a) the detonation pressure, psia; (b) the detonation temperature, °R; (c) the burned gas density, lbm/ft^3; (d) the detonation wave velocity, ft/sec; and (e) the burned gas velocity, ft/sec.

7.19 Organic dust inside confined spaces and suspended in air, such as grain elevators and silos, can produce explosive reactions if accidentally ignited. Consider a stoichiometric cellulose-air mixture as a model for the dust. For these conditions, calculate (a) the ideal detonation temperature, °R; (b) the ideal detonation pressure, atm; (c) the burned-gas density, lbm/ft^3; (d) the detonation wave velocity, ft/sec; and (e) the burned-gas velocity, ft/sec.

8

Chemical Kinetics

8.1 INTRODUCTION

Heat- and/or work-producing chemical energy conversion processes have been described previously using overall chemical reaction mechanisms and time-independent energy arguments alone. Actually, combustion is a result of dynamic, or time-dependent, events that occur on a molecular level among atoms, molecules, radicals, and solid boundaries. Many important fuel-engine interface characteristics, therefore, cannot be understood without first introducing additional time-dependent principles of physical chemistry. The dynamic study of molecular chemistry, or *chemical kinetics,* covers subjects including kinetic theory of gases, statistical thermodynamics, quantum chemistry, and elementary reactions and reaction rate theory. In this chapter, a general but simplified overview of this area of science will be presented in order to help the combustion engineer better understand kinetic factors that play a major role in efficiently burning any fuel for the purpose of producing heat and/or power while minimizing the formation of incomplete combustion products.

8.2 KINETIC THEORY OF GASES

A simple model for the microscopic world of matter in the gas phase developed using kinetic theory assumes that:

Matter exists as discrete particles or *molecules*.
Molecules can be treated ideally as small spheres of diameter σ.
Mean distance between molecules $>> \sigma$.
Molecules are in continuous three-dimensional motion.
Each molecule moves in a random direction through space at a different speed.
No appreciable interatomic forces exist between gas molecules except when they collide (zero attractive and infinite repulsive force at contact).
Speed and directional characteristics for any molecule will remain constant until it interacts with another particle or a solid boundary.

Kinetic theory of gases provides both scientists and engineers with useful molecular descriptions of important ideal-gas mixture properties, including density, pressure, temperature, and internal energy. At equilibrium, any overall mixture property of an isolated inert gas mixture, i.e., P, T, $\bar{u}$, $\bar{h}$, or $\bar{\rho}$, will remain constant, even though the kinetic characteristics of all molecules will be continuously changing. Density, or mass per unit volume, is simply total molecular mass associated with molecules contained within a given space divided by that same volume. Continuum and molecular calculations may yield quite different values for density, however, if the number of molecules and/or the volume is quite small, an important physical condition that can occur in vacuum environments or in the earth's upper atmosphere.

Consider a differential cube $dx\ dy\ dz$, shown in Figure 8.1, having characteristic dimensions greater than molecular diameter σ and containing N_A molecules of inert gas A described by a general velocity $\vec{V}$

$$\vec{V}<x,y,z> = X\vec{i} + Y\vec{i} + Z\vec{k}$$

Since the direction and speed of any particular molecule will be changing with time, all molecules do not move at the same velocity $\vec{V}$. It is more appropriate then to describe molecular motion in terms of a statistical speed distribution function Ψ, developed by Maxwell in 1860 and given as

$$\Psi<V> = 4\pi\left[\frac{m}{2\pi kT}\right]^{3/2} V^2 \exp\left\{\frac{-mV^2}{2kT}\right\} \tag{8.1}$$

where

m = molecular mass of A, $m = MW/N_0$, g mole/molecule
V = molecular speed, cm/sec
N_0 = Avogadro's constant = 6.02×10^{23} molecules/mole
k = Boltzmann's constant = 1.3804×10^{-16} g cm^2/sec^2 K

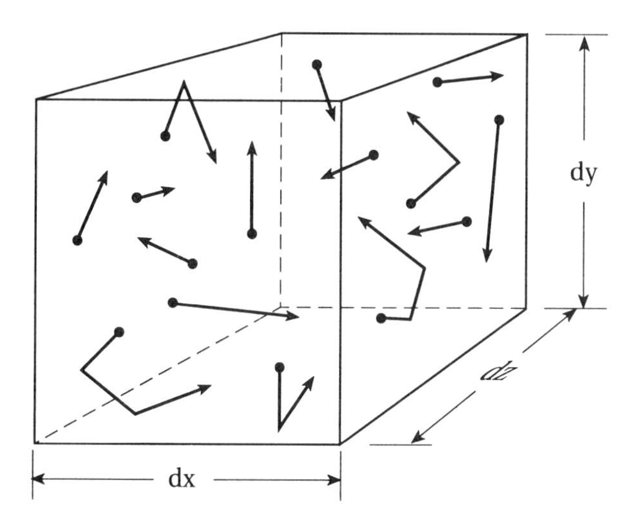

Figure 8.1 Kinetic motion of inert gas molecules within differential molecular cube.

The speed distribution function Ψ yields distributed molecular speeds that are functions of molecular mass m, absolute temperature T, and Boltzmann's constant k , i.e., universal gas constant per molecule. The most probable speed, the velocity magnitude associated with the maximum of the Maxwell distribution function, is found to equal

$$V_{mp} = \left[\frac{2\ T}{m} \right]^{1/2} \quad \text{cm/sec} \tag{8.2}$$

while mean speed for *all* molecules is given as

$$V_m = \left[\frac{8\ T}{\pi m} \right]^{1/2} \quad \text{cm/sec} \tag{8.3}$$

and mean square speed is found to be equal to

$$\overline{V^2}_m = \left[\frac{3\ T}{m} \right]^{1/2} \quad \text{cm/sec} \tag{8.4}$$

Consider a single A molecule moving through the differential volume at a mean speed V_m. The volume swept out per unit time by this molecule, i.e., base × height, would equal

$$\text{volume} = \pi\sigma^2 V_m \tag{8.5}$$

$$= \pi\sigma^2 \left[\frac{8kT}{\pi m}\right]^{1/2} = \sigma^2 \left[\frac{8\pi kT}{m}\right]^{1/2} \quad \text{cc/sec} \tag{8.6}$$

The ideal number of molecular collisions of an A molecule per unit time with all other A molecules in this swept volume is then equal to the volume swept out per unit time multiplied by the number of molecules of A within the volume, or

$$\text{No. A collisions} = \sigma^2 \left[\frac{8\pi kT}{m}\right]^{1/2} \left[\frac{N_A}{dx\,dy\,dz}\right]$$

$$= \sigma^2 \left[\frac{8\pi kT}{m}\right]^{1/2} [\overline{\text{A}}] \quad \text{collisions/sec} \tag{8.7}$$

where $[\overline{\text{A}}]$ is the molar concentration of A or

$$[\overline{\text{A}}] = \left[\frac{N_A}{dx\,dy\,dz}\right] \quad \text{molecules/cc}$$

The collision frequency Z, or total frequency of collisions between all molecules of A, then equals the number of A molecules per unit volume mutliplied by the rate of collisions per molecule of A.

$$Z_{AA} = [\overline{\text{A}}][\overline{\text{A}}]\sigma^2 \left[\frac{8\pi\ \ T}{m}\right]^{1/2}$$

The *mean free path* λ, or average distance traveled by any molecule between collisions, is determined knowing the distance traveled per unit time and dividing by the number of collisions per unit time that this particle experiences or

$$\lambda = \frac{V_m}{\pi\sigma^2 V_m[\overline{\text{A}}]} = \frac{1}{\pi\sigma^2[\overline{\text{A}}]} \quad \text{cm/collisions} \tag{8.8}$$

A more precise analysis of the collisional process that accounts for movement of all molecules yields a corrected expression for mean free path as

$$\lambda = \frac{1}{\sqrt{2}\,\pi\sigma^2[\overline{\text{A}}]} \quad \text{cm/collisions} \tag{8.9}$$

The *bimolecular collision frequency Z*, or total frequency of collisions between molecules of species A and B per unit volume of gas, can be obtained from a rigorous analysis of the two-body collision process, with results found to equal

$$Z_{AB} = [\overline{A}][\overline{B}]\frac{\sigma_{AB}^2}{\delta}\left[\frac{8\pi kT}{m_{AB}^*}\right]^{1/2} \tag{8.10}$$

where

σ_{AB} = average diameter = $(\sigma_A + \sigma_B)/2$
m_{AB}^* = reduced molecular mass = $[(m_A + m_B)/m_A m_B]$
δ = symmetry factors
= 1 if A $\neq$ B
= 2 if A = B

EXAMPLE 8.1 Kinetic theory of gases predicts a molecular diameter of 4 $\times 10^{-8}$ cm for air at 0°C and 1-atm pressure. Using this information, determine the following: (a) mass of an air molecule, g/molecule; (b) molecular density, number of molecules/cc; (c) mean molecular speed, cm/sec; (d) mean free path, cm/collision; and (e) collision frequency, number of collisions/sec.

Solution:

1. Molecular mass:

$$m = \frac{MW}{N_0} = \frac{28.97 \text{ g/g mole}}{6.02 \times 10^{23} \text{ molecules/g mole}}$$

a. $m = 4.81 \times 10^{-23}$ g/molecule

2. Molecular density:

$$\rho = \frac{P}{RT} = \frac{(101{,}000 \text{ N/m}^2)\,(28.97 \text{ kg/kg mole})\,(1000 \text{ g/kg})}{(8314 \text{ m K/kg mole K})\,(273\text{K})\,(100 \text{ cm/m})^3}$$

$$= 1.29 \times 10^{-3} \text{ g/cm}^3$$

$$n_0 = \frac{\rho}{m} = \frac{1.289 \times 10^{-3} \text{ g/cm}^3}{4.81 \times 10^{-23} \text{ g/molecule}}$$

b. $n_0 = 2.68 \times 10^{19}$ molecules/cm^3

3. Mean molecular speed:

$$V_m = \frac{8kT}{\pi m}^{1/2} = \left[\frac{(8)(1.38 \times 10^{-16} \text{ cm}^2 \text{ g/sec}^2 \text{ K})(273\text{K})}{(\pi)(4.81 \times 10^{-23} \text{ g/molecule})}\right]^{1/2}$$

c. $V_m = 4.47 \times 10^4$ cm/sec

4. Mean free path:

$$\lambda = \frac{1}{\sqrt{2}\,\pi\sigma^2 n_0}$$

$$= \frac{1}{\sqrt{2}\,(\pi)\,(4 \times 10^{-8})^2(2.68 \times 10^{19}\ \text{molecules/cm}^2)}$$

d. $\lambda = 2.1 \times 10^{-5}$ cm/molecular collision

5. Collision frequency Z:

$$Z = \frac{V_m <\text{cm/sec}>}{<\text{cm/molecular collision}>}$$

$$= \frac{4.47 \times 10^4\ \text{cm/sec}}{2.1 \times 10^{-5}\ \text{cm/molecular collision}}$$

e. $Z = 2.13 \times 10^9$ molecular collisions/sec

Pressure can be interpreted kinetically as the net normal force resulting from all individual momentum exchange processes occurring at solid boundaries as a result of the molecular motion within the same volume. Note that the total kinetic energy and momentum of gas molecules within this volume will remain constant since all collisions are perfectly elastic. Consider again the motion of a single A molecule of mass m traveling through the original differential cube $dx\,dy\,dz$ with the mean velocity $\vec{V}_m$,

$$\vec{V}_m = X_m\vec{i} + Y_m\vec{j} + Z_m\vec{k} \tag{8.11}$$

The normal force exerted on the $dy\,dz$ wall by this A molecule is equal to its change in x momentum from $+MX_m$ to $-MX_m$ during each collision or

$$|F_X|_A = 2mX_m \tag{8.12}$$

Table 8.1 Molecular Diameter σ and Mean Free Path λ for Several Gases at STP

Gas	σ, m $\times 10^{-10}$	λ, m $\times 10^{-8}$
Argon	2.90	10.9
Helium	2.00	22.9
Nitrogen	3.50	7.46
Oxygen	2.95	10.5
Carbon dioxide	3.30	8.39
Ammonia	3.00	10.2

An A molecule will travel the distance dx in a time interval dx/X_m, and the time between repetitive collisions with the $dy\,dz$ face is equal to $2\,dx/X_m$. The number of collisions of an A molecule with the $dy\,dz$ face is, therefore, equal to $X_m/2\,dx$, and the change in x momentum per unit time is then the product of the number of collisions per unit time and momentum exchange per molecule or

$$\left.mV_x\right)_A = \left[\frac{X_m}{2\,dx}\right]\left[2mX_m\right] = \frac{2mX_m^{\,2}}{2\,dx}$$

$$\left.mV_x\right)_A = \frac{mX_m^{\,2}}{dx} \tag{8.13}$$

The pressure, i.e., force per unit area, due to an A molecule is equal to

$$\left.P_x\right)_A = \frac{mX_m^{\,2}}{dx\,dy\,dz} \tag{8.14}$$

and total pressure due to all A molecules is then given as

$$P_x = \frac{\Sigma mX_m^{\,2}}{dx\,dy\,dz} = \frac{1}{dx\,dy\,dz}\,\Sigma mX_x^{\,2} \tag{8.15}$$

Since pressure at equilibrium must be the same for all three faces, it follows that

$$P = P_x = P_y = P_z = \frac{1}{3\,dx\,dy\,dz}\,\Sigma m\left[X_m^{\,2} + Y_m^{\,2} + Z_m^{\,2}\right]$$

$$= \frac{1}{3V}\Sigma mV_m^{\,2}$$

and

$$PV = \frac{2}{3}\left[\frac{1}{2}\Sigma mV_m^{\,2}\right] \tag{8.16}$$

A simple kinetic model for internal energy of an inert ideal gas mixture of A molecules would equal the total sum of kinetic energy of all translating A molecules or

$$\overline{U} = \Sigma\overline{E}_{\text{trans}} = \frac{1}{2}\Sigma mV_m^{\,2} \tag{8.17}$$

The ideal-gas law found in Chapter 1 was written as

$$PV = N\overline{R}T \tag{1.4}$$

Combining Equations (8.16) and (1.4) yields

$$\frac{2}{3}\left[\frac{1}{2}\Sigma mV_m^{\ 2}\right]=N_A\overline{R}T$$

or

$$T=\frac{2}{3}\left(\frac{1}{N_A\overline{R}}\right)\Sigma mV_m^{\ 2} \tag{8.18}$$

Equation (8.18) provides a kinetic definition for the temperature of a monatomic gas. Increasing the mean molecular speed will raise both the gas temperature and the internal energy. Note also that since all collisions are elastic, two systems of the same gas at the same temperature but different pressure and/or volume could have quite different collision frequencies but would still have the same internal energy per unit mass since internal energy of an ideal gas is only a function of temperature.

Combustion chemistry involves nonmonatomic gaseous constituents that do not remain inert and, in fact, many species undergo significant molecular rearrangement and energy transformation during a reaction process. At temperatures associated with the process of combustion, molecules can store quantized energy in rotational, vibrational, and electronic modes in addition to that of simple translation kinetic energy; see Figure 8.2. Thus, molecular structure of reactive constituents is also an important factor in gas-phase chemical conversion processes. A complete understanding of molecular physical chemistry and combustion cannot be described using kinetic theory of gases and will require additional knowledge of subjects including statistical thermodynamics and quantum chemistry. Energy associated with an overall combustion process is not simply the kinetic

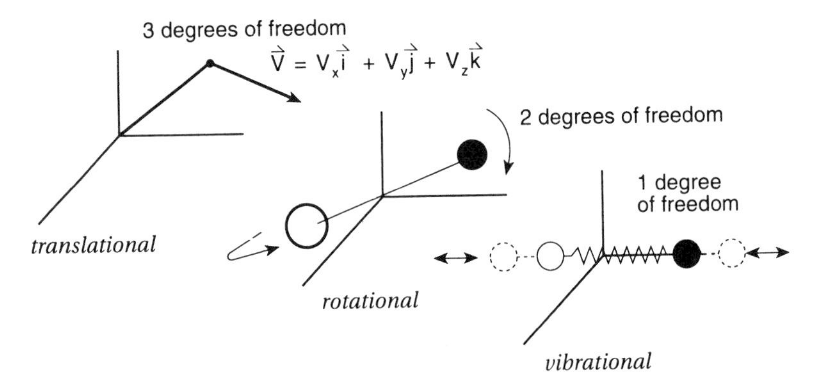

Figure 8.2 Internal energy storage modes of a gaseous diatomic molecule.

Table 8.2 C/H/N/O Molecular Bond Strengths

Bond	kca/mole
C−C	85.5
C−H	98.1
C=C	143.0
C≡C	194.3
C−C	78.3
C−N	81.0
C−O	85.8
C−S	63.7
C=O	167 (formaldehyde)
	172 (higher aldehydes)
	183 (ketones)
C≡N	210.6
H−H	104.2
N−N	60.0
N−H	88.0
N≡N	225.5
O−O	33.1
O−H	110.6

Source: Penner, S., *Chemistry Problems in Jet Propulsion* (Pergamon Press, 1957).

molecular energy but rather a result of effective collisions between reactive molecules that can cause a disruption of molecular structure and release energy stored within molecular bonds, see Table 8.2.

EXAMPLE 8.2 Acetylene, C_2H_2, is an unsaturated hydrocarbon compound having the structure H−C≡C−H. Using the standard state reactions and bond energies found in Table 8.2, estimate the heat of formation of acetylene.

Solution:

1. Standard state reaction:

$$2C_S + H_2 \xrightarrow{STP} C_2H_2 + \Delta H_f$$

2. Carbon gas formation reaction energy, from Table B.1 of in the appendixes:

$$C_S \xrightarrow{STP} C_G + 171.7 \text{ kcal}$$

3. Hydrogen formation reaction energy, from Table B.1 in the appendixes:

$$H_2 \rightarrow 2H + 104.2 \text{ kcal}$$

4. Using bond energies then for the reaction

$$2C + 2H \rightarrow C_2H_2 + \Delta H_f$$

where

$$\begin{aligned}\Delta H_f &= 2 \times \text{C–H bond energy} + \text{C}\equiv\text{C bond energy}\\ &= (2)(98.1) + (194.3)\\ &= 390.5 \text{ kcal}\end{aligned}$$

5. The standard state reaction, 1 above, can be written as

$$2C_G + 2H \rightarrow C_2H_2 - 390.5$$
$$2C_S \rightarrow 2C_G + 343.4$$
$$H_2 \rightarrow 2H + 104.2$$

or

$$\Delta H_f = 390.5 - 343.4 - 104.2 = -57.1 \text{ kcal}$$

This value is in reasonable agreement with the value of 54.194 found in Appendix B.

8.3 COLLISION THEORY AND CHEMICAL REACTIONS

Chemical kinetics provides a means of describing dynamic events effecting changes that molecular components undergo within a reactive mixture. Reaction equations that describe actual molecular activity occurring on the microscopic level are called *elemental reactions*. For example, the elemental reaction given by Equation (8.19) below shows that two hydrogen atoms are produced from the dissociation of a single hydrogen molecule.

$$H_2 + M \rightarrow H + H + M \tag{8.19}$$

Equation (8.20), an overall reaction equation, however, is not an example of an elemental reaction since interaction of a single methane molecule with two oxygen molecules is not the actual kinetic mechanism for forming carbon dioxide and water.

$$CH_4 + 2O_2 \rightarrow CO_2 + 2H_2O \tag{8.20}$$

A generalized elemental reaction between atomic species A and B to form products C and D is expressed in Equation (8.21) as

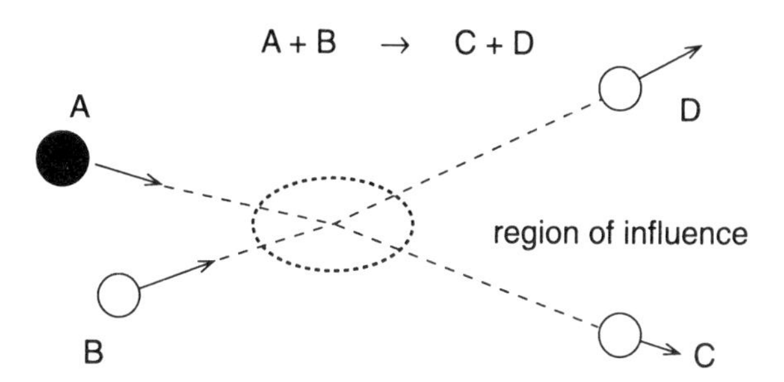

Figure 8.3 Kinetic model for molecular rearrangement via elemental reaction collisions.

$$aA + bB \rightarrow cC + dD \tag{8.21}$$

Note that atomic species A, B, C, and D may be molecules, radicals, and/or atoms. Equation (8.21) shows that only interatomic interactions between reactant species A and B occur in order to form the products C and D. Furthermore, the elemental reaction above assumes that reactants A and B are inert until they approach each other, at which point rearrangement of molecular structures to yield new species C and D occurs only when these constituents are in near one another. After elemental product species C and D are formed, they separate from a molecular conversion site and remain stable until each, in turn, comes into close contact with additional molecules, atoms, radicals, and/or solid boundaries, see Figure 8.3.

A plot of species concentration versus reaction time is useful in the study of the nature of an elemental reaction rate, see Figure 8.4. The *rate of reaction R*, a change in concentration of reactive species *i* with respect to time for an elemental reaction, can be expressed as

$$R_i \equiv -\frac{d[^-]_i}{dt} \tag{8.22}$$

where

R = reaction rate
$[^-]$ = molar concentration, g mole/cc
t = time, sec

Reaction rates can be thought of as a chemical velocity of reaction; i.e., higher speeds imply greater rates of conversion. The general relationships between rate of reaction of A and B and reactant species molar concentra-

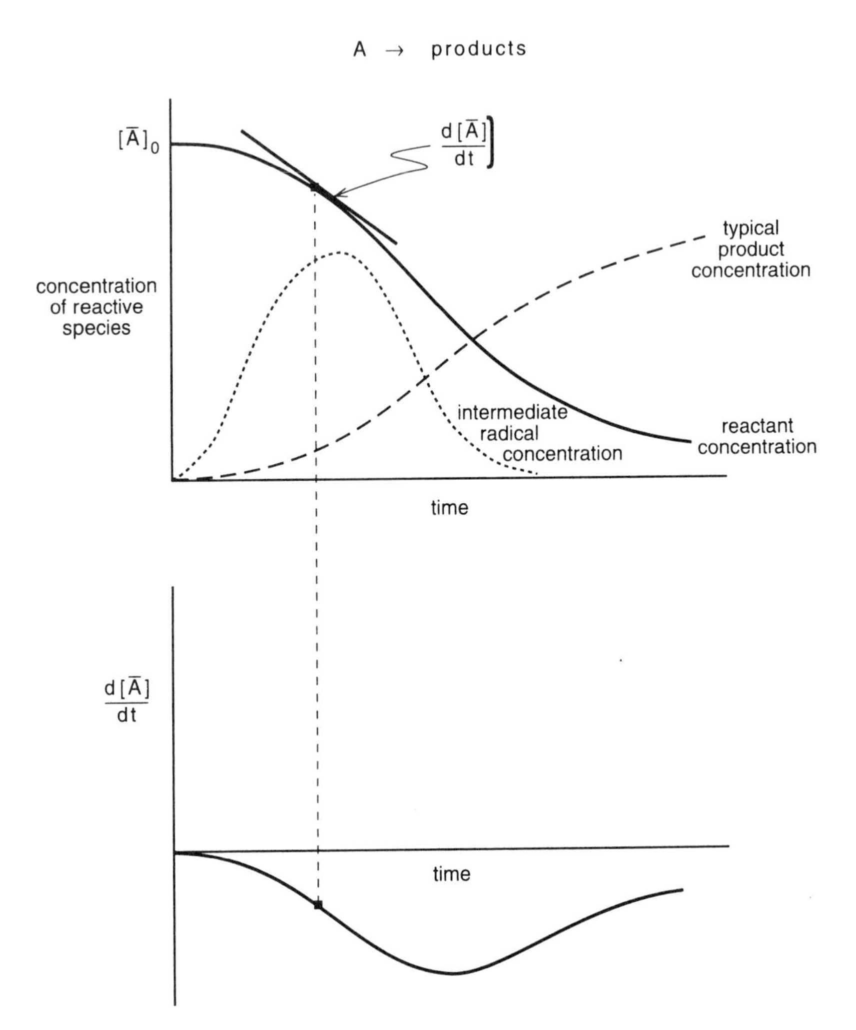

Figure 8.4 Composition-time schematic for decomposition.

tions for the elemental reaction given by Equation (8.21) has been found to satisfy the expressions

$$R_A = \frac{-1}{a}\frac{d[\overline{A}]}{dt} = k\,[\overline{A}]^a[\overline{B}]^b \tag{8.23a}$$

$$R_B = \frac{-1}{b}\frac{d[\overline{\mathrm{B}}]}{dt} = k[\overline{\mathrm{A}}]^a[\overline{\mathrm{B}}]^b \tag{8.23b}$$

where

k = elemental rate constant

$a,b, \ldots$ = reaction integers indicating the number of atomic members of A, B, . . . participating in the atomic reaction (usually 1 or 2).

The *molecularity,* or order, of an elemental reaction is the number of atomic members that participate in a reaction process, see Table 8.3 and Figure 8.5. For the general elemental reaction given by Equation (8.21),

a ≡ order of reaction with respect to A

b ≡ order of reaction with respect to B

$a + b$ ≡ overall reaction order

A reaction in which the slope of concentration versus time yields a straight line is termed a zeroth-order reaction, while a reaction in which the slope satisfies the expression

$$R_A = \frac{d[\overline{\mathrm{A}}]}{dt} = -k\,[\overline{\mathrm{A}}] \tag{8.24}$$

is called a first-order or unimolecular reaction with respect to A. Equations (8.25a) and (8.25b) are examples of bimolecular or second-order reactions, predominant reactions that occur in high-temperature combustion.

$$R_A = \frac{d[\overline{\mathrm{A}}]}{dt} = -k\,[\overline{\mathrm{A}}][\overline{\mathrm{B}}] \tag{8.25a}$$

or

$$R_A = \frac{d[\overline{\mathrm{A}}]}{dt} = -k\,[\overline{\mathrm{A}}][\overline{\mathrm{A}}] \tag{8.25b}$$

Note that if B is insert during a bimolecular reaction, a second-order reaction may appear to be first-order in A; i.e., $[\overline{\mathrm{B}}]$ = const.

Equations (8.23a) and (8.23b) show that chemical velocity R depends on concentrations of reactive species A and B, reaction integers a and b, and a rate constant k. Elemental reaction rates do depend on absolute reaction temperature; i.e., higher reaction temperatures produce higher reaction rates, while lower temperatures result in lower reaction rates. Many experimentally measured elemental reactions will double their kinetic speed with each 10K (10°R) increase in reaction temperature.

Table 8.3 Characteristics of Different-Order Reactions

Reaction order	Differential equation	Rate constant k_A from integrated equation	Half-life, $t_{1/2}$
0	$-\frac{d[\bar{A}]}{dt} = k_A$	$\frac{[\bar{A}]_0 - [\bar{A}]}{t}$	$\frac{[\bar{A}]_0}{2k_A}$
1	$-\frac{d[\bar{A}]}{dt} = k_A[\bar{A}]$	$\frac{1}{t} \ln \frac{[\bar{A}]_0}{[\bar{A}]}$	$\frac{1}{k_A} \ln 2$
2 Type I	$-\frac{d[\bar{A}]}{dt} = k_A[\bar{A}]^2$	$\frac{1}{t}\left(\frac{1}{[\bar{A}]} - \frac{1}{[\bar{A}]_0}\right)$	$\frac{1}{k_A[\bar{A}]_0}$
Type II	$-\frac{d[\bar{A}]}{dt} = k_A[\bar{A}][\bar{B}]$	$\frac{a \ln ([\bar{A}]/[\bar{A}]_0) \times [\bar{B}]_0/[\bar{B}])}{t(b[\bar{A}]_0 - a[\bar{B}]_0)}$	$\frac{a}{k_A(a[\bar{B}]_0 - b[\bar{A}]_0)} \ln \left(2 - \frac{b[\bar{A}]_0}{a[\bar{B}]_0}\right)$
3 Type I	$-\frac{d[\bar{A}]}{dt} = k_A[\bar{A}]^3$	$\frac{1}{2t}\left(\frac{1}{[\bar{A}]^2} - \frac{1}{[\bar{A}]_0^2}\right)$	$\frac{3}{2k_A[\bar{A}]_0^2}$
Type II	$-\frac{d[\bar{A}]}{dt} = k_A[\bar{A}]^2[\bar{B}]$	$\frac{-a}{t(b[\bar{A}]_0 - a[\bar{B}]_0)}\left(\frac{1}{[\bar{A}]} - \frac{1}{[\bar{A}]_0}\right) + \frac{ab}{t(b[\bar{A}]_0 - a[\bar{B}]_0)^2} \ln \left(\frac{[\bar{A}][\bar{B}]_0}{[\bar{A}]_0[\bar{B}]_0}\right)$	$\frac{-a}{k_A[\bar{A}]_0(b[\bar{A}]_0 - a[\bar{B}]_0)} - \frac{ab \ln (2 - b[\bar{A}]_0/a[\bar{B}]_0)}{k_A(b[\bar{A}]_0 - a[\bar{B}]_0)^2}$
Type III	$-\frac{d[\bar{A}]}{dt} = k_A[\bar{A}][\bar{B}][\bar{C}]$		
n	$-\frac{d[\bar{A}]}{dt} = k_A[\bar{A}]^n$	$\frac{1}{(n-1)t}\left(\frac{1}{[\bar{A}]^{n-1}} - \frac{1}{[\bar{A}]_0^{n-1}}\right)$	$\frac{(2^{n-1} - 1)}{k_A[\bar{A}]_0^{n-1}(n-1)}$

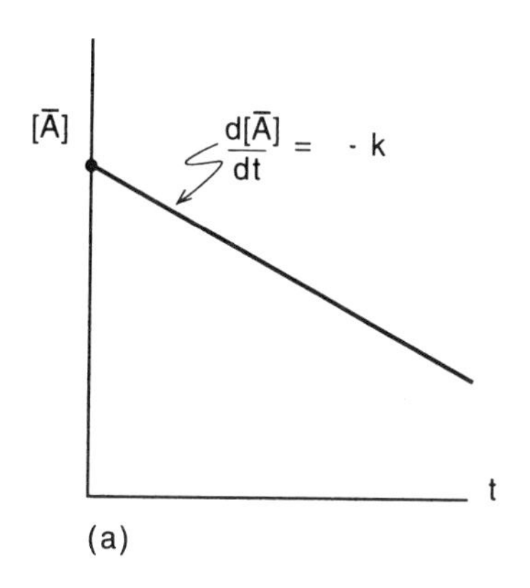

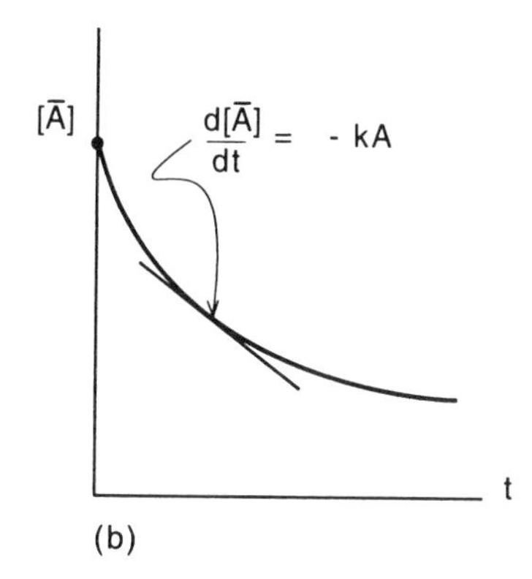

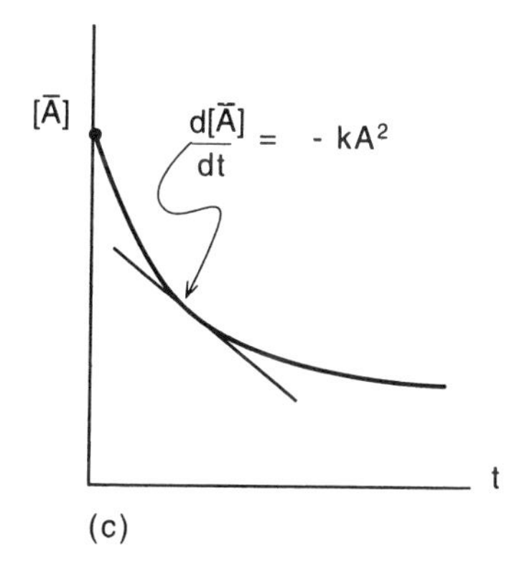

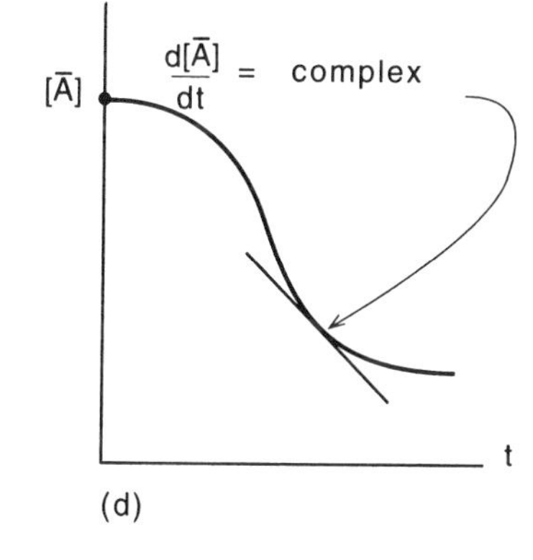

Figure 8.5 Concentration vs. time diagrams for various chemical reactions: (a) zeroth-order reaction, (b) first-order reaction, (c) second-order reaction, (d) autocatalytic oxidation reaction.

EXAMPLE 8.3 The high-temperature reaction of H_2 and O_2 involves the elementary initial reaction

$$H_2 + M \rightarrow H + H + M$$

where

$[\overline{H}_2]$ = diatomic hydrogen concentration
$[\overline{H}]$ = monotomic hydrogen concentration
$[\overline{M}]$ = total mixture concentration

(a) Write the reaction rate expression of H_2; (b) determine the relationship between $[H_2]$ and time; (c) repeat part (b) for [H]; and (d) evaluate an expression for the time it takes the concentration of H_2 to be reduced by one-half by this reaction.

Solution:

1. Reaction rate:

$$\frac{1}{1}\frac{d[\overline{H}_2]}{dt} = -k[\overline{H}_2]\,[\overline{M}]$$

$$\frac{d[\overline{H}_2]}{[\overline{H}_2]} = -k[\overline{M}]\,dt$$

Assume that

k, $[\overline{M}]$ are constant $\qquad [\overline{M}] >> [\overline{H}_2]$

Integrating yields

$$1\text{n}\,\frac{[\overline{H}_2]}{[\overline{H}_2]_0} = -k[\overline{M}]\,(t - t_0)$$

where

$[\overline{H}_2]_0$ = H_2 concentration at $t = t_0$
$[\overline{H}_2]$ = H_2 concentration at t
t_0 = initial reaction time

a. $$\frac{[\overline{H}_2]}{[\overline{H}_2]_0} = \exp\{-k[\overline{M}]t\}$$

2. Concentration versus time:

Let

$$[\overline{H}_2] = [\overline{H}_2]_0 - x$$

where

x = extent of reaction

$$[\overline{H}] = 2x$$

$$\frac{[\overline{H}]}{[\overline{H}_2]_0} = 2\left\{\frac{[\overline{H}_2]_0 - [\overline{H}_2]}{[\overline{H}_2]_0}\right\}$$

$$= 2\left\{1 - \frac{[\overline{H}_2]}{[\overline{H}_2]_0}\right\}$$

b. $$= 2\,[1 - \exp(-k[\overline{M}]t)]$$

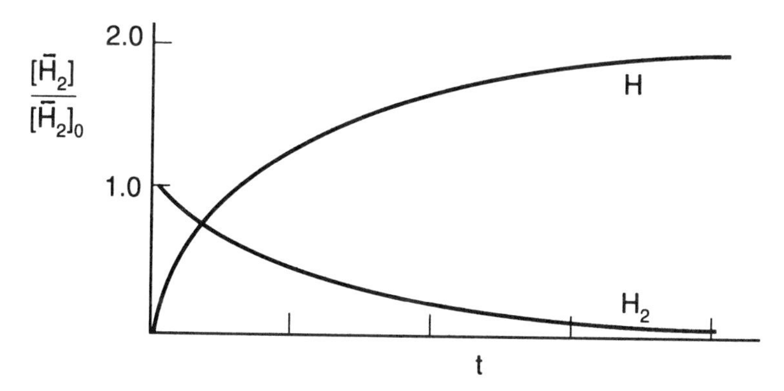

3. $[\overline{H}_2]$ half-life:

$$\frac{[\overline{H}_2]}{[\overline{H}_2]_0} = 0.5 = \exp\left\{-k[\overline{M}]t_{1/2}\right\}$$

c. $$t_{1/2} = \frac{-\ln 0.5}{k[M]} = \frac{0.6431}{k[M]}$$

Comments: This problem illustrates how an elementary reaction can be integrated to predict the concentration versus time in the simplest case.

Several theoretical models have been proposed to predict the observed behavior of an elemental rate constant k, including collision theory, statistical thermodynamics, and quantum mechanics. The following arguments will be based on the simple collision theory presented in Section 8.2. At first approximation, it is assumed that an elemental reaction occurs whenever appropriate reactant molecules and/or atoms collide. Recall from Section 8.2 that the frequency of collisions between two species A and B was equal to

$$Z_{A-B} = \sigma_{A-B}^2\left\{8\pi \ell N_0 T\left[\frac{m_A + m_B}{m_A m_B}\right]\right\}^{1/2}[\overline{A}][\overline{B}] \tag{8.10}$$

In this case, the ideal reaction rate for atomic species A colliding with species B would equal

$$R_A = Z_{A-B} \tag{8.26}$$

Combining Equations (8.7a), (8.8), and (8.9), and rearranging yield a collision expression for the rate constant k as

$$k = \sigma^2_{A-B} \left\{ 8\pi k N_0 T \left[\frac{m_A + m_B}{m_A m_B} \right] \right\}^{1/2} \tag{8.27}$$

The assumption that each collision results in chemical reaction yields rate constants that are unrealistically high. Collisions do occur between species, in fact, that do not result in chemical reaction. An engineering approximation still based on collision theory is to assume that only a fraction of these molecular collisions are capable of producing chemical reactions or

$$R_{A-B} = f \times Z_{A-B} \tag{8.28}$$

where f = fraction of collisions that result in chemical reactions; $f << 1.0$.

Arrhenius (1889) proposed that the temperature dependency of the rate constant k could be expressed as

$$k = C \times \exp \left\{ \frac{-\Delta \overline{E}}{\overline{R}T} \right\} \tag{8.29}$$

or

$$\ln k = \ln C - \Delta \overline{E}/\overline{R}T$$

where

C = effective conversion frequency factor
$\Delta \overline{E}$ = reaction activation energy
$\overline{R}$ = universal gas constant
T = absolute reaction temperature

Combining Equations (8.27–8.29) yields the following relationship:

$$C \exp \left\{ \frac{-\Delta \overline{E}}{\overline{R}T} \right\} = f \sigma^2_{AB} \left\{ 8\pi k N_0 T \left[\frac{m_A + m_B}{m_A m_B} \right] \right\}^{1/2} \tag{8.30}$$

The effective conversion collision frequency factor C is assumed to equal the collision frequency expression or

$$C = \sigma^2_{A-B} \left\{ 8\pi k N_0 T \left[\frac{m_A + m}{m_A m_B} \right] \right\}^{1/2} \tag{8.31}$$

with the fraction of collisions that result in chemical reactions then equal to

$$f = \exp \left\{ \frac{-\Delta \overline{E}}{\overline{R}T} \right\} \tag{8.32}$$

EXAMPLE 8.4 Consider the elementary reaction

A + A → products

Assume that the molecular weight of A is 50 g/g mole, the effective molecular diameter is 3.5×10^{-8}cm, and the activation energy for the reaction is 40 kcal/g mole. Determine the Arrhenius rate constant k and half-life $t_{1/2}$ using simple collision theory for (a) 500K, (b) 1000K, (c) 1500K, and (d) 2000K. Let $[A]_0 = 1$ mole g/cc.

Solution:

1. Elementary reaction:

A + A → products

$$\frac{d[A]}{dt} = -k[A][A] = = -k[A]^2$$

$$\frac{d[A]}{[A]^2} = -k\,dt$$

$$\int_{[A]_0}^{[A]} \frac{d[A]}{[A]^2} = -\int_0^t k\,dt = -k\int_0^t dt$$

$$\frac{1}{[A]} - \frac{1}{[A]_0} = -kt$$

$$t = \frac{1}{k}\left[\frac{1}{[A]} - \frac{1}{[A]_0}\right]$$

2. Half-life $t_{1/2}$:

for $t_{1/2}$

$$[A] = 0.5[A]_0$$

$$t_{1/2} = \frac{1}{k}\left[\frac{1}{0.5[A]_0} - \frac{1}{[A]_0}\right] = \frac{1}{k[A]_0}$$

3. Arrhenius rate constant:

$$k = \sigma_A{}^2\left\{8\pi \not{k} N_0 T\left[\frac{2m_A}{m^2{}_A}\right]\right\}^{1/2} \exp\left\{\frac{-\Delta\bar{E}}{RT}\right\}$$

or

$$k = (3.5 \times 10^{-8})^2\left\{8\pi(1.38 \times 10^{-16})(6.02 \times 10^{23})\right.$$

$$\left.\left(\frac{2}{50}\right)T^{1/2}\right\}\exp\left\{\frac{-40{,}000}{(1.987)\,(T)}\right\}$$

$$= 1.1195 \times 10^{-11} [T^{1/2}] \exp\left\{\frac{-20131}{T}\right\} \quad \text{cc/sec molecule}$$

or

$$6.02 \times 10^{23} \text{ molecules/mole}$$
$$k = 6.739 \times 10^{12} \times T^{1/2} \exp\left\{-20131/T\right\}$$

and

$$t_{1/2} = 1/k$$

T, K	k, cc/sec mole	$t_{1/2}$, sec
500	4.93×10^{-4}	2.03×10^{3}
1000	3.85×10^{5}	2.60×10^{-6}
1500	3.87×10^{8}	2.58×10^{-9}
2000	1.28×10^{10}	7.8×10^{-11}

Figure 8.6 shows the energy versus reaction path plot for an ideal endothermic and an exothermic elementary reaction. In both instances, heat of reaction is the difference in chemical potential energy between reactant and product states. The activation energy is seen as an energy barrier that prevents all collisions from resulting in chemical reactions. Stable molecules have intermolecular forces associated with their chemical bonding that are strong enough to hold the molecular complex together. Successful elemental reactions have collisional energy in excess of the activation energy that, when redistributed through the colliding complex at the instant of interaction, can cause dissociation or association of single-, double-, or triple-bonded molecules. Table 8.4 lists magnitudes of various elementary C/H/N/O reaction rate constants.

The rate constant k is not a true constant but is dependent on reaction temperature, and its behavior must be experimentally obtained. For example, the rate constant for the general elemental reaction given by Equation (8.3) can be determined using an initial slope of a concentration versus time plot, in which case the initial reaction temperature and the initial molar concentrations $[\overline{A}]_0$ and $[\overline{B}]_0$ are known. The initial reaction rate then equals

$$R_A = \left.\frac{\Delta[\overline{A}]}{\Delta t}\right) t=0 \tag{8.33}$$

and

$$R_A = k[\overline{A}]_0^a \, [\overline{B}]_0^b \tag{8.34}$$

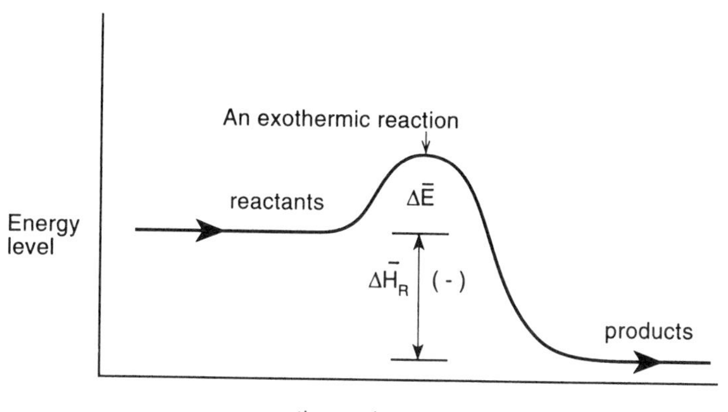

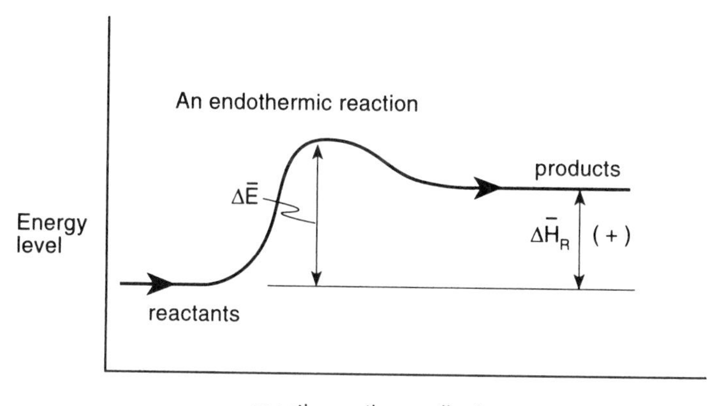

where:

$\Delta\bar{E}$ = activation energy

$\Delta\bar{H}_R$ = heat of reaction

Figure 8.6 Energy reaction path schematic for an elementary reaction.

Combining Equations (8.33) and (8.34) and solving for the rate constant k yield the expression

$$k = \frac{1}{[\overline{A}]_0[\overline{B}]_0} \left.\frac{\Delta[\overline{A}]}{\Delta t}\right)_{t=0} \tag{8.35}$$

Table 8.4 Elementary C/H/N/O Reaction Rate Constants, $k = AT^n \exp(-E/RT)$

Reaction	A	n	E	Reaction	A	n	E
1. $H + CO_2 = CO + OH$	5.6×10^{13}	0	23,500	29. $O + N_2 + M = N_2O + M$	6.3×10^{14}	0	56,800
2. $O + N_2 = NO + N$	1.44×10^{14}	0	75,580	30. $H + O_2 + M = HO_2 + M$	1.5×10^{15}	0	−1,000
3. $O + NO = N + O_2$	4.1×10^{9}	1	38,340	31. $H + HO_2 = OH + OH$	2.5×10^{14}	0	1,900
4. $N + OH = NO + H$	4.21×10^{13}	0	0	32. $OH + HO_2 = H_2O + O_2$	1.2×10^{13}	0	1,000
5. $OH + H_2 = H_2O + H$	2.2×10^{13}	0	5,150	33. $O + HO_2 = OH + O_2$	5.0×10^{13}	0	1,000
6. $OH + OH = O + H_2O$	5.75×10^{12}	0	780	34. $H + HO_2 = H_2 + O_2$	2.5×10^{13}	0	700
7. $O + H_3 = H + OH$	1.74×10^{13}	0	9,450	35. $H + HO_2 = O + H_2O$	1.0×10^{13}	0	1,000
8. $H + O_2 = O + OH$	1.56×10^{14}	0	16,633	36. $H_2 + HO_2 = H + H_2O_2$	1.9×10^{13}	0	24,000
9. $O + H + M = OH + M$	3.6×10^{17}	−1	0	37. $H_2O_2 + M = 2OH + M$	7.1×10^{14}	0	−5,100
10. $O + O + M = O_2 + M$	3.6×10^{17}	−1	0	38. $CH_4 + M \rightarrow CH_3 + H + M$	1.5×10^{19}	0	99,960
11. $H + H + M = H_2 + M$	1.81×10^{18}	−1	0	39. $CH_4 + H = CH_3 + H_2$	5.1×10^{13}	0	12,900
12. $H + OH + M = H_2O + M$	7.3×10^{18}	−1		40. $CH_4 + OH = CH_3 + H_2O$	2.85×10^{13}	0	4,968
13. $O + CO_2 = CO + O_2$	1.9×10^{13}	0	54,150	41. $CH_4 + O = CH_3 + OH$	1.7×10^{13}	0	8,700
14. $CO + O + M = CO_2 + M$	6.0×10^{17}	−1	2,484	42. $CH_3 + O_2 = HCO + H_2O$	1.0×10^{11}	0	0
15. $N_2 + O_2 = N + NO_2$	2.7×10^{14}	−1	120,428	43. $HCO + OH = CO + H_2O$	3.0×10^{13}	0	0
16. $N_2 + O_2 = NO + NO$	4.2×10^{14}	0	119,100	44. $H + CO + M = HCO + M$	1.0×10^{17}	1	0
17. $NO + NO = N + NO_2$	3.0×10^{11}	0	0	45. $HO_2 + HO_2 = H_2O_2 + O_2$	6.5×10^{13}	0	0
18. $NO + M = O + NO + M$	2.27×10^{17}	−0.5	148,846	46. $H_2O_2 + H = H_2O + OH$	3.18×10^{14}		9,000
19. $O + NO + M = NO_2 + M$	1.05×10^{15}	0	−1,870	47. $CH_2 + \frac{1}{2} O_2 \rightarrow CO + H_2^+$	$5.52 \times 10^{8}\ p^{-0.825}$	1	24,642
20. $N + O_2 + M = NO_2 + M$	7.0×10^{11}	−1		48. $H + NO + M = HNO + M$	4.8×10^{15}		0
21. $O + NO_2 = NO + O_2$	1.0×10^{13}	0	600	49. $HNO + H + M = H_2NO + M$			
22. $H + NO_2 = NO + OH$	7.25×10^{14}	0	1,930	50. $HNO + HNO + M \rightarrow H_2O + N_2O + M$			
23. $N + CO_2 = CO + NO$	2.0×10^{11}	−0.5	7,950	51. $R + NO + M \rightarrow RNO + M$			
24. $CO + NO_2 = NO + CO_2$	2.0×10^{11}	−0.5	4,968	52. $R + RNO + M \rightarrow R_2NO + M$			
25. $H + N_2O = OH + N_2$	3.01×10^{13}	0	10,800	53. $R + R_2NO + M \rightarrow R_2NOR + M$			
26. $O + N_2O = O_2 + N_2$	3.61×10^{13}	0	24,800	54. $H + HNO = H_2 + NO$	1.4×10^{11}	0.5	
27. $N_2 + NO_2 = NO + N_2O$	1.41×10^{14}	0	83,000	55. $H + HCO = H_2 + CO$	2.0×10^{13}	0	0
28. $NO + HO_2 = OH + NO_2$	6.0×10^{11}	0	0				

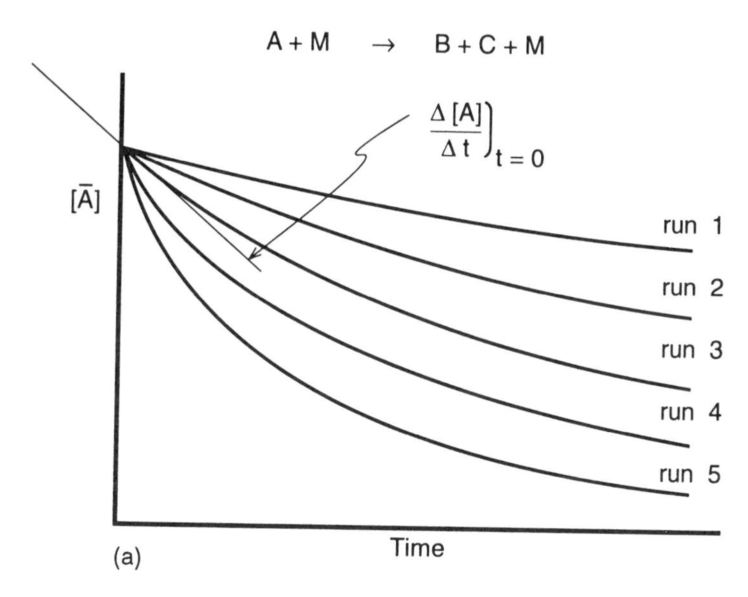

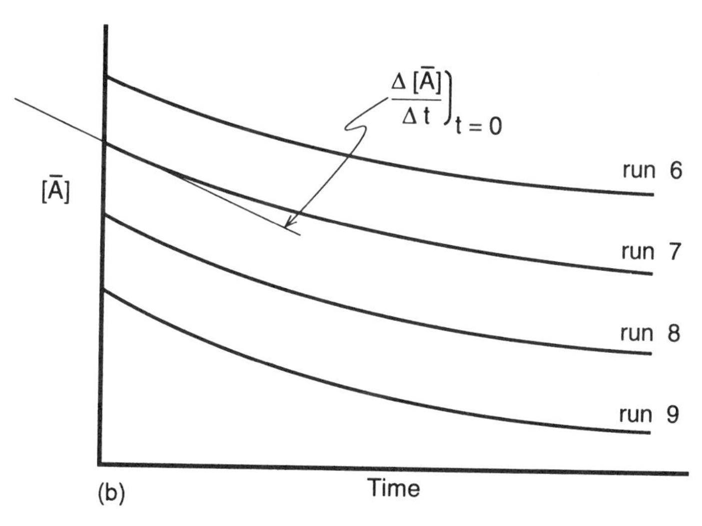

Figure 8.7. Concentration vs. time: (a) fixed initial concentration, variable initial temperatures; (b) fixed initial temperature, variable initial concentration.

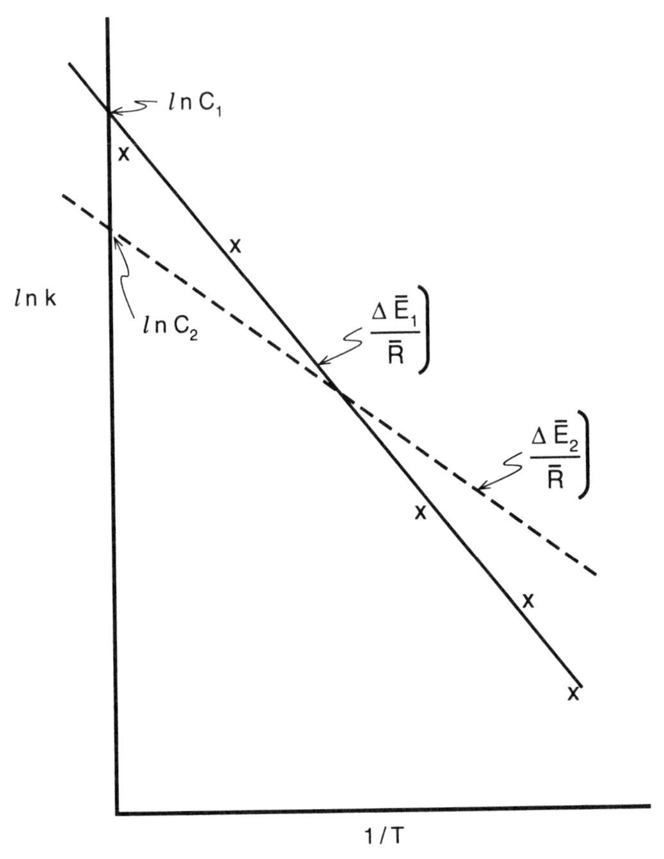

Figure 8. 8 Elementary Reaction Arrhenius Rate Constant vs. Temperature (Schematic Diagram)

Figure 8.7 illustrates an initial slope analysis of this general elemental reaction to determine the effect of varying the concentration and reaction temperature on the rate constant k. Experimentally obtained elementary rate constants, such as those found using Equation (8.35), should plot as a straight line on natural log versus inverse temperature coordinates, as illustrated by Figure 8.8. The activation energy can then be obtained from the slope of the curve-fitted data and frequency factor from the abscissa intercept.

Overall rate expressions are often written to correlate combustion of a given fuel as

$$\frac{d[\text{fuel}]}{dt} = -\,k\,[\text{fuel}]^{\alpha}[\text{oxidant}]^{\beta} \tag{8.36}$$

Equation 8.36, however, is an empirical expression based on fitted experimental data and not on elemental reaction analysis. Recall Table 8.4 lists Arrhenius rate constants for several important elemental reactions.

8.4 COMPLEX CHEMICAL KINETICS MECHANISMS

Gas-phase chemical kinetics associated with most operating combustion machinery is compounded by both the complexity of the many concurrent elemental molecular reactions associated with an energy conversion process and the heterogeneous boundaries within which these chemical events are confined. No effort will be made to discuss the fuel-engine chemical kinetics interface for any particular combustion device in this chapter. Instead, only basic dynamic elements of a combustion process will be discussed. In the following section, bimolecular abstraction, association, branching, and dissociation reactions will be described, kinetic events that play a major role in high-temperature gas-phase oxidation chemistry.

EXAMPLE 8.5 Certain gas-phase chemical reactions can be experimentally studied by measuring the changes in total reaction pressure with time by using a constant-volume reaction vessel. Consider the dissociation reaction

$$A_2 \xrightarrow{k} 2A$$

(a) Determine an expression for $[A_2]$ and $[A]$ in terms of the reaction temperature and total pressure; (b) determine the rate of dissociation of A_2 using part a; and (c) evaluate the rate constant k using parts (a) and (b).

Solution:

1. Elementary reaction:

$$A_2 \xrightarrow{k} 2A$$

where

$$\frac{d[A_2]}{dt} = -k[A_2]$$

$$\frac{d[A_2]}{dt} = -k\,dt$$

or

$$\ln\frac{[A_2]}{[A_2]_0} = -kt$$

Let

$$[A_2] = [A_2]_0 - Z$$
$$[A] = [A]_0 + 2Z$$

2. The total concentration at any time is given by

$$[A_2] + [A] = \{[A_2]_0 - [A_2]_0 + [A_2]\} + \{2[A_2]_0 - 2[A]\}$$
$$= 2[A_2]_0 - [A_2]$$

3. Assuming an ideal-gas behavior, then,

$$P_{tot}V = N_{tot}\bar{R}T$$

or

$$\frac{N_{tot}}{V} = [tot] = \frac{P_{tot}}{\bar{R}T}$$

4. Combining items 2 and 3,

$$2[A_2]_0 - [A_2] = \frac{P_{tot}}{\bar{R}T}$$

or

$$[A_2] = 2[A_2]_0 - \frac{P_{tot}}{\bar{R}T}$$

$$\text{at } t = 0 \qquad P_{tot} = P_0$$

a. $$[A_2] = \frac{2P_0 - P_{tot}}{\bar{R}T}$$

$$[A] = \frac{P_{tot} - P_0}{\bar{R}T}$$

5. Substituting the above into item 1 yields

$$\ln\left[\frac{(2P_0 - P_{tot})\,\bar{R}T}{P_0\,\bar{R}T}\right] = -kt$$

$$\ln\left[2 - \frac{P_{tot}}{P_0}\right] = -kt$$

$$k = \frac{-1}{t}\left[\ln 2 - \frac{P_{tot}}{P_0}\right]$$

Comment: The above analysis suggests that the total pressure versus time data for a reaction could be used to evaluate the rate constant k. The above expression is valid only if the reaction is the sole kinetic step and the mixture behaves as an ideal gas.

Homogeneous gas-phase combustion involves a large number and variety of different elemental reactions that: (1) begin the kinetic process; (2) produce short-lived, reactive intermediate radicals that propagate the overall mechanism; and (3) terminate the kinetic process by forming stable products of combustion, see Figure 8.9. Many of these individual reactions occur simultaneously, while others, termed *chain reactions,* are consecutive. The short-lived reactive species that propagate these interactive reactions are commonly referred to as *chain carriers.*

High-temperature oxidation chemistry in premixed homogeneous fuel-air mixtures is initiated by elemental fuel pyrolysis reactions. Equation (8.37) is an example of an endothermic bimolecular *dissociation* reaction of methane in which the methyl radical CH_3 and a hydrogen atom are produced.

$$CH_4 + M \rightarrow CH_3 + H + M \tag{8.37}$$

A period of apparent inactivity can exist in the initial stages of fuel dissociation, during which significant concentrations of those active radicals necessary to further propagate the overall combustion process are produced.

Equations (8.38a) and (8.38b) are examples of bimolecular *abstraction* reactions, delayed kinetic processes that require radicals such as H and O atoms and an OH radical in order to occur.

$$H + CH_4 \rightarrow H_2 + CH_3 \tag{8.38a}$$

$$OH + CH_4 \rightarrow HOH + CH_3 \tag{8.38b}$$

The reaction described by Equation (8.38a) shows a hydrogen atom abstracted by the collision of a methane molecule with a hydrogen atom to produce diatomic hydrogen and a methyl radical. In Equation (8.38b), a hydrogen atom is abstracted from the collision of a hydroxyl radical with a methane molecule to produce water and a methyl radical. Activation energies for abstraction reactions, in general, are not as great as those for bimolecular dissociation reactions.

Equations (8.39a) and (8.39b) are examples of *branching reactions,* reactions that produce more active radicals than they consume.

$$H + O_2 \rightarrow OH + O \tag{8.39a}$$

$$O + H_2 \rightarrow OH + H \tag{8.39b}$$

In Equation (8.39a), an oxygen atom and a hydroxyl radical are produced from the collision of a hydrogen atom with an oxygen molecule, whereas a hydrogen atom and a hydroxyl radical are generated from the collision of an oxygen atom with a hydrogen molecule in Equation (8.39b). Branching reactions are the chief reactions that actually control the overall rate of an oxidation process.

Equation (8.40) is an example of an *association* reaction, a bimolecular radical recombination reaction that forms a stable species, while Equation (8.41) is an example of a *termolecular,* or three-body, association reaction.

$$CH_3 + H \rightarrow CH_4 \tag{8.40}$$

$$H + H + M \rightarrow H_2 + M \tag{8.41}$$

Since the probability of a successful three-body collision is less than for a bimolecular collision, these types of reactions will have slower reaction rates but are major mechanisms for high-temperature radical recombination reactions.

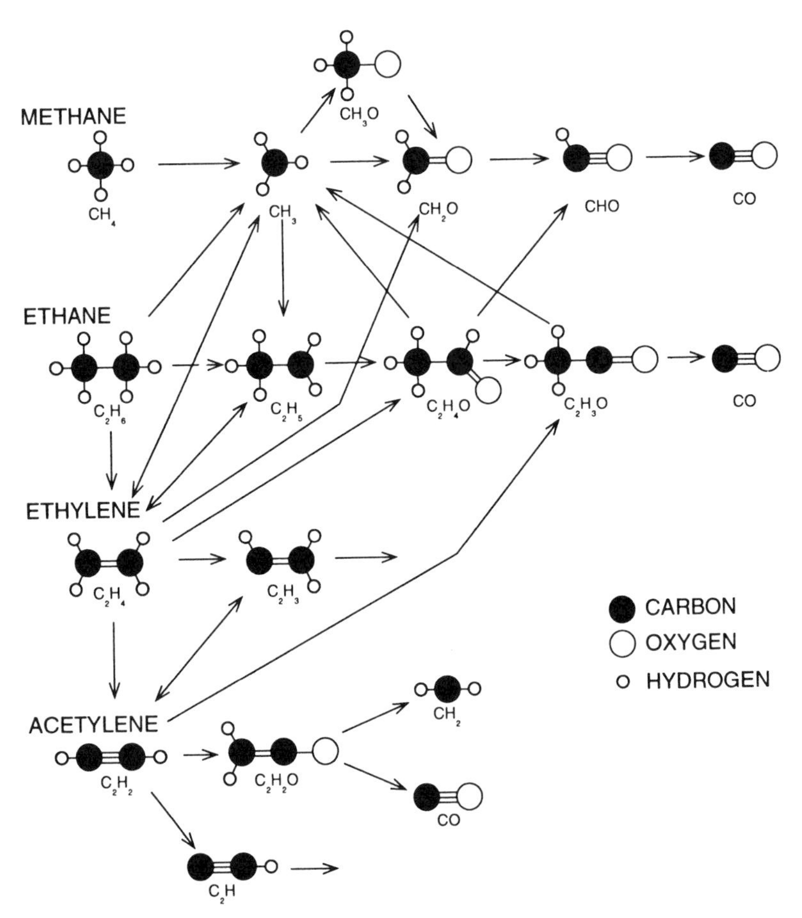

Figure 8.9 Complex elementary C/H/O reaction paths.

EXAMPLE 8.6 Many oxidation reactions involve the dissociation of the fuel molecule to form an active radical. Consider the reaction system,

$$A_2 \underset{k_2}{\overset{k_1}{\rightleftarrows}} 2A$$

For this process, write an expression for (a) the rate of change of A_2, (b) the ratio of k_1/k_2 at equilibrium, and (c) the relationship among k_1, k_2, and the equilibrium constant K_p.

Solution:

1. Rate of change in A_2:

$$\frac{d[A_2]}{dt} = -k_1[A_2] + k_2[A]^2$$

Now, let

$$[A_2] = [A_2]_0 - x \qquad [A] = 2x$$

$$\frac{d}{dt}\left\{[A_2]_0 - x\right\} = -k_1\left\{[A_2]_0 - x\right\} + k_2(2x)^2$$

a. $$\frac{-dx}{dt} = -k[A_2]_0 + k_1x + 4k_2x^2$$

2. At equilibrium,

$$\frac{d[A_2]}{dt} = 0 \qquad \text{or} \qquad 0 = -k_1[A_2] + k_2[A]^2$$

and

$$\frac{k_1}{k_2} = \frac{[A]^2}{[A_2]}$$

3. Equilibrium constant k_c:

Define

$$k_c \equiv \frac{[A]^2/[A_0]^2}{[A_2]/[A_{2_0}]}$$

where

$$[A_{2x}]_0 = [A]_0 \equiv 1.0$$

$$K_c = \frac{[A]^2}{[A_2]} = \frac{k_1}{k_2} \qquad \frac{k_1}{k_2} = K_c$$

4. Equilibrium constant k_p:

$$K_p = \frac{(P/P_0)_A^2}{(P/P_0)_{A2}}$$

where

$$(P_0)_{A,A2} = 1.0 \text{ atm}$$

$$K_p = \frac{P_A^{\ 2}}{P_{A_2}}$$

and

$$P_A + P_{A_2} = P_{\text{tot}}$$

5. Ideal gas

$$P_A = [\text{A}]\overline{R}T \qquad P_{A_2} = [\text{A}_2]\overline{R}T$$

$$K_p = \frac{P_A^{\ 2}}{P_{A_2}} = \frac{\{[\text{A}]\overline{R}T\}^2}{\{[\text{A}_2]\overline{R}T\}} = \frac{[\text{A}]^2}{[\text{A}_2]}\overline{R}T$$

$$K_p = K_c\overline{R}T$$

where $\overline{R} = 82.057$ cc atm/g mole K

$$K_c = \frac{K_p}{\overline{R}T}$$

$$\frac{k_1}{k_2} = \frac{K_p}{\overline{R}T}$$

where $K_p \Rightarrow$ JANAF data.

The salient features of homogeneous gas-phase hydrocarbon-air combustion, i.e., initiation, branching, and termination, can be illustrated by the less complex but well-studied H_2-O_2 reaction system. Initiation of the H_2-O_2 reaction mechanism begins with the pyrolysis of hydrogen. At low temperatures, dissociation of the hydrogen molecule will be due to heterogeneous wall collisions or

$$H_2 + W \rightarrow HW + H \tag{8.42}$$

At higher temperatures, gas-phase pyrolysis of the hydrogen molecule will become important, or

$$H_2 + M \rightarrow H + H + M \tag{8.43}$$

High-temperature endothermic dissociation of the oxygen molecule will produce oxygen atoms as expressed in Equation (8.26).

$$O_2 + M \rightarrow O + O + M \tag{8.44}$$

Kinetics associated with the initial stages of an oxidation process and induction delay period are a function of initial reactant concentrations, temperature, and pressure but are not rate-controlling steps of the overall kinetic mechanisms.

Equations (8.45a) and (8.45b) are chain-branching reactions that are responsible for the rapid propagation of the H_2-O_2 reaction mechanism. These reactions are often referred to as being autocatalytic in that they create more intermediate radicals than they consume, thereby self-accelerating the reaction process.

$$H_2 + O \rightarrow OH + H \tag{8.45a}$$

$$O_2 + H \rightarrow OH + O \tag{8.45b}$$

Equations (8.46a) and (8.46b) are propagation reactions that result in the formation of stable products of combustion but require the presence of an intermediate radical species in order to occur.

$$OH + H_2 \rightarrow HOH + H \tag{8.46a}$$

$$HO_2 + H_2 \rightarrow HOH + OH \tag{8.46b}$$

EXAMPLE 8.7 Complex oxidation mechanisms can involve many postulated elementary reactions. These reactions result in the formation of reactive radical species. An important assumption often applied to such a scheme is that these intermediate radicals reach a steady state, i.e., don't change with time. Consider the reaction set below for the H_2-O_2 system.

$$\begin{aligned}
\text{wall} + H_2 &\overset{1}{\rightarrow} 2H \\
H + O_2 &\overset{2}{\rightarrow} OH + O \\
O + H_2 &\overset{3}{\rightarrow} OH + H \\
OH + H_2 &\overset{4}{\rightarrow} HOH + H \\
HOH + H &\overset{5}{\rightarrow} OH + H_2 \\
\text{wall} + H &\overset{6}{\rightarrow} \text{removal at surface} \\
\text{wall} + OH &\overset{7}{\rightarrow} \text{removal at surface}
\end{aligned}$$

Using these reactions (a) write an expression for the formation of H_2O; (b) for H atoms, (c) OH radical, and (d) O atoms. (e) Using the steady-state approximation for (b) → (d), express part (a) in terms of H_2 and O_2 and H steady state.

Solution:

1. Formation of H_2O:

a. $$\frac{d[H_2O]}{dt} = k_4[OH][H_2] - k_5[H_2O][H]$$

2. Formation of H:

b. $$\frac{d[H]}{dt} = k_1[H_2][W] - k_2[H][O_2] + k_3[O][H_2] + k_4[OH][H_2] - k_5[H][H_2O] - k_6[H][W]$$

3. Formation of OH:

c. $$\frac{d[OH]}{dt} = k_2[H][O_2] + k_3[O][H_2] - k_4[OH][H_2] + k_5[H_2O][H] - k_7[OH][W]$$

4. Formation of O:

d. $$\frac{d[O]}{dt} = -k_2[H][O_2] - K_3[O][H_2]$$

5. The steady-state assumption assumes that the concentration of certain radicals remains constant during reaction, or

$$\frac{d[H]}{dt} = \frac{d[OH]}{dt} = \frac{d[O]}{dt} = 0$$

6. H-atom steady state:

$$O = k_1[H_2][W] - k_2[H][O_2] + k_3[O][H_2] + k_4[OH][H_2] - k_5[H][H_2O] - k_6[H][W]$$

7. OH steady state:

$$0 = k_2[H][O_2] + k_3[O][H_2] - k_4[OH][H_2] + k_5[H_2O][H] - k_7[OH][W]$$

8. O-atom steady state:

$$0 = k_2[H][O_2] - k_3[O][H_2]$$

or

$$\left.\frac{[H]}{[O]}\right)_{ss} = \frac{k_3[H_2]}{k_2[O_2]}$$

9. Substituting part 7 into part 6,

$$0 = k_2[\mathrm{H}][\mathrm{O_2}] + k_3\left(\frac{k_2[\mathrm{O_2}][H]}{k_3[\mathrm{H_2}]}\right)[\mathrm{H_2}]$$

$$= k_4[\mathrm{OH}][\mathrm{H_2}] + k_5[\mathrm{H_2O}][\mathrm{H}] - k_7[\mathrm{OH}][\mathrm{W}]$$

$$[\mathrm{OH}]\{k_4[\mathrm{H_2}] + k_7[\mathrm{W}]\} = [\mathrm{H}]\{2k_2[\mathrm{O_2}] + k_5[\mathrm{H_2O}]\}$$

$$\left.\frac{[\mathrm{OH}]}{[\mathrm{H}]}\right)_{\mathrm{SS}} = \frac{2k_2[\mathrm{O_2}] + k_5[\mathrm{H_2O}]}{k_4[\mathrm{H_2}] + k_7[\mathrm{W}]}$$

10. Substituting parts 7 and 8 into part 5,

$$0 = k_1[\mathrm{H_2}][\mathrm{W}] - k_1[\mathrm{H}][\mathrm{O_2}] + k_2[\mathrm{H}][\mathrm{O_2}]$$

$$- k_5[\mathrm{H}][\mathrm{H_2O}] - k_6[\mathrm{H}][\mathrm{W}]$$

$$+ k_4[\mathrm{H_2}]\left\{\frac{2\,k_2[\mathrm{O_2}] + k_5[\mathrm{H_2O}]}{k_4[\mathrm{H_2}] + k_7[\mathrm{W}]}\right\}[H]$$

and

$$[\mathrm{H}]_{\mathrm{SS}} = \frac{k_1[\mathrm{H_2}][\mathrm{W}]}{k_5[\mathrm{H_2O}] + k_6[\mathrm{W}]\dfrac{-k_4[\mathrm{H_2}]\,\{2k_2[\mathrm{O_2}] + k_5[\mathrm{H_2O}]\}}{k_4[\mathrm{H_2}] + k_7[\mathrm{W}]}}$$

11. Formation of H_2O:

$$\frac{d[\mathrm{H_2O}]}{dt} = k_4[\mathrm{OH}]_{\mathrm{SS}}[\mathrm{H_2}] - k_5[\mathrm{H_2O}][\mathrm{H}]_{\mathrm{SS}}$$

$$= \left[\,k_4[\mathrm{H_2}]\left\{\frac{2k_2[\mathrm{O_2}] + k_5[\mathrm{H_2O}]}{k_4[\mathrm{H_2}] + k_7[\mathrm{W}]}\right\} - k_5[\mathrm{H_2O}]\right][\mathrm{H}]_{\mathrm{SS}}$$

Comments: This problem illustrates that even the simplest kinetic mechanisms are highly complex. Numerical techniques are necessary to predict the concentration-time profiles for even the simplest cases. In addition, the solutions to many practical combustion kinetics are complicated by the thermodynamic, fluid mechanic, and heat transfer coupling. Boundary conditions at the walls further influence the actual physics of high-temperature chemical kinetics.

Table 8.5 lists the mechanism for H_2-O_2 combustion postulated by Hinshelwood.

In Chapter 7, two types of ideal combustion processes were predicted that could propagate through a homogeneous premixed combustible mixture: detonations, or high-pressure supersonic reactions, and deflagrations,

Table 8.5 H_2-O_2 Reaction Mechanism

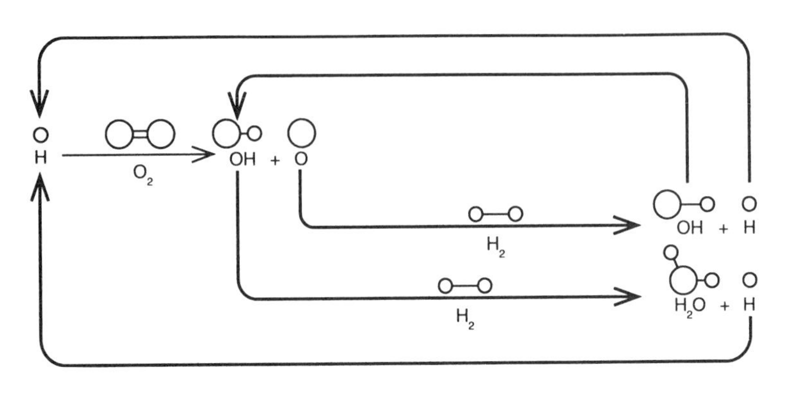

H/O Chain Branching Reactions

$$\mathrm{H_2 + wall} \longrightarrow \mathrm{WH + H}$$

$$\mathrm{H + O_2} \xrightarrow{1} \mathrm{HO + O}$$

$$\mathrm{O + H_2} \xrightarrow{2} \mathrm{HO + H}$$

$$\mathrm{HO + H_2} \underset{4}{\overset{3}{\rightleftarrows}} \mathrm{HOH + H}$$

$$\mathrm{H + O_2 + M} \xrightarrow{5} \mathrm{HOH + H}$$

$$\mathrm{H + wall} \xrightarrow{6} \text{stable species}$$

$$\mathrm{HO_2 + wall} \xrightarrow{7} \text{stable species}$$

$$\mathrm{HO + wall} \xrightarrow{8} \text{stable species}$$

$$\mathrm{HO_2 + H_2} \xrightarrow{9} \mathrm{H_2O + OH}$$

$$\mathrm{OH + H_2} \xrightarrow{10} \mathrm{H_2O + H}$$

$$\mathrm{O + H_2O} \underset{12}{\overset{11}{\rightleftarrows}} \mathrm{OH + OH}$$

or low-pressure subsonic reactions. The H_2-O_2 kinetic mechanisms described in this chapter can be used to better understand the physical difference between deflagration and detonation wave propagations. Figure 8.10 illustrates a thermal explosion plot for the hydrogen-oxygen system. For H_2-O_2 mixtures below 400°C (752°F), no rapid homogeneous reaction

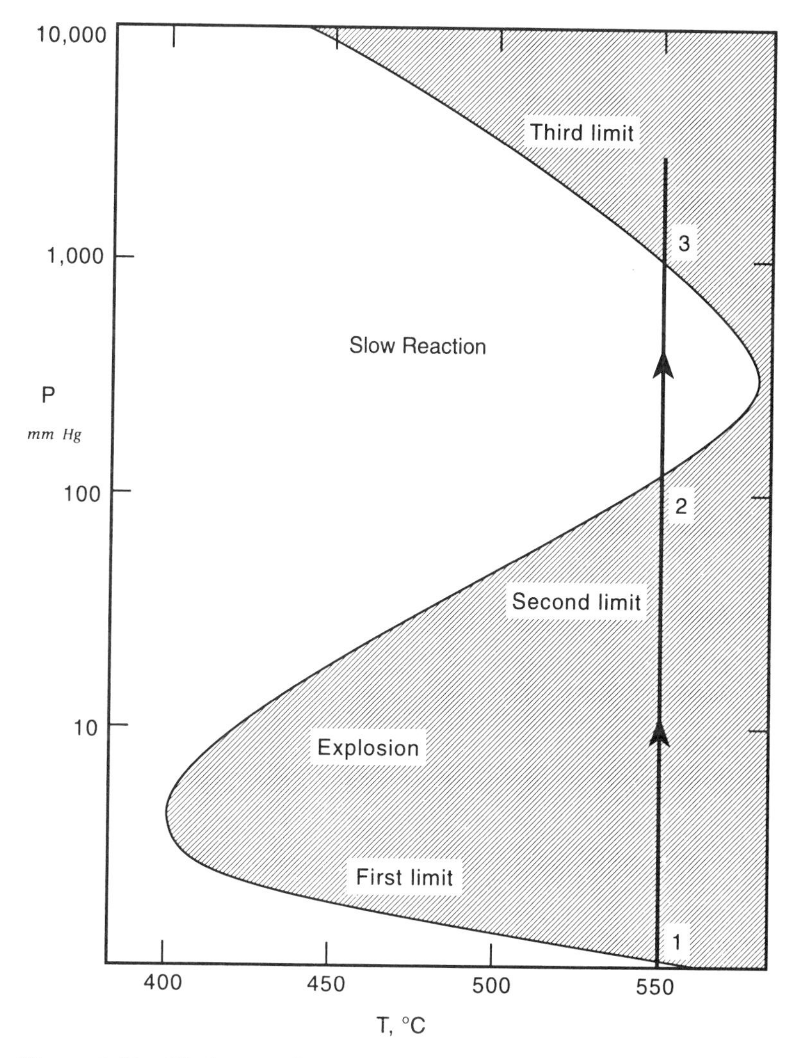

Figure 8.10 Hydrogen-Oxygen Thermal Explosion Limits

occurs without external energy input. In this region, slow heterogeneous surface reactions dominate, and overall reaction rate is a function of material and its surface area–reaction volume ratio. For H_2-O_2 mixtures above 600°C (1112°F), thermal explosions spontaneously occur in this region since chain-branching autocalytic reactions cause the combustion process to run away rapidly in a fashion similar to the physics of a nuclear explosion. Between 400 and 600°C (752–1112°F), a very different chemistry can be observed. Consider the fixed temperature line shown on Figure 8.10. Increasing pressure, i.e., collisions, will increase the overall rate of reactions but, below the first explosion limit, heat loss to walls controls the process, and radical quenching by recombination at the wall dominates below the first explosion limit. Further increases in pressure at fixed temperature raise the number of gas collisions, and the rate of energy release by gas-phase reactions becomes greater than the rate of energy loss to the walls. Increased radical formation above the first explosion limit then causes an autocatalytic explosion. Further increases in pressure will cause gas-phase radical recombination reactions to quench the process above the second explosion limit. Above the third explosion limit, additional thermolecular reactions cause the reaction again to explode spontaneously.

Kinetic mechanisms similar to those described above can be used to model fuel ignition delay, emissions formation, and other aspects of high-temperature oxidation thermochemistry. Detailed kinetic mechanisms are developed for only a few simple combustion reactions, and most of this work is done using experimental apparatus that is decoupled from any operating internal or external combustion system.

8.5 BASIC FLAME THEORY

Autocatalytic chain-branching reaction systems, as illustrated previously by the H_2-O_2 mechanism in Section 8.4, play a major role in the physics of propagating supersonic explosions or detonation waves. A subsonic reaction front that passes through a reactive gas mixture, termed a deflagration in Chapter 7, will now be referred to as a *flame*. The physics of flames is described in many ways, including specific flame reaction mechanisms, flame structures, and thermochemistry within a reactive fluid flow. Most important, flames associated with gas-phase combustion can be classified as being either *premixed,* i.e., flames that propagate through homogeneous fuel-air mixtures, or *diffusion,* i.e., flames that separate regions of pure fuel from pure oxidant.

An engineering analysis of flame physics will apply much of the thermochemical material covered in earlier chapters of this text. For example, overall flame temperature and corresponding heat release will depend

Table 8.6 Approximate Ignition Temperatures in 1-atm Air

Combustile	Formula	Temperature °F	Temperature °C
Sulfur	S	470	245
Charcoal	C	650	345
Fixed carbon (bituminous coal)	C	765	405
Fixed carbon (semibituminous coal)	C	870	465
Fixed carbon (anthracite)	C	840–1115	450–600
Acetylene	C_2H_2	580–825	305–440
Ethane	C_2H_6	880–1165	470–630
Ethylene	C_2H_4	900–1020	480–550
Hydrogen	H_2	1065–1095	575–590
Methane	CH_4	1170–1380	630–765
Carbon monoxide	CO	1130–1215	610–665
Kerosene	—	490–560	255–295
Gasoline	—	500–800	260–425

Source: Steam: Its Generation and Use, B & W Co. (1963)

on, among other factors, the nature and phase of the particular fuel being burned, total gas pressure, and reactant stoichiometry, as well as the overall reactive product gas composition. Actual peak flame temperatures closely correspond to trends predicted by adiabatic flame calculations with maximum values near stoichiometric reactant conditions. A description of physics within flames, however, is based in part on the kinetic material presented in this chapter. A bimolecular kinetic mechanism for chemistry occurring within a flame, for example, predicts characteristic reaction times on the order of bimolecular collision frequency, as well as a characteristic reaction dimension, or *flame thickness,* on the order of the molecular mean free path. Two factors will cause an actual flame to differ from the bimolecular kinetic flame model: occurrences of three-body collisions and energy loss from the flame as a result of radiation heat transfer.

Both thermal and radical-generated electromagnetic waves are produced by propagating reaction zones and often result in visible or luminous flame emissions. This apparent flame "color" is strongly affected by thermochemical properties such as the temperature, pressure, overall carbon–hydrogen atom ratio, and type of intermediate radicals produced within the flame. The major source of light emitted from within the primary reaction zone of a premixed flame is due to high concentrations of intermediate radicals present in the flame front and, therefore, is chemical kinetic in

Table 8.7 Flammability Limits in Air at Ambient Temperature and 1-atm Pressure

Full	Leanest %	Richest %
Acetone (C_3H_6O)	3.10	11.15
Acetylene (C_2H_2)	2.50	80.00
Ammonia (NH_3)	16.10	26.60
Benzene (C_6H_6)	1.41	7.10
Butane (C_4H_{10})	1.86	8.41
Butyl ($C_4H_{10}O$)	1.45	11.25
Butylene (C_4H_8)	1.98	9.65
Carbon monoxide (CO)	12.50	74.20
Ethane (C_2H_6)	3.22	12.45
Ethyl (C_2H_6O)	4.25	18.95
Ethylene (C_2H_4)	3.05	28.60
Heptane (C_7H_{16})	1.00	6.70
Hexane (C_6H_{14})	1.27	6.90
Hydrogen (H_2)	4.00	74.20
Methane (CH_4)	5.00	14.00
Methyl (CH_4O)	7.10	36.50
Octane (C_8H_{18})	0.95	—
Pentane (C_5H_{12})	1.42	7.80
Propane (C_3H_8)	2.37	9.50
Propyl (C_3H_8O)	2.15	13.50
Propylene (C_3H_6)	2.40	10.30
Propylene oxide (C_3H_6O)	2.10	21.50
Toluene (C_7H_8)	1.45	6.75

Adapted from Bureau of Mines, Bulletin 503, 1952, *Limits of Flammability of Gases and Vapors,* by H. F. Coward and G. W. Jones.

nature. Postflame emission is thermal rather than kinetic, and Planck's law for blackbody radiation generally characterizes the temperature and wavelength properties of this type of thermally excited emission.

H_2-O_2 flames produce little or no visual emission although H/O radical species such as the OH radical do emit electromagnetic energy in the ultraviolet region of the spectrum. Hydrocarbon flames will, in addition, emit electromagnetic energy in the infrared region as a result of the presence of water vapor and C/O species such as CO and CO_2. Stoichiometric and lean hydrocarbon flames often appear blue in the primary zone (kinetic), while fuel-rich flames are green. Flames with regions of high–carbon particulate concentrations will appear yellow (thermal). Soot-producing flames, having high concentrations of carbon with thermally excited blackbody radiant heat transfer, may be beneficial in certain heating and steam generation applications that require high heat transfer rates from combustion gases. Heat loss from such a flame, on the other hand, may be detrimental to the performance of propulsion machinery requiring adiabatic expansion of these high-pressure and high-temperature gases in order to produce power.

Premixed flames can assume many geometric structures, including flat, conical, spherical, and even cellular-shaped reaction zones. Central ignition of a quiescent gaseous fuel-air mixture contained within a spherical bomb calorimeter will produce a propagating, spherically shaped premixed flame that will travel through the reactive mixture toward the wall with unburned gases ahead of the reaction zone and burned products of combustion behind. An example of a stabilized premixed flame can be found in the inner core of the reaction zone produced by a bunsen burner in which fuel gas, supplied to the burner base, mixes with air entering the burner tube through the side parts and base, to produce a homogeneous mixture, which then exits at the top of the burner tube.

An important premixed flame parameter is *flame speed,* the propagation rate of a flame front relative to the homogeneous reactants. Premixed flames travel in a direction normal to their flame front, burning toward unburned reactants and away from burned products. Magnitudes for a subsonic deflagration front velocity can vary anywhere from a fraction of a centimeter per second to that of the local speed of sound. Maximum flame speeds occur in stoichiometric mixtures, with speed falling off for the same fluid conditions but with leaner and/or richer mixtures. Flame speeds can be steady or unsteady or can even appear to have a zero velocity, in which event the reactive flow passes through a stabilized flame thermally attached to an appropriate flameholder. Many factors influence flame speed, including initial thermodynamic state, overall fuel-air thermochemistry, and chemical kinetics, as well as the aerodynamics of flow. Basically, premixed flame speed is kinetically controlled; that is, flame speed is a function of chemical composition and of complex mechanisms and reaction rates occurring within the flame. Additional factors will also influence flame speed, such as whether the flame is traveling within a laminar or turbulent flow; whether boundary layer interactions occur, including heat transfer and friction losses at a solid wall; and whether the geometric configuration of the flame front is one-, two-, or three-dimensional. Combustion velocities must be experimentally measured; however, experimentally reported values obtained in a combustion laboratory facility for a particular fuel-air ratio may differ from those actually produced by operational combustion heat and/or power machinery.

The minimum overall temperature at which any fuel-air mixture begins to burn as its *ignition temperature,* see Table 8.6. On a microscopic level, flame initiation and propagation occur in a very local region of a homogeneous mixture when sufficient external energy is provided, such as that supplied by an electric discharge or a spark, which raises local temperature and, hence, collision frequency in excess of that associated with the ignition temperature. *Ignitability* is a means of assessing the relative ability to initiate ignition within a given fuel-air mixture. The self-ignition temperature, or temperature at which spontaneous ignition occurs, can be considerably different from temperatures within a flame. Reactions that begin at the self-ignition temperature, in fact, often show an induction delay, i.e., time period between initiation and observable reaction (~ 1 sec), while reaction time within flames is much shorter (~ 10^{-5}sec). Ignition temperatures and ignition energy measurements are extremely sensitive to experimental facilities and techniques used to obtain these parameters, and experimentally reported values, like flame speed, obtained in a combustion laboratory facility may differ from those actually produced by heat and/or power combustion machinery. *Flammability limits* bracket the rich-to-lean fuel-air mixture range beyond which flame propagation cannot occur after an ignition source is removed, even if the mixture is at its ignition temperature, see Table 8.7. Heat loss to a solid surface and loss of active radical species can cause a premixed flame to be extinguished. The inability to completely burn a gaseous charge at the walls of a closed vessel and the inability to ignite a gas-fired burner properly using improper electrodes are two examples of *flame quenching.* Quenching characteristics of premixed flames are reported in terms of a *quenching diameter,* the minimum diameter of a tube containing a stationary gaseous mixture through which a flame can pass without extinction.

A reaction zone that separates regions of pure fuel from regions of pure oxidant and that requires, in order to burn, mixing of these reactants with active radicals within the combustion zone is called a *diffusion flame.* Recall that heat transfer is a result of a temperature gradient and that mass flow is a result of a pressure gradient, whereas diffusion of species is a result of a concentration gradient. Burning coal, wood fires, and candles, as well as diesel fuel droplet combustion, are examples of combustion processes that produce diffusion flames. It is difficult to burn a single log or lump of coal because radiation heat loss from burning solids is quite high, and two or more reradiating surfaces are needed to reduce these heat losses. Heat from the flame of a candle vaporizes candle wax, which then mixes by diffusion with the surrounding air to burn in a diffusion flame.

Also, burned gases within the inner core of a bunsen burner flame mix with the surrounding air to produce a diffusion flame at the outer core of the bunsen burner flame.

Diffusion flames are more dependent on mixing rate between reactants than on kinetic mechanisms, as is the case with premixed flames. No single parameter, such as a burning velocity, can be associated with diffusion flames because the physics for diffusion flames is considerably more complex. One-dimensional premixed flame models have been developed, but the three-dimensional nature of a diffusion flame, with its associated temperature, pressure, and concentration gradients, does not lend itself to simple modeling. Low-speed laminar flow diffusion flames, such as those produced by a candle, are controlled by molecular diffusion. High-speed diffusion flames associated with industrial burners and gas turbine combustors depend on turbulent mixing, in which case fuel is introduced as discrete droplets and the aerodynamics of flow are critical to a proper burn.

In this chapter, basic principles of gaseous chemical kinetics and flame chemistry have been introduced. Most practical combustion machinery utilizes liquid and solid fuels. In these instances, atomization and vaporization of liquid fuels and reduction in size and dispersion of solid fuels become essential. These issues, as well as the heterogeneous chemistry of soot formation, cool flames, and safety topics such as flashback and blowoff, have not been treated. A more complete discussion of these and other chemical kinetic subjects can be found in specific literature.

PROBLEMS

8.1 Equal volumes of ideal gases at the same temperature and total pressure contain equal numbers of molecules. Using the additional fact that one mole of a gas at *STP* will occupy 22.4 liters, determine the universal gas constant for an ideal gas in units of (a) atm liters/g mole K; (b) ft lbf/lb mole °R; (c) Btu/lb mole °R; and (d) N m/kg mole K.

8.2 Calculate the mean free path for the gases found in Table 8.1 at (a) 298K and 10-atm pressure; (b) 298K and 0.1-atm pressure; and (c) 1800K and 1-atm pressure.

8.3 Consider 1 ft^3 of air at *STP*. Determine (a) the acoustic velocity, ft/sec; (b) the most probable speed, ft/sec; and (c) the mean square speed, ft/sec. Repeat parts (a)–(c) for $T = 900$ °R.

8.4 Plot the speed distribution function Ψ as a function of molecular velocity V for helium at (a) 0°C; (b) 300°C; and (c) 900°C. From your results, show that by increasing temperature, the distribution will become broader, yielding more atoms having higher velocities and, hence, greater molecular energy.

8.5 A 0.5-m^3 tank conntains N_2 at 101 kPa and 298K. Calculate (a) the molecular density, molecules/cc; (b) the root mean square velocity, cm/sec; and (c) the mean molecular kinetic energy, N m/molecule.

8.6 The Loschmidt number is defined as the number of molecules per unit volume of any gas at a given temperature and pressure. Show that this number is the same for any gas that obeys the ideal-gas law. Calculate the standard Loschmidt number for 1-atm pressure and 25°C, no./cc.

8.7 Show that the mean molecular speed can be related to the sonic velocity, c as

$$V_m = \left[\frac{8}{\pi\gamma}\right]^{1/2} c$$

8.8 The internal energy of a monatomic gas can be treated as having an $\overline{R}T/2$ contribution for each directional degree of freedom. Using this kinetic energy model, calculate (a) the constant-volume molar specific heat, kJ/kg mole K; (b) the constant-pressure molar specific heat, kJ/kg mole K; and (c) the molar specific heat ratio for a monatomic gas.

8.9 The internal energy of a diatomic gas can be treated as having an $\overline{R}T/2$ contribution for each directional degree of freedom plus an $\overline{R}T/2$ contribution for each rotational degree of freedom. Using this kinetic energy model, calculate (a) the constant-volume molar specific heat, kJ/kg mole K; (b) the constant-pressure molar heat, kJ/kg mole K; and (c) the molar specific heat ratio for a low-temperature diatomic gas.

8.10 At high temperatures, a diatomic gas can also have an $\overline{R}T$ contribution from a vibrational energy contribution. Using this kinetic energy model, calculate (a) the constant-volume molar specific heat, kJ/kg mole K; (b) the constant-pressure molar specific heat, kJ/kg mole K; and (c) the molar specific heat ratio for a high-temperature diatomic gas.

8.11 Using the bond energy values in Table 8.2, calculate ΔH_f^0 for the following gaseous compounds: (a) methane, CH_4; (b) methylene, C_2H_4; (c) n-hexane, C_6H_{14}; and (d) methanol, CH_3OH, kcal/g mole.

8.12 Calculate the bimolecular collision frequency between N_2 and He as compared to that for the collision frequency between N_2 and O_2, no./sec.

8.13 Determine the mean speeds for the gases found in Table 8.1 at (a) 0°C; (b) 300°C; (c) 900°C; and (d) 1200°C, m/sec.

8.14 The initiation step in most high-temperature oxidation reactions involves the thermal decomposition of the fuel. Let the pyrolysis of a fuel A be represented by the elementary reaction

$$A \rightarrow B + C$$

where

[A]	=	fuel concentration
[B]	=	radical product B concentration
[C]	=	radical product C concentration

For an initial concentration of A of $[A]_0$, find an expression for (a) the rate of dissociation of A; (b) the rate constant for the reaction k; and (c) the half-life for the process $t_{1/2}$.

8.15 Consider the elementary reaction

$$A + A \rightarrow \text{products}$$

Assume a molecular weight of 50 g/g mole for A, the effective molecular diameter is 3.5×10^{-8} cm, and the activation energy for the reaction is 40,000 cal/g mole. Determine the temperature rise needed to double the rate of reaction for an initial temperature equal to (a) 500K; (b) 1000K; (c) 1500K; and (d) 2000K. (Assume $[A]_0 = 1$ g mole/cc).

8.16 At 500K, the rate of a bimolecular reaction is 10 times the rate at 400K. Using collision theory, evaluate the activation energy ΔE of this reaction at these conditions.

8.17 The elementary reaction

$$O_2 + M \underset{k_2}{\overset{k_1}{\rightleftarrows}} O + O + M$$

has a forward rate constant

$$k_1 = 3.61 \times 10^{18}\, T^{-10} \exp -\left\{ \frac{118028}{\overline{R}T} \right\}$$

For a reaction temperature of 2000K, calculate (a) the equilibrium constant K_p; (b) the equilibrium constant K_c; and the reverse rate constant k_2.

8.18 The gas-phase thermal decomposition of nitrous oxide N_2O at 1030K was studied in a constant-volume vessel at various initial pressures of N_2O. The half-life data obtained for the reactions were found to be as given below:

P_0 mm Hg	52.5	139	290	260
$t_{1/2}$sec	860	470	255	212

Determine a rate constant equation that fits these data.

8.19 The thermal decomposition of nitrogen dixoide can be determined via the elementary initiation reaction

$$NO_2 + NO_2 \xrightarrow{k} NO + O + NO_2$$

Using the rate constants for the NO_2 reaction given below, determine the following: (a) the differential equation for NO_2 rate of decomposition; (b) the integrated expression for $[NO_2]$ versus time; (c) the reaction activation energy, cal/g mole; (d) the effective NO_2 molecular diameter, m.

$$k<592\text{K}> = 522 \text{ cm}^3/\text{sec g mole}$$
$$k<656\text{K}> = 5030 \text{ cm}^3/\text{sec g mole}$$

8.20 The formation of nitrogen oxide in the high-temperature H_2-O_2 system has been postulated to occur via the chain reaction

$$O_2 + N_2 \underset{k_2}{\overset{k_1}{\rightleftarrows}} O + O + N_2$$

$$N_2 + O \underset{k_4}{\overset{k_3}{\rightleftarrows}} NO + N$$

$$O_2 + N \underset{k_6}{\overset{k_5}{\rightleftarrows}} NO + O$$

(a) Write an expression for the rate of formation of [NO], [O], and [N].
(b) Determine the steady-state concentration of the radicals [O] and [N].
(c) Develop an expression for the rate of formation of [NO] in terms of the steady-state concentration in (b).

8.21 The formation of many important radical species involved in certain oxidation reactions can be experimentally studied using radiation techniques. Beer's law expresses the decrease in radiation intensity of wavelength υ after it passes through an absorbing medium as

$$\frac{I_l^\upsilon}{I_0^\upsilon} = \exp\{-[A]\, l\, \epsilon_\upsilon\}$$

where

I_0 = incident radiation intensity
I_l = transmitted radiation intensity
[A] = concentration of radical, g moles/cc
υ = radiation wavelength, Å
l = radiation pathlength, cm
ϵ_υ = molar extinction coefficient

Consider the formation of radical A via the elementary reaction

$$A_2 + M \xrightarrow{k} 2A + M$$

Use Beer's law to develop (a) an equation for the rate of formation of the radical A, and (b) an expression for k in terms of the initial concentrations of $[A_2]_0$ and $[A]_0$.

8.22 The maximum spectral emissive power for blackbody thermal radiation is given by Wien's displacement law

$$\lambda_{max} T = 2897 \ \mu\text{m K}$$

Determine the temperature required for maximum emission in the (a) ultraviolet ($\lambda = 0.3\ \mu$m), K; (b) visible ($\lambda = 0.5\ \mu$m), K; and (c) infrared ($\lambda = 1.2\ \mu$m), K.

8.23 Laminar premixed flame speeds S_L for fuel-air mixtures are found to fit the relationship empirically:

$$S_L = a + bT^n_{init} \qquad \text{m/sec} \qquad 200 < T_{init} < 600\text{K}$$

where

Gas	a	b	n
methane	0.08	1.6×10^{-6}	2.11
propane	0.10	3.4×10^{-6}	2.00
benzene	0.30	7.9×10^{-9}	2.92
isooctane	0.12	8.4×10^{-7}	2.19

Calculate the laminar flame speed for methane for an initial temperature of (a) 200K; (b) 400K; and (c) 600K, m/sec.

8.24 Repeat 8.23 for (a) propane, (b) benzene, and (c) isooctane.

8.25 The pressure dependency for a laminar premixed flame is found to satisfy the relationship

$$\ln S_L = c + d \ln P \qquad \text{m/sec}$$

where

$S_L = 0.225$ m/sec at 150 kPa
$S_L = 0.223$ m/sec at 405 kPa

Plot (a) $\ln S_L$ versus $\ln P$, and (b) S_L versus P; (c) determine the values of c and d.

9

Combustion Engine Testing

9.1 INTRODUCTION

In the first eight chapters, considerable time was devoted to several significant engineering aspects of combustion science. Obviously, a description of high-temperature thermochemistry is essential to understanding the general performance of any internal combustion engine. Even so, proper design and operation of spark-ignition, compression-ignition, and gas turbine engines cannot be based on theoretical and analytical modeling alone. Much of what is known today concerning these devices has been demonstrated by experimentation. Before turning to the fuel-engine energy characteristics of particular *internal combustion* (IC) engines in coming chapters, the general subject of engine testing will first be covered. In this chapter, therefore, useful definitions and standard engine testing terminology will be reviewed. Emphasis will be given to piston-cylinder devices.

9.2 INTERNAL COMBUSTION ENGINE NOMENCLATURE

Successful commercial development of IC engines has been done, in part, to the reciprocating piston-cylinder configuration; see Figure 9.1. Also, the development of potential future prime movers, such as the Stirling engine,

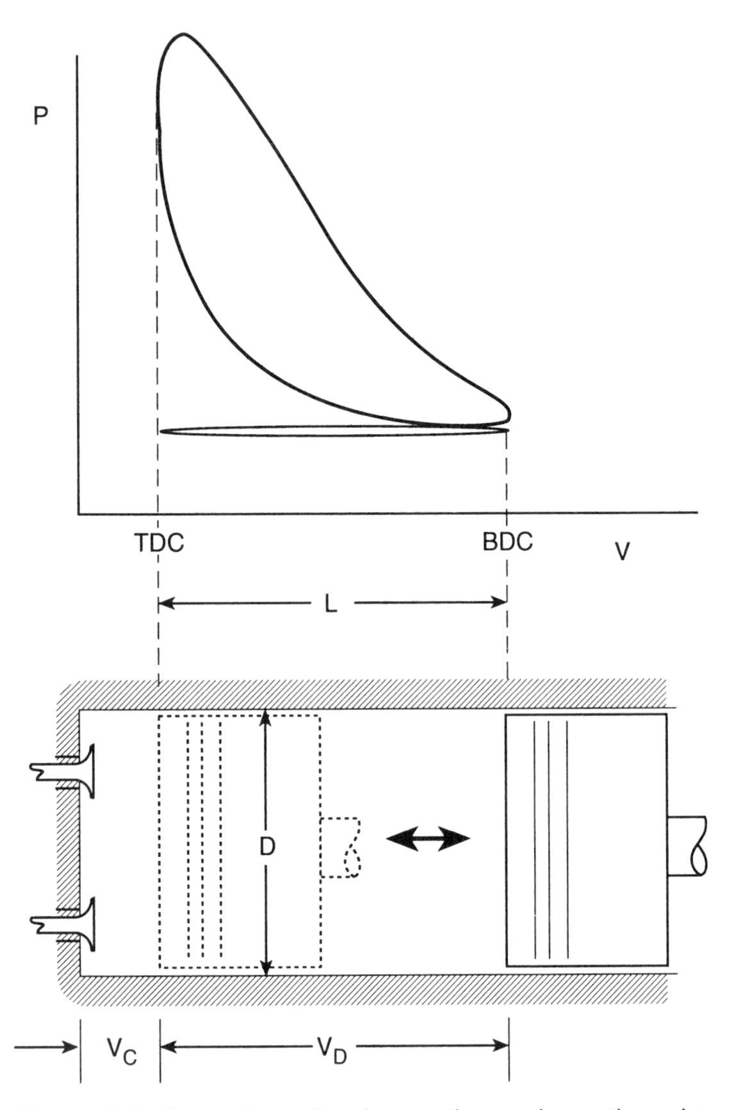

Figure 9.1 Internal combustion engine reciprocating piston cylinder configuration.

will utilize a positive displacement geometry. By referring to Figure 9.1, some useful terms pertinent to such machines can better be understood.

The piston diameter D is termed the *bore,* while the piston travel in one direction, L, is called the *stroke.* An expansion stroke motion requires that a piston travel from its piston upper limit of travel, i.e., *top dead center* (TDC), to its piston lower limit of travel, i.e., *bottom dead center* (BDC). Common engineering practice is to state a given engine geometry in terms of its bore and stroke as $D \times L$. The volume swept by a piston during one stroke, V_D, is referred to as cylinder *displacement volume,* described in terms of bore and stroke as

$$V_D = \frac{\pi}{4} D^2 L \qquad \frac{\text{m}^3}{\text{stroke}} \left(\frac{\text{ft}^3}{\text{stroke}} \right) \tag{9.1}$$

Volume above the piston at the TDC location is referred to as the *clearance volume,* V_C, and is often expressed as a percentage of the displacement volume c, where $0 < c < 1.0$:

$$V_C = cV_D \qquad \text{m}^3 \ (\text{ft}^3) \tag{9.2}$$

The geometric ratio of the volume at BDC to that at TDC is termed *compression ratio,* r_v, or

$$r_v = \frac{\text{volume at BDC}}{\text{volume at TDC}} > 1.0 \tag{9.3a}$$

and, in terms of the clearance and displacement volumes, is expressed as

$$r_v = \frac{V_C + V_D}{V_C} = 1 + \frac{V_D}{V_C} = 1 + \frac{1}{c} \tag{9.3b}$$

Another useful parameter is the *pressure ratio* r_p which, for the piston-cylinder case, is given by the expression

$$r_p = \frac{\text{pressure at TDC}}{\text{pressure at BDC}} > 1.0 \tag{9.4}$$

EXAMPLE 9.1 A 4 × 5-in. single-cylinder engine is being used as a research engine. The engine, a four-stroke IC engine, is running at 1800 rpm with inlet conditions of 73°F and 14.5 psia. Determine (a) the displacement volume V_D, ft^3; (b) the clearance volume V_C for an 8:1 compression ratio, ft^3; (c) the ideal volumetric flow rate $\dot{V}$, ft^3/min. and (d) the ideal air mass flow rate $\dot{m}$, lbm/min.

Solution:

1. Displacement volume:

a. $V_D = \dfrac{\pi D^2}{4} \times L = \dfrac{\pi}{4}\left(\dfrac{4}{12}\,\text{ft}\right)^2 \times \left(\dfrac{5}{12}\,\text{ft}\right) = 0.0364\ \text{ft}^3/\text{stroke}$

2. Clearance volume:

b. $V_C = \dfrac{V_D}{r_v - 1} = \dfrac{0.0364}{7} = 0.00520\ \text{ft}^3/\text{stroke}$

3. Ideal volumetric flow rate:

$$\dot{V} = \frac{V_D < \text{ft}^3/\text{power stroke} > N < \text{rev/min} >}{n < \text{revolutions/power stroke} >} = \frac{(0.0364)(1800)}{2}$$

c. $\dot{V} = 32.76\ \text{ft}^3/\text{min}$

4. Mass flow rate, assuming induced charge is air alone:

$$\dot{m} = \frac{P\dot{V}}{RT} = \rho < \frac{\text{lbm}}{\text{ft}^3} > \times\ \dot{V} < \frac{\text{ft}^3}{\text{min}} >$$

d. $\dot{m} = \dfrac{(14.5\ \text{lbf/in.}^2)(144\ \text{in.}^2/\text{ft}^2)(32.76\ \text{ft}^3/\text{min})}{(1545/28.96\ \text{ft lbf/lbm°R})(533°\text{R})} = 2.41\ \text{lbm/min}$

Comments: In an actual engine, the irreversible nature of the exhaust and intake processes will prevent the ideal induction process from occurring. This discrepancy is due to thermal and fluid effects, such as residual gas composition, temperature and pressure; intake air-fuel ratio; inlet pressure and temperature; fluid viscosity; and losses due to mixing and fluid flow through valve openings.

Each mechanical cycle of a reciprocating IC engine involves an intake, compression, combustion, expansion, and exhaust of a (fuel)-air mixture. In a *two-stroke engine,* shown in Figure 9.2, two strokes, i.e., an intake-compression stroke and a power-exhaust stroke, repeat each mechanical cycle. In a *four-stroke engine,* shown in Figure 9.3, each mechanical cycle consists of separate intake, compression, power, and exhaust strokes.

At the TDC position after an exhaust stroke, it is evident that it will be impossible to fill the entire volume on intake with a fresh charge. This is due in part to the fact that each cylinder will contain trapped products of combustion, called *residuals,* in the clearance volume. A larger clearance volume would therefore increase the dilution of the intake and, because of gas dynamic and thermal effects, the actual volumetric intake may be less than the ideal volumetric intake. The *volumetric efficiency* η_v relates the actual and ideal induction performance of an engine intake process as

$$\eta_v = \frac{\text{mass (volume) actually induced at inlet } P\ \&\ T}{\text{engine displacement mass (volume)}} \times 100 \qquad (9.5)$$

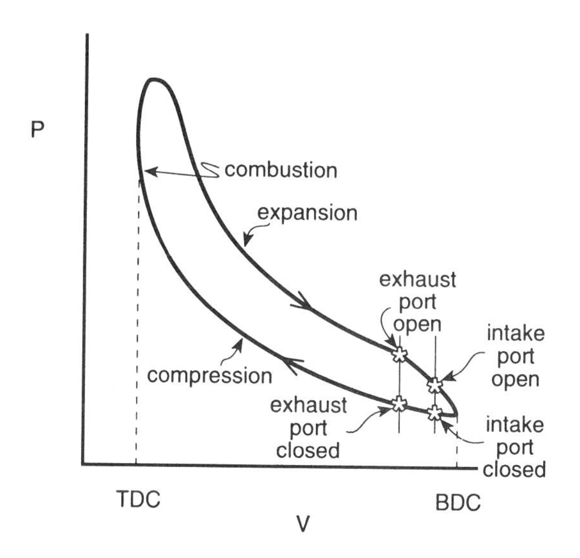

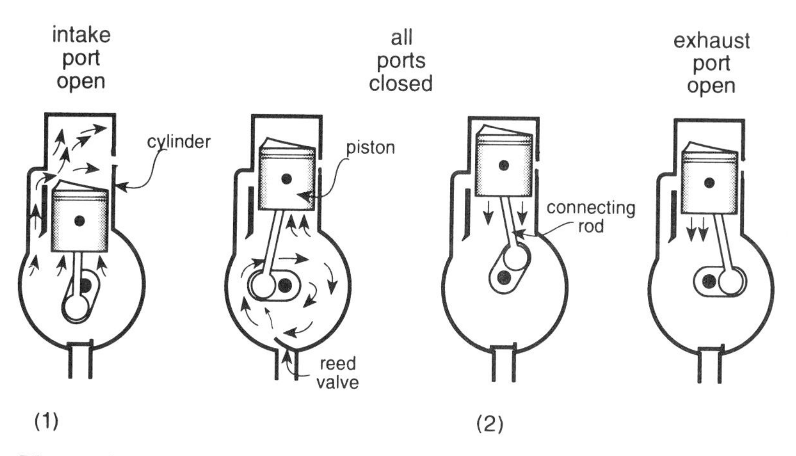

Figure 9.2 Two-stroke IC engine: (1) intake-compression stroke; (2) power-exhaust stroke.

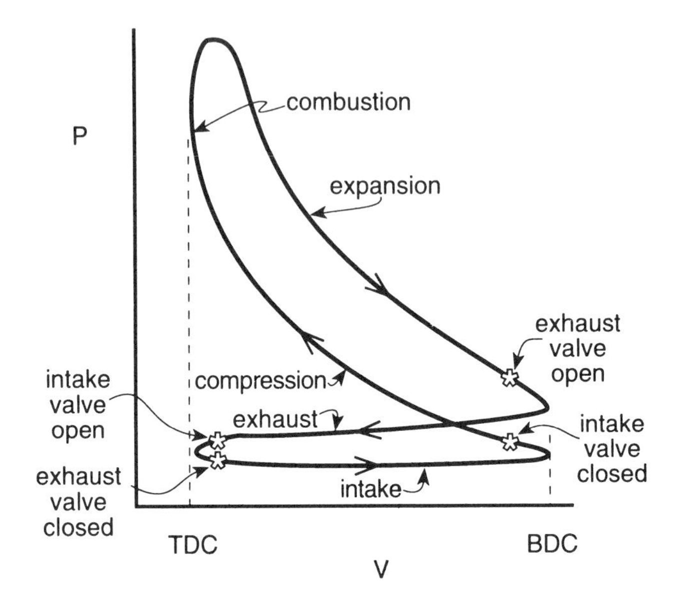

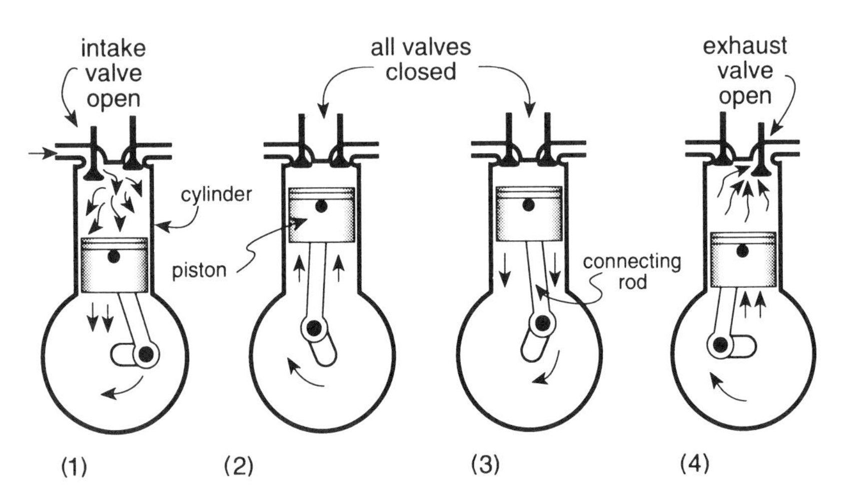

Figure 9.3 Four-stroke IC engine: (1) intake stroke; (2) compression stroke; (3) power stroke; and (4) exhaust stroke.

Several of these definitions have direct bearing on the energy characteristics of reciprocating piston-cylinder geometries as well as rotating machinery.

9.3 INDICATED ENGINE PERFORMANCE

When describing the energy transfers associated with a particular IC engine, it is essential to specify clearly the system boundaries chosen for analysis. For example, considering the (fuel)-air mixture as a system, the thermodynamic description is termed an *indicated engine analysis* and deals with the heat/work transfers to/from the gases within an engine. Consideration of the complete engine or vehicle as a thermodynamic system is termed a *brake engine analysis.*

Indicated performance predictions ideally can be based on thermodynamic arguments alone. Results based on an indicated *air standard cycle analysis* can be found in most engineering thermodynamic textbooks. In this approach, a closed system consisting of only air, i.e., air standard cycle, replaces the actual fuel-air mixture, and the intake and exhaust strokes are replaced by heat transfer processes. A *cold air* cycle approach uses a room-temperature specific heat ratio $\gamma = 1.4$ in performance calculations, while a *hot air* cycle with $\gamma = 1.3$ is often used to represent actual IC engine energetics more realistically.

Thermodynamic analysis of air standard cycles often utilizes a pressure-volume, or *P-V,* diagram to visualize the engine-energy transfer. Theoretical or *ideal indicated P-V* diagrams can be drawn from first principles for Otto, Diesel, and/or Stirling engines. From a basic energy analysis, ideal peak pressure, net work and thermal efficiency for those classic heat engines can be predicted. Direct experimental measurement using *P-V* history of an actual engine, either by mechanical or electronic techniques, produces an *indicator diagram.* Figures 9.2 and 9.3 show generalized *indicated P-V* diagrams for typical IC engines. Obviously, considerable differences will distinguish between ideal and actual indicator diagrams for both spark- and compression-ignition engines.

Boundary expansion work for closed systems can be evaluated in terms of pressure and volume as a summation of the infinitesimal force through distances during the cycle as

$$W_I = \int P\,dV \qquad \frac{\text{N m}}{\text{stroke}}\left(\frac{\text{ft lbf}}{\text{stroke}}\right) \tag{9.6}$$

where

W_I = indicated work	N m/stroke (ft lbf/stroke)
P = cylinder pressure	N/m^2 (lbf/ft^2)
V = cylinder volume	m^3/stroke (ft^3/stroke)

Defining an average or *mean effective pressure* $\overline{P}$, *one can calculate indicated cycle work during the power stroke as*

$$W_I = \int \overline{P}\, dV = \overline{P} \int dV = \overline{P}(V_{\mathrm{BDC}} - V_{\mathrm{TDC}}) = \overline{P} V_D \qquad (9.7a)$$

or

$$\overline{P} = \frac{W_I}{V_D} \qquad \frac{\mathrm{N/m^2}}{\mathrm{stroke}} \left(\frac{\mathrm{lbf/ft^2}}{\mathrm{stroke}} \right) \qquad (9.7b)$$

Indicated mean effective pressure (*IMEP*) is thus seen as an "average" pressure that, if acting over a power stroke, would produce the same work as the actual cycle. Using the definition for displacement volume, indicated work per cylinder can be expressed as

$$W_I = \overline{P} \left(\frac{\pi}{4} \right) D^2 L = \overline{P} A L \qquad (9.8)$$

Note that an *ideal* indicated mean effective pressure can be determined from classic cycle analysis, whereas an *actual* indicated mean effective pressure must be obtained directly from numeral integration of an operating engine indicator card.

EXAMPLE 9.2 The mechanical or electronic measurement of an actual IC engine pressure-volume diagram is termed an engine indicator diagram. The recorded experimental information can be used to determine the indicated power produced by the combustion gases in the cylinder. Consider the idealized indicator card diagram of a 3-in. × 5-in. four-stroke engine operating at 1400 rpm shown below.
Using the diagram shown, calculate (a) peak cycle pressure, psi; (b) the indicated mean effective pressure, psi; and (c) the indicated engine power, ihp.

Solution:

1. Peak pressure from the diagram:

   a. $P_{max} = 400$ psia

2. Net work/power stroke from the diagram:

   $W_{net} = \oint P\, dV$

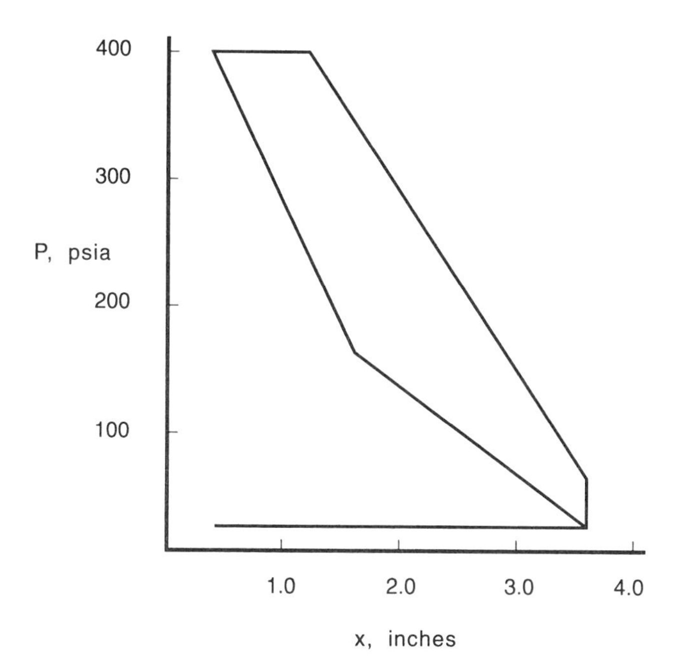

The net work is given by the enclosed area.

$$A_{\text{net}} = (400 - 20 \text{ psia}) \times (3.6 - 0.4 \text{ in.})$$
$$- \tfrac{1}{2}(400 - 60) \times (3.6 - 1.2) - \tfrac{1}{2}(160 - 20) \times (3.6 - 1.6)$$
$$- \tfrac{1}{2}(400 - 160) \times (1.6 - 0.4) - (160 - 20) \times (1.6 - 0.4)$$
$$= 356 \text{ psi in.}$$

This area is assumed to equal the product of the mean effective pressure and the *diagram* displacement length.

$$A_{\text{net}} = \overline{P} X_{\text{base}} = 356 \text{ psi in.}$$

or

b. $$\overline{P} = \frac{356 \text{ psi in.}}{(3.6 - 0.4) \text{ in.}} = 111.25 \text{ psi}$$

3. Ideal engine power:

$$\dot{W}_I = \frac{\overline{P} < \text{lbf/ft}^2 > L < \text{ft} > A < \text{ft}^2 > N \text{ (rev/min) } C < \text{no. cylinders} >}{(33{,}000 \text{ ft lbf/hp min}) \; n < \text{rev/power stroke} >}$$

and

$$\dot{W}_I = \frac{(111.25 \text{ lbf/in.}^2)(144 \text{ in.}^2/\text{ft}^2)(5/12)(\pi/4)(3/12)^2(1400)}{(33{,}000)(2)}$$

c. $= 6.95$ ihp/cylinder

Comments: An experimentally recorded indicator card can be integrated by digital or planimetric techniques to determine the actual indicated mean effective pressure. Note also that the mechanically or electrically produced illustrations require a linear vertical scale, "pressure per unit vertical displacement," and a horizontal linear scale, "linear displacement per unit volume." Such a figure allows the *IMEP* to be obtained without recourse to thermodynamic or engine information.

Practical considerations of an IC engine more frequently are in terms of the *rate* at which work is being produced, i.e., power. The *indicated engine power* can be obtained from the *IMEP* in SI as:

$$\dot{W}_I = \frac{\overline{P} < \dfrac{\text{N/m}^2}{\text{stroke}} > L < \text{m} > A < \text{m}^2 > N < \dfrac{\text{rev}}{\text{min}} > C < \text{no. cylinders} >}{\left(60 \dfrac{\text{sec}}{\text{min}}\right)\left(1000 \dfrac{\text{N m}}{\text{kW sec}}\right) n < \dfrac{\text{revolutions}}{\text{power stroke}} >} \text{ kW} \tag{9.9a}$$

Indicated power is expressed in Engineers' units as:

$$\dot{W}_I = \text{IHP} = \frac{\overline{P} < \dfrac{\text{lbf/ft}^2}{\text{stroke}} > L < \text{ft} > A < \text{ft}^2 > N < \dfrac{\text{rev}}{\text{min}} > C < \text{no. cylinders} >}{\left(33{,}000 \dfrac{\text{ft lbf}}{\text{hp min}}\right) n < \dfrac{\text{revolutions}}{\text{power stroke}} >} \text{ hp} \tag{9.9b}$$

Engine power is a result of energy released during particular chemical reactions occurring near TDC. In thermodynamic considerations of ideal indicated cycle performance, heat addition is used to approximate this process. The maximum or ideal theoretical heat added to any cycle can be expressed in terms of a fuel heating value as

$$\dot{Q} = \dot{m}_{\text{fuel}} \times LHV_{\text{fuel}} \quad \frac{\text{kJ}}{\text{sec}}\left(\frac{\text{Btu}}{\text{min}}\right) \tag{9.10}$$

where

$$\dot{m}_{\text{fuel}} = FA \times \dot{m}_{\text{air}} \quad \frac{\text{kg fuel}}{\text{sec}}\left(\frac{\text{lbm}}{\text{min}}\right) \tag{9.11}$$

with

$$\dot{m}_{\text{air}} = \left(\frac{P\dot{V}_D}{R_{\text{air}}T}\right)_{\text{inlet}} \quad \frac{\text{kg air}}{\text{sec}}\left(\frac{\text{lbm}}{\text{min}}\right) \tag{9.12}$$

Indicated thermal efficiency η_{thermal} for an air standard thermodynamic cycle can be written as

$$\eta_{\text{thermal}} = \frac{\text{net work (power) output}}{\text{net heat (flux) addition}} \tag{9.13a}$$

In SI:

$$\eta_{\text{thermal}} = \frac{\dot{W}_I \; <\text{kW}>}{\dot{m}_{\text{fuel}} \; <\text{kg/sec}> \; LHV \; <\text{kJ/kg}>} \times 100 \tag{9.13b}$$

and in Engineers' units:

$$\eta_{\text{thermal}} = \frac{IHP \; <\text{hp}> \; (2545 \text{ Btu/hp hr})}{\dot{m}_{\text{fuel}} \; <\text{lb/hr}> \; LHV \; <\text{Btu/lbm}>} \times 100 \tag{9.13c}$$

Indicated heat rate (IHR) expresses the heat addition required per unit of power produced and, as such, is as "inverse" of thermal efficiency, or

$$IHR = \frac{\dot{Q}}{\dot{W}_I} = \frac{\dot{m}_{\text{fuel}} HV}{\dot{W}_I} \quad \frac{\text{kJ}}{\text{kW sec}} \left(\frac{\text{Btu}}{\text{hp min}} \right) \tag{9.14}$$

The *indicated specific fuel consumption* ($ISFC$) determines a normalized fuel consumption rate to the indicated engine power output as

$$ISFC = \frac{\dot{m}_f}{\dot{W}_I} \quad \frac{\text{kg fuel}}{\text{kW sec}} \left(\frac{\text{lbm}}{\text{hp min}} \right) \tag{9.15}$$

Specific fuel consumption can be considered as a statement of fuel consumption per unit of engine output. Smaller values for this expression imply that less fuel is necessary to produce a unit of power output.

It is essential to recognize that the central aspect of any engine operation or design begins with understanding fundamental processes that occur within the engine combustion chamber. The magnitude of each of these indicated engine performance parameters will vary as a function of engine type, load, fuel-air ratio, fuel type, and a variety of additional variables.

9.4 BRAKE ENGINE PERFORMANCE

The second law of thermodynamics predicts that energy in transition degrades as a result of irreversible processes. From such principles of irreversibility as entropy production and friction, it is apparent that any energy transfer associated with gases within the piston, i.e., indicated engine performance, will be considerably different from those resulting energy transfers to the engine drive train, i.e., *brake engine performance*. Because of mechanical and thermal energy losses within and from an engine, its brake work, or power, will therefore be less than the corresponding indicated work or power

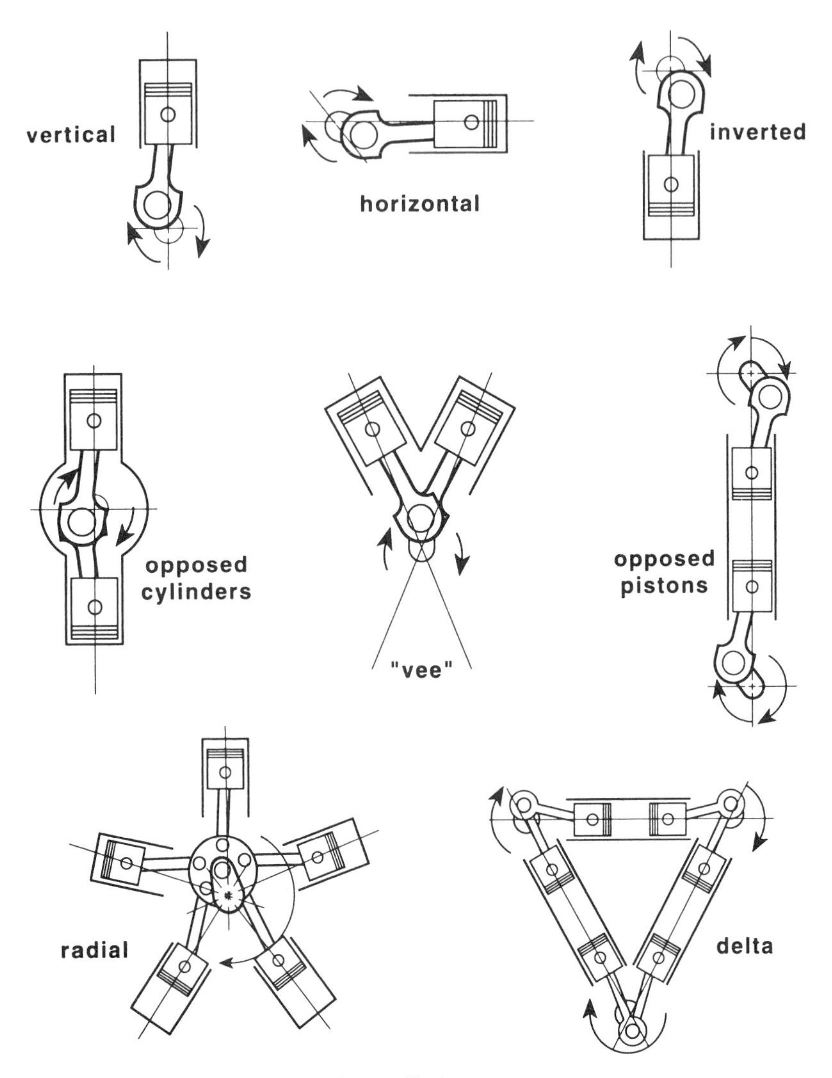

Figure 9.4 Various IC engine cylinder arrangements.

$$\dot{W}_I > \dot{W}_B \qquad \text{kW (hp)} \tag{9.16}$$

and

$$\dot{W}_I = \dot{W}_B + \dot{W}_F \qquad \text{kW (hp)} \tag{9.17}$$

where

$\dot{W}_I$ = indicated power (generated)
$\dot{W}_B$ = brake power (delivered)
$\dot{W}_F$ = friction power (lost)

The engine *mechanical efficiency,* η_{mech}, provides a means of determining how effective a particular engine is at reducing internal losses.

$$\eta_{\text{mech}} = \frac{\dot{W}_B}{\dot{W}_I} \times 100 \tag{9.18}$$

Most commercial IC engines, except for certain research models, are produced in multicylinder configurations. Figure 9.4 illustrates a few of the specific arrangements that have been used to develop engines that satisfy particular design factors, such as vibrational, inertial, and frictional influences while, at the same time optimizing items such as power, torque, and/or engine size.

Internal combustion engine power is often transmitted through mechanical gearing, electric generators, or propellor shafting. These added energy transfers will result in further reductions in actual delivered power. Such losses can be accounted for by introducing transmission, generator, and/or propellor efficiencies.

Brake engine performance, unlike ideal indicated performance, cannot be determined from basic theory and therefore requires experimental laboratory measurements. Brake characteristics are often obtained by means of a power absorption device termed a *dynamometer.* The simplest and earliest means of measuring brake performance was by use of a *friction* dynamometer or brake, shown in Figure 9.5. Engine power is dissipated by friction loading the prony brake absorption band. Since mechanical work can be described as the product of a force acting through a distance, the brake work during one revolution is given by the expression

$$W_B = \text{distance} \times \text{force} = (2\pi r) \times f \tag{9.19}$$

where

W_B = brake working during one revolution, N m (ft lbf)
r = drive shaft wheel radius, m (ft)
f = friction force, N (lbf)

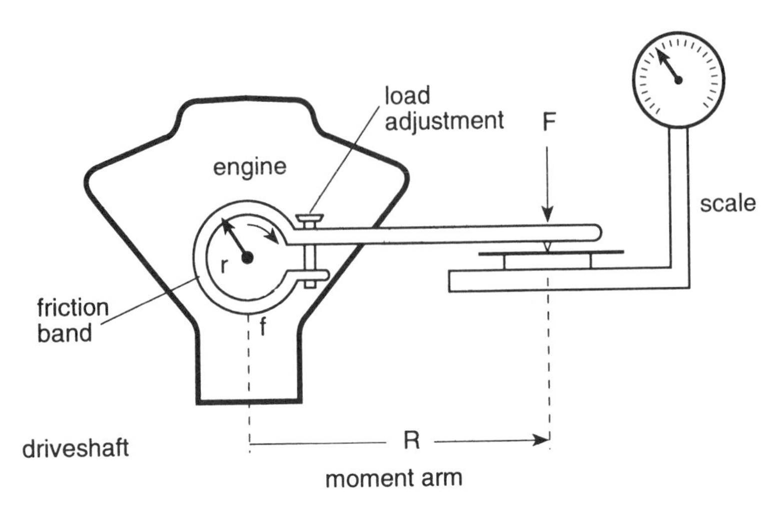

Figure 9.5 Friction band engine dynamometer schematic.

Brake torque τ is the product of a moment arm and corresponding brake force or

$$\tau = rf \qquad \text{N m (ft lbf)} \tag{9.20a}$$

The drive shaft torque is absorbed by an opposing dynamometer measuring torque and at dynamic equilibrium,

$$\tau = rf = RF \qquad \text{N m (ft lbf)} \tag{9.20b}$$

where

R = measuring moment arm, m (ft)

F = measuring load as determined by a scale, N (lbf)

Torque is a statement of an amount of work a prime mover can produce and is therefore a measure of how large a load a given engine is capable of pulling. Since engine power is a *rate* of work term, it will specify how fast a particular engine will pull a given load. *Brake power* $\dot{W}_B$ can be determined from torque in SI as:

$$\dot{W}_B = \frac{2\pi R <\text{m/rev}> F <\text{N}> N <\text{rev/sec}>}{(1000 \text{ N m/kW sec})} \qquad \text{kW} \tag{9.21a}$$

In Engineers' units brake power is expressed as:

$$\dot{W}_B = BHP = \frac{2\pi R <\text{ft/rev}> F <\text{lbf}> N <\text{rev/min}>}{(33{,}000 \text{ ft lbf/hp min})} \qquad \text{hp} \tag{9.21b}$$

Notice that brake power, unlike indicated power, is not modified by the number of power strokes per revolution or number of cylinders.

A *brake mean effective pressure* ($BMEP$) can be obtained from the brake power equation (9.21) and the indicated power equation (9.9) in SI as:

$$BMEP = \frac{\dot{W}_B <\text{kW}> \left(1000 \dfrac{\text{N m}}{\text{kW sec}}\right) \left(60 \dfrac{\text{sec}}{\text{min}}\right) n < \dfrac{\text{rev}}{\text{power stroke}} >}{L < \dfrac{\text{m}}{\text{stroke}} > A < \text{m}^2 > N < \dfrac{\text{rev}}{\text{min}} > C < \text{no. cylinders} >} \ \text{N/m}^2 \tag{9.22a}$$

Engineers' units:

$$BMEP = \frac{\dot{W}_B <\text{hp}> \left(33{,}000 \dfrac{\text{ft lbf}}{\text{hp min}}\right) n < \dfrac{\text{rev}}{\text{power stroke}} >}{L < \dfrac{\text{ft}}{\text{stroke}} > A < \text{ft}^2 > N < \dfrac{\text{rev}}{\text{min}} > C < \text{no. cylinders} >} \ \text{lbf/ft}^2 \tag{9.22b}$$

Combining Equations (9.9), (9.22), and (9.17), mechanical efficiency can be expressed in terms of mean effective pressures as

$$\eta_{\text{mech}} = \frac{\dot{W}_B}{\dot{W}_I} \times 100 = \frac{BMEP}{IMEP} \times 100 \tag{9.23}$$

Brake thermal efficiency, η_{brake}, for an engine can be determined using Equation (9.13a) in terms of engine brake power output.

SI:

$$\eta_{\text{brake}} = \frac{(1000 \text{ N m/kW})\ \dot{W}_B <\text{kW}>}{\dot{m}_{\text{fuel}} <\text{kg/sec}> HV <\text{kJ/kg}>} \times 100 \tag{9.24a}$$

Engineers' units:

$$\eta_{\text{brake}} = \frac{(2545 \text{ Btu/hp hr})\ BHP <\text{hp}>}{\dot{m}_{\text{fuel}} <\text{lbm/hr}>\ HV <\text{Btu/lbm}>} \times 100 \tag{9.24b}$$

Brake heat rate (BHR), an inverse of brake thermal efficiency, expresses the ideal fuel-air energy addition required by an engine per unit of brake power output as

$$BHR = \frac{\dot{Q}}{\dot{W}_B} = \frac{\dot{m}_{\text{fuel}} HV}{\dot{W}_B} \qquad \frac{\text{kJ}}{\text{kW sec}} \left(\frac{\text{Btu}}{\text{hp min}}\right) \tag{9.25}$$

Brake specific fuel consumption ($BSFC$) expresses the normalized fuel consumption to the brake engine power output as

$$BSFC = \frac{\dot{m}_f}{\dot{W}_B} \qquad \frac{\text{kg fuel}}{\text{kW sec}} \left(\frac{\text{lbm}}{\text{hp min}}\right) \tag{9.26}$$

EXAMPLE 9.3 An 11.5 × 12.75-cm six-cylinder, two-stroke marine diesel delivers 170 bkW at 2100 rpm. The engine burns 50.9 kg fuel/hr. The indicated engine power is 205 ikW. For these conditions, calculate (a) the engine torque, N m; (b) *BMEP*, kPa; (c) brake specific fuel consumption, kg/bkW hr; (d) *IMEP*, kPa; (e) indicated specific fuel consumption, kg/ikW hr; (f) mechanical efficiency; and (g) friction power, fkW.

Solution:

1. Brake engine torque:

$$\text{a. } \tau_B = \frac{(1000 \text{ N m/kW sec})(170 \text{ bkW})(60 \text{ sec/min})}{2\pi(2100 \text{ rev/min})} = 773 \text{ N m}$$

2. *BMEP*:

$$\overline{P} = \frac{W_B\, 1000\, n}{LANC} = \frac{(170 \text{ bkW})(1000 \text{ N m/kW sec})(1)(60 \text{ sec/min})}{(0.1275 \text{ m})\ \pi/4\ (0.115 \text{ m})^2\ (2100 \text{ rev/min})\ 6}$$

$$\text{b. } \overline{P}_B = 6.113 \times 10^5 \text{ N/m}^2 = 611 \text{ kPa}$$

3. Brake specific fuel consumption:

$$\text{c. } BSFC = \frac{50.9 \text{ kg fuel/hr}}{170 \text{ bkW}} = 0.299 \text{ kg/bkW hr}$$

4. *IMEP*:

$$\text{d. } \frac{IMEP}{BMEP} = \frac{\dot{W}_I}{\dot{W}_B} \quad \text{or } \overline{P}_I = \frac{(205)}{(170)}(611) = 737 \text{ kPa}$$

5. Indicated specific fuel consumption:

$$\text{e. } ISFC = \frac{50.9 \text{ kg fuel/hr}}{205 \text{ ikW}} = 0.248 \text{ kg/ikW hr}$$

6. Mechanical efficiency:

$$\text{f. } \eta_{\text{mech}} = \frac{\dot{W}_B}{\dot{W}_I} = \frac{170}{205} = 0.829 = 83\%$$

7. Friction power:

$$\text{g. } \dot{W}_F = \dot{W}_I - \dot{W}_B = 205 - 170 = 35 \text{ fkW}$$

As discussed earlier, a dynamometer can be used to investigate a given engine experimentally under various operational conditions. A brief description of engine dynamometers is given in the following section. A more complete discussion of these devices can be found in the engineering literature.

The *prony brake* is the simplest and most inexpensive measuring system; see Figure 9.5. A brake drum is attached to an engine, and a friction band lined with asbestos is then used to load the engine. A constant resisting torque is developed for a given friction band pressure. Engine load can be changed by varying band pressure. The sliding resistance between the brake liner and the brake drum dissipates the engine power as heat. Too severe or too rapid a change in this mechanical braking can result in a loss of engine speed or even cause the engine to stall. This type of dynamometer is difficult to maintain at a constant load over a long testing period because of the heat generated by friction loading. Brake cooling is required and is provided by water. If water gets on the brake drum lining, it can influence brake performance and ensuing engine measurements. In general, the characteristics of the prony brake dynamometer limit its use to testing small, low-speed engines.

The *water brake* dynamometer is a hydraulic or fluid power absorption system. Basically, this type of engine utilizes a pump driven by an engine. Engine torque is transmitted through the pump because of the viscous nature of water. As the pump impeller rotates, a torque will be transmitted through the water and will be resisted by the pump casing. The pump casing is allowed to move or rotate, permitting casing torque to be measured using a brake arm and scale technique similar to the prony brake mentioned previously. Constant water flow rate will maintain a constant load as long as the viscosity (water temperature) is kept constant, whereas variations in loading can be accomplished by adjusting the water flow rate through the brake. This type of dynamometer is compatible with a wide range of engine speeds and sizes. It is a more stable system than a prony brake and will not stall as easily as the prony when a load is varied.

A *fan brake* dynamometer is also a fluid power absorption system. The viscous resistance of air, as opposed to water, is used in fan dynamometers to load a given IC engine. A fan is attached to an engine, and air resistance to the rotating fan blades absorbs the engine power. In this device, a specially calibrated fan is used to load the engine. Adjustments for load variations require change in fan blade radius, diameter and/or pitch. Typically, a particular blade is used for a specific load and then replaced for different loads. Fan dynamometers are not very accurate and are most frequently used to test an engine in a very long test over several hours at a given condition.

Several dynamometers are available for engine testing that use electrical as opposed to mechanical means of loading. The *electric generator* dynamometer is a power absorption system in which the engine is connected to an electric motor/generator set. An engine-generated torque is developed by a rotating stator and a fixed armature, which is resisted by an electromagnetic field. This electrically generated torque can be measured by allowing the generator housing to move or rotate, and connection of this to a brake arm and scale permits engine power measurement similar to that by prony brake and water brake systems. This type of dynamometer is governed by its electric generator efficiency. In addition, generated power may be dissipated through joule heating of resistance coils or a light bulb load bank, but a generator can also be run as a motor to measure engine friction power. These units are expensive and are most frequently matched with low-power and high-speed IC engines.

The *electric current* dynamometer is similar to an electric generator system. It is an electromagnetic power absorption device in which an engine drives a disk or rotor inside an electric coil. The engine torque is resisted by a torque produced by the electromagnetic field. Brake load is varied by changing the DC current in the coil.

The *engine chasis* dynamometer is another type of dynamometer. These devices place an entire vehicle on a treadmill for the purpose of loading the vehicle for testing. This is more precisely termed a *vehicle* versus an engine dynamometer.

9.5 ENGINE PERFORMANCE TESTING

In practice, IC engines are designed for use over a wide operational range and therefore require testing to determine their particular performance characteristics. Several reasons come to mind for such sophisticated testing, including: evaluating an engine to determine whether it meets design parameters predicted by the manufacturer; research and development programs involving modified or new engine concepts; and laboratory exercises used to illustrate engine principles to undergraduate engineering students. Standard tests have been established and maintained by engineering societies such as the Society of Automotive Engineers (SAE), American Society of Mechanical Engineers (ASME), and the American Society of Testing Materials (ASTM).

Engine testing typically provides information such as indicated and/or brake performance parameters, fuel-engine interface sensitivity, and/or exhaust gas emission characteristics. In general, this type of information can be quite complex and often is unit-sensitive, which is to say, a function of engine type, size, load, speed, fuel, and ambient environment.

EXAMPLE 9.4 The following two-stroke diesel engine data were obtained during a standard engine test.

1. Fraction of full load, %	25	50	75	85	100
2. 100-cc fuel consumption time, sec	77	65	52	47	43
3. Fuel specific gravity at 70°F, °API	45	45	45	45	45
4. Air supply mass flow rate, lbm/min	11	11	11	11	11
5. Air supply inlet temperature, °F	70	71	71	71	72
6. Cooling water mass flow rate, lbm/min	13.5	13.5	13.5	13.5	13.5
7. Cooling water inlet temperature, °F	60	60	60	60	60
8. Cooling water outlet temperature, °F	115	130	150	160	170
9. Exhaust gas temperature, °F	315	392	465	476	560
10. Dynamometer load voltage, V	120	120	120	120	120
11. Dynamometer load current, A	40	70	125	140	165

Determine the following: (a) rate of energy being supplied by the fuel, Btu/lbm; (b) rate of energy delivered by the engine to the dynamometer, Btu/lbm; (c) rate of energy lost to coolant, Btu/lbm; (d) rate of energy lost to exhaust gases, Btu/lbm; and (e) rate of unaccounted-for energy lost by radiation, oil coolant, etc., Btu/lbm.

Solution:

1. Energy balance, fuel input:

   Fuel specific gravity:

$$API\langle 60\rangle = [0.002(60-70) + 1]\,45°\text{API} \qquad (5.7)$$

$$= 44°\text{API}$$

   From Appendix D,

   At 44°API $\quad SG = 0.8063 \quad LHV = 18{,}600$ Btu/lbm

Fuel density:

$$\rho_{\text{fuel}} = (SG)\,(\rho_{H_2O})$$
$$= (0.8063)(62.4\ \text{lbm/ft}^3) = 50.313\ \text{lbm/ft}^3$$

Fuel mass flow, 25% load:

$$\dot{m}_{\text{fuel}} = \frac{(50.313\ \text{lbm/ft}^3)(100\ \text{cc})(60\ \text{sec/min})}{(77\ \text{sec})[(2.54\ \text{cm/in.})(12\ \text{in./ft})]^3} = 0.138\ \text{lbm fuel/min}$$

Fuel energy supply to engine, 25% load:

a. $|\dot{Q}_{\text{fuel}}| = \dot{m}_{\text{fuel}} \times LHV = (0.138\ \text{lbm/min})\ (18{,}600\ \text{Btu/lbm})$
$= 2567\ \text{Btu/min}$

2. Energy balance, dynamometer load:

$$\dot{W}_{\text{brake}} = I^2R = VR$$

Brake load, 25%:

b. $\dot{W}_{\text{brake}} = (120\ \text{V})(40\ \text{A}) = 4800\ \text{W}$
$= (4.8\ \text{kW})(56.87\ \text{Btu/kW min}) = 273\ \text{Btu/min}$

3. Energy balance, cooling water:

$$|\dot{Q}_{H_2O}| = \left(\dot{m}_{H_2O} C_P \Delta T\right)_{H_2O}$$

Coolant loss, 25% load:

c. $|\dot{Q}_{H_2O}| = (13.5\ \text{lbm/min})(1.0\ \text{Btu/lbm°R})(115 - 60\text{°R})$
$= 742.5\ \text{Btu/min}$

4. Energy balance, combustion gases:

$$|\dot{Q}_{\text{gas}}| = (\dot{m}_{\text{fuel}} + \dot{m}_{\text{air}})C_{p/\text{gas}}T_{\text{gas}}\ (-\ \dot{m}C_pT)_{\text{air}} - (\dot{m}C_pT)_{\text{fuel}}$$

The specific heats for the air and exhaust gases will depend on both temperature and composition. For this calculation, use the properties of air for first estimation.

From Appendix A

$$C_p\ \langle \text{Btu/lbm°R} \rangle = \frac{\overline{C}_p\ \langle \text{cal/g mole K} \rangle \times 1.8001}{MW_{\text{air}} \times (9\text{K}/5\text{°R})}$$

T,°R	$\overline{C}$, cal/g mole K	C, Btu/lbm°R
530	6.947	0.240
775	7.044	0.243
852	7.092	0.245
925	7.142	0.246
936	7.151	0.247
1020	7.219	0.249

Gas loss, 25% load:

d. $|\dot{Q}_{gas}| = (0.138 + 11 \text{ lbm/min})(0.243 \text{ Btu/lbm°R})(775\text{°R}) - (11)(0.240)(530) = 698 \text{ Btu/min}$

5. Energy balance, miscellaneous losses:

$$\dot{E}_{fuel} = \dot{W}_{brake} + \dot{Q}_{H_2O} + \dot{Q}_{gas} + \dot{Q}_{misc}$$

or

$$\dot{Q}_{misc} = \dot{E}_{fuel} - \dot{W}_{brake} - Q_{H_2O} - \dot{Q}_{gas}$$

Miscellaneous losses, 25% load:

e. $\dot{Q}_{misc} = 2567 - 273 - 742.5 - 698 = 853.5 \text{ Btu/min}$

6. Energy balance summary:

% Load	$\dot{Q}_{fuel}$	$\dot{W}_B$, Btu/min	$\dot{Q}_{H_2O}$, Btu/min	$\dot{Q}_{gas}$	$\dot{Q}_{misc}$
25	2567	273	742.5	698	854
50	3050	478	945.0	931	696
75	3813	853	1215.0	1150	595
85	4222	955	1350.0	1196	721
100	4613	1126	1485.0	1458	544

7. Diesel energy utilization on a percentage basis, or

$$\% \text{ power} = \frac{\dot{W}_B}{\dot{E}_{fuel}} \times 100 \qquad \% \text{ Exhaust} = \frac{\dot{Q}_{gas}}{\dot{E}_{fuel}} \times 100$$

$$\% \text{ Coolant} = \frac{\dot{Q}_{H_2O}}{\dot{E}_{fuel}} \times 100 \qquad \% \text{ Misc} = \frac{\dot{Q}_{misc}}{\dot{E}_{fuel}} \times 100$$

% Load	% Power	% Coolant	% Exhaust	% Misc
25	10.63	28.92	27.20	33.25
50	15.67	30.98	30.52	22.82
75	22.37	31.86	30.16	15.60
85	22.62	31.98	28.33	17.03
100	24.40	32.19	31.61	11.79

Evaluation of engines involves many critical measurements, including fuel and air mass (volume) flow rates, cylinder pressure, power output, speed, and emission concentrations. Pressure indicators, useful in indicated performance analysis, can provide cylinder pressure-volume and/or pressure-time (crank angle) diagrams. Low-speed engines are compatible with piston spring measurements, while pressure transducers are useful with high-speed engine movements. Air volumetric flow rates can be determined using air box and orifice techniques, while fuel flow rate can be measured using a gravity-feed or weighing scale system. Engine speed can be monitored using flywheel measurements such as tachometers, magnetic signals, or strobiscopic techniques. Brake power output can be absorbed using engine dynamometers, as discussed in Section 9.4.

Engine performance maps provide a means of representing complete engine characteristics over a wide range of speed, power, and torque. The general nature of these curves can be ascertained from relations developed in this chapter.

Internal combustion engine power is directly related to its fuel-air mixture and, for a particular engine geometry and specific fuel-air ratio, the ideal charge per cylinder per cycle should be independent of speed. In Section 9.2, the ideal and actual induction processes were related by the use of a volumetric efficiency η_v. Volumetric efficiency is strongly influenced by heat transfer, fuel parameters, the thermodynamic state of both intake charge and residual gases, and engine parameters such as intake and exhaust manifolds and valve design. For a particular condition of engine load, valve timing, and fuel-air input, the actual volumetric efficiency versus engine speed yields the characteristic curve shown in Figure 9.6. At low engine (piston) speed, an incoming charge has little kinetic energy influence on the intake process, i.e., *ramming.* At higher speeds, inertial ramming will increase the charge input to a maximum valve but, as speed is increased, frictional effects will tend to reduce the magnitude of η_v from its midspeed optimum valve. Changing engine operational regime may shift the volumetric efficiency curve.

At constant load, Equation (9.20) shows that, ideally, engine torque, $\tau = rf$, is independent of speed. Since power results from the combustion of a fuel-air mixture, an actual torque curve, like volumetric efficiency, will fall off at high and low engine speeds; see Figure 9.7. Equation (9.21) relates brake power, torque, and engine speed as

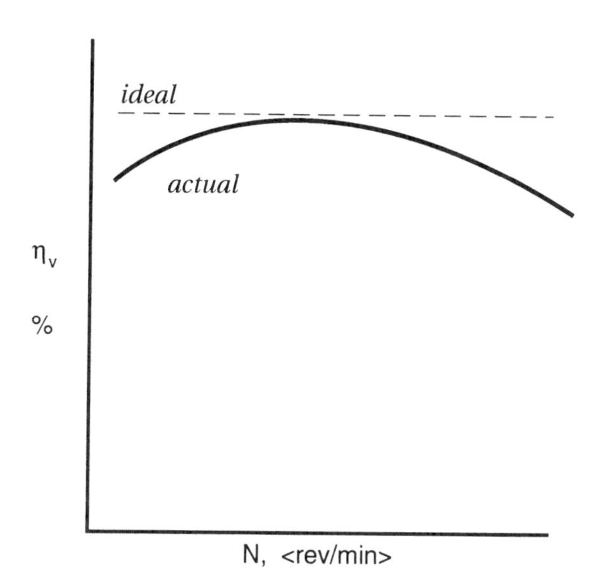

Figure 9.6 Volumetric efficiency vs. engine speed.

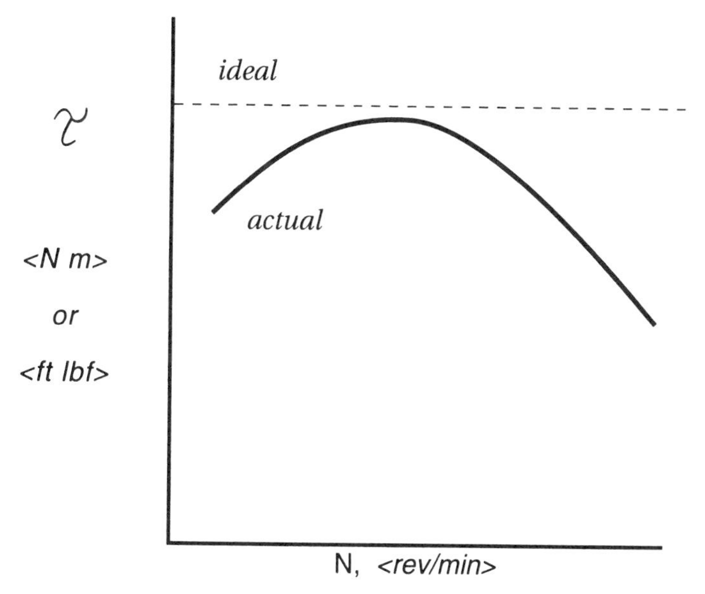

Figure 9.7 Engine torque vs. engine speed.

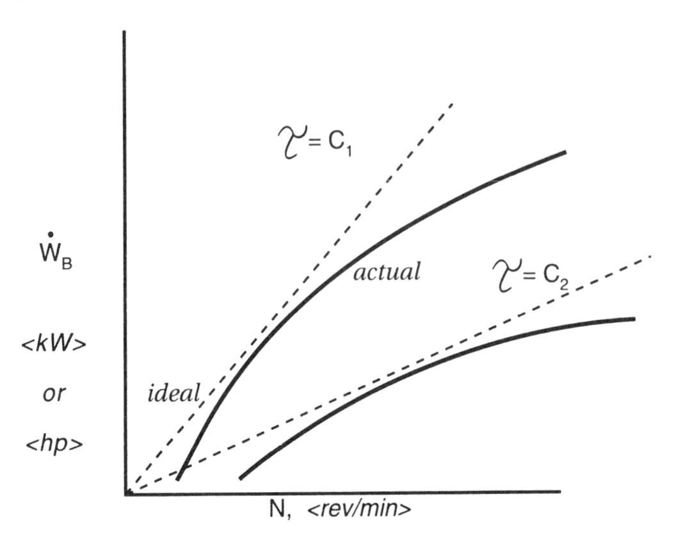

Figure 9.8 Brake power vs. engine speed.

$$\dot{W}_B = C_1 \times \tau \times N \tag{9.27}$$

Brake power versus engine speed for fixed load should ideally yield a straight line having a constant slope; see Figure 9.8. Actual brake power will fall off at low and high speeds because of the speed dependency of the torque curve.

Brake mean effective pressure, *BMEP*, is calculated knowing brake power and speed, using Equation (9.22) or

$$BMEP = C_2 \times \frac{\dot{W}_B}{N} \tag{9.28}$$

or combining Equations (9.27) and 9.(28)

$$BMEP = C_2 \times C_1 \times \tau \times \frac{N}{N} = C_3 \times \tau \tag{9.29}$$

Ideally, *BMEP* versus speed plots as a straight line having a positive slope and passes through the origin; see Figure 9.9. Again, as with brake power, a speed-dependent torque relation will produce an actual *BMEP* curve having low- and high-speed fall off regions.

Brake specific fuel consumption, *BSFC*, Equation (9.26), a measure of fuel consumed per unit of brake power can be related to the induced air and brake power as

$$BSFC = \frac{\dot{m}_f}{\dot{W}_B} = \frac{\dot{m}_{\text{air}}\, FA}{\dot{W}_B} = C_4 \times \frac{\dot{m}_{\text{air}}}{\dot{W}_B} \tag{9.30}$$

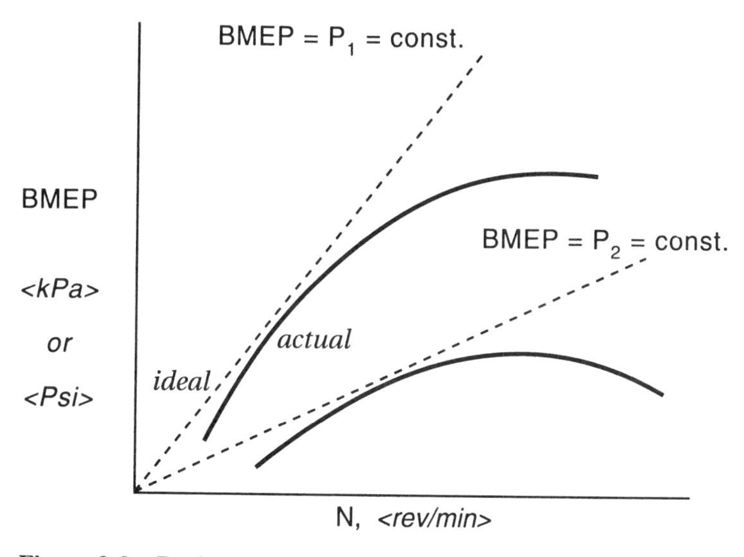

Figure 9.9 Brake mean effective pressure vs. engine speed.

For a fixed load and air-fuel ratio, the volumetric efficiency and brake power curves show that specific fuel consumption versus engine rpms would produce large values at low and high rpms, with a minimum rather than a maximum valve at midrange speed; see Figure 9.10. Brake thermal efficiency, Equation (9.24), i.e., the ratio of brake power to fuel consumption, should be inversely proportional to the brake specific consumption curve as shown in Figure 9.10.

As stated earlier, friction power $\dot{W}_F$, a measure of engine losses, increases with speed. Mechanical efficiency, Equation (9.18), defined as the ratio of brake to indicated power,

$$\eta_{\text{mech}} = \frac{\dot{W}_B}{\dot{W}_I} = \frac{\dot{W}_B}{\dot{W}_B + \dot{W}_F} \tag{9.18}$$

will tend to decrease with speed since friction power tends to increase at a greater rate than the sum of brake and friction power; see Figure 9.10.

Additional aspects of IC engine testing, such as matching load and output or correcting for nonstandard barometric testing, can be found in many standard IC engine textbooks.

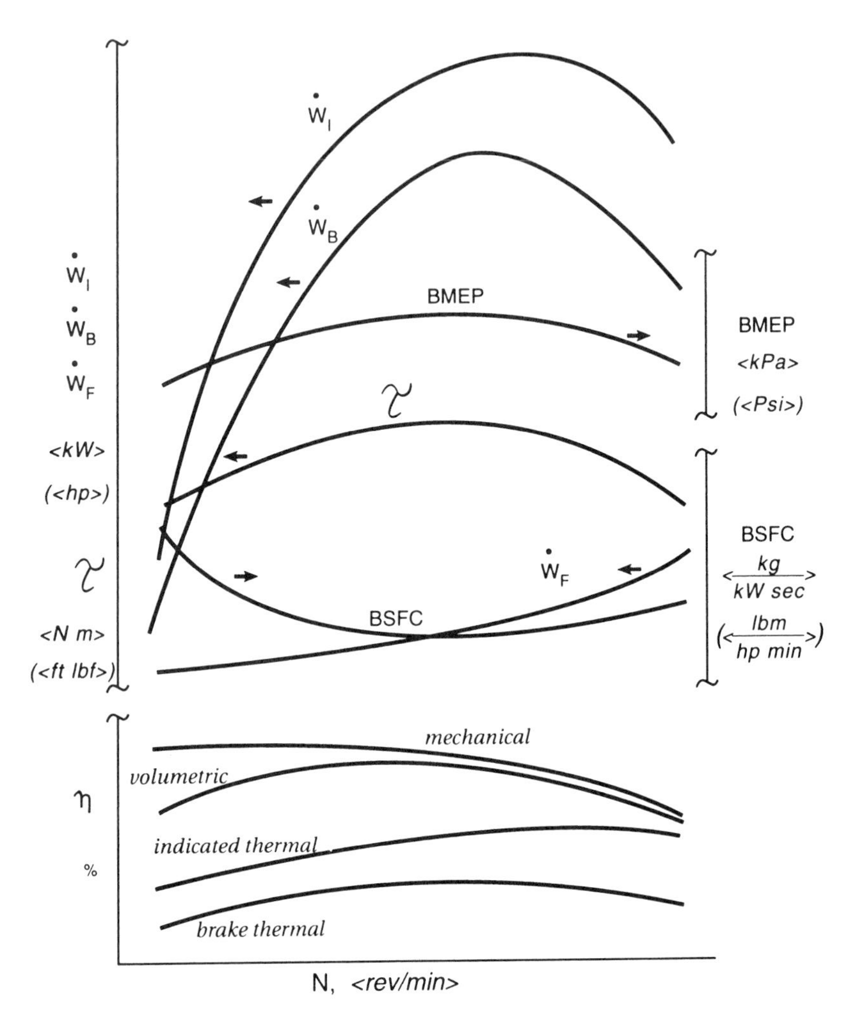

Figure 9.10 Typical IC engine performance map.

9.6 THE COOPERATIVE FUEL RESEARCH (CFR) ENGINE

The design and development of more efficient internal combustion engines will require a fuller understanding of the general characteristics of fuel-engine compatibility. As described in previous sections of this chapter, many important brake and indicated performance parameters for a variety of engine types and sizes are determined in the engine test cell.

One significant fuel-engine interaction not mentioned in earlier sections but worthy of attention is the tendency of fuel-air mixtures to detonate at certain conditions in both spark-ignition (SI) and compression-ignition (CI) engines. This explosive nature is termed *autoignition,* or *knock.* Knock characteristics are a function of the chemical nature of a fuel and the combustion processes in a particular engine. Since knock performance of fuels must be determined experimentally, a variety of standardized tests for both SI and CI fuels have been established by the American Society for Testing Materials (ASTM). These tests involve the use of specific engines and include research, motor, or vehicle tests.

The research method for testing SI and CI fuels for knock uses a single-cylinder *Cooperative Fuel Research* (CFR) engine. This unique 3.25 in × 4.5 in. four-stroke engine has several parameters that can be varied while the engine is in operation, including the compression ratio (3:1 to 15:1), air-fuel ratio, and/or fuel type. Variable compression is accomplished by means of a hand crank and worm gear mechanism, which allows the cylinder and head assembly to be raised or lowered with respect to the crankshaft. A unique overhead valve design maintains constant-volume clearance with varying compression ratio.

Three separate gravity-fed fuel bowls are connected to a horizontal draft, air-bled jet carburetor, allowing any of three different fuels to be used and/or interchanged while the engine is running. The fuel-air ratio can also be changed by simply raising or lowering each of the three fuel bowls.

Engine speed is maintained at 900 rpm by means of an engine-belted, synchronous AC power generator. Also, particular SI versions of the CFR engine have components that allow spark timing to be changed, while CI configurations have the capability of varying fuel injection rate and timing.

Recall from Chapter 5 that liquid hydrocarbon fuel characteristics are a function of their distillation curves. Since most commercial fuels are a mixture of many compounds of varying volatility, the knock characteristics will, in general, relate to their distillation curves. Detonation, or knock, in an engine will result in a rapid rise in peak engine pressure. Cylinder pressure in the research test is measured using a pressure transducer with filtered voltage output displayed on a knock meter. The knock rating of a particular fuel is obtained by matching the knock intensity of a sample at fixed operating conditions to that produced by blends of reference fuels.

Four primary fuels are used as standards for making mixtures of reference blends. Spark-ignition engine fuel ratings are measured in terms of an *octane* rating, while compression-ignition engine fuel ratings are specified by a *cetane* scale. Octane numbers are based on isooctane having an octane value of 100 and *n*-heptane having an assigned value of 0 octane. Thus, a 90-octane reference blend would be 90% isooctane and 10% *n*-heptane mixture by volume or

$$\text{Reference octane no.} = \%\ \text{isooctane} + \%\ n\text{-heptane}$$

Cetane numbers are based on *n*-cetane (nexadecane) having a cetane value of 100 and *n*-methylnaphthalene ($C_{11}H_{10}$) having an assigned value of 0 cetane. The cetane scale for a reference blend by volume is given as

$$\text{Reference cetane no.} = \%\ n\text{-cetane} + \%\ C_{11}H_{10}$$

In general, the higher the octane rating, the lower the tendency for the fuel-air mixture to autoignite in an SI engine. Also, the higher the cetane rating of a fuel, the greater the tendency for autoignition in a CI engine. Thus, a high-octane fuel will be a low-cetane fuel, and a low-octane fuel will have a high cetane index.

Knock rating of IC engine fuels based on the research test method is often different from values obtained using the motor method. Octane and/or cetane rating variations according to knock testing by research and motor methods are specified by *fuel sensitivity*. In general, the research method yields octane ratings higher than those obtained from the motor test. Current practice requires that automotive fuels be based on an average of the research and motor octane numbers.

Fuel	Old scale	New scale
Unleaded	91	87
Regular	94	89
Midpremium	96	91.5
Premium	99	94

Additives, such as tetraethyl lead (TEL), have been used in the past to raise the octane values of gasoline, but environmental impact has forced the development of unleaded fuels and alternate octane boosters, such as the recent efforts with anhydrous ethanol–gasoline blends, or gasohol. In addition, some fuels may actually yield octane numbers in excess of 100.

9.7 ENGINE EMISSIONS TESTING

In Section 9.6, standard tests necessary to determine IC engine knock characteristics, one of several fuel-engine critical interfaces, were discussed. Current practice requires that the engineer also understand the true nature of exhaust gases emitted by an engine. Over 90% of the typical exhaust gas composition for most newer automotive engines, on a volumetric basis, consist chiefly of CO_2, H_2O, and N_2. In addition, over 200 specific trace chemical compounds have been identified in engine exhaust, including carbon monoxide, CO; nitric oxides, NO_x; unburned hydrocarbons, UHC; partially oxidized hydrocarbons, PHC; sulfur dioxide, SO_2; and particulates and smoke; see Figure 9.11 and Table 9.1. Combustion engine testing and performance evaluation now include emissions testing. Exhaust gas, or *emissions,* analysis is an important engine performance activity because of the increasing number of IC engines in operation, the geographical and unique local climates where they are concentrated, as well as the variety of specific engines in use.

The concentrations of engine pollutants, i.e., exhaust products other than O_2, H_2O, and N_2, are very small and cannot be measured using the boiler or furnace stack sampling techniques discussed in earlier chapters, such as the Orsat analyzer. Engine exhaust gas instrumentation must generally be designed to detect 1 part per million (ppm) of a particular species in a mixture.

Table 9.1 Internal Combustion Engine Exhaust Gas Emissions

Carbon monoxide (CO): CO, ordorless, toxic gas (0.3% CO in air is lethal within 30 min), product of incomplete combustion.
Unburned Hydrocarbons (UHC): C_xH_y, paraffins, olefins, and aromatics (odor and carcinogenic constituents), products of incomplete combustion.
Partially Burned Hydrocarbons (PHC): $C_xH_y \cdot CHO$, aldehydes; $C_xH_y \cdot CO$, ketones; and $C_xH_y \cdot COOH$, carboxylic acids; quenched products of low-temperature combustion.
Nitric oxides (NO_x): NO, nitric oxide, colorless and odorless gas; NO_2, nitrogen dioxide, reddish orange gas, corrosive and toxic; N_2O, nitrous oxide, colorless, odorless laughing gas. High-temperature by-products of lean combustion dissociation in excess air, major contributor to photochemical smog.
Sulfur Oxides (SO_x): SO_2, sulfur dioxide, nonflammable colorless gas, source of sulfuric acid (H_2SO_4) and acid rain, product of combustion of sulfur impurities in hydrocarbon fuels.

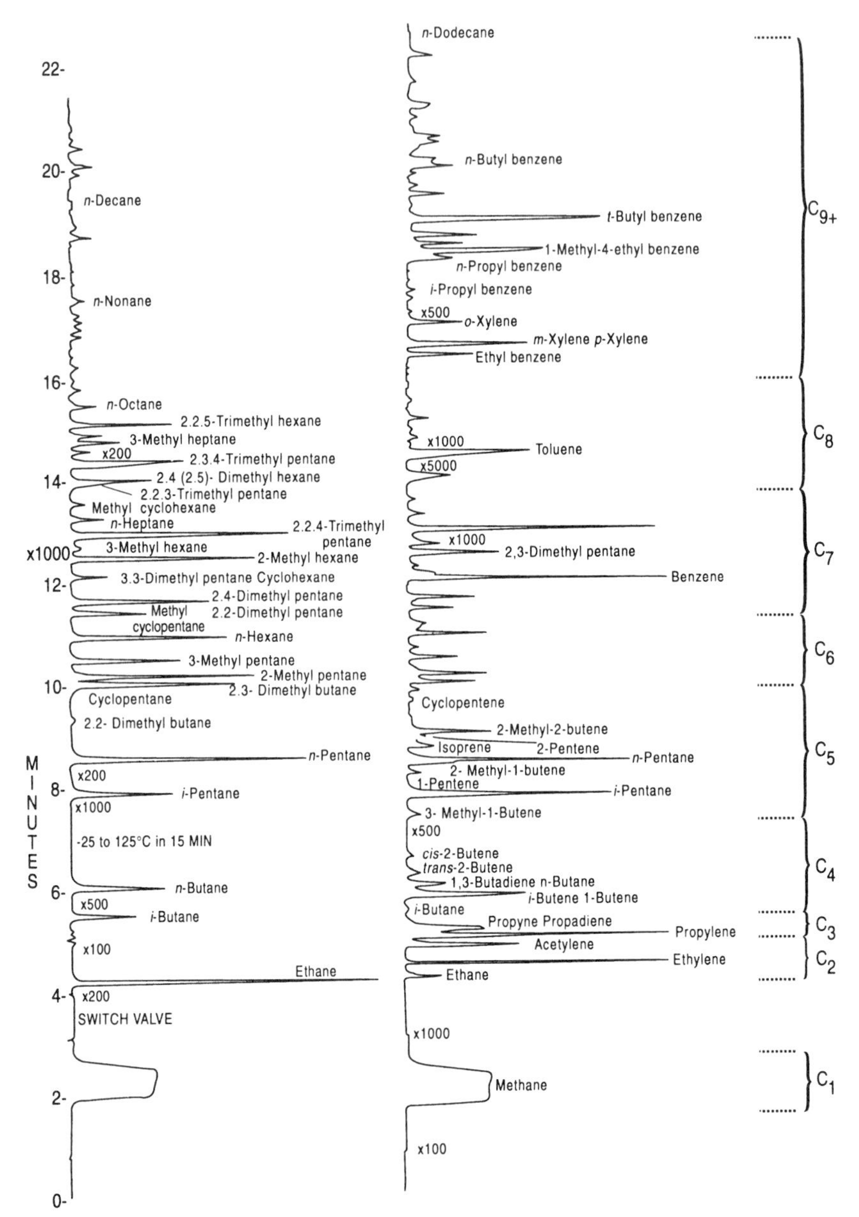

Figure 9.11 Typical gas chromatograph of exhaust gas emissions.

The ability to absorb or emit radiation is one means of characterizing or identifying particular chemical compounds. Spectroscopy characterizes the interactions of electromagnetic radiation and matter. Absorption spectroscopy utilizes the ability of matter to absorb radiation, whereas emission spectroscopy involves the radiant energy emitted by matter. No chemical species absorbs or emits radiation throughout the entire electromagnetic spectrum from ultraviolet to infrared. The distinct absorption or emission spectrum for a given compound is determined by its molecular structure and quantized internal energy levels. Energy is stored in a molecule as translational, rotational, vibrational, electronic, and chemical energy. The infrared spectrum can be related, for example, to rotational and vibrational state changes in a molecule, whereas ultraviolet radiation can be associated with electronic state changes in a molecule.

Electromagnetic radiation via quantum mechanics principles is converted by the spectroscopic emission detector into a measurable quantity, such as an electric current or voltage. Among the requirements for a useful spectroscopic detector are: (1) a high sensitivity to measured radiation with a low background noise level, (2) a fast time response to changes in radiation flux, (3) a low fatigue level, and (4) a measurable detection signal.

Ultraviolet and visible photons can cause photoejection of electrons from certain specially treated surfaces. Incorporation of these photosensitive materials into the construction of certain detectors, termed *photoelectric detectors,* allows conversion of radiant photon flux into an electric current. Infrared radiation can cause a decrease in the resistance of certain materials that are poor electric conductors. Application of these materials in designing certain exhaust gas detectors allows conversion of an infrared photon flux into measurable voltages. This internal photoconductive effect, since no electron ejection occurs as in phototube-type detectors, is found in certain metallic sulfides, certain element semiconductors such as silicon or germanium, and many compound semiconductors such as indium antimonide (InSb).

EXAMPLE 9.5 An automobile consumes 3.33 gal/hr of fuel while running at a constant speed of 50 mph on an engine dynamometer test facility. The spark-ignition engine burns a rich isooctane mixture. Exhaust gas analysis, on a dry volumetric basis, gives the following results.

9.54%	CO_2
4.77%	CO
3.15%	CH_4
0.84%	H_2
2.8%	O_2
78.9%	N_2

The specific gravity of isooctane is 0.702. Calculate (a) the fuel-air equivalence ratio; (b) the air-fuel ratio on a mass basis; (c) the fuel density, lbm/gal; (d) the fuel consumed, mpg; and (e) the grams of CO produced per mile.

Solution

1. Reaction equation:

Stoichiometric:

$$C_8H_{18} + 12.5\,[O_2 + 3.76N_2] \rightarrow 8CO_2 + 9H_2O + 47N_2$$

Actual:

$$\Phi\, C_8H_{18} + 12.5\,[O_2 + 3.76\,N_2] \rightarrow$$
$$a\,[0.0954CO_2 + 0.0477CO + 0.0315CH_4 + 0.0084H_2$$
$$+ 0.028O_2 + 0.789N_2] + bH_2O$$

Nitrogen balance:

$$(25)(3.76) = (0.789)(2)a \qquad a = 59.569$$

Carbon balance:

$$8\Phi = [0.0954 + 0.0477 + 0.0315](59.569)$$

a. $$\Phi = 1.30$$

Oxygen balance:

$$25 = (59.569)[(2)(0.0954) + 0.0477 + 0.056] + b$$
$$b = 7.457$$

Hydrogen balance (check):

$$(18)(1.30) = 23.4$$
$$(59.569[(4)(0.0315) + 0.016] + (2)(7.457) = 23.4$$

2. Mass air-fuel ratio:

b. $$AF = \frac{(12.5)(4.76)(28.97)\text{ lbm air}}{(1.3)(114\text{ lbm fuel})} = 11.6\,\frac{\text{lbm air}}{\text{lbm fuel}}$$

3. Mass of fuel consumed per hour:

$$\rho_{fuel} = (0.702)(62.4\text{ lbm/ft}^3) = 43.8\text{ lbm fuel/ft}^3$$
$$= (43.8\text{ lbm fuel/ft}^3)\,(231\text{ in.}^3\text{/gal})(1728\text{ in.}^3\text{/ft}^3)^{-1}$$

c. $$\rho_{fuel} = 5.855\text{ lbm fuel/gal}$$

$$\dot{M}_{fuel} = (5.855\text{ lbm fuel/ gal})(3.33\text{ gal/hr}) = 19.5\text{ lbm fuel/hr}$$

where

$$Z<\text{mpg}> = (50\ \text{mph})(3.33\ \text{gal/hr})^{-1}$$

d. $Z = 15$ mpg

4. Grams of CO produced per mile:

$$\frac{M_{CO}}{M_{fuel}} = \frac{(59.569)(0.0477)(28)\ \text{lbm CO}}{(1.3)(114)\ \text{lbm fuel}} = 0.5368\ \frac{\text{lbm CO}}{\text{lbm fuel}}$$

or

$$\dot{M}_{CO} = \frac{\left(0.5368\ \dfrac{\text{lbm CO}}{\text{lbm fuel}}\right)\left(19.5\ \dfrac{\text{lbm fuel}}{\text{hr}}\right)\left(0.45359\ \dfrac{\text{kg}}{\text{lbm}}\right)\left(100\ \dfrac{\text{g}}{\text{kg}}\right)}{(50\ \text{mph})}$$

e. $\dot{M}_{CO} = 94.96 = 955$ g CO/mi

It is often necessary to separate various exhaust gas emission species from each other prior to their detection and analysis. Gas chromatography is a material separation phenomenon in which given exhaust gas/vapor mixture species are segregated into pure samples contaminated only by an inert carrier gas. These segregated compounds are then detected, giving a quantitative analysis of the gas mixture. The column, the heart of the chromatograph, is a component that isolates and partitions the various chemical species in a given exhaust sample. The detector provides the means of quantizing the column effluent. Several types of detectors are currently used, among them the infrared analyzer, the mass spectrometer, the flame ionization detector, and the thermal conductivity cell. The detector is designed to produce a signal that is proportional to the mass of the unknown sample. The voltage generated by the detector, often after suitable amplification, is recorded as a function of time; see Figure 9.11.

Standard tests, such as the California, U.S. Federal, and Europa programs, are used to evaluate the toxic pollutant emission levels of new vehicles. Vehicles are tested in laboratories appropriately equipped with chassis dynamometers and suitable emission instrumentation. Most current tests are based on a constant-volume sampling (CVS) system in which exhaust gases are accumulated in a bag during a programmed test period. The vehicle is run using a standard test fuel, such as indolene (C_7H_{13}), and is made to operate over a specified driving cycle consisting of various cruise, acceleration, and deceleration modes; see Figure 9.12. An experimental recording of the engine speed versus test time is made and compared to the

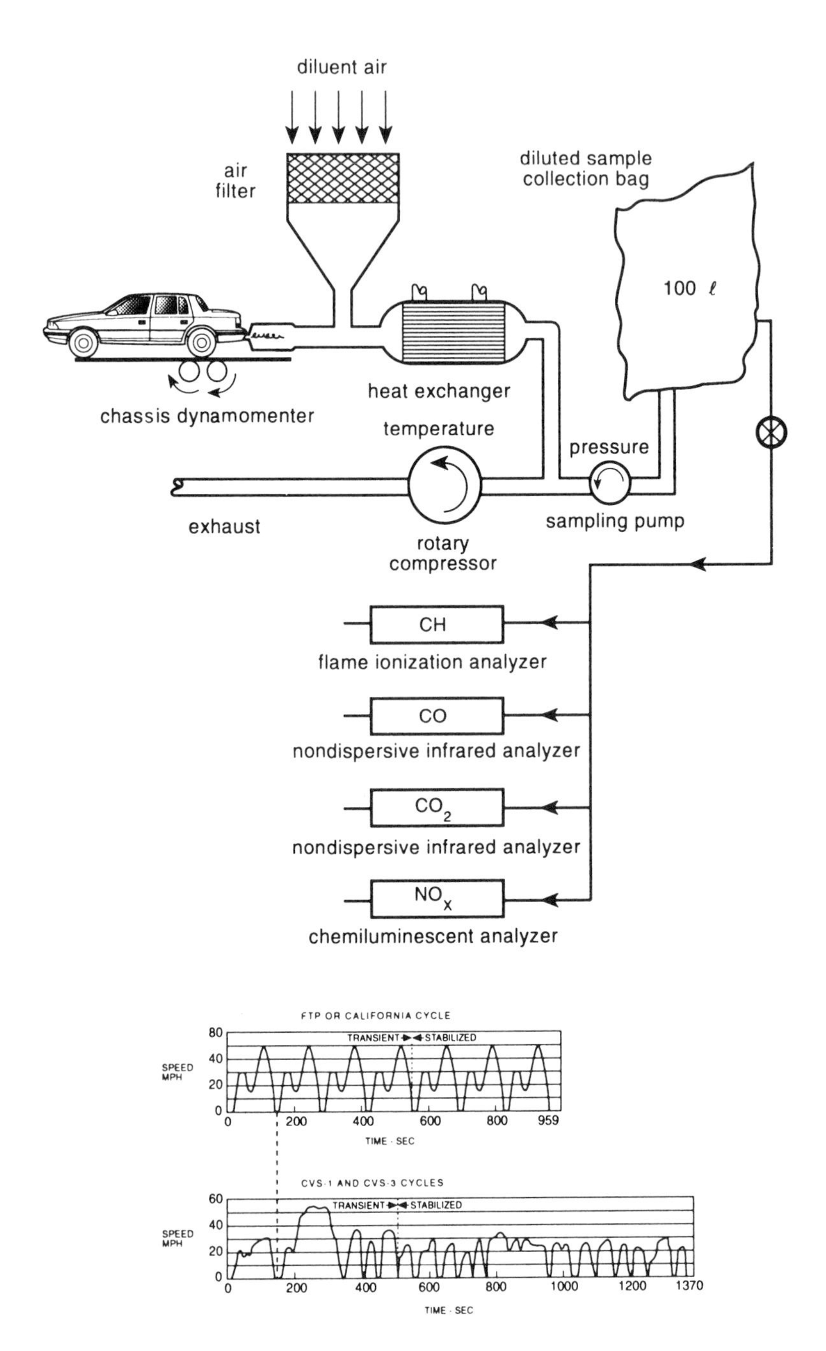

Figure 9.12 Constant-volume engine emission system schematic.

Table 9.2 U.S. Emission and Fuel Economy Standards

	Emission standards g/mi						Fuel Economy Standards
	Federal			California			
	HC[a]	CO[b]	NO_x[c]	HC[a]	CO[b]	NO_x[c]	mpg
1979	1.5	15	2.0	0.41	9	1.5	19
1980	0.41	7	2.0	0.41	9	1.0	20
1981	0.41	3.4[d]	1.0	0.41	3.4	1.0	22
				0.41	7	0.7 (Opt.)	
1982	0.4	3.4[d]	1.0	0.41	7	0.4	24
				0.41	7	0.7 (Opt.)	
1983	0.41	3.4	1.0	0.41	7	0.4	26
1984	0.41	3.4	1.0	0.41	7	0.4	27
1985	0.41	3.4	1.0	0.41	7	0.4	27.5

[a]HC unburned hydrocarbons. [b]CO carbon monoxide.
[c]NO_x nitric oxides. [d]Possible waiver to 7 g/mi CO.
Source: U.S. Environmental Protection Agency (EPA) and California Air Resources Board (CARB)

precise speed versus time requirement of the standard test. Exhaust gases collected during the driving cycle are diluted with filtered ambient air supplied by a compressor running at a constant delivery rate. Dilution with excess air ensures that the water in the exhaust will not condense. After driving the cycle, a small fixed sample is withdrawn from the collection bag and analyzed. Emissions analysis is then reported, usually in terms of a million moles of product or grams of sample per mile. Typical emission standards for U.S. spark-ignition engines are given in Table 9.2.

PROBLEMS

9.1 The Cooperative Fuel Research (CFR) engine is used in the octane rating of gasolines. This unique four-stroke single-cylinder engine runs at a constant speed of 900 rpm and has a variable compression ratio that can be adjusted from 3:1 to 15:1 while the engine is in operation. The 3.25-in. × 4.5-in. engine geometry maintains a constant displacement volume while the clearance volume can be adjusted. Calculate (a) the displacement volume, ft^3; (b) the clearance volume, ft^3; (c) the ideal volumetric airflow rate, ft^3/min; and (d) the ideal mass flow rate of air for the engine as a function of compression ratio, lbm/min. Assume that air enters the engine at *STP* conditions.

9.2 The surface-to-volume ratio at TDC is important in achieving the optimum combustion process in any IC engine. Four possible geometries are shown below, and each configuration has a bore and stroke of 3 in. × 4 in.

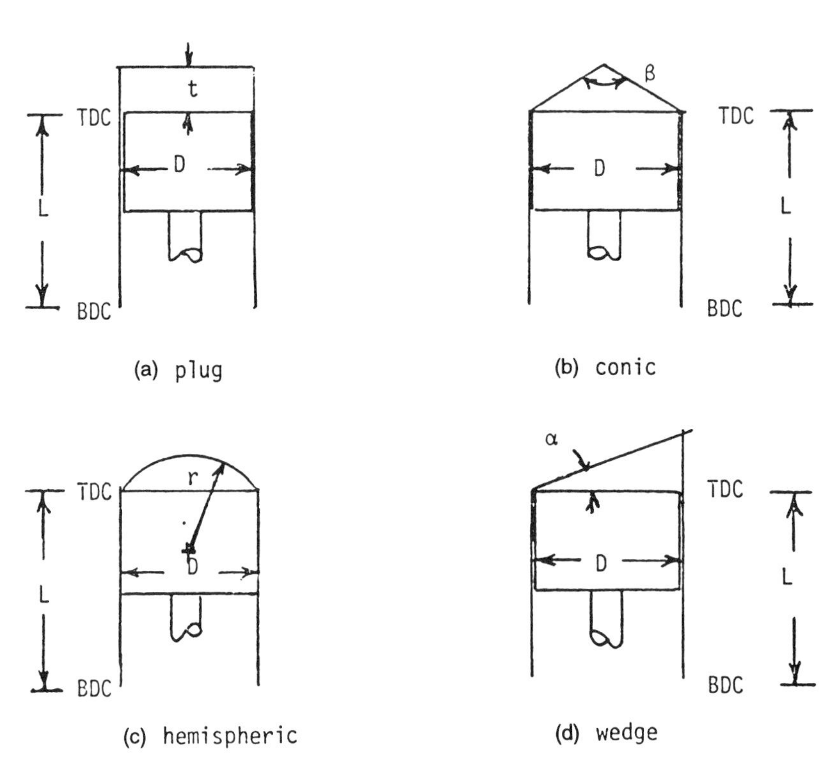

For these geometries, calculate, for an 8:1 compression ratio, (a) geometric parameters for the combustion chambers; (b) the surface-to-volume ratios at TDC; and (c) the mass of air at *STP* that could occupy the four clearance volumes, kg.

9.3 An internal combustion engine cannot be fully charged with fresh reactants during the intake stroke. At BDC, the displacement volume of an engine would ideally be filled with the fresh charge while the clearance volume would contain residual exhaust gases. For this condition, (a) show that the mass of gas in the displacement volume, M_c, can be determined from the total mass of air that could fill the entire volume, M_t, and the compression ratio, r_v, as

$$M_c = M_t\left[1 - \frac{1}{r_v}\right]$$

(b) Show that the ideal mass flow rate of air induced by the engine in terms of engine speed N and strokes per cycle n would become

$$\dot{M}_a = (M_t N/n)\left[1 - \frac{1}{r_v}\right]$$

(c) Show that the ideal mass flow rate of fuel induced by the engine in terms of engine speed N and strokes per cycle n would become

$$\dot{M}_f = (M_t N/nAF)\left[1 - \frac{1}{r_v}\right]$$

9.4 A six-cylinder, four-stroke spark-ignition engine is running at 1800 rpm. The 10-cm × 11.5-cm engine has an 8:1 compression ratio and burns an isooctane air mixture having an equivalence ratio of 0.8. Inlet conditions are 23°C and 100 kPa abs. Determine the ideal mass flow rate of fuel-air mixture to the engine for a volumetric efficiency of (a) 100%, (b) 95%, (c) 90%, and (d) 85%, kg/min.

9.5 Internal combustion engine compression and expansion processes are thermodynamically modeled using the polytropic process relationship

$$Pv^n = \text{const}$$

The value of n for a particular engine can be obtained by plotting the experimental expansion and compression P-V data on a log P versus log V graph, which results in straight line relationships; the slope is n. Demonstrate the technique by considering a 3-in. × 5-in. diesel engine with an 18:1 compression ratio and inlet conditions of 14.8 psi and 72°F. Plot the polytropic process log P versus log V diagram for n = (a) 1.4, (b) 1.35, and (c) 1.30.

9.6 The indicated mean effective pressure and indicated work per cycle can be obtained from an experimental measurement and integration of the cylinder pressure versus swept cylinder volume diagram for an engine. An equivalent indicated heat transfer to the working fluid within the cylinder as a result of combustion can be obtained from the pressure and crank data. Show that the first law for a closed system, i.e., the working fluid in the cylinder, can be related to the crank angle θ as

$$\frac{\delta Q<\theta>}{d\theta} = P<\theta>\frac{dV}{d\theta} + \frac{dU}{d\theta}$$

If the gases within the engine can be treated as an ideal gas, show that

$$\frac{R\,dT}{d\theta} = \frac{P\,dV}{d\theta} + \frac{V\,dP}{d\theta}$$

Assuming constant specific heats for the gas within the cylinder, show that the heat transfer as a function of crank angle can be written as

$$\frac{\delta Q<\theta>}{d\theta} = \frac{\gamma}{\gamma - 1} P<\theta> \frac{dV}{d\theta} + \frac{1}{\gamma - 1} V<\theta> \frac{dP}{d\theta}$$

9.7 A four-stroke, six-cylinder IC engine operates at a constant speed of 1600 rpm. The 1.2-cm × 2.0-cm engine produces a mechanical indicator card having a total area of 3.25 cm^2 and a length of 1.0 cm. The spring constant for the indicator is given as 1082 kPa/cm of deflection. Calculate (a) the indicated mean effective pressure per cylinder, kPa; (b) the indicator card height for a maximum cylinder pressure of 12,120 kPa, cm; and (c) the indicated engine power, kW.

9.8 A 4-in. × 5-in. single-cylinder engine is being used as a research engine. The engine, a four-stroke IC engine, is running at 1800 rpm, with inlet conditions of 73°F and 14.5 psia. Find (a) the displacement volume, $in.^3$; (b) the clearance volume, $in.^3$; (c) the ideal volumetric flow rate, ft^3/min; and (d) the ideal mass flow rate for an engine with an 8:1 compression ratio, lbm/min.

9.9 A four-stroke CI engine is delivering 600 bhp while running at 2400 rpm. The total engine piston displacement is 1650 $in.^3$, and the mass AF ratio is 22:1. The engine uses 69.5 lbm air/min. Determine (a) the brake mean effective pressure, psi; (b) the brake torque, ft lbf; and (c) the brake specific fuel consumption, lbm fuel/hp hr.

9.10 A six-cylinder IC engine is being performance-tested on an electric dynamometer. At full load, the experimental data obtained are as follows:

Engine rpm	N	1500 rpm
Dynamometer load	F	64.5 lbf
Brake arm	R	2 ft
Friction power	FHP	2.5 hp
Fuel flow rate	M_f	0.245 lbm fuel/min
Heating value	LHV	12,000 Btu/lbm

Determine (a) the brake power delivered to the dynanometer, hp; (b) the engine mechanical efficiency, %; and (c) the engine brake thermal efficiency, %.

9.11 A four-stroke 4.75-in. × 6.5-in. diesel engine generates the following test data:

Engine rpm	N	1160 rpm
Dynamometer load	F	120 lbf

Brake arm	*R*	1.75 ft
Friction power	*FHP*	57.3 hp
Mass *AF* ratio		13.5:1
Fuel oil specific gravity		0.82
Ambient air conditions		14.7 psia, 68°F
Fuel flow rate		21 lbm/hr

Calculate (a) the fuel lower heating value, Btu/gal; (b) the engine *BHP*, *FHP*, and *IHP*, hp; (c) the engine mechanical efficiency, %; (d) the brake specific fuel consumption, lbm/hp hr; (e) the brake mean effective pressure, psi; (f) the engine brake torque, ft lbf; and (g) the brake thermal efficiency, %.

9.12 An engine is using 9.1 lbm air/min while operating at 1200 rpm. The engine requires 0.352 lbm fuel/hr to produce 1 hp of indicated power. The *AF* ratio is 14:1, and the indicated thermal efficiency is 82%. Find (a) the engine brake power, hp; and (b) the fuel heating value, Btu/lbm fuel.

9.13 A four-cylinder, 20-cm × 30-cm two-cycle engine has an indicated mean effective pressure of 610 kPa and a mechanical efficiency of 83%. For operation at 2100 rpm, determine (a) the brake power, kW; (b) the brake torque, N m; and (c) the brake specific fuel consumption for a 78% volumetric efficiency for the engine, kg/kW sec.

9.14 A compression-ignition engine has a brake power output of 2900 kW and burns diesel fuel having an *API* value of 30. For a brake thermal efficiency of 37.6%, calculate (a) the lower heating value of the fuel, kJ/kg fuel; and (b) the brake specific fuel consumption, kg/kW sec.

9.15 A six-cylinder, four-stroke IC engine is running at 2680 rpm. The 11.5 cm × 13 cm produced an indicator card diagram having an upper power loop area of 7.25 cm^2 and a lower pumping loop area of 0.38 cm^2. The indicator card has a displacement length of 7.45 cm and a vertical peak deflection of 3.5 cm. A spring scale of 100 kPa/cm displacement is required for the pressure measurement. From this information, find (a) the peak pressure, kPa; (b) the indicated mean effective pressure, kPa; and (c) the indicated power, kPa.

9.16 A performance map for a four-stroke IC engine shows a minimum *BSFC* of 0.338 kg/kW hr when operating at 80% of the rated engine speed. Brake power output of 128 kW, volumetric efficiency of 95%, and an *AF* ratio of 15:1 are measured at this condition. If the engine is operated at 120% of its rated speed, the brake power output is found to be 158 kW, with a *BSFC* of 0.398 kg/kW hr. If the volumetric efficiency remains the same, what is the new *AF* ratio? Determine the volumetric efficiency if the same *AF* ratio is maintained.

9.17 A six-cylinder, 4.3-in. × 4.125-in. automotive engine burns 25.5 lbm/hr of gasoline while running at 3000 rpm and producing half-load on a dynamometer. If this engine is used in a vehicle at the same rpm and load and the engine–to–rear-axle speed ratio is 3.75 to 1, determine (a) the vehicle speed, mph, and (b) the fuel consumption, mpg. (Assume a rear-tire radius of 14 in.)

10
Spark-Ignition Engine Combustion

10.1 INTRODUCTION

Early internal combustion engine technology was a result of efforts by many individuals from around the world as well as a variety of important historical factors. Several names are frequently associated with the development of the spark-ignition IC engine, including Lenoir, who patented the first commercially successful internal combustion engine in 1860. Nickolaus Otto has been credited with producing the four-stroke, spark-ignition gas engine in 1876; however, his contribution was based on a cycle proposed 14 years earlier by Beau de Rochas.

In 1895, Charles E. and J. Frank Duryea built the first successful gasoline automobiles in the United States but, by 1898, there were still less than 1000 automobiles in use. Ten years later, Henry Ford offered his first Model T Ford and, by 1911, there were more than a half-million cars in the country. When the highly successful Model T was withdrawn from production in 1927, over 15 million units had been produced, and its unit cost had dropped from an initial price of $825 to only $290. The early growth of a U.S. automotive industry was due to part to the concurrence of several unique historical factors, including: (1) a westward population migration that required an ability to move easily between vastly separated and dispersed low-density population centers; (2) the greater per capita income and purchasing power of Americans relative to the rest of the world; and (3) the availability of gasoline, a suitable and relatively cheap motor fuel.

The question: Which came first—better engines or better fuels? is reminiscent of the proverbial dilemma of the chicken and the egg. Actually, the history of automotive combustion engine development cannot be separated from the growing and developing U.S. oil industry. At the end of the nineteenth century, kerosene, which was used chiefly to enrich coal gas, began to lose its value as a result of Edison's invention of the electric light. By the turn of the century, the emerging gasoline-powered automotive engine was still competing with electric- and even steam-powered vehicles, but the gasoline engine won out in America, in large measure because of the needs of an expanding and mobile population and the American oil industry's ability to shift from supplying its kerosene product to cheap gasoline. In 1911, the Standard Oil Companies, for the first time, sold more gasoline than kerosene.

Early straight-run low-octane gasolines, however, knocked in the 4.1-to-1 compression ratio engines of that day. As early as 1913, efforts were under way to find compounds that would improve gasoline octane ratings. In 1923, after testing over 30,000 compounds, Drs. Midgley and Boyd found that a teaspoon of tetraethyl lead mixed with a gallon of gasoline raised the fuel octane rating from 75 to 85, which satisfactorily eliminated knock in these low–compression ratio engines. Modern spark-ignition (SI) engines need gasoline having an ignition quality, or *octane rating,* that is in the mid to upper 90 range; see Chapter 9.

Professor R. F. Sawyer from University of California, Berkeley, stated, in an essay found in the *Future of the Automobile,* that "the automobile engine persisted through 100 years with no significant changes from its original design." From its very start, the growth of the American automotive industry was based on a successful marketing strategy that did not require scientific innovation to nurture basic spark-ignition combustion research and development. For example, most major engine technology that distinguished post–World War II gasoline engines from the turn-of-the-century engines had been achieved by the late 1920s.

Today, since the environmental and fuel crises of the 1960s and 1970s, worldwide combustion research activity is at its highest level. Critical fuel-engine interactions for SI engines are illustrated in Figure 10.1. A resurgence in fundamental and applied combustion work is directing efforts toward the following goals:

To develop engines that will show a substantial increase in useful power output and a decrease in required fuel consumption

To design engines that will improve combustion efficiency and reduce generation of harmful pollutant by-products

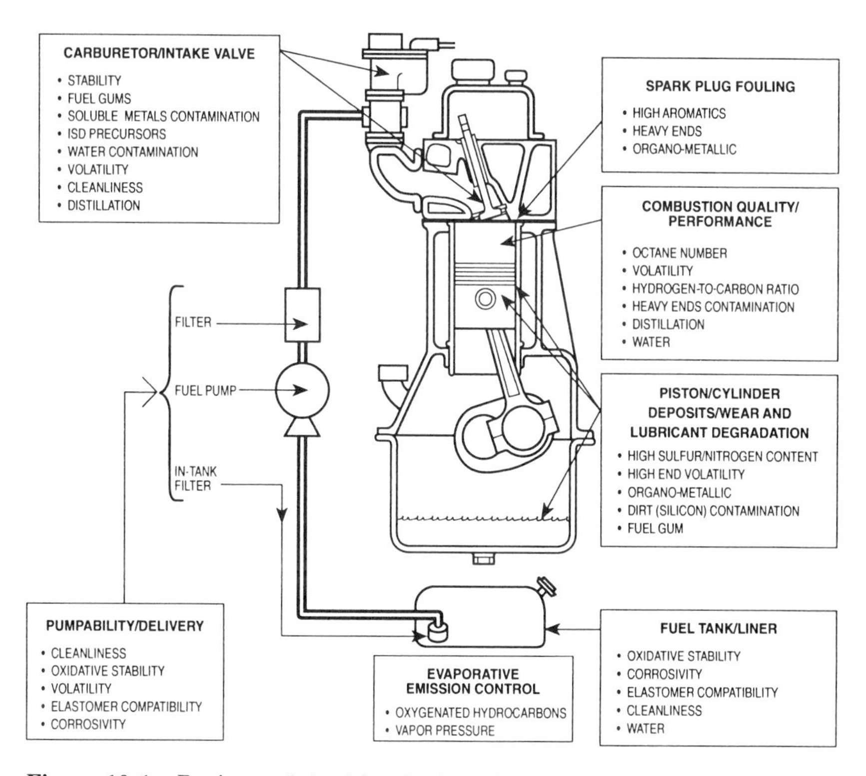

Figure 10.1 Basic spark-ignition fuel-engine interface.

To produce spark-ignition engines having wider tolerances to a variety of fuel alternatives

10.2 THERMODYNAMICS AND SPARK-IGNITION ENGINE MODELING

Spark-ignition, or SI, engines are internal combustion engines in which fuel-air mixture chemical kinetics are initiated by a local electric discharge. Successful ignition occurs only if the spark process provides sufficient energy within a very small region of the fuel-air mixture to ensure that (1) this local charge is above its self-ignition temperature and (2) initial local heat release by combustion within this local charge is greater than any instantaneous heat loss. The original spark-ignition engine of Otto required ignition near TDC, with rapid combustion taking place prior to an expansion stroke.

Several levels of thermochemical modeling can be used to predict theoretical energy conversion characteristics of SI engines, including air standard thermodynamic cycle analysis, fuel-air cycle analysis, and full engine simulations. Professors D. Foster and P. Myers in an SAE paper entitled "Can Paper Engines Stand the Heat?" indicated that IC engine models:

Are a mathematical catalogue of the most current physical information of IC engine energy characteristics
Provide a framework to systemize thinking about fundamental variables and mechanisms affecting engine performance
Give physical insight into engine chemistry and identify areas where fundamental information is needed
Provide a quantitative description of *intermediate* steps between measured data points
Cannot be the sole basis for designing engines but can go hand in hand with experimentation to achieve this end.

Air standard Otto cycle analysis, a zeroth-order SI engine model analysis, is covered in most classical thermodynamics texts, in which case the charge is assumed to consist of a fixed amount of air alone. Since this mass is constant, combustion and exhaust processes are replaced with heat transfer and, using classical thermodynamics, with either constant or variable specific heats; calculations can be made to predict ideal cycle indicated state parameters such as pressure, temperature, and density, as well as indicated energy parameters, including work, heat transfer, and thermal efficiency. An air standard Otto cycle consists of the following four ideal processes:

1–2 Isentropic compression from BDC to TDC
2–3 Constant-volume heat addition at TDC
3–4 Isentropic expansion from TDC to BDC
4–1 Constant-volume heat rejection at BDC

A dimensionless analysis based on a technique used by Walker in his Stirling engine work can be written for an ideal Otto cycle and utilizes the following dimensionless parameters:

τ $= T_3/T_1$ = maximum cycle temperature ratio
r_v $= V_1/V_2$ = compression ratio
r_e $= V_4/V_3$ = expansion ratio
α $= T_3/T_2 = P_3/P_2$ constant volume heating temperature ratio
ρ $= T_4/T_1 = P_4/P_1$ constant volume cooling temperature ratio
$IMEP$ = indicated mean effective pressure
ξ $= IMEP/P_1$ = dimensionless specific work

For a constant specific heat analysis, and using the dimensionless parameters listed above, the following expressions for temperature can be written:

$$T_2 = T_1 r_v^{\gamma-1}$$
$$T_3 = \alpha T_2 = \alpha r_v^{\gamma-1} T_1 = \tau T_1$$
$$T_4 = \rho T_1 = (1/r_e)^{\gamma-1} T_3$$

Now, for an Otto cycle, $v_2 = v_3$ and $v_4 = v_1$, which means that compression and expansion ratios are equal or

$$r_v = v_1/v_2 = v_4/v_3 = r_e$$

and

$$\rho = \tau / r_v^{\gamma-1}$$

where

$$T_4 = (\tau / r_v^{\gamma-1}) T_1$$

The net cycle work on a unit mass basis is given as

$$\begin{aligned} w_{\text{net}} &= {}_1w_2 + \cancel{{}_2w_3} + {}_3w_4 + \cancel{{}_4w_1} = u_1 - u_2 + u_3 - u_4 \\ &= C_v(T_1 - T_2 + T_3 - T_4) \end{aligned}$$

or

$$w_{\text{net}} = \frac{RT_1}{(\gamma-1)}\left[1 - r_v^{\gamma-1} + \tau - \frac{\tau}{r_v^{\gamma-1}}\right] \tag{10.1}$$

The net work divided by displacement volume, i.e., indicated mean effective pressure (*IMEP*) is then equal to

$$\begin{aligned} IMEP &= \frac{w_{\text{net}}}{v_1 - v_2} = \frac{w_{\text{net}}}{v_1 - (v_1/r_v)} = \frac{w_{\text{net}}}{v_1[1 - (1/r_v)]} \\ &= \frac{RT_1[(1 - r_v^{\gamma-1}) + \tau - \tau/r_v^{\gamma-1}]}{v_1(\gamma-1)[1 - 1/r_v]} \end{aligned} \tag{10.2}$$

Specific work ξ, *IMEP* in dimensionless form, is written as

$$\xi = \frac{IMEP}{P_1} = \frac{[(1-r_v^{\gamma-1}) + \tau - \tau/r_v^{\gamma-1}]}{(\gamma-1)[1 - 1/r_v]} \tag{10.3}$$

External heat addition per unit mass, q_H, is

$$q_H = {}_2q_3 = C_v[T_3 - T_2] = C_v T_1[\tau - r_v^{\gamma-1}]$$

$$\frac{q_H}{C_v T_1} = \tau - r_v^{\gamma-1} \tag{10.4}$$

Indicated thermal efficiency η is, therefore, given as

$$\eta = \frac{w_{net}}{q_H} = \frac{[RT_1/(\gamma-1)][1 - r_v^{\gamma-1} + \tau - \tau/r_v^{\gamma-1}]}{C_v T_1 [\tau - r_v^{\gamma-1}]}$$

$$= \frac{1 - r_v^{\gamma-1} + \tau(1 - 1/r_v^{\gamma-1})}{\tau - r_v^{\gamma-1}} = \frac{\tau - r_v^{\gamma-1} + 1 - \tau/r_v^{\gamma-1}}{\tau - r_v^{\gamma-1}}$$

$$= 1 + \frac{1 - \tau/r_v^{\gamma-1}}{\tau - r_v^{\gamma-1}} = 1 + \frac{(r_v^{\gamma-1})[1 - \tau/r_v^{\gamma-1}]}{(r_v^{\gamma-1})[\tau - r_v^{\gamma-1}]}$$

$$= 1 + \frac{1}{r_v^{\gamma-1}} \frac{[r_v^{\gamma-1} - \tau]}{[\tau - r_v^{\gamma-1}]} = 1 - \left(\frac{1}{r_v}\right)^{\gamma-1} \tag{10.5}$$

EXAMPLE 10.1 Using the dimensionless Otto cycle analysis developed in this chapter, determine the value for specific work as a function of compression ratio for the following constraints: (a) fixed peak temperature ratio, $\tau = 10$; (b) fixed peak pressure ratio, $p_3/p_1 = 30$, and (c) fixed heat addition per unit mass, $q_H/C_v T_1 = 10$. For each of these conditions, determine (d) the corresponding values of τ, p_3/p_1, and $q_H/C_v T_1$. Assume that $\gamma = 1.4$.

Solution

1. Specific work:

$$\xi = \frac{[(1 - r_v^{\gamma-1}) + \tau - \tau/r_v^{\gamma-1}]}{(\gamma - 1)[1 - (1/r_v)]}$$

2. External heat addition per unit mass:

$$\frac{q_H}{C_v T_1} = \tau - r_v^{\gamma-1}$$

3. Maximum pressure ratio, P_3/P_1:

$$\frac{P_3}{P_1} = \left(\frac{P_3}{P_2}\right)\left(\frac{P_2}{P_1}\right) = \left(\frac{T_3}{T_2}\right) r_v^{\gamma} = \alpha r_v^{\gamma}$$

4. $V = c$ heating temperature ratio α:

$$\alpha = \tau/r_v^{\gamma-1}$$

5. Fixed peak temperature ratio τ:

a. r_v	τ	q_H/C_vT_1	P_3/P_1	ξ
2	10	8.6805[a]	20[a]	10.5095[a]
3	10	8.4482	30	11.2658
4	10	8.2589	40	11.7180
5	10	8.0963	50	12.0103
6	10	7.9523	60	12.2062
7	10	7.8221	70	12.3390
8	10	7.7026	80	12.4281
9	10	7.5918	90	12.4856
10	10	7.4881	100	12.5196

[a] $\dfrac{q_H}{C_vT_1} = 10 - 2^{0.4} = 8.6805$

$$\frac{P_3}{P_1} = \alpha r_v^{\gamma} = \left(\frac{\tau}{r_v^{\gamma-1}}\right) r_v^{\gamma} = \tau r_v = (10)(2) = 20$$

$$\xi = \frac{1 - 2^{0.4} + 10 - (10/2^{0.4})}{(0.4[1 - (1/2)]} = 10.5095$$

6. Fixed peak pressure ratio P_3/P_1:

b. r_v	P_3/P_1	τ	q_H/C_vT_1	ξ
2	30[a]	15[a]	13.6805	16.5631[a]
3	30	10	8.4482	11.2658
4	30	7.5	5.7589	8.1709
5	30	6.0	4.0963	6.0766
6	30	5.0	2.9523	4.5316
7	30	4.286	2.1078	3.3250
8	30	3.75	1.4526	2.3438
9	30	3.33	0.9251	1.5215
10	30	3.0	0.4881	0.8161

[a] $\dfrac{P_3}{P_1} = \alpha r_v^{\gamma} = \left(\dfrac{\tau}{r_v^{\gamma-1}}\right) r_v^{\gamma} = \tau r_v$

$$\tau = \frac{P_3/P_1}{r_v} = \frac{30}{2} = 15$$

$$\frac{q_H}{C_vT_1} = 15 - 2^{0.4} = 13.6805$$

$$\xi = \frac{1 - 2^{0.4} + 15 - (15/2^{0.4})}{(0.4)[1 - (1/2)]} = 16.5631$$

7. Fixed specific heat addition, q_H/C_vT_1:

c.

r_v	q_H/C_vT_1	τ	P_3/P_1	ξ
2	10	11.3195[a]	22.6390[a]	12.1071[a]
3	10	11.5518	34.6555	13.3352
4	10	11.7411	46.9644	14.1884
5	10	11.9037	59.5183	14.8342
6	10	12.0477	72.2860	15.3492
7	10	12.1780	85.2453	15.7746
8	10	12.2974	98.3792	16.1350
9	10	12.4082	111.6740	16.4463
10	10	12.5119	125.1189	16.7192

[a] $\tau = (q_H/C_vT_1) + r_v^{\gamma-1} = 10 + 2^{0.4} = 11.3195$

$P_3/P_1 = \tau r_v = (11.3195)(2) = 22.6390$

$$\xi = \frac{1 - 2^{0.4} + 11.3195 - (11.3195/2^{0.4})}{(0.4)[1 - (1/2)]} = 12.1071$$

Comment This problem illustrates the use of dimensionless Otto cycle relationships in the study of variations in a constant specific heat SI engine model. Results from this parametric design study analysis can be plotted in dimensionless form as shown in Figure 10.2.

Figure 10.2 illustrates dimensionless characteristics of a basic constant specific heat air standard Otto cycle. Any air standard Otto cycle predictions would appear to be of limited value since any operating SI engine: (1) does not operate as a closed system, (2) does not actually consume air but requires a fuel-air mixture, and (3) does not operate with an external heat transfer but utilizes complex combustion processes that occur within the engine. Thermodynamics arguments are based on time-independent equilibrium principles, and they cannot be expected to address many important time-related engine effects such as engine friction, intake and exhaust dynamics, spark timing, rate of heat release, heat transfer losses, and chemical kinetics of engine emissions. Thermodynamic predictions for indicated efficiency and indicated specific work as functions of compression ratio, however, do, in fact, follow the general performance trends of SI engines.

The zeroth-order air standard Otto model can be modified to include an ideal exhaust process, in which case exhaust valve dynamics and blowdown of hot gases through an exhaust orifice will be neglected. It will be assumed, however, that the working substance discharged by an engine is air that has undergone isentropic compression from BDC to TDC (1–2),

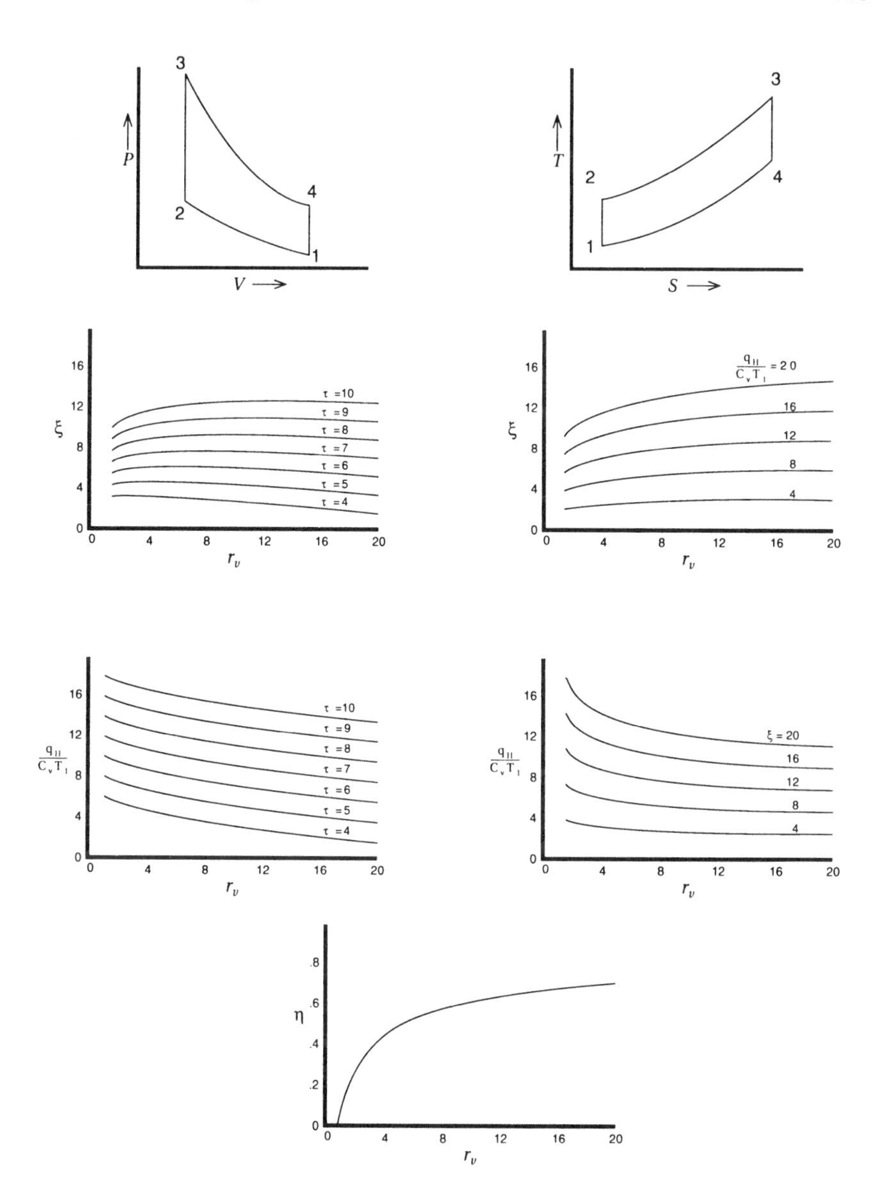

Figure 10.2 Ideal air standard Otto cycle thermodynamic characteristics.

external heat addition at TDC (2–3), and isentropic expansion from TDC to BDC (3–4). For the four-stroke engine, an ideal exhaust process would begin after a power stroke expansion to BDC and would continue until the piston again reaches TDC position 5, at which point it is assumed that both exhaust and intake pressures are equal. Ideally, blowdown would occur with the piston remaining in its BDC position. At TDC after the exhaust stroke, there will be a certain quantity of trapped exhaust gases at 5 remaining in the cylinder, described by a *residual factor* f, and defined as the mass of these clearance product gases at the end of an exhaust stroke divided by the total mass of residual and intake gases in the engine at the beginning of the compression stroke. Referring to Figure 10.3 and assuming a total unit mass of charge confined during compression, combustion, and expansion, it follows that

$$m_{tot} = m_{air} + m_{res} = m_1 = m_2 = m_3 = m_4 = 1.0$$

with

$$f = \frac{m_5}{m_1} < 1.0$$

For the exhaust blowdown, then,

Initial state	Final state
$P = P_4$	$P = P_5 \quad = P_1$
$T = T_4$	$T = T_5 \quad \neq T_6 \neq T_1$
$V = V_{BDC}$	$V = V_{TDC}$
$m = m_{tot} \quad = 1$	$m = m_{res} \quad = f$

During the exhaust process, no mass enters, one exhaust stream exits, and the conservation of mass for a control volume around the piston-cylinder configuration can be expressed using Equation (7.2a) as

$$\left.\frac{Dm}{Dt}\right)_{CM} = \left.\frac{\partial m}{\partial t}\right)_{CV} + \sum_{out} \dot{m}_j - \cancel{\sum_{in} \dot{m}_i} = 0 \qquad (7.2a)$$

$$\left.\frac{dm}{dt}\right)_{CV} = -\dot{m}_{exh}$$

or

$$dm = -\dot{m}_{exh}\, dt \qquad (10.6)$$

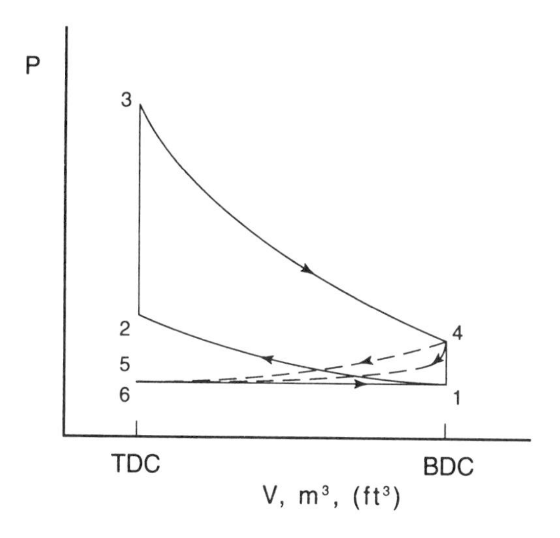

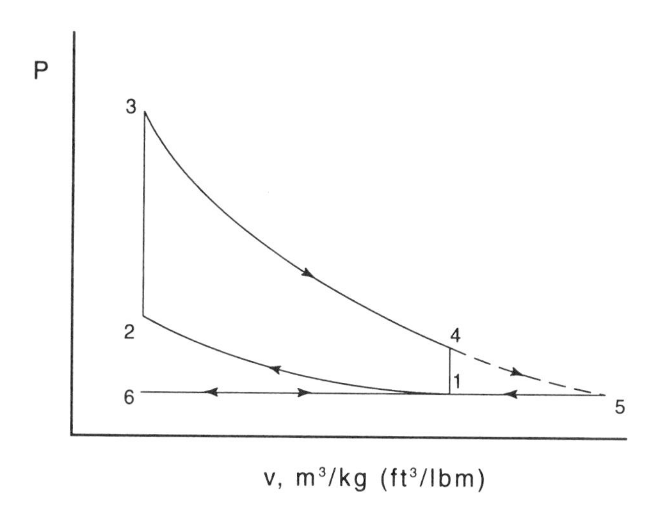

Figure 10.3 Indicator diagram for the four-stroke Otto cycle.

The conservation of energy for the control volume can be expressed using Equation (7.9) as

$$\left.\frac{DE}{Dt}\right)_{CV} = \left.\frac{\partial E}{\partial t}\right)_{CV} + \sum_{out} \dot{m}_j e_j - \sum_{in} \dot{m}_i e_i = \dot{Q} - \dot{W} \tag{7.9}$$

which, for adiabatic exhaust blowdown, is expressed simply as

$$\left.\frac{d(me)}{dt}\right)_{CV} = -\dot{W} - \dot{m}_{exh}e_{exh}$$

and

$$\left.\frac{d(me)}{dt}\right)_{CV} = \left.\frac{d(mu)}{dt}\right)_{CV} = m\left.\frac{du}{dt}\right)_{CV} + u\left.\frac{dm}{dt}\right)_{CV}$$

$$= m\left.\frac{du}{dt}\right)_{CV} - \dot{m}_{exh}u$$

or

$$m\left.\frac{du}{dt}\right)_{CV} - \dot{m}_{exh}u = -\dot{W} - \dot{m}_{exh}e_{exh} = -\dot{W} - \dot{m}_{exh}h_{exh}$$

$$m\,du = -dW + \dot{m}_{exh}(u - h)\,dt$$

$$m\,du = -P\,dV + \dot{m}_{exh}(u - h)\,dt$$

$$m\,du = -P\,dV - (u - h)dm \tag{10.7}$$

For an ideal gas having constant specific heats, Equation (10.7) becomes

$$mC_v\,dT = -P\,dV - (C_v - C_p)T\,dm = -P\,dV + RT\,dm$$

where

$$PV = mRT \tag{1.3}$$

$$P\,dV + V\,dP = RT\,dm + Rm\,dT$$

Substituting $RT\,dm$ from this expression into Equation (10.7) yields

$$mC_v\,dT = -P\,dV + [P\,dV + V\,dP - Rm\,dT] = V\,dP - Rm\,dT$$

which, rearranged, gives

$$m(C_v + R)\,dT = +V\,dP$$

$$\left[\frac{PV}{RT}\right]C_p\,dT = +V\,dP$$

or

$$\left(\frac{C_P}{R}\right)\frac{dT}{T} = \frac{dP}{P} \tag{10.8}$$

$$\left(\frac{T_{fin}}{T_{init}}\right)^{C_p/R} = \left(\frac{P_{fin}}{P_{init}}\right)$$

and

$$\frac{T_{\text{fin}}}{T_{\text{init}}} = \left(\frac{P_{\text{fin}}}{P_{\text{init}}}\right)^{(\gamma-1)/\gamma} = \left(\frac{V_{\text{fin}}}{V_{\text{init}}}\right)^{1-\gamma} \tag{10.9}$$

Equation (10.9) shows that the ideal exhaust blowdown process is isentropic, in which case gas temperature at the end of an exhaust stroke is given as

$$T_5 = \left(\frac{1}{r_v}\right)^{1-\gamma} T_4 = (r_v)^{\gamma-1}\, T_4 \tag{10.10}$$

and a residual fraction f can then be expressed as

$$f = \frac{m_5}{m_4} = \left(\frac{P_5V_5}{RT_5}\right)\left(\frac{RT_4}{P_4V_4}\right) = \left(\frac{V_5}{V_4}\right)\left(\frac{P_5}{P_4}\right)\left(\frac{T_4}{T_5}\right)$$

$$= \left(\frac{1}{r_v}\right)\left(\frac{P_5}{P_4}\right)\left(\frac{1}{r_v}\right)^{\gamma-1} = \left(\frac{1}{r_v}\right)^{\gamma}\left(\frac{P_5}{P_4}\right)$$

$$f = \left(\frac{1}{r_v}\right)^{\gamma}\left(\frac{P_1}{P_4}\right) \tag{10.11}$$

For an ideal isobaric and adiabatic intake process, the piston begins an intake stroke at TDC, with ambient air at P_6 and T_6 entering the piston-cylinder configuration and mixing with the remaining residual gases at P_5 and T_5. As was done for the exhaust analysis, inlet valve dynamics and gas dynamics of flow through an intake orifice will be neglected. The intake process ends at BDC, with conditions in the control volume identified as state 1. Thus, for an ideal intake process,

Initial state	Final state
$P = P_5 \quad = P_1$	$P = P_1$
$T = T_5 \quad \neq T_1$	$T = T_1$
$V = V_{\text{TDC}}$	$V = V_{\text{BDC}}$
$m = m_{\text{res}} \quad = f$	$m = m_{\text{tot}} \neq m_{\text{res}} = 1$

The conservation of mass for a control volume around the piston-cylinder configuration during an adiabatic intake process for one inlet stream can be written as

$$\left.\frac{Dm}{Dt}\right)_{\text{CM}} = \left.\frac{\partial m}{\partial t}\right)_{\text{CV}} + \sum_{\text{out}} \dot{m}_j - \sum_{\text{in}} \dot{m}_i = 0 \tag{7.2a}$$

$$\left.\frac{dm}{dt}\right)_{\text{CV}} = +\, \dot{m}_{\text{int}}$$

or

$$dm = + \dot{m}_{\text{int}}\, dt$$

The conservation of energy for the adiabatic intake process can be stated as

$$\left.\frac{DE}{Dt}\right)_{CM} = \left.\frac{\partial E}{\partial t}\right)_{CV} + \sum_{\text{out}} \dot{m}_j e_j - \sum_{\text{in}} \dot{m}_i e_i = \dot{Q} - \dot{W}$$

$$\left.\frac{d(me)}{dt}\right)_{CV} = -\dot{W} + \dot{m}_{\text{int}} e_{\text{int}}$$

or

$$d(me)_{CV} = -\dot{W}\, dt + dm_{CV} h_{\text{intake}} \tag{10.12}$$

Integration of Equation (10.12) for a constant-pressure intake process yields

$$(m_{\text{fin}} u_{\text{fin}} - m_{\text{init}} u_{\text{init}}) = -P[V_{\text{BDC}} - V_{\text{TDC}}]$$
$$+ (m_1 - m_5) h_6$$

or

$$u_1 - f u_5 = P[V_{\text{TDC}} - V_{\text{BDC}}] + (1-f) h_6$$
$$u_1 - f u_5 = fRT_5 - RT_1 + (1-f) h_6$$
$$u_1 + RT_1 = f(u_5 + RT_5) + (1-f) h_6$$
$$h_1 = f h_5 + (1-f) h_6 \tag{10.13}$$

Equation (10.13) requires that conditions at both 5 and 6 be known in order to determine conditions at 1 prior to compression. Note that residual gas conditions at 5 and the intake state at 6 will influence properties at BDC prior to compression at 1. However, to determine state point 5 at the end of the exhaust stroke, one must know conditions at states 4, 3, 2, and 1; hence, a trial-and-error solution is necessary to calculate clearance gas weight fraction f and temperature T_1 at the beginning of compression.

An ideal open system SI engine analysis in which inlet and exhaust pressures are equal is commonly referred to as an *unthrottled* engine. *Supercharging,* a technique for power-boosting a given engine design, raises charge delivery pressure above ambient by the use of an engine-driven compressor. *Turbocharging,* on the other hand, is a particular method of engine supercharging using hot-gas expansion through an exhaust gas–driven turbine, charge-compression compressor set to provide the required charge preparation. Figure 10.4 shows ideal indicator diagrams for SI engines.

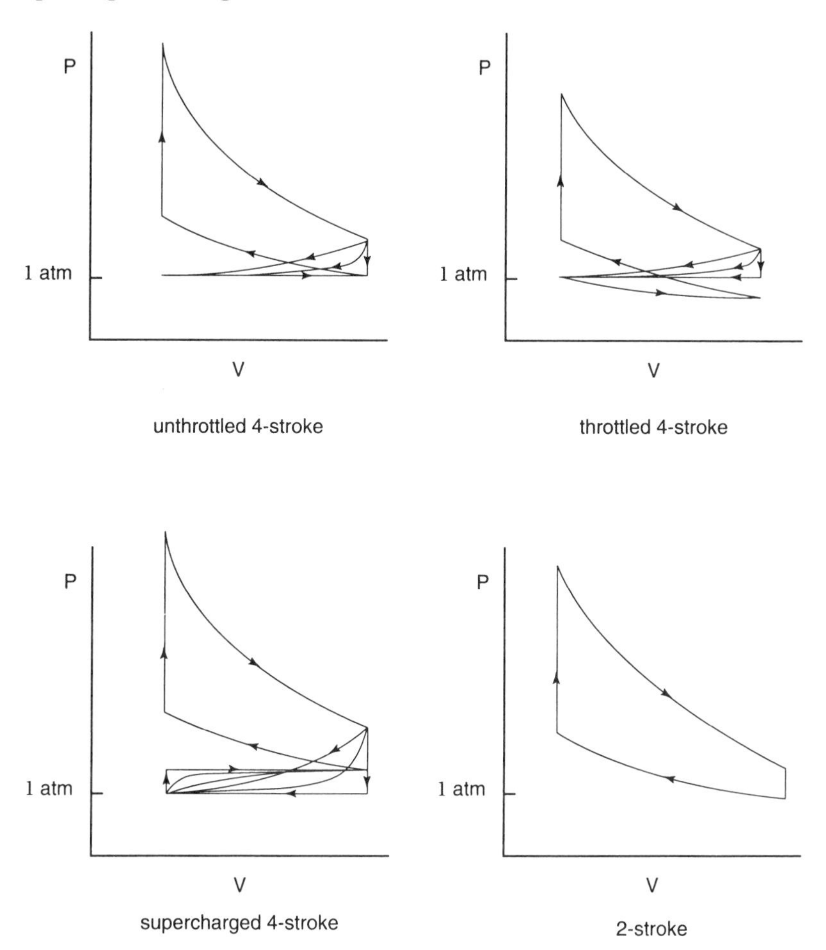

Figure 10.4 Various indicator diagrams for SI engine operation.

Extended air standard Otto cycle analysis will not predict how inducted fuel-air charge and burned product gas composition thermochemistry or internal energy conservation via combustion influence indicated engine performance predictions. Development of a fuel-air cycle analysis, however, allows some important fuel-air physics to be included in IC engine thermochemical modeling. At this level, real gas stoichiometry, specific heat variations, and dissociation effects are considered. Performance calculations approximate a finite burn rate with an instantaneous combustion process. Suitable energy-entropy charts with particular equivalent ratios for several useful hydrocarbon fuels are frequently used. Both Hottell and Starkmann

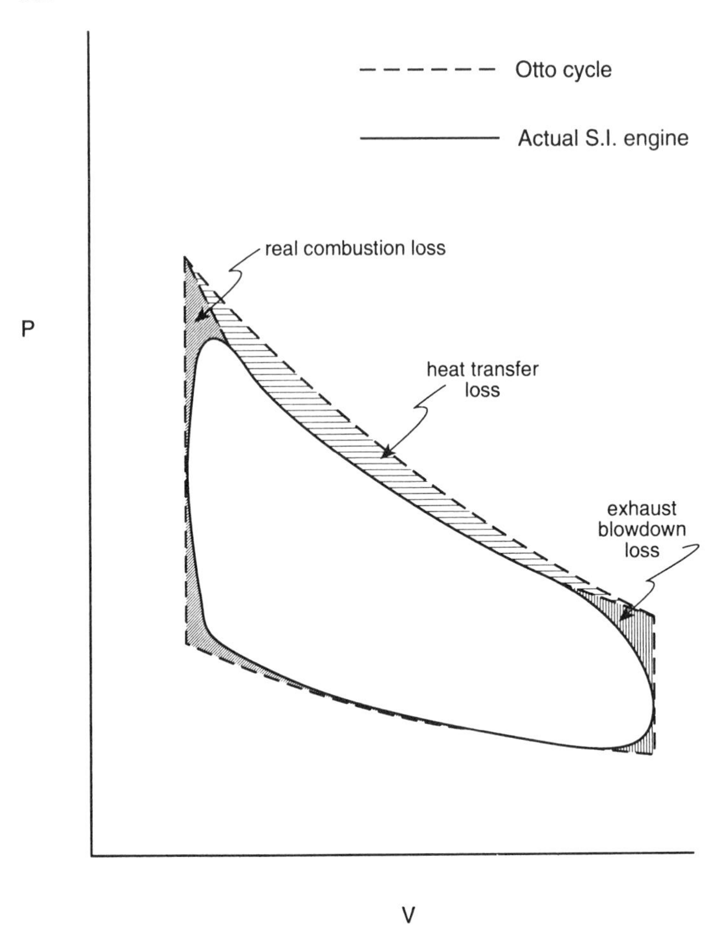

Figure 10.5 Effects of time-related losses on SI indicator card.

published charts for rich, lean, and stoichiometric frozen isooctane-air mixtures, as well as corresponding charts for their equilibrium products of combustion; see Chapter 3. Property values for charted fuel vapor-dry air mixtures are based on ideal-gas mixture predictions, assuming variable specific heats and no residual gas present. Property values for product charts are based on frozen complete combustion composition for temperatures below 1700 K (3060°R) and equilibrium composition for temperatures above 1700 K (3060°R). Several texts, such as that by C.F. Taylor (1966), give a more complete discussion of the use of these charts in IC engine calculations.

For fuel-air Otto analysis, compression of a fuel-air mixture assumes frozen charge composition and an isentropic process. Combustion at TDC is approximated as a constant-volume adiabatic flame process in which products of combustion are assumed to have either a frozen complete reaction composition for temperatures less than 1700K (3060°R) or an equilibrium reaction composition for conditions above 1700K (3060°R). Expansion from TDC is treated as an isentropic and frozen composition expansion process. Reactant and product charts can also be used in a fuel-air analysis that includes exhaust residual gas recirculation.

Finite piston motion and speed, real intake and exhaust gas exchange rates, heat transfer, friction, and transient engine requirements are but a few of the dynamic engine effects that influence SI engine operation (see Figure 10.5). The probabilistic and spatially resolved physics of periodic ignition, the complex chemical kinetics of combustion, and the generation of pollution by-products are examples of dynamic chemistry that also impact SI engine operation. With the advent of high-speed computers, programs and codes can now be written and are available that allow a wider selection and specification of SI fuel alternatives, greater range and choice of *FA* ratio, and a more general thermochemical analysis and that can begin to include some of the time-dependent physics of actual engines.

10.3 FUEL THERMOCHEMISTRY AND SPARK-IGNITION ENGINES

Today, proper charge preparation is essential to good SI engine performance. Precise fuel-air control allows SI engines to run over a wide range of loads and speeds. Fuel metering and control are most frequently accomplished by carburetion; see Figure 10.6. Both manifold and throttle-body fuel injection techniques are now being pursued in an effort to develop engines having better fuel economy, power output, emissions control, and/or a broader tolerance to fuel alternatives. Note, however, that charge preparation and mixture control are quite different for carbureted and fuel-injected engines. Power and torque demands for SI engines are met by varying the mixture fuel-air ratio. For example, engines built to operate on gasoline can have an air-fuel ratio that may vary from approximately 8:1 (rich limits) to 20:1 (lean limits) while running. Too lean a mixture makes it difficult to ignite, while too rich a mixture yields unburned hydrocarbon and carbon monoxide levels that may be in excess of acceptable limits. Fuel-air limits usually fall within fuel-air mixture flammability limits.

Many basic fuel thermochemical properties are important to proper charge preparation and engine operation. For example, complete vaporization and mixing of a liquid fuel in air prior to compression is an essential element of classic SI charge preparation. Fuel-air delivery systems, such as

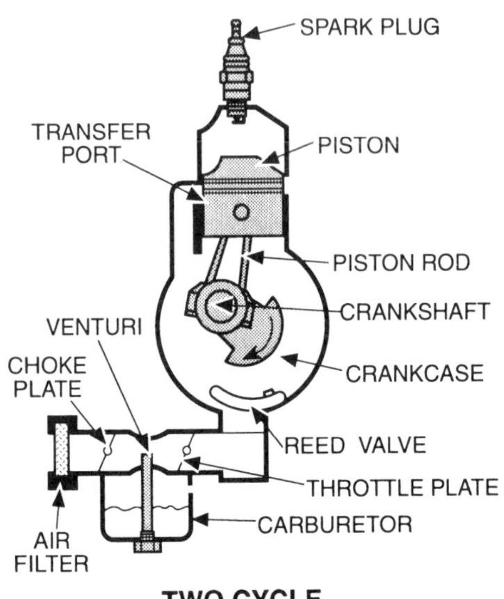

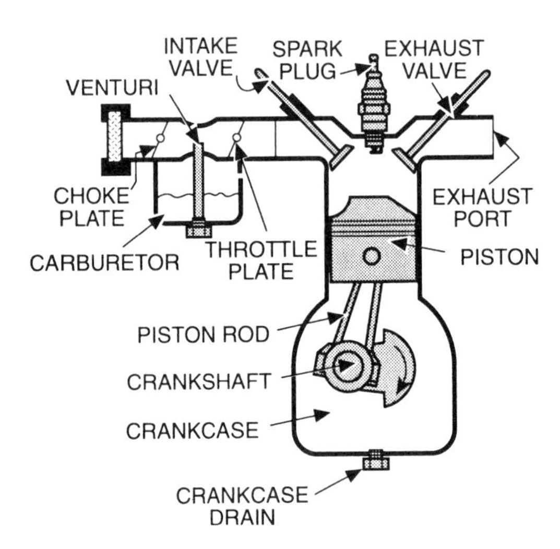

Figure 10.6 Charge preparation and the carbureted SI engine.

those used in standard carbureted induction mechanisms, require the use of liquid fuel blends having appropriate low-temperature distillation curves; recall Chapter 5. Note that this fuel-engine nature of volatility will mean that a lower effective fuel distillation temperature is needed during start-up and warm-up than when an engine is running. A distillation temperature that is too low may in some instances, cause fuel to vaporize in the fuel delivery system, resulting in *vapor lock,* a severe situation in which liquid fuel flow to an engine is blocked. Vapor lock can be crucial in aircraft fuel applications where high-altitude flight will mean a lower fuel boiling point temperature than at sea level. Seasonal atmospheric temperature fluctuations create a need for supplying both summer and winter fuel blends. Note also that unvaporized fuel-air mixtures admit liquid fuel droplets into an engine, diluting and scrubbing lubricating oil film from cylinder walls and increasing friction. Preheating an intake manifold can help to reduce the amount of liquid fuel that reaches cylinder walls, but this would also reduce indicated power output due to the expansion of, and reduction in, the heated charge actually inducted into the engine.

The more volatile a fuel, the greater will be the amount vaporized, and the potentially larger will be the temperature drop within a carburetor, a condition that may cause any moisture in air to condense and ice the carburetor. Less volatile fuels will prevent icing but can make warm-up of the engine more difficult. Volatility requirements for an SI engine designed to run on pure compounds, such as alcohols or propane, are even more unmanageable during cold start since these fuels are not blends but rather are substances that have fixed saturation temperatures for a given saturation pressure. Cold starts often necessitate some additional fuel enrichment to ensure sufficient vaporized fuel for proper ignitability, which is traditionally provided in carbureted engines by the choke. Choking can also prevent hesitation, stumble, stall, and backfire during cold starting. Uniform fuel-air distribution is hard to achieve in multicylinder carbureted engines. Complete vaporization is formidable, and liquid droplets may enter the intake manifold. The momentum of airflow and that of liquid fuel droplets are considerably different, resulting in uneven fuel distribution between cylinders. Cylinder-to-cylinder *FA* ratio variations can also occur under transient operation, causing a time variation, or lag, during charge induction. The poor part-load efficiency of SI engines is primarily an inability to maintain cylinder-to-cylinder mixture quality control. Improvements in fuel control and distribution may result from the development of direct cylinder-injected engines.

Energy is required to vaporize liquid fuel during charge preparation and induction prior to combustion. Both latent heat of vaporization and ignition energy of SI fuels are functions of mixture composition within any cylinder.

Mixture ignition temperatures must be lower than their fuel-air self-ignition temperatures to prevent preignition during compression. Internal combustion engine combustion produces gases in the exhaust gas stream at temperatures in which water will most probably be a vapor and, thus, the heat of combustion for a particular fuel should be based on its lower heating value. Condensation of this moisture in the exhaust line occurs when the engine is shut off, which can cause rusting of engine exhaust components. There also can be an appreciable quantity of dissociated products of combustion in the burned charge resulting from high-temperature endothermic reactions. These energy-absorbing reactions will lower the total overall exothermic energy released by the fuel, and the actual temperature rise at TDC will be less than that predicted if dissociation is ignored.

Full power means heavy loads and rich mixture combustion. Maximum

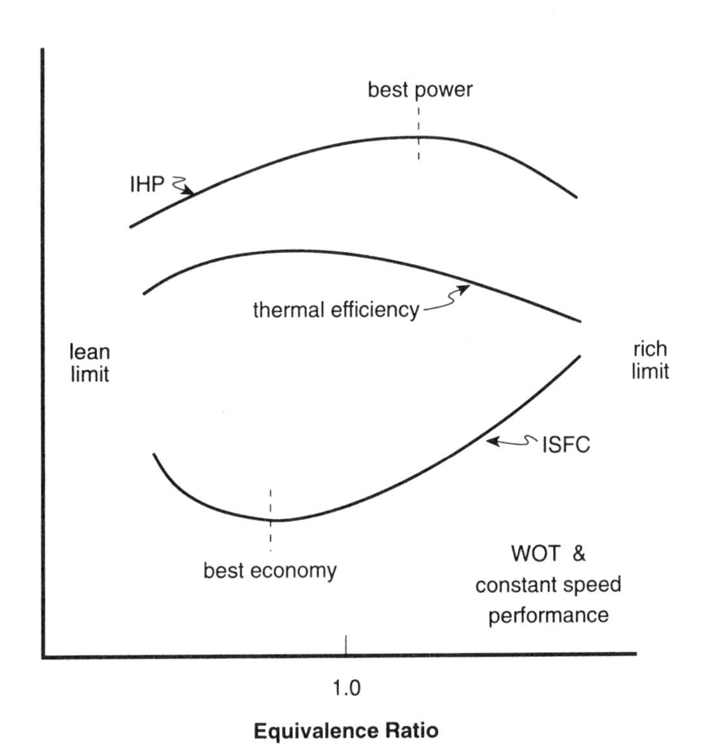

Figure 10.7 Stoichiometry and SI indicated engine performance.

air mass must be inducted into the cylinders. Wide open throttle, or WOT, ensures conditions of minimum intake resistance for a carbureted engine. The influences of residual gases remaining from a previous mechanical cycle are low when operating at WOT. This high power output regime utilizes a rich mixture in each cylinder and produces the greatest values for both peak and mean effective pressures. Excess fuel in the cylinders also can help to prevent overheating of exhaust valves and inhibit the tendency of SI fuels to knock during combustion. Engines are run at full power at their best power point (see Figure 10.7).

Idle, or no-load, operation also requires rich mixtures in order to overcome friction and sustain power. With a carbureted engine, this is accomplished with a nearly closed throttle setting, but throttling produces a lower induction cylinder pressure (vacuum) and less air intake. This means that the quantity and condition of hot residual gases left in the clearance volume have a more significant influence on charge induction and preparation. With a lower total charge, peak pressure and indicated mean effective pressures will be greatly reduced.

Cruise performance means a partially opened throttle setting and lean stoichiometry. The intake charge inducted has less dilution by hot residuals. Only enough energy is needed from combustion to overcome both engine friction and vehicle drag. This reduction in power output means that steady cruise operation can burn a lean mixture and operation of the engine is at the point of best economy (again see Figure 10.7).

EXAMPLE 10.2 A four-cylinder SI engine with a total displacement of 820 cc and having 8:1 compression ratio is to be analyzed using an air standard Otto cycle analysis. Conditions at BDC are assumed to be 105 kPa abs and 27°C. For a total heat addition of 2.5 kJ, calculate the following, assuming constant specific heats: (a) the maximum cycle temperature and pressure, K and kPa; (b) the ideal indicated thermal efficiency, %; and (c) the indicated mean effective pressure, kPa. Repeat parts (a)–(c), assuming variable specific heats and using JANAF data for air.

Solution

1. Conditions at BDC state 1, assuming charge is air alone:

$$V_D = \frac{V_{\text{tot}}}{4} = V_1 - V_2 = V_1\left[1 - \left(\frac{1}{r_v}\right)\right]$$

$$V_1 = \left(\frac{8}{7}\right)\left(\frac{0.00082\ \text{m}^3}{4}\right) = 0.000234\ \text{m}^3/\text{cyl}$$

$$m = \frac{P_1 V_1}{R T_1} = \frac{(105{,}000 \text{ N/m}^2)(0.000234 \text{ m}^3/\text{cyl})}{(8314 \text{ N m/kg K}/28.97)(300\text{K})}$$

$$= 0.000285 \text{ kg/cyl}$$

2. Process 1 → 2 isentropic compression $S_1 = S_2$ for constant specific heats, from Appendix B for air at 300K:

$$C_p\Big)_{\text{air}} = \frac{(6.949 \text{ cal/g mole K})\left(4.187 \dfrac{\text{kJ/kg mole}}{\text{cal/g mole}}\right)}{(28.97 \text{ kg/kg mole})} = 1.0043 \text{ kJ/kg K}$$

$$C_v = C_p - R = 1.0043 - 0.287 = 0.7173 \text{ kJ/kg}$$

$$\gamma = \frac{C_p}{C_v} = \frac{1.0043}{0.7173} = 1.400$$

Constant specific heats (Cold air standard)

$$r_v = \frac{V_{\text{BDC}}}{V_{\text{TDC}}} = \left(\frac{T_2}{T_1}\right)^{1/\gamma-1}$$

$$T_2 = (8)^{0.4}(300) = 689\text{K}$$

Variable specific heats (Use Example 3.2)

$$r_v = \frac{V_{\text{BDC}}}{V_{\text{TDC}}} = \frac{V_r<T_1>}{V_r<T_2>}$$

$$V_r<T_2> = (18{,}020)/8 = 2252$$

$$T_2 = 678.5\text{K}$$

3. Process 2 → 3, V = constant heat addition:

$${}_2Q_3 = m(u_3 - u_2)$$

Constant specific heats:

a. $$T_3 = T_2 + \frac{{}_2Q_3}{mC_v} = 689 + \frac{(2.5 \text{ kJ}/4)}{(0.000285 \text{ kg})(0.7173 \text{ kJ/kg K})} = 3746\text{K}$$

$$P_3 = \frac{mRT_3}{V_3} = \frac{(0.000285 \text{ kg})(8314 \text{ N m/kg K})(3746\text{K})(8)}{(28.97)(0.000\,234 \text{ m}^3)}$$

$$P_3 = 10{,}474{,}861 \text{ N/m}^2 = 10{,}475 \text{ kN/m}^2$$

Variable specific heats:

$${}_2\bar{q}_3 = [\Delta\bar{h}<T_3> - \bar{R}T_3] - [\Delta\bar{h}<T_2> - \bar{R}T_2]$$

where

$$\bar{q} = \frac{(2.5 \text{ kJ}/4)(28.97 \text{ kg/kg mole})}{\left(4.187 \dfrac{\text{kJ/kg mole}}{\text{cal/g mole}}\right)(0.000285 \text{ kg})} = 15{,}173 \text{ cal/g mole}$$

$$T_2 = 678.5\text{K} \qquad \Delta\bar{h}<T_2> = 2735 \text{ cal/g mole}$$

or

$$\Delta\bar{h}<T_3> - 1.987\ T_3 = 15{,}173 + [2735 - (1.987)(678.5)]$$
$$= 16{,}560 \text{ cal/g mole}$$

by trial and error, $T_3 = 3023\text{K}$

$$\Delta\bar{h}<T_3> = 22{,}567 \text{ cal/g mole}$$

$$P_3 = \frac{(0.000285)(8314)(8)(3023)}{(28.97)(0.000234)(1000)} = 8453 \text{ kPa}$$

4. $3 \rightarrow 4$, isentropic expansion $S_3 - S_4$:

Constant specific heats

$$\frac{T_4}{T_3} = \left(\frac{V_{\text{TDC}}}{V_{\text{BDC}}}\right)^{\gamma-1}$$

$$T_4 = (1/8)^{0.4}(3746)$$
$$= 1630\text{K}$$

Variable specific heats (Use Example 3.2)

$$V_r<T_4> = 8V_r<T_3> = (8)(20.06)$$
$$V_r<T_4> = 160.48$$
$$T_4 = 1630\text{K}$$

5. Indicated thermal efficiency:

$$\eta = \frac{\text{desired energy output}}{\text{required energy output}} = \frac{w_{\text{net}}}{q_{\text{add}}} = 1 + \frac{{}_4q_1}{{}_2q_3}$$

Constant specific heats:

b. $$\eta = 1 - \left(\frac{T_1}{T_2}\right) = 1 - \left(\frac{300}{689}\right) = 0.565 = 56.5\%$$

Variable specific heats:

$$\eta = 1 - \left[\frac{\bar{u}<T_4> - \bar{u}<T_1>}{\bar{u}<T_3> - \bar{u}<T_2>}\right] = 1 - \left[\frac{\Delta\bar{h}<T_4> - \Delta\bar{h}<T_1> - \bar{R}(T_4 - T_1)}{\Delta\bar{h}<T_3> - \Delta\bar{h}<T_2> - \bar{R}(T_3 - T_2)}\right]$$

$$= 1 - \left[\frac{10{,}363 - 35 - (1.987)(1630 - 300)}{22{,}567 - 2735 - (1.987)(3023 - 679)}\right] = 0.494 - 49.4\%$$

6. Indicated mean effective pressure:

$$IMEP = \frac{W_{\text{net}}}{V_D} = \frac{\eta_2 Q_3}{[V_1 - V_2]}$$

Constant specific heats:

c. $$IMEP = \left[\frac{(0.565)(2.5/4 \text{ kN m})}{(0.00082 \text{ m}^3/4)}\right] = 1722.6 = 1723 \text{ kPa}$$

Variable specific heats:

$$\text{c.}\quad IMEP = \left[\frac{(0.494)(2.5/4)}{(0.00082/4)}\right] = 1506.1 = 1506\ \text{kPa}$$

Comment The influence of specific heat on the predicted air standard Otto cycle is seen in this example. For the same geometry and heat input, the assumption of constant heats yields too large an indicated thermal efficiency, indicated mean effective pressure, and indicated peak temperature. The hot-air standard value of 1.3 gives values that are closer to those obtained using the air tables. Further realism results from fuel-air/product gas thermochemistry, open system analysis, and incomplete combustion.

10.4 SPARK-IGNITION INTERNAL COMBUSTION ENGINE COMBUSTION

Spark-Ignition combustion processes can be classified as normal, i.e., a spark-initiated process; abnormal, i.e., preignition via hot-spot processes, such as carbon deposits or hot spark plugs; and self-ignition, i.e., autocatalylic detonation processes. Historic SI engine combustion requires a homogeneous combustion process, while current efforts are directed toward nonhomogeneous or stratified charge engine chemistry.

Basic design objectives of classic spark-ignition combustion include:

Developing a high level of turbulence within the charge
Promoting a rapid but smooth rise in pressure versus time (crank angle) during burn
Achieving peak cylinder pressure as close to TDC as possible
Establishing the maximum premixed flame speed
Burning the greatest portion of the charge early in the reaction process
Attaining complete fuel-air charge mixture combustion
Precluding the occurrence of detonation or knock
Minimizing combustion heat loss

Today, most SI engines are designed to operate within the 1000–6000-rpm range with spark firing required from 8 to 50 times each second. Engine power results from spark-initiated chemical kinetic and related thermochemical events occurring with a given piston-cylinder head configuration. Spark-ignition engine output, therefore, is strongly influenced by the TDC combustion chamber geometry. Some of these more obvious shape factors include:

TDC surface to volume configuration characteristics that allow heat loss and reduce power output

Combustion chamber configurations during burn that can promote the development of detonation waves and combustion knock
Particular configurations with crevice and cold wall quench zones that cause incomplete combustion

Flame speed, a characteristic of fuel-engine interactions, is a maximum near stoichiometric proportions that for gasoline engines, can achieve values in the range of 20–40 m/sec (66–131 ft/sec). Increasing engine speed will promote greater turbulence and higher flame speeds, but spark-ignition must be advanced to maintain combustion near the TDC position. Spark advance is also required at part-load operation, a regime characterized by lower power output, lower total energy release, and lower flame speed. Part-load performance, with its greater residual gas charge and hotter running cylinders, makes the engine more sensitive to spark knock. Flame speed for extremely lean burn can be so slow that a flame may still be progressing through the chamber when inlet valves open, leading to backfire into the inlet manifold. Spark advance is set at minimum ignition advance for best torque (MBT).

A spherical combustion chamber having a radius r_0 can be used to describe essential elements of SI premixed homogeneous charge combustion. Assume that such a geometry contains reactants initially at uniform conditions T_2 and P_2. Central ignition of this fuel-air mixture produces a spherically shaped premixed flame front that burns outward toward the combustion chamber wall. Recall from Chapter 8 that premixed flame speed depends on the initial thermodynamic state, the fuel-air ratio, the chemical kinetics of the fuel-air mixture, and the aerodynamics of flow.

Initial rates of chemical reactions and flame propagation at the center of the constant-volume chamber will be relatively slow because of low turbulence, temperature, and pressure within the original charge. This is why efforts are made to initiate swirl and turbulence within a cylinder during charge induction and prior to charge ignition. Too much turbulence, however, could contribute to increased convective heat transfer loss, flame extinction, and/or reduction in indicated power output.

As the flame front progressively burns through the charge, burned gas product temperature T_3, pressure P_3, and specific volume v_3 behind the flame front will be greater, causing thermal compression of the remaining unburned charge. In addition, heat transfer and diffusion of some reactive radical elements across the traveling reaction zone from the burned product gases will raise local temperature and reactivity in the charge ahead of the flame. This means that flame speed will increase as it moves into these gases. Note that this differs from the one-dimensional steady-state and steady-flow subsonic reaction propagation process described in Chapter 7.

In Chapter 7, a reaction front was burning in an open duct with constant upstream conditions whereas, in an SI engine, flame propagation occurs within confined spaces having upstream conditions that are not constant.

The highest kinetic rates, gas temperatures, and hence flame speeds are achieved in the middle region of the sphere. Since flame speeds are greatest in this area, the largest part of the charge is burned during this stage of combustion. Thermal compression of the unburned charge ahead of the flame, however, continues during this portion of the reaction.

Flame extinction and termination of combustion occur near the wall. High heat transfer rates between the moving flame front and cold wall causes rapid reduction in gas temperature and flame speed approaching the cavity wall. Quenching of the flame occurs prior to its arrival at the wall, leaving a boundary layer region of incomplete reaction within the volume. Note that flame quench in narrow gaps and crevice geometries of an actual chamber can play a major role in emissions formation in an operating engine.

Recall from Chapter 8 that increasing values of reaction temperature and pressure caused the hydrogen-oxygen system to shift from a subsonic flame to a supersonic detonation wave. Higher temperatures and pressures are found in the unburned charge at the end of burn as a result of thermal compression and heating. The potential for knock, or detonation, therefore, exists in a spherical cavity near the end of burn, where conditions may allow gas phase autocatalytic reactions to propagate at a rate that causes oxidation to self-accelerate. This phenomenon of engine knock can be observed by the use of a cylinder pressure-crank-angle diagram; see Figure 10.8.

Central ignition and combustion of a homogeneous fuel-air charge using a constant-volume spherical chamber produces the following general traits:

Low initial burn rate with little flame expansion and turbulence

An intermediate burn rate having the greatest flame speed, percent of burn, and distance of travel

Terminal burn rate having the maximum flame front area and a thermally compressed end charge that can potentially detonate

Figure 10.9 graphically illustrates these results. The flame front-time plot is often referred to as the classic premixed S combustion curve.

Control or limitation of knock in SI engines is a key goal of engine research and development. A given CFR octane-rated fuel will burn differently in the many types of engines that are now available. It should be noted that volume is not necessarily constant during combustion in an actual SI engine. Efforts are being made to deal with this problem through combustion chamber design and/or basic fuel science research to develop

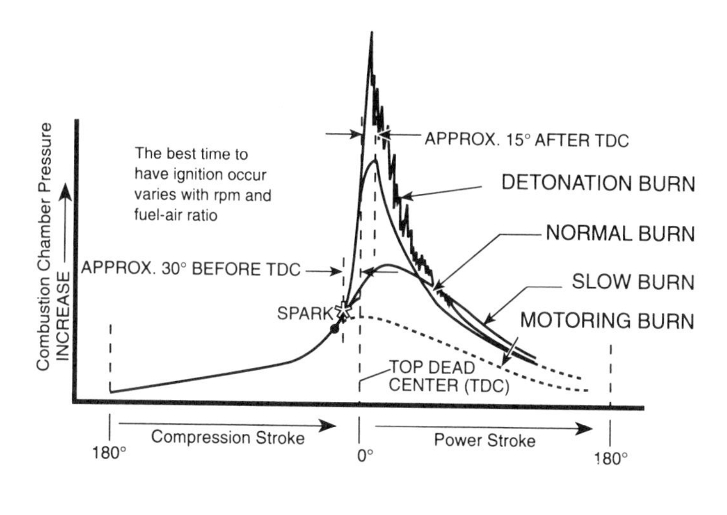

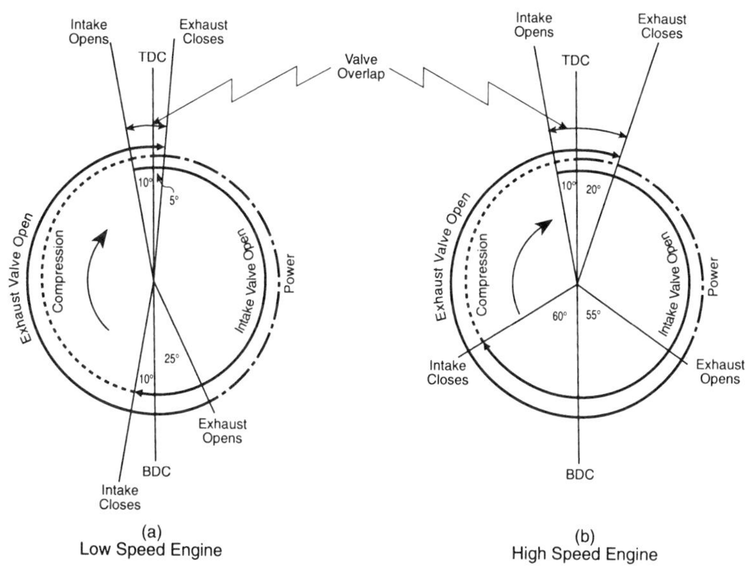

Figure 10.8 Spark-ignition combustion-pressure-crank angle diagram.

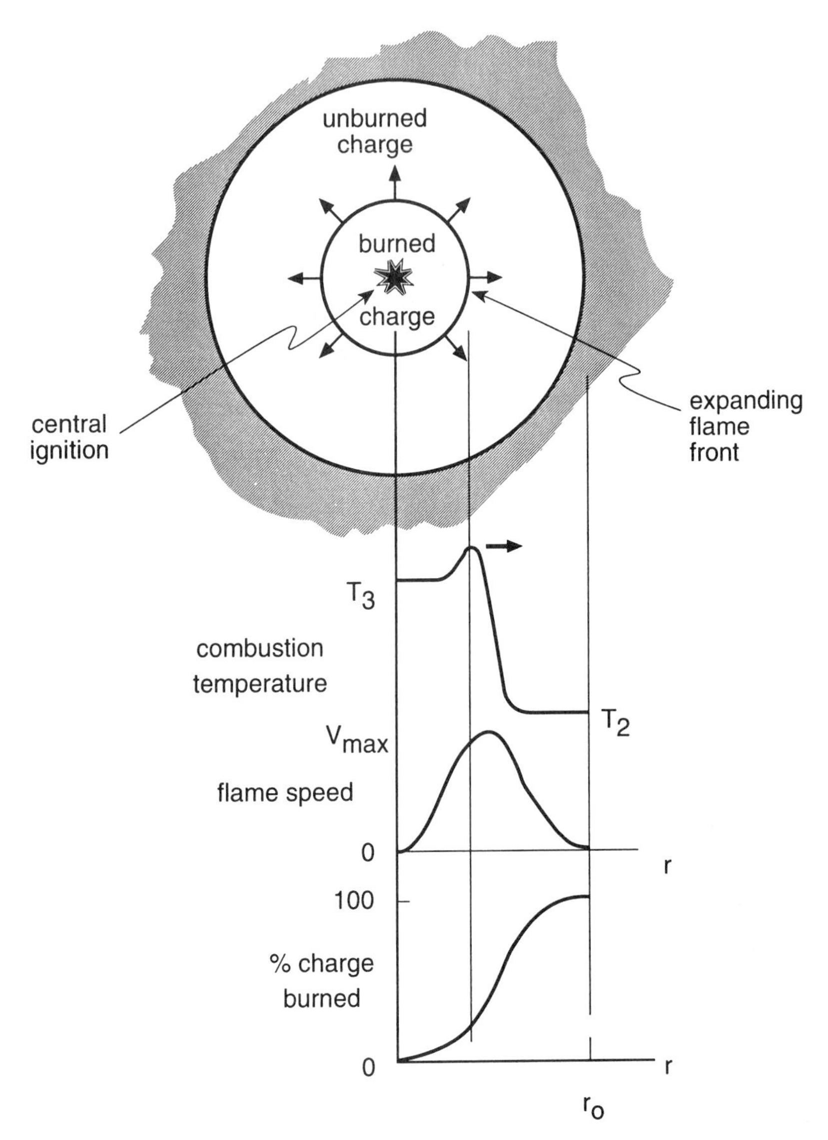

Figure 10.9 Spark-ignition constant-volume combustion characteristics.

fuels having higher octane ratings. To illustrate, consider side ignition of the spherical charge previously described. The final flame front area is reduced, but the distance the flame would have to travel is increased, i.e., $2r_0$ versus r_0. This added distance could increase thermal compression of end gases and promote knock. A simple cone or wedge-shaped chamber, rather than a cylinder, with spark ignition at the base would yield an initially large burn rate and flame front, with reduced area and mass at the end of burn. This is a technique used to control knock, but the total charge in this instance may be less than for a sphere. Many geometric configurations and SI combustion techniques are pursued in an attempt to achieve optimal engine performance (see Figure 10.11).

Thermodynamic and time-related factors both play a role in controlling SI engine knock. Lower-reaction thermal conditions and shorter periods of time for reaction will reduce the tendency for spark knock. Hence, thermodynamic factors that can reduce the potential for knock include:

Reducing the charge by throttling (heat release)
Raising the combustion chamber wall temperature (heat loss)
Operating with an excessively rich or lean mixture (lower peak temperature)
Charge dilution
Lowering the inlet air temperature
Increasing the air inlet humidity
Raising the octane number of the fuel

Time factors that can reduce the potential for knock include the following:

Increasing cylinder turbulence (flame speed)
Increasing engine speed
Decreasing flame travel distance
Retarding spark timing

Indirect methods that can help control knock include:

Promotion of combustion chamber swirl (turbulence)
Exhaust gas recirculation (temperature)
Manifold and/or cylinder-injected combustion
Nonhomogeneous combustion

Efforts are in progress to develop engines that *stratify* the charge, i.e., replace the homogeneous charge with a mixture that is richer during the middle stages of reaction and leaner during termination of flame propagation. This type of burn could not only help control knock but would promote more complete combustion of a fuel-air charge and reduce fuel consumption. Development of suitable additives to increase the octane rating of gasoline could also be a means of decreasing the fuel-engine knock sensitivity.

Incomplete combustion is a natural consequence of SI engine operation and results from the unique nature of each particular fuel-engine interface; i.e., the quality of fuel used, the specific geometry of an engine's combustion chamber, and the particular means of ignition and burn utilized to produce power. The major by-products of incomplete combustion, or *engine pollutants,* associated with SI chemistry are CO, or carbon monoxide; UHC, or unburned hydrocarbons; and NOx, or nitric oxides. Equilibrium product predictions using first and second law analysis as described in Chapter 3 do not predict levels of engine emissions measured using techniques described in Chapter 9. Equilibrium thermochemistry is independent of time, but the process of combustion in an SI engine is, of necessity, time-bounded. An understanding of emissions cannot be based on simple thermochemical calculations but rather requires a knowledge of both fuel and engine characteristics.

EXAMPLE 10.3 An ideal open system air standard Otto engine has a compression ratio of 9 to 1. Charge intake conditions are given as 80°F and 14.7 psi with combustion approximated by a heat addition of 650 Btu/lbm. Using a constant specific heat analysis and residual gas recirculation, determine the following: (a) charge temperature at BDC prior to compression; (b) peak cycle temperature; (c) residual gas temperature at TDC after exhaust; and (d) residual gas fraction f. Compare these results to those for an equivalent Otto cycle.

Solution

1. Properties of air, from Appendix B at *STP*:

$$\overline{C}_p\langle T_0\rangle = 6.947 \text{ cal/g mole K}$$

or

$$C_p = \frac{(6.947)(1.8001 \text{ Btu/lb mole K})\ (5\text{K})}{(9°\text{R})(28.97 \text{ lbm/lb mole})} = 0.24\ \frac{\text{Btu}}{\text{lbm °R}}$$

$$\overline{C}_v = \overline{C}_p - \overline{R} = 6.947 - 1.987 = 4.960 \text{ cal/g mole K}$$

$$C_v = \frac{(4.960)(1.8001)(5)}{(9)(28.97)} = 0.171\ \frac{\text{Btu}}{\text{lbm °R}}$$

$$\gamma = \frac{C_p}{C_v} = \frac{0.24}{0.171} = 1.4$$

2. Otto cycle analysis, no intake/exhaust/recirculation:

$$T_1 = 80°\text{F} \qquad P_1 = 14.7 \text{ psi}$$

$$v_1 = \frac{(53.34 \text{ ft lbf/lbm °R})\ (540°\text{R})}{(14.7 \text{ lbf/in.}^2)(144 \text{ in.}^2/\text{ft}^2)} = 13.607\ \frac{\text{ft}^3}{\text{lbm}}$$

$$T_2 = T_1\left(\frac{v_1}{v_2}\right)^{\gamma-1} = 540\ (9)^{0.4} = 1300.4°\text{R}$$

$$T_3 = T_2 + \frac{q}{C_v} = 1300 + (650/0.171) = 5101.2°\text{R}$$

$$T_4 = T_3\left(\frac{v_3}{v_4}\right)^{\gamma-1} = 5101\left(\frac{1}{9}\right)^{0.4} = 2118.1°\text{R}$$

3. Open cycle analysis—assume $T_1 = 560°\text{R}, f = 0.03$:

$$v_1 = \frac{(53.34)\ (560)}{(14.7)\ (144)} = 14.11\ \text{ft}^3/\text{lbm}$$

4. Compression process:

$$T_2 = T_1\left(\frac{v_1}{v_2}\right)^{\gamma-1} = 560(9)^{0.4} = 1348.6°\text{R}$$

$$P_2 = P_1\left(\frac{v_1}{v_2}\right)^{\gamma} = (14.7)(9)^{1.4} = 318.6\ \text{psia}$$

5. Heat addition, $q(1 - f) = C_v(T_3 - T_2)$:

$$T_3 = T_2 + (1 - f)q/C_v = 1348.6 + \frac{(1.0 - 0.03)\ (650)}{0.171}$$

$$T_3 = 5035.7°\text{R}$$

$$P_3 = \frac{(53.34)(5035.7)(9)}{(14.11)(144)} = 1189.8\ \text{psi}$$

6. Expansion and exhaust:

$$f = \left(\frac{v_3}{v_5}\right) = \left(\frac{P_5}{P_3}\right)^{1/\gamma} = \left(\frac{14.7}{1189.8}\right)^{1/1.4} = 0.043$$

7. Residual gas temperature:

$$T_5 = T_3\left(\frac{v_3}{v_5}\right)^{\gamma-1} = (5035.7)(0.043)^{0.4} = 1430.4°\text{R}$$

8. Intake and mixing process, $P = c$ (adiabatic):

$$C_pT_1 = fC_pT_5 + (1 - f)C_pT_6$$

$$T_1 = fT_5 + (1 - f)T_6$$

$$= (0.043)(1430.4) + (1 - 0.043)(540)$$

$$= 578.3°\text{R}$$

9. Return to step 3 and choose a new T_1 and f, and repeat steps 3–8. By iteration, one finds that

Part		Otto cycle
a	$T_1 = 580°R$	(540°R)
b	$T_3 = 5030.7°R$	(5101.2°R)
c	$T_5 = 1442°R$	
d	$f = 0.044$	

Comments This problem illustrates how the residual gas influences SI engine thermochemistry. These calculations did not include the temperature dependency of γ, the impact fuel-air composition, and losses such as heat, friction, and pumping volumetric efficiency.

Carbon monoxide, an intermediate product of combustion of any hydrocarbon fuel, is a rich SI combustion product produced at both full-load and idle operation. At idle, 0.3 vol % CO is lethal and can cause death in confined spaces within 30 min. The major source of CO is a result of chemical kinetics within the bulk gas; however, CO is also produced by partial oxidation of UHC during the exhaust stroke as well as dissociation of CO_2 produced during combustion. The rapid drop in gas temperature during expansion will freeze CO concentrations at levels different from those predicted on the basis of equal temperatures and equilibrium composition calculations. Control of CO is chiefly by improved *FA* management and lean burn, such as with a stratified engine. Additional CO reduction is obtained using exhaust gas afterburning management via catalyst.

Unburned hydrocarbons, or UHC, can be found in the boundary layer around the combustion walls, where convective and radiative heat transfer have quenched the reaction at the end of flame travel. Additional sources of UHC can include fuel pyrolyzed during burn, lean operation misfire, rich combustion during start-up, and *FA* variation between cylinders. Noncombustion UHC emissions include desorbed or outgassed fuel that had undergone oil film loading early in the charge induction process and gases trapped in small quench zones and crevice volumes formed by the piston crown, top ring, and cylinder walls. Nonengine UHC emissions include fuel tank, crankcase, and carburetor evaporative losses. Initial efforts to reduce and control UHC in SI engines used improved carburetion, spark timing, and careful attention to minimize the surface-to-volume ratio in the design of a combustion chamber. Lean burn combustion, a technique for CO management, has also been used as a technique for controlling UHC emissions. Recall that stratified combustion is based on an ability to burn rich during that portion of the reaction in which the greatest fraction of charge is consumed. A lean burn design approach produced engines having poor fuel economy with unacceptable performance, which required increased maintenance. Most current UHC control systems today incorporate catalytic UHC reduction of the exhaust. Oxidizing catalytic converters containing materials such as platinum reduce both CO and UHC without interfering with engine operation but require the use of unleaded gasoline to prevent catalyst poisoning. Figure 10.10 illustrates the general relationship between CO, UHC, and NOx emission concentrations and fuel-air mixture equivalence ratio for an SI engine and hydrocarbon fuel such as gasoline.

Figure 10.11 shows several SI combustion chamber geometries that have been used in an attempt to achieve a specific design goal such as controlling detonation (wedge), increasing power (hemispherical), and reducing emissions (stratified charge). Other chambers provide examples of unique SI combustion processes (jet valve and pulse burn) that have also been developed in an effort to improve SI engine performance. The cylinder head

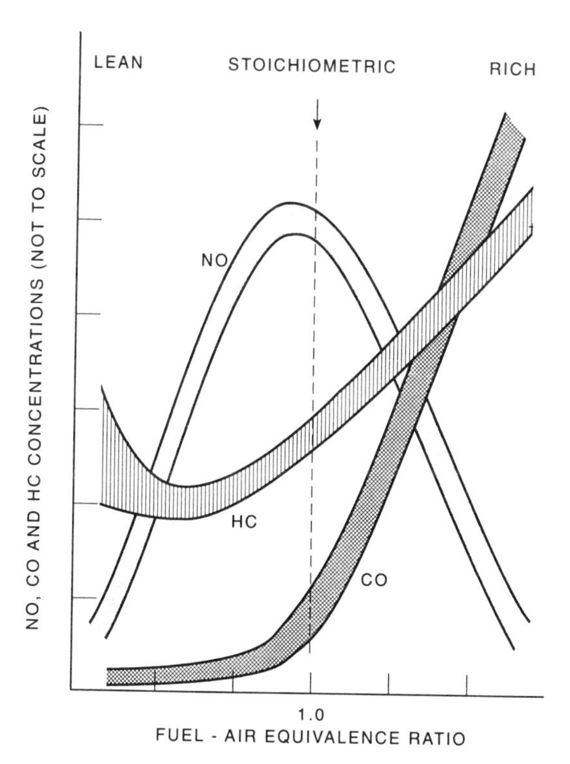

Figure 10.10 Spark-ignition emissions vs. equivalence ratio.

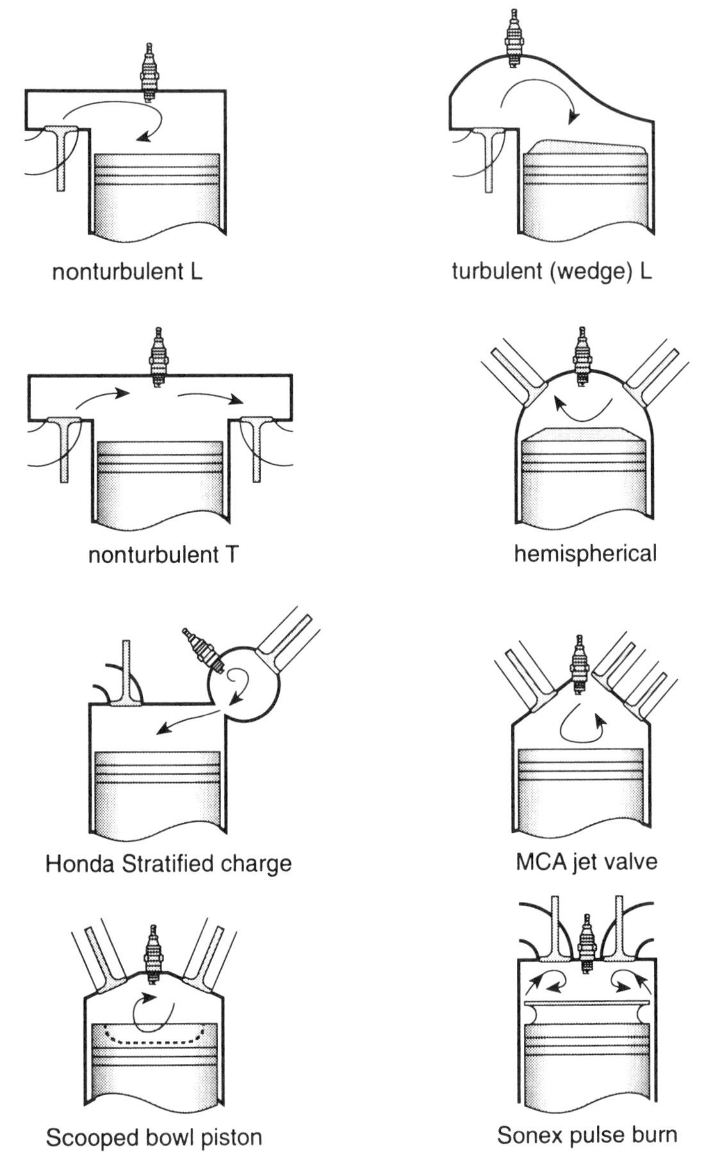

Figure 10.11 Various SI combustion chamber designs.

must provide more than just a cover for the piston crown at TDC and definition for the chamber volume. For example, the spark plug and the exhaust valve are most frequently located near each other and away from the end gas, while the end gas is located in a cool region of the chamber to minimize detonation.

EXAMPLE 10.4 An automotive SI engine burns liquid octane at the rate of 0.0006 kg/sec. Both air and fuel are assumed to enter the engine at 25°C and 1 atm. The exhaust gases leave the engine at 900K and 1 atm. Steady-state operation results from an equivalence ratio of 0.833, with 88% of the carbon in the fuel burning to form carbon dioxide and the remaining 12% forming carbon monoxide. Measurements show that the indicated heat loss from the engine is equal to 80% of the indicated work output of the engine. Using this information, obtain the following: (a) fuel-air ratio, kg fuel/kg air; (b) combustion product composition, ppm; (c) exhaust gas dew point temperature, K; (d) indicated engine power output, ikW; and (e) indicated specific fuel consumption, kg/ikW sec.

Solution

1. Stoichiometric equation:

$$C_8H_{18} + 12.5[O_2 + 3.76N_2] \rightarrow 8CO_2 + 9H_2O + (12.5)(3.76)N_2$$

2. Actual reaction:

$$0.833C_8H_{18} + 12.5[O_2 + 3.76N_2] \rightarrow$$
$$0.833[(0.88CO_2 + 0.12CO)8 + 9H_2O] + 2.49O_2 + (12.5)(3.76)N_2$$

3. Fuel-air ratio:

$$FA = \frac{(0.833 \text{ kg/mole})(114 \text{ kg/kg mole})}{(12.5)(4.76 \text{ kg mole})(28.97 \text{ kg/kg mole})}$$

a. $$= 0.055 \text{ kg fuel/kg air}$$

4. Exhaust gas analysis:

$$\bar{x}_{CO_2} = \frac{(0.833)(8)(0.88)}{14.16 + 2.49 + 47} = 0.0921 \text{ (92,100 ppm)}$$

$$\bar{x}_{CO} = \frac{(0.833)(8)(0.12)}{63.65} = 0.0126 \text{ (12,600 ppm)}$$

$$\bar{x}_{H_2O} = \frac{(0.833)(9)}{63.65} = 0.1178 \text{ (117,800 ppm)}$$

b. $\bar{x}_{O_2} = \dfrac{2.49}{63.65} = 0.0391$ (39,100 ppm)

$\bar{x}_{N_2} = \dfrac{(12.5)(3.76)}{63.65} = 0.7384$ (738,400 ppm)

5. Exhaust gas dew point T:

$$P_{H_2O} = \bar{x}_{H_2O}\, P_{tot} = (0.1178)(101\text{kPa})$$
$$= 11.89 = 11.9 \text{ kPa}$$

c. $T_{sat} \langle 11.9 \text{ kPa} \rangle = 322\text{K}$

6. Energy balance:

$$\cancel{\frac{dE}{dt}} = \frac{\delta Q}{dt} - \frac{\delta W}{dt} + \sum \dot{N}_i \bar{e}_i - \sum \dot{N}_j \bar{e}_j$$

or

$$0 = Q - W + \sum N_i \bar{e}_i - \sum N_j \bar{e}_j$$
$$W - (0.8)(W) = \sum N_i \bar{e}_i - \sum N_j \bar{e}_j$$
$$0.2W = \{(12.5)(3.76)[\bar{h}_f^0 + \Delta\bar{h}\langle 900 \rangle\}_{N_2} + (2.49)[\bar{h}_f^0 + \Delta\bar{h}\langle 900 \rangle]_{O_2}$$
$$+ (0.833)(9)[\bar{h}_f^0 + \Delta\bar{h}\langle 900 \rangle]_{H_2O_g} + (0.833)(8)(0.12)[\bar{h}_f^0$$
$$+ \Delta\bar{h}\langle 900 \rangle]_{CO} + (0.833)(8)(.88)[\bar{h}_f^0 + \Delta\bar{h}\langle 900 \rangle]_{CO_2}\}$$
$$- \{ (12.5)[\bar{h}_f^0 + \Delta\bar{h}\langle 298 \rangle]_{O_2} + (12.5)(3.76)\ [\bar{h}_f^0 + \Delta\bar{h}\langle 298 \rangle]_{N_2}$$
$$+ (0.833)[\bar{h}_f^0 + \Delta\bar{h}\langle 298 \rangle]_{C_8H_{18}}\}$$
$$0.2W = \{(12.5)(3.76)[4355] + (2.49)[4600]$$
$$+ (0.833)(9)[-57{,}798 + 5240] + (0.833)(8)(0.12)[-26{,}417 + 4397]$$
$$+ (0.833)(8)(.88)[-94{,}054 + 6702]\} - \{(0.833)[-59{,}740]\}$$
$$0.2W = -657{,}993.9 \text{ cal/g mole}$$
$$W = -3{,}289{,}969 = -3{,}290{,}000 \text{ cal/g mole}$$

and

$$\dot{W} = \frac{(3.29 \times 10^6 \text{ cal/g mole})(4.187\,\dfrac{\text{kJ/kg mole}}{\text{cal/g mole}})(0.0006 \text{ kg/sec})}{(114 \text{ kg/kg mole})}$$

d. $\dot{W} = 72.5 \text{ kJ/sec} = 72.5\text{ikW}$

7. Indicated specific fuel consumption:

$$ISFC = \frac{\dot{m}_{\text{fuel}}}{\dot{W}_I} = \frac{0.0006 \text{ kg/sec}}{72.5 \text{ ikW}}$$

e. $= 8.28 \times 10^{-6} = 8.3 \times 10^{-6}$ kg/ikW sec

10.5 SPARK-IGNITION ENGINE FUEL ALTERNATIVES

Mobility in today's world is due, in part, to the historic and successful relationship that has existed between gasoline and the SI engine. Future SI engine technology will be required to provide improved engines that can utilize a wider range of fuel alternatives than just leaded gasoline, including the petroleum-based distillate fuel fractions between gasoline and diesel, non-petroleum-based gasolines produced from resources such as tar sands and shale oil, coal-derived liquid fuels, biomass compounds, and alcohols, as well as alcohol-petroleum blends. Table 10.1 summarizes nominal properties for certain of these fuel choices. Dr. Gordon Millar, former president of SAE, has observed that an omnivorous engine must be produced in order to use these fuels effectively, a design goal that will raise many new challenges in materials-fuel interactions for the engine designers and manufacturers who will build these improved machines. Among the compound and interactive factors that will influence the success or failure of these new prime movers are: economics; a viable industrial base for providing suitable fuel alternatives; acceptable performance characteristics for the next generation of mobility propulsion hardware, including better fuel economy and emissions control; and overall fuel-engine system durability, reliability, and efficiency.

The ideal energy characteristics for an SI engine fuel should include:

Storage as a stable liquid at ambient temperature

Low vapor pressure to prevent vapor lock but with light fractions for easy vaporization during cold start

High energy content per unit mass and a fuel density at ambient conditions for both good fuel economy and power

High octane number to allow increased compression ratio operation for better thermal efficiency

Chemical constituents that minimize the generation of polluting engine emissions

Environmentally safe fuel-handling characteristics

Antioxidant, anti-icing, ignition control, cleansers, antiknock, and other required fuel additivies.

In the past, leaded gasoline, which was the principal commercial spark-ignition engine fuel, best met the greatest number of these important fuel

Table 10.1 Nominal Properties for Certain SI Fuels[a]

Compound	Methane	Propane	Hydrogen	Methanol	Ethanol	"Gasoline"
Formula	CH_{4_l}	$C_3H_{8_l}$	H_{2_l}	CH_3OH	C_2H_5OH	C_8H_{17}
Molecular weight	16.0	44.0	2.0	32.0	46.0	Variable
Stoichiometric AF	17.24	15.67	34.47	6.45	9.01	14.5–15.1
Specific gravity	0.3	0.5		0.794	0.785	0.7
h_{fg} kJ/kg	509	426	447	1168	921	349
(Btu/lbm)	(219)	(183)	(192)	(502)	(396)	(150)
Freezing pt., °C	−182	−188		−98	−117	−57
(°F)	(−296)	(−306)	(—)	(−144)	(−179)	(−70)
Boiling pt. °C	−162	−42	−253	65	78	27–227
(°F)	(−259)	(−44)	(−423)	(149)	(172)	(80–440)
LHV kJ/kg	49,537	45,980	116,114	20,106	26,991	47,125
(Btu/lbm)	(21,297)	(19,768)	(49,920)	(8,644)	(11,604)	(20,260)
Autoignition T, °C	732			470	363	221–260
(°F)	(1350)	(—)		(878)	(685)	(430–500)
Octane number	120	105		107	107	91–105

[a]Data from: ASTM Special Technical Publications No. 109A, 1963, and No. 225, 1958; Phillips Petroleum Co: *Reference Data for Hydrocarbons*, 1962.

criteria. today, however, concern about the impact of lead on the environment requires that unleaded gasolines be produced and utilized. Vigorous efforts are now under way to discover suitable octane boosters for unleaded gasoline and to extend and expand current crude oil gasoline resources in an attempt to continue producing a world supply of a usable unleaded fuel. Gasolines can also be distilled from several of the solid and liquid fuel resources discussed in Chapters 4 and 5. Gasoline fractions from shale oil and coderived distillates may have distillation curves similar to those produced from crude oil but may differ widely in their specific properties, such as carbon-to-hydrogen ratio, ignition temperature, viscosity, ultimate analysis, and even molecular structure. Oil shale gasoline, for example, is highly paraffinic and has a low octane number, whereas coal-derived distillate gasoline is generally highly aromatic and has a high octane number. Germany produced 10,000 bbl/day of high octane fuel from coal during World War II, a technology being used today in South Africa; see Chapter 5. Hydrogenation of these gasolines may be necessary to make them more suitable for use with present SI engine designs. The performance of engines burning these various gasolines will depend strongly on how closely they compare to the current grades of gasoline being refined from crude oil.

The cleanest-burning SI fuel is hydrogen, i.e., no CO, CO_2, or UHC. Products of hydrogen-air engine combustion have less dissociation and a higher specific heat ratio than those for gasoline-air chemistry. Indicated thermal efficiency of a gasoline engine when it is switched to operate on hydrogen is comparable to that using gasoline, but the engine requires some minor modifications. For example, a piston-ring redesign may be necessary to limit hydrogen blowby gas. Hydrogen combustion has some unique characteristics that would influence its use as an SI fuel, including wider flammability limits, higher flame speed, lower ignition temperature, and higher detonation tendency than gasoline-air mixtures. Spark-ignition engines that burn hydrogen also have a greater tendency to preignite, backfire, and spark-knock during stoichiometric operation than when running on gasoline. Smaller quenching distances are found for hydrogen-air combustion. This allows a flame front to pass through smaller areas when valves open, causing backfiring into the intake manifold in carbureted engines. Conversion of a given engine to stoichiometric hydrogen combustion will result in a lower indicated power output than when the engine is burning gasoline, but an ability to run lean (wider flammability limits), along with increasing compression ratio, can recover this loss in power. The high NOx emissions level, peak temperature, and flame speed can all be reduced by lean operation, use of exhaust gas recirculation (EGR), or recycling of the exhaust moisture. Efficient fuel storage in vehicle applications is a major disadvantage since hydrogen must be stored either as a liquid or as a gas using metal hydride absorption, but these solutions introduce severe weight and volume penalties to the vehicle. In addition, hydrogen is not naturally occurring and is, therefore, truly an exotic SI fuel option that may be practical in the near term only in large metropolitan areas where severe environmental constraints on pollution levels now exist.

Methane is also a cleaner-burning fuel than gasoline. Charge preparation techniques must be modified when burning methane in order to accommodate its larger required fuel volumetric flow rate and the resulting reduction in volumetric efficiency. Charge preparation is somewhat simplified in that methane is already a vapor, a benefit when cold-starting an engine. One hundred cubic feet of methane gas is approximately equal to one gallon of gasoline in terms of stored chemical energy content, but operation with methane does require ignition timing adjustments because of its lower flame speeds. Methane has a high octane number and is easily ignited but has a higher ignition temperature than gasoline.

Liquefied natural gas (LNG), compressed natural gas (CNG), synthetic natural gas (SNG), and even producer gas can all be used as SI fuels. Sulfur impurities in these fuels, however, can cause extensive engine component corrosion. Liquefied petroleum gas (LPG), such as butane and/or propane, are also suitable automotive fuels that have been used in certain limited dual-fuel applications. High-pressure gas can be stored in heavy cylinders and is then throttled, metered, and supplied by means of a pressure regulator, fuel flow gauge, and fuel-air mixer. A fuel solenoid shutoff valve is necessary for safety when the engine is not running.

Alcohol is another resource that has been used as an SI engine fuel. The two most practical power alcohol fuels are methanol, which can be produced from coal and/or natural gas, and ethanol, which can be produced from fermentation of certain biomass materials; see Chapter 5. American farmers burned a grain alcohol fuel called argol in their tractors during the depression, while Germany used alcohol as a military fuel during World War II when gasoline supplies were cut off. Post–World War II French vehicles burned a 50/50 gasoline-alcohol blend and, during the energy crisis of the 1970s, Americans were encouraged to use gasohol, a 90% gasoline-10% ethanol fuel blend in which the alcohol acted chiefly as a political additive. Activities in Brazil, where ethanol production can be based on its sugarcane industry, and in Canada, where methanol produced from natural gas is feasible, strongly suggest that alcohol fuels may be a reasonable choice for the future mix of worldwide SI fuel options.

EXAMPLE 10.5 The influence of variable specific heats and gas composition is incorporated in the fuel-air thermochemical SI engine analysis. Consider a stoichiometric mixture of gaseous octane-air and, using JANAF data found in Appendix B, find (a) the mixture relative pressure P_r and (b) the mixture relative volume v_r as functions of temperature. Repeat parts (a) and (b) for (c) ideal complete combustion products.

Solution

1. Stoichiometric equation:

$$C_8H_{18} + 12.5(O_2 + 3.76N_2) \rightarrow 8CO_2 + 9H_2O + 47N_2$$

2. Reactant mole fractions:

$$\bar{x}_{C_8H_{18}} = \frac{1}{1 + (12.5)(4.76)} = 0.01653$$

$$\bar{x}_{O_2} = \frac{12.5}{1 + (12.5)(4.76)} = 0.20661$$

$$\bar{x}_{N_2} = \frac{(12.5)(3.76)}{1 + (12.5)(4.76)} = 0.77686$$

3. Product mole fractions:

$$\bar{x}_{CO_2} = \frac{8}{17 + 47} = 0.12500$$

$$\bar{x}_{H_2O} = \frac{9}{17 + 47} = 0.14063$$

$$\bar{x}_{N_2} = \frac{47}{17 + 47} = 0.73437$$

4. Reactant mixture entropy:

$$\bar{s}^0\langle T\rangle = \Sigma \bar{x}_i \bar{s}_i^0\langle T\rangle$$

T		cal/g mole K			
K	°R	$\bar{s}^0_{C_8H_{18}}$	$\bar{s}^0_{O_2}$	$\bar{s}^0_{N_2}$	$\bar{s}^0\langle T\rangle$
298	536	111.807	49.004	45.770	47.5298
300	540	111.853	49.047	45.815	47.5744
400	720	126.555	51.091	47.818	49.7960
500	900	140.574	52.722	49.386	51.5826
600	1080	153.816	54.098	50.685	53.0949
700	1260	166.387	55.297	51.806	54.4213
800	1440	178.320	56.361	52.798	55.6090
900	1620	189.540	57.320	53.692	56.6870
1000	1800	200.202	58.192	54.507	57.6767

5. Reactant relative pressure and volume:

$$P_r<T> = \exp\{\bar{s}^0<T>/\bar{R}\} \qquad v_r<T> = \bar{R}T/P_r<T>$$

T			
K	°R	$P_r \times 10^{-10}$	$v_r \times 10^{+10}$
298	536	2.4462	242.1
300	540	2.5017	238.3
400	720	7.6526	103.86
500	900	18.806	52.829
600	1080	40.257	29.615
700	1260	78.479	17.723
800	1440	142.67	11.142
900	1620	245.45	7.286
1000	1800	403.90	4.920

6. Product mixture entropy:

$$\bar{s}^0<T> = \Sigma \bar{x}_j \bar{s}_j^0<T>$$

T		cal/g mole K			
K	°R	$\bar{s}^0_{CO_2}$	$\bar{s}^0_{H_2O}$	$\bar{s}^0_{N_2}$	$\bar{s}^0<T>$
1800	3240	72.391	61.965	59.320	61.3258
1900	3420	73.165	62.612	59.782	61.8529
2000	3600	73.903	63.234	60.222	62.3557
2100	3780	74.608	63.834	60.642	62.8366
2200	3960	75.284	64.412	61.045	63.2984
2300	4140	75.931	64.971	61.431	63.7413
2400	4320	76.554	65.511	61.802	64.1676
2500	4500	77.153	66.034	62.159	64.5782
2600	4680	77.730	66.541	62.503	64.9742
2700	4860	78.286	67.032	62.835	65.3566
2800	5040	78.824	67.508	63.155	65.7258
2900	5220	79.344	67.971	63.465	66.0836
3000	5400	79.848	68.421	63.765	66.4301
3100	5580	80.336	68.858	64.055	66.7656
3200	5760	80.810	69.284	64.337	67.0918

7. Product relative pressure and volume:

T			
K	°R	$P_r \times 10^{-10}$	$v_r \times 10^{+10}$
1800	3240	2,534.3	1.411
1900	3420	3,304.2	1.143
2000	3600	4,255.6	0.9338
2100	3780	5,420.8	0.7698
2200	3960	6,839.1	0.6392
2300	4140	8,546.8	0.5347
2400	4320	10,592	0.4502
2500	4500	13,023	0.3814
2600	4680	15,896	0.3250
2700	4860	19,269	0.2784
2800	5040	23,203	0.2398
2900	5220	27,782	0.2074
3000	5400	33,074	0.1802
3100	5580	39,158	0.1573
3200	5760	46,144	0.1378

Several thermochemical properties of neat alcohol do make it unsuitable for direct use with current engine technology, a situation that cannot be easily circumvented without some major redesign. Alcohols have a higher latent heat of vaporization and lower vapor pressure than gasoline and will require a modified intake and charge preparation technique for successful consumption. Alcohol, unlike gasoline blends, has a simple saturation temperature-pressure relationship and, therefore, preheating is essential to eliminate cold-start difficulties. These same characteristics make alcohol fuels less likely to vapor-lock and, with their greater latent heats of vaporization, yield higher volumetric efficiencies as a result of charge cooling during induction. Alcohol-fueled racing engines have used this property successfully to provide a greater mass of charge and thus increase the indicated torque and power output of their vehicles. These engines run rich, an impractical condition for consumer operation given the current stringent requirements on emissions and fuel economy. Efforts to improve fuel economy are further aggravated with alcohol since the heating value, on a mass basis, for both methanol and ethanol is less than that of gasoline or, in other words, more gallons of fuel are burned per mile. Also, stoichiometric fuel-air ratios for alcohol combustion are greater than those for gasoline-air, meaning that, for equal amounts of air, more alcohol fuel is needed. A standard carburetor cannot meter sufficient fuel for stoichiometric combustion and, in fact, will cause the engine to run lean. Thermochemically, Table 10.1 would suggest that ethanol is more favorable than methanol, i.e., has higher heat of combustion and lower latent heat of vaporization. However, it should be recalled that the successful fuel-engine interface design involves more issues than just thermodynamic properties of a given fuel. Alcohol-air mixtures have wide flammability limits, high flame speed, and low flame luminosity, i.e., less radiation heat loss, which suggests potentially better part-load performance. Octane numbers of ethanol and methanol are higher than gasoline and have even been considered as blending agents to raise the octane number of unleaded gasoline. The higher octane number for pure ethanol may allow an engine to be designed having a 12:1 compression ratio. Alcohol motor fuels are more susceptible to hot-spot preignition, but this can be alleviated somewhat by using a spark plug having a lower heat range. Emissions of UHC, CO, and NOx should be lower for alcohol-fueled engines, but partially oxidized aldehyde emissions will generally be greater because of their lower charge temperature and lean operation limits. Alcohol motor fuels will not degrade over time as certain gasoline constituents do, but they do absorb water, and alcohol-water mixtures will separate when stored. Alcohols are also more corrosive when in contact with metals such as copper and brass, in addition to certain plastics and gasket materials. Alcohol provides less natural lubricity than gasolines and results in lower engine component durability, greater wear, and possibly greater rust because of its affinity for water.

EXAMPLE 10.6 A 4-in. × 4-in. four-stroke IC engine having a 0.571-in. clearance operates at a constant piston speed of 1300 rpm. The six-cylinder engine burns a stoichiometric mixture of octane and air. Conditions at BDC after intake can be taken to a vaporized fuel-air mixture at 14.7 psi and 80°F. Based on the fuel air Otto cycle and JANAF data for the engine thermochemistry, determine (a) the ideal indicated net work, ft lbf; (b) the ideal indicated mean effective pressure, psi; (c) the ideal indicated thermal efficiency; and (d) the ideal indicated specific fuel consumption.

Solution:

1. Engine geometry:

$$V_{\text{clearance}} = \frac{\pi}{4}\left(\frac{0.571}{12}\ \text{ft}\right)\left(\frac{4}{12}\ \text{ft}\right)^2 = 0.0041524\ \text{ft}^3$$

$$V_{\text{displacement}} = \frac{\pi}{4}\left(\frac{4}{12}\ \text{ft}\right)\left(\frac{4}{12}\ \text{ft}\right)^2 = 0.0290888\ \text{ft}^3$$

or

$$V_{\text{TDC}} = 0.0041524\ \text{ft}^3$$

$$V_{BDC} = 0.0041524 + 0.0290888 = 0.0332412 \text{ ft}^3$$

$$r_v = \frac{V_{BDC}}{V_{TDC}} = \frac{0.0332412}{0.0041524} = 8.00{:}1$$

2. Stoichiometric equation, ideal combustion:

$$C_8H_{18} + 12.5[0_2 + 3.76\ N_2] \rightarrow 8CO_2 + 9H_2O + (12.5)(3.76)N_2$$

Reactant mole fractions:

$$\bar{x}_{C_8H_{18}} = \frac{1.0}{1.0 + (12.5)(4.76)} = 0.01653$$

$$\bar{x}_{0_2} = \frac{12.5}{60.5} = 0.20661 \qquad \bar{x}_{N_2} = \frac{(12.5)(3.76)}{60.5} = 0.77686$$

Product mole fractions:

$$\bar{x}_{CO_2} = \frac{8}{8 + 9 + (12.5)(3.76)} = 0.1250$$

$$\bar{x}_{H_2O} = \frac{9}{64} = 0.1406 \qquad \bar{x}_{N_2} = \frac{47}{64} = 0.7344$$

3. Conditions at BDC, state 1:

$$T_1 = 540\ °R = 300K$$

$$N_{tot} = \frac{P_1V_1}{\overline{R}T_1} = \frac{(14.7\text{ lbf/in.}^2)(144\text{ in.}^2/\text{ft}^2)(0.0332413\text{ ft}^3)}{(1545\text{ ft lbf/lb mole °R})(540\text{ °R})}$$

$$= 8.434 \times 10^{-5}\text{ lb moles}$$

$$\overline{U}_{tot}<T_1> = \sum_{i=1} \bar{x}_i N_{tot}[\bar{h}_f^0 + \Delta\bar{h}<T_1> - \overline{R}T_1]_i$$

Using JANAF data from Appendix B,

$$U_{tot}<300K> = \left\{(8.434 \times 10^{-5}\text{ lb mole})\left[1.8001\frac{\text{Btu/lb mole}}{\text{cal/g mole}}\right]\right\}\times$$

$$\{(0.20661)[0 + 13 - (1.987)(300)]_{O_2}$$
$$+ (0.77686)[0 + 13 - (1.987)(300)]_{N_2}$$
$$+ (0.01653)[-49{,}820 + 82 - (1.987)(300)]_{C_8H_{18}}\}\text{ cal/g mole}$$

$$U_1<300K> = -\ 0.21338\text{ Btu}$$

4. Process 1–2 isentropic compression of reactants:

$$\frac{V_2}{V_1} = \frac{v_r<T_2>}{v_r<T_1>} = \frac{1}{8}$$

From Example 10.5,

$$v_r\langle 300\text{K}\rangle = 238.3 \times 10^{-10}$$

or

$$v_r\langle T_2\rangle = \frac{238.3 \quad 10^{-10}}{8} = 29.79 \times 10^{-10}$$

and

$$T_2 = 600\text{K} = 1080\ °\text{R}$$

5. Conditions at TDC, state 2:

$$T_2 = 600\text{K} = 1080\ °\text{R}$$

$$P_2 = \frac{N_{tot}\bar{R}T_2}{V_2} = \frac{(8.434 \times 10^{-5}\text{ lb moles})(1545\text{ ft lbf/lb mole °R})(1080\text{ °R})}{(0.0041524\text{ ft}^3)(144\text{ in.}^2/\text{ft}^2)}$$

$$= 235.36\text{ psi}$$

$$U_{tot}\langle T_2\rangle = \sum_{i=1} \bar{x}_i N_{tot}[\bar{h}_f^0 + \Delta\bar{h}\langle T_2\rangle - \bar{R}T_2]_i$$

Using JANAF data from Appendix B,

$$U_{tot}\langle 600\text{K}\rangle = \{(8.434 \times 10^{-5})(1.8001)\} \times$$
$$\{(0.20661)[0 + 2{,}210 - (1.987)(600)]_{O_2}$$
$$+ (0.77686)[0 + 2{,}125 - (1.987)(600)]_{N_2}$$
$$+ (0.01653)[-49{,}820 + 18{,}813 - (1.987)(600)]_{C_8H_{18}}\}$$
$$U_2\langle 600\text{K}\rangle = 0.061137\text{ Btu}$$

6. Process 2–3 V = constant adiabatic flame; assume complete combustion:

$$\delta Q - P\,dV = dU \qquad U\langle T_3\rangle = U\langle T_2\rangle$$

$$U\langle T_2\rangle = \sum_{react} \bar{x}_i N_{tot}\,[\bar{h}_f^0 + \Delta\bar{h} - \bar{R}T]_i = 0.61148\text{ Btu}$$
$$U\langle T_3\rangle = \sum_{prod} N_j[\bar{h}_f^0 + \Delta\bar{h} - \bar{R}T]_j$$

or

$$0.061137 = \{0.1394 \times 10^{-5}\text{ lb mole } C_8H_{18})(1.8001)\} \times$$
$$\{[-94{,}054 + \Delta\bar{h}\langle T_3\rangle - 1.987\,T_3]_{CO_2}(8\text{ lb mole } CO_2/\text{lb mole } C_8H_{18})$$
$$+ [-57{,}798 + \Delta\bar{h}\langle T_3\rangle - 1.987\,T_3]_{H_2O}(9\text{ lb mole } H_2O/\text{lb mole } C_8H_{18})$$
$$+ [0 + \Delta\bar{h}\langle T_3\rangle - 1.987\,T_3]_{N_2}(12.5)(3.76\text{ lb mole } N_2/\text{lb mole } C_8H_{18})\}$$

Rearranging yields

$(8\Delta\bar{h}<T_3>)_{CO_2} + (9\Delta\bar{h}<T_3>)_{H_2O} + 47\Delta\bar{H}<T_3>_{N_2} - 127.168\ T_3 =$
$1{,}296{,}978$

By trial and error,

$$T_3 = 3137\text{K} = 5647°\text{R}$$

Also,

$$P_3 = \frac{(0.1394 \times 10^{-5})(8 + 9 + 47)(1545)(5647)}{(0.0041524)(144)} = 1302 \text{ psi}$$

7. Process 3–4 isentropic expansion of products:

$$\frac{V_4}{V_3} = \frac{v_r<T_4>}{v_r<T_3>} = 8$$

Again, using the results from Example 10.5,

$$v_r<T_3> = 0.150 \times 10^{-10}$$

with

$$v_r<T_4> = (8)(0.150 \times 10^{-10}) = 1.20 \times 10^{-10}$$

and

$$T_4 = 1880\text{K} = 3384°\text{R}$$

8. Conditions at BDC, state 4:

$$T_4 = 1880\text{K} = 3384°\text{R}$$

$$P_4 = \frac{(0.1394 \times 10^{-5})(8 + 9 + 47)(1545)(3384)}{(0.0332413)(144)}$$

$$= 97.445 \text{ psi}$$

$$U_{tot}<T_4> = \Sigma N_j[\bar{h}_f^0 + \Delta\bar{h}<T_4> - \bar{R}T_4]_j$$

Using JANAF data from Appendix B yields

$$U_4<1880\text{K}> = \{(0.1394 \times 10^{-5})(1.8001)\}$$
$$\times \{(8)[-94{,}054 + 20{,}132 - (1.987)(1880)]_{CO_2}$$
$$+ (9)[-57{,}798 + 15{,}921 - (1.987)(1880)]_{H_2O}$$
$$+ (12.5)(3.76)[12{,}389 - (1.987)(1880)]_{N_2}\}$$
$$= -1.5685 \text{ Btu}$$

9. Ideal indicated net work per cylinder:

$$W_{net} = {}_1W_2 + \cancel{{}_2W_3} + {}_3W_4 + \cancel{{}_4W_1}$$

With process $i \rightarrow j$:

$$\delta Q - \delta W = dU$$

$${}_1W_2 = U\langle T_1\rangle - U\langle T_2\rangle$$

$${}_3W_4 = U\langle T_3\rangle - U\langle T_4\rangle$$

$$W_{\text{net}} = U\langle T_1\rangle - U\langle T_2\rangle + U\langle T_3\rangle - U\langle T_4\rangle$$

but

$$U\langle T_3\rangle = U\langle T_2\rangle \qquad \text{(see step 6)}$$

$$W_{\text{net}} = -0.21338 + 1.5685 = 1.355 \text{ Btu/cyl}$$

a. $W_{\text{net}} = (1.355)(778) = 1054 \text{ ft lbf/cyl}$

10. Ideal indicated mean effective pressure:

$$\text{b. } IMEP = \frac{W_{\text{net}}}{V_D} = \frac{(1054 \text{ ft lbf/cyl})}{(0.0290888 \text{ ft}^3)(144 \text{ in.}^2/\text{ft}^2)} = 252 \text{ psi/cyl}$$

11. Ideal indicated thermal efficiency:

$$\eta = \frac{W_{\text{net}}}{\dot{M}_{\text{fuel}} HV}$$

$$= \frac{1.355 \text{ Btu/cyl}}{(1{,}317{,}440 \times 1.8001 \text{ Btu/lb mole fuel})(0.1394 \times 10^{-5} \text{ lb mole fuel/cyl})}$$

c. $\eta = 0.410 = 41.0\%$

12. Ideal indicated engine power, Equation (9.9b):

$$\dot{W}_I = \frac{\overline{P}LANc}{33{,}000\, n} = \frac{(252 \text{ psi/cyl})(4/12 \text{ ft/stroke})(\pi)(4 \text{ in.})^2(1300 \text{ rpm})\ (6 \text{ cyl})}{(4)(33{,}000 \text{ ft lbf/hp min})(2 \text{ rev/power stroke})}$$

$$= 125 \text{ ihp}$$

13. Ideal indicated specific fuel consumption:

$$\dot{M}_{\text{fuel}} = \left(114 \times 0.1394 \times 10^{-5} \frac{\text{lbm fuel}}{\text{intake}}\right)\left(1300 \frac{\text{rev}}{\text{min}}\right)\left(\frac{1 \text{ intake}}{2 \text{ rev}}\right)\left(6 \text{ cyl}\right)$$

$$= 0.6198 \text{ lbm fuel/min}$$

and

$$ISFC = \frac{\dot{M}_{\text{fuel}}}{W_I} = \frac{(0.6198 \text{ lbm fuel/min})(60 \text{ min/hr})}{(125 \text{ ihp})}$$

d. $= 0.2975 \text{ lbm/ihp hr}$

10.6 THE WANKEL ROTARY ENGINE

The Wankel rotary engine design, developed and patented by Felix Wankel in Germany (ca. 1960), was the result of his efforts to perfect rotary valve sealing. Mechanically, the Wankel rotary engine (see Figure 10.12) is quite different from the reciprocating piston-cylinder positive displacement engine geometry normally used in SI internal combustion engine applica-

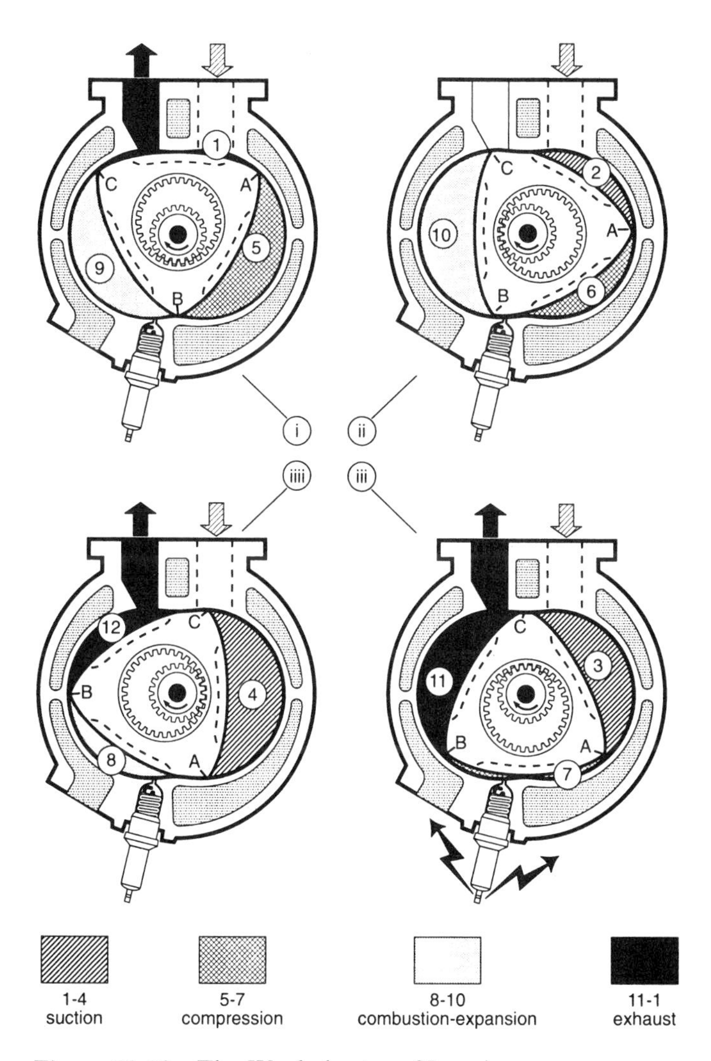

Figure 10.12 The Wankel rotary SI engine.

tions. Rather than a piston, the Wankel has a three-sided rotor that rotates within an apitrochoidal (somewhat oval) housing. The working fluid is confined within three spaces between faces of the rotors and housing by apex and side seals on the rotor. Three expansion processes per rotor revolution occur since each rotation of the rotor allows three separate charges to move through the engine per lobe.

The Wankel rotary engine is, in effect, a "four-stroke" spark-ignition engine that requires the following events per rotor rotation (see Figure 10.12):

1. Intake (1–4): Volume increases because of rotor rotation and the resulting fuel-air suction through the intake port
2. Compression (5–7): Volume sealing and reduction, with resulting increased confined gas pressure and temperature
3. Combustion (6–7): Electrical spark ignition of fuel-air mixture at minimum volume and rapid heat release, raising both pressure and gas temperature
4. Power (8–10): Volume increase with rotor rotation because of work done by the expanding high-pressure gas on the rotor lobe
5. Exhaust (11–1): Volume decrease and exhaust resulting from rotor rotation and the exposure of the exhaust port

The Wankel configuration offers several potential advantages over the piston engine, some of which are:

Considerably fewer parts. For example, a typical V8 engine has over 1000 total parts, of which 390 are moving parts. A Curtis-Wright 185-hp Wankel has only 633 parts, of which 154 are moving. This offers advantages such as lighter and smaller engines, increased power density, i.e., power output per unit engine weight, and lower production costs.

Less friction. Fewer moving parts suggest less friction. Less power is lost to friction since fewer parts are subject to wear. This allows a rotary engine to operate at high speeds and, therefore, higher power output.

Less vibration. Rotary engine motion is rotational rather than reciprocating; thus, a rotary-engine vehicle generally runs more smoothly than a piston-engine vehicle, providing smooth output torque characteristics.

Advantages of a two-stroke engine with a four-step (stroke) power sequence. The Wankel is designed so that its rotor rotates once for every three shaft rotations, and the three-sided rotor allows three complete power cycles with each rotor rotation. Each lobe therefore experiences a power cycle on every rotation much like a single-cylinder two-stroke piston engine.

Lower NO_X emissions. Although an advantage in itself, this is actually a result of poor engine combustion efficiency and lower peak temperature. (Recall that NO_X formation is a result of high-temperature $N_x \rightleftarrows O_2$ thermochemistry.)

Greater knock insensitivity. This, too, is a result of the poor combustion for these engines but does allow the use of lower octane fuels than piston engines.

The Wankel has some disadvantages when compared to the piston engine. These include:

Poor long-term rotor-sealing integrity. The rotor needs an assembly of apex and side seals that are subject to significant wear. Note that, in reciprocating engines, ring velocities are low at TDC during combustion, whereas rotor speed remains high during the development of peak pressure in a Wankel engine. As a result, gas seals tend to fail after a relatively short time. A simple and effective solution, such as improved ring designs for reciprocating piston engines, has not yet been fully developed for the Wankel.

Greater heat transfer losses. Because of combustion chamber geometry, the Wankel has a much higher surface area–to–volume ratio than a piston engine. As a result, more energy released by combustion is lost as heat, resulting in lower engine efficiencies. This heat loss is a contributor to a Wankel's poor combustion efficiency.

Inefficient combustion. The Wankel produces emissions that are high in unburned hydrocarbons (UHCs). The fuel rate and oil consumption for Wankel engines are somewhat greater than for comparable reciprocating engines.

Unlike the piston engine, the same area of the Wankel housing is always exposed to hot combustion gases. This area never gets cooled by fuel-air intake charge, as is the case in a reciprocating engine, and it therefore remains hot while other parts of the housing are cooler. This uneven heating causes engine cooling distortion and material problems.

PROBLEMS

10.1 The Lenoir cycle, a forerunner of the modern spark-ignition engine, had combustion without compression of the charge. The air standard cycle for this engine consists of the following ideal processes:

1–2	$P = c$	cooling from BDC to TDC
2–3	$V = c$	heat and addition at TDC
3–1	$s = c$	expansion from TDC to BDC

Assuming constant specific heats, derive expressions for the following in terms of γ, T_1, T_2, and T_3: (a) compression ratio, (b) net *IMEP*, (c) heat additions, and (d) indicated thermal efficiency.

10.2 Repeat Example, 10.1 for (a) monatomic gas, $\gamma = 1.67$; (b) $\gamma = 1.3$; and (c) $\gamma = 1.2$.

10.3 Conditions at the beginning of compression in an SI engine with an 8.5:1 compression ratio are 14.7 psia and 77°F. Determine the conditions at the end of ideal compression, assuming properties for the charge to be (a) air with constant specific heats at T_0; (b) air with constant specific heats at the average temperature for part a; (c) air with variable specific heats; (d) stoichiometric C_8H_{18}-air mixture with constant specific heats at T_0; and (e) stoichiometric C_8H_{18}-air mixture with variable specific heats.

10.4 Use the results of Problem 10.3, and calculate the ideal compression work per unit mass of charge, assuming the properties for the charge to be (a) air with constant specific heats at T_0; (b) air with constant specific heats at the average temperature for part a; (c) air with variable specific heats; (d) stoichiometric C_8H_{18}-air mixture with constant specific heats at T_0; and (e) stoichiometric C_8H_{18}-air mixture with variable specific heats via combustion charts.

10.5 Conditions at the beginning of expansion in an SI engine having an 8.5:1 compression ratio are 4320°R and 920 psia. Determine the conditions at the end of ideal expansion, assuming properties for the burned gases to be (a) air with constant specific heats at T_0; (b) air with constant specific heats at the average temperature for part a; (c) air with variable specific heats; (d) stoichiometric C_8H_{18}-air complete combustion products and constant specific heat at T_0; and (e) stoichiometric C_8H_{18}-air products of combustion via combustion charts.

10.6 Using the results of Problem 10.5, calculate the ideal expansion work per unit mass of charge, assuming properties for the burned gases to be (a) air with constant specific heats at T_0; (b) air with constant specific heats at the average temperature for part a; (c) air with variable specific heats; (d) stoichiometric C_8H_{18}-air complete combustion products and constant specific heat at T_0; and (e) stoichiometric C_8H_{18}-air products of combustion via the combustion charts.

10.7 Consider an Otto cycle model of an SI engine with an 9:1 compression ratio in which 700 Btu/lbm of heat are added at TDC. Conditions at BDC prior to compression are 15 psia and 80°F. Early combustion can be modeled ideally as a constant-volume heat addition process that occurs prior to reaching TDC, followed by compression to TDC of the combustion gases prior to expansion. Assume that early heat addition occurs at a volume equal to twice the volume at TDC, and calculate the ratio of (a) peak pressures, (b) thermal efficiencies, and (c) *IMEPS* for this extended Otto cycle to the basic Otto cycle. Use a constant specific heat air standard cycle analysis.

10.8 Repeat Problem 10.7 for the condition of late combustion, in which case constant-volume heat addition occurs after compression to TDC and the beginning of expansion. Assume that late heat addition occurs at a volume equal to twice the volume at TDC, and calculate the ratio of (a) peak pressures, (b) thermal efficiencies, and (c) *IMEPS* for this extended Otto cycle to the basis Otto cycle. Use a constant specific heat air standard cycle analysis.

10.9 The peak brake power for an automobile engine operating at 4500 rpm is rated as 250 brake hp. Brake thermal efficiency at this condition is 28%. Assuming a lower heating value of 19,000 Btu/lbm for the fuel, calculate (a) peak load fuel consumption, lbm/h; (b) peak load torque, ft-lbf; and (c) peak load brake specific fuel consumption, lbm/hp-h.

10.10 A prototype spark-ignition engine maintains a vehicle at 35 mph on a chassis dynamometer. The engine consumes 2.32 gal/h of isooctane while the measure volumetric flow rate of air during the test is found to be 50 ft^3/min at *STP*. For these conditions, determine (a) fuel consumption, lbm fuel/h; (b) *FA* ratio, lbm fuel/lbm air; and (c) fuel economy, mph.

10.11 A six-cylinder, four-stroke SI engine has been converted to burn liquid methanol and air, with an equivalence ratio of 0.9 at an 8:1 compression ratio. The indicated power developed by the engine is 150 ihp when running at 3300 rpm. Engine volumetric efficiency is 87%, and conditions on intake can be taken as 14.6 psia and 85°F. For these conditions, calculate (a) the actual combustion equation *AF* ratio; (b) molar *AF* ratio; (c) fuel-air mixture mass flow rate; (d) heat release per mass of fuel-air mixture at *STP*; (e) indicated thermal efficiency; and (f) indicated specific fuel consumption.

10.12 A spark-ignition internal combustion engine is proposed to burn a blend of methanol and ethanol. Intake conditions are 14.5 psia and 50°F. At 50°F, the intake vapor pressure of methanol is 1.08 psia, while that of ethanol is 0.45 psia. Assume that the air-fuel mixture delivered to the engine is a homogeneous vapor, and find: (a) the mixture molar *AF* ratio; (b) the mass *FA* ratio; (c) the mixture equivalence ratio; and (d) the percent theoretical air.

10.13 An Otto cycle model for a spark-ignition internal combustion engine has a 5:1 compression ratio and burns a stoichiometric mixture of isooctane vapor and air. Conditions at the beginning of the compression process are 14.7 psia and 594°R. The residual fraction is assumed to be 0.047. For these conditions and using combustion charts, determine: (a) P, T, v, and u around the cycle; (b) net cycle work, Btu/lbm air; (c) indicated thermal efficiency; (d) exhaust gas temperature at BDC, °R; and (e) residual fraction f.

10.14 An Otto cycle design is based on the combustion of methanol in 15% excess air. The conditions after compression of the air-fuel mixture are given as 270 psia and 1250°R. Constant-volume combustion is assumed to occur after spark ignition of the compressed charge, with the resulting temperature after reaction being tripled. Assuming complete combustion, determine: (a) the balanced reaction equation; (b) the molar FA ratio; (c) the complete combustion product mole fractions; and (d) the peak cycle pressure.

10.15 A four-cylinder, four-stroke IC engine has been modified to run on hydrogen. The 2.5-cm × 2.75-cm engine is running at 1200 rpm. The engine is designed to burn lean with an equivalence ratio of 0.8. Using JANAF data for the H_2-O_2 reaction system, find the following for a fuel-air Otto cycle analysis: (a) net work, N m; (b) indicated MEP, kPa; (c) indicated thermal efficiency, %; (d) indicated power, ikW; and (e) indicated specific fuel consumption, kg/kW hr.

10.16 An automobile consumes 3.33 gal of fuel/h while running at a constant speed of 50 mph on a chassis dynamometer. The spark-ignition engine burns a rich isooctane mixture and exhaust gas analysis on a dry volumetric basis is as follows:

9.54%	CO_2	0.84%	H_2
4.77%	CO	2.80%	O_2
3.15%	CH_4	78.90%	N_2

The specific gravity of isooctane is 0.702. Determine (a) FA equivalence ratio; (b) AF ratio on a mass basis; (c) fuel consumption, mph; and (d) CO production, g/mi.

10.17 A four-cylinder 6.25-cm × 7.5-cm four-stroke spark-ignition engine running at 1800 rpm burns a mixture of indolene and 20% excess air. The volumetric efficiency of the engine with inlet conditions of 105 kPa and 28°C is 92%. One of every 12 ignitions results in a misfire. Assuming ideal combustion otherwise with all unburned hydrocarbons produced by misfire, find (a) fuel mass flow rate, g/hr; (b) carbon monoxide mass flow rate, g/hr; and (c) unburned hydrocarbon mass flow rate, g/hr.

10.18 A spark-ignition engine design is required to maintain a 28.2% brake thermal efficiency when burning a variety of fuel alternatives.

Determine the brake specific fuel consumption required when burning (a) normal liquid octane, (b) propane, (c) methanol, and (d) hydrogen

10.19 An SI engine with a 9.5:1 compression ratio runs on a mixture of *n*-octane and 125% theoretical air. Conditions at BDC can be taken as 104 kPa and 27°C. Using a JANAF fuel-air Otto cycle analysis, determine (a) conditions at the end of compression; (b) adiabatic flame temperature for complete combustion at TDC, K; and (c) peak cylinder pressure, kPa. Repeat parts b and c, assuming equilibrium conditions at the end of combustion.

10.20 A six-cylinder 6.5-cm × 11-cm four-stroke SI engine running at 2400 rpm burns a mixture of CH_4 in 20% excess air. Assuming a volumetric efficiency of 87% and inlet conditions of 105 kPa and 30°C, determine (a) fuel mass flow rate, g/sec; (b) charge volumetric flowrate, cc/sec; and (c) ideal carbon monoxide exhaust mass flow rate, g/sec.

10.21 A spark-ignition engine is designed such that flame propagation is across the cylinder diameter. When the engine is operating at 6000 rpm, a nominal flame speed of 9.2 m/sec travels across the combustion volume in 9.6 msec. For these conditions, calculate (a) the distance of flame travel, cm; (b) the crank angle of rotation during combustion; (c) the flame speed required at 1600 rpm to travel the same crank angle as for 6000 rpm; and (d) the crank angle at 1600 rpm, assuming the same flame speed as for 6000 rpm.

10.22 A four-cylinder 2.5-in. × 3-in. four-stroke SI engine operates at 1800 rpm, with an octane–air rich mixture having an air-fuel ratio of 15 to 1. Calculate (a) fuel mass flow rate. lbm/hr; (b) ideal CO_2 mass flow rate, lbm/hr; (c) ideal CO mass flow rate for 98% combustion of carbon, lbm/hr; and (d) the mass of UHC if all the emissions resulted from misfire every 12 ignitions, i.e., no combustion, lbm/hr.

10.23 A small air-cooled SI engine is found to produce 32 ikW of power output. Exhaust analysis of the dry products of combustion are found to be equal to

CO_2	8.2%
CO	0.8%
O_2	4.5%
N_2	86.5%

The fuel is methanol liquid and enters the engine at 25°C, with a fuel flow rate 3.5 g/sec. For these conditions, find (a) the heat loss from the engine, kW; (b) the indicated thermal efficiency, %; and (c) the mass flow rate of CO_2, g/sec.

10.24 Indolene, C_7H_{13}, is used extensively in CVC emissions testing of SI engines. Fuel economy can be calculated using a carbon balance on the measured engine emissions. For indolene, determine (a) the H/C mass ratio; (b) the grams of carbon per gallon of fuel (SG = 0.7404); and (c) the mass fraction of carbon in indolene, CO_2, and CO. Using parts a, b, and CO, show also that the miles per gallon of fuel consumed during testing can be determined to equal

$$\text{MPG} = \frac{2426 < \text{g°C/gal fuel} >}{\{0.866[\text{HC}] + 0.273[\text{CO}_2] + 0.429[\text{CO}]\} < \text{g°C/mi} >}$$

where [ ] = measured emission in g/mi

10.25 A CVC test on a four-cylinder, four-stroke engine yields 1.52 g/mi UHC, 2.92 g/mi NO_X, 10.54 g/mi CO, and 565.17 g/mi CO_2. Using the results of Problem 10.21, estimate fuel economy for the test run, mpg.

11

Compression-Ignition Engine Combustion

11.1 INTRODUCTION

Several early pioneers in the development of the spark-ignition internal combustion engine were mentioned briefly in Chapter 10. The modern compression-ignition, or CI, internal combustion engine, however, is the result of the initial efforts by one man, Rudolf Christian Karl Diesel (1858–1913). Diesel's goal was to design and develop a unique reciprocating piston-cylinder combustion engine that would burn considerably less fuel per unit power output than available alternative prime movers of his day. He is credited also with being the first inventor of his time to rely on the use of first principles of thermodynamics rather than experimentation alone to achieve the goal of developing a new heat engine. Much of Diesel's early work was published in his *Rational Heat Power* (1892). Early design work and performance predictions were based on analytical models developed prior to fabricating and testing an engine. His original concept was that of a constant-temperature internal combustion (IC) process that would allow his engine to approach the ideal heat input of a Carnot cycle heat engine. Diesel later modified his initial engine and thermodynamic model to that of his now famous constant-pressure heat addition thermodynamic cycle.

Diesel built an early prototype machine in Augsburg, Germany, which, in February 1894, ran for approximately 1 min at 88 rpm while producing 13.2-ihp (indicated horsepower) output. By 1895, an operating engine was developed that produced 20 ihp at approximately 165 rpm. Diesel technology was introduced to the general public by 1898, when three commercial diesel engines, one from Krupp, Deutz, and Maschinenfabrik Augsburg Works (known today as M.A.N.), respectively, were displayed at an exposition of power plants in Munich, Germany.

The simplicity of the diesel's ancillary charge preparation and ignition systems, with its ability to run on far cruder fuels, have made the diesel a successful IC engine. Compression-ignition engines have found use in a variety of areas, including small-scale stationary power generation; heavy-duty mobility propulsion applications such as locomotive, road transport, and off-road earth-moving equipment; and marine power, where longevity and reliability are essential. The first diesel-powered automobile was built in 1929 by Clessie Cummins, with Mercedes-Benz following suit seven years later in Europe. General Motors began full production of diesel cars in 1977 and, today, automotive and small truck diesels are being built in America, Asia, and Europe.

This chapter will emphasize basic and concurrent fuel engine interface characteristics of compression-ignition engines (see Figure 11.1) and

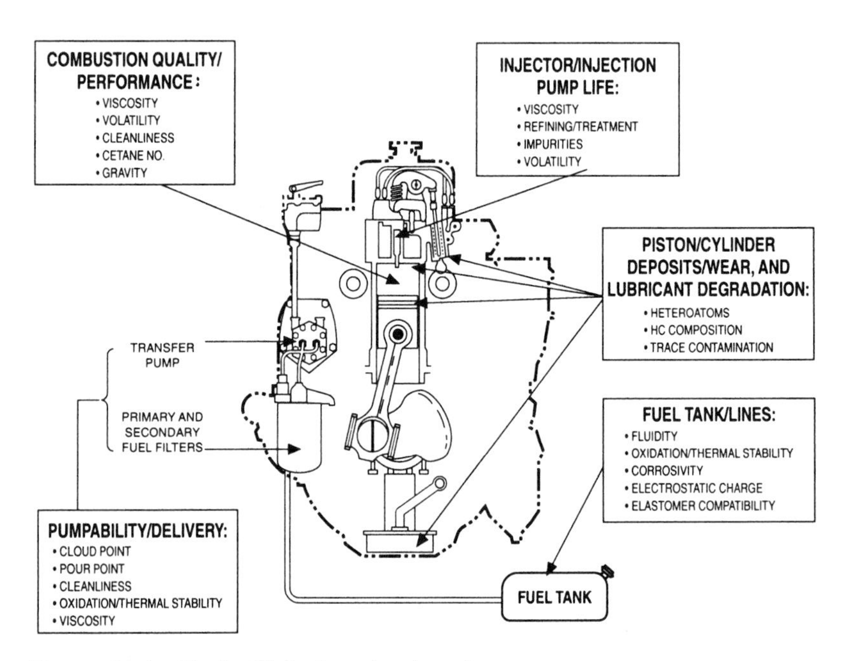

Figure 11.1 Basic CI fuel-engine interface.

thereby complement the material dealing with spark-ignition fuel-engine interface characteristics presented in Chapter 10. Chapter coverage will include general thermochemical modeling, compatible fuel characteristics, combustion processes, and emission associated with compression-ignition engines.

11.2 THERMODYNAMICS AND COMPRESSION-IGNITION ENGINE MODELING

Dimensionless thermodynamic models for CI engines can be developed as was done for SI engines in Chapter 10. The classic thermodynamic Otto ($V=c$ heat addition) and/or Diesel ($P=c$ heat addition) cycles are not appropriate for predicting the energy limits of today's intermediate- and high-speed diesel engines. The Dual or combined cycle (V and $P = c$ heat additions) is a more proper thermodynamic model for a modern CI engine. An air standard Dual cycle consisting of the following ideal processes can be used to predict the general energetics of CI engines:

1–2 Isentropic compression from BDC to TDC
2–3 Constant-volume partial heat addition at TDC (valid for high-speed CI engines)
3–4 Constant-pressure partial heat addition during expansion (valid for all CI engines)
4–5 Isentropic expansion to BDC
5–1 Constant-volume heat rejection at BDC

A dimensionless analysis of a compression-ignition engine will be based on the following dimensionless parameters:

$\tau = T_4/T_1$ = maximum cycle temperature ratio
$r_v = V_1/V_2$ = compression ratio
$r_e = V_4/V_5$ = expansion ratio $\neq r_v$
$\alpha = T_3/T_2$ = constant-volume heating temperature ratio
$\beta = T_4/T_3$ = constant-pressure heating temperature ratio
$\rho = T_5/T_1$ = constant-volume cooling temperature ratio
η = indicated thermal efficiency
$\xi = IMEP/P_1$ = dimensionless specific work

Using constant specific heat analysis and the dimensionless parameters listed above, the following dimensionless relationships for temperature now can be written for the Dual cycle:

$$T_2 = T_1 r_v^{\gamma-1}$$

where, from 1–2, $Pv^{\gamma} = \text{const}$

$$T_3 = \alpha T_2 = \alpha T_1 r_v^{\gamma-1}$$

$$T_4 = \beta T_3 = \beta\alpha T_2 = \beta\alpha T_1 r_v^{\gamma-1} = \tau T_1 \tag{11.1}$$

For the Dual-cycle process 3–4, heat addition is at constant pressure during expansion, or

$$P_4 = \frac{RT_4}{v_4} = \frac{RT_3}{v_3} = P_3$$

$$\frac{v_4}{v_3} = \frac{T_4}{T_3} = \beta$$

and, since $v_5 = v_1$,

$$\frac{v_4}{v_5} = \frac{v_4}{v_1} = \left(\frac{v_4}{v_3}\right)\left(\frac{v_3}{v_1}\right) = \left(\frac{v_4}{v_3}\right)\left(\frac{v_2}{v_1}\right) = \frac{\beta}{r_v}$$

$$\frac{T_5}{T_4} = \left(\frac{v_4}{v_5}\right)^{\gamma-1} = \left(\frac{\beta}{r_v}\right)^{\gamma-1}$$

$$\frac{T_5}{T_1} = \left(\frac{T_5}{T_4}\right)\left(\frac{T_4}{T_1}\right) = \frac{\tau\beta^{\gamma-1}}{r_v^{\gamma-1}} = \rho \tag{11.2}$$

Using Equations (11.1) and (11.2), it follows that

$$\alpha\beta r_v^{\gamma-1} = \tau$$

$$\frac{[\alpha\beta r_v^{\gamma-1}]\beta^{\gamma-1}}{r_v^{\gamma-1}} = \rho$$

and thus

$$\rho = \beta^{\gamma}\alpha \tag{11.3}$$

The net work for a Dual cycle w_{net} can be written as

$$\begin{aligned} w_{\text{net}} &= {}_1w_2 + \cancel{{}_2w_3} + {}_3w_4 + {}_4w_5 + \cancel{{}_5w_1} \\ &= u_1 - u_2 + P_3(v_4 - v_3) + u_4 - u_5 \\ &= C_v(T_1 - T_2) + C_v(\gamma - 1)(T_4 - T_3) + C_v(T_4 - T_5) \end{aligned}$$

Using the dimensionless parameters cited earlier, w_{net} is then

$$\begin{aligned} w_{\text{net}} &= C_v T_1\left[1 - \left(\frac{T_2}{T_1}\right) + (\gamma - 1)\left\{\left(\frac{T_4}{T_1}\right) - \left(\frac{T_3}{T_1}\right)\right\} + \left(\frac{T_4}{T_1}\right) - \left(\frac{T_5}{T_1}\right)\right] \\ &= C_v T_1\left[1 - r_v^{\gamma-1} + \gamma\left(\frac{T_4}{T_1}\right) - \gamma\left(\frac{T_3}{T_1}\right) - \left(\frac{T_4}{T_1}\right) + \left(\frac{T_3}{T_1}\right) + \left(\frac{T_4}{T_1}\right) - \left(\frac{T_5}{T_1}\right)\right] \end{aligned}$$

or

$$w_{\text{net}} = \frac{RT_1}{(\gamma - 1)}\left[1 - r_v^{\gamma-1} + \gamma(\tau - \alpha r_v^{\gamma-1}) + \alpha\left\{r_v^{\gamma-1} - \left(\frac{\tau}{\alpha r_v^{\gamma-1}}\right)^{\gamma}\right\}\right] \tag{11.4}$$

Ideal indicated mean effective pressure, *IMEP*, for a Dual cycle, defined as the net work divided by the displacement volume, is then equal to

$$IMEP = \frac{w_{\text{net}}}{v_1[1 - (1/r_v)]} = \frac{P_1[1 - r_v^{\gamma-1} + \gamma(\tau - \alpha r_v^{\gamma-1}) + \alpha\{r_v^{\gamma-1} - (\tau/\alpha r_v^{\gamma-1})^{\gamma}\}]}{(\gamma - 1)[1 - 1/r_v]}$$

which, written in dimensionless form as specific work ξ, is given as

$$\xi = \frac{IMEP}{P_1} = \frac{[1 - r_v^{\gamma-1} + \gamma(\tau - \alpha r_v^{\gamma-1}) + \alpha\{r_v^{\gamma-1} - (\tau/\alpha r_v^{\gamma-1})^{\gamma}\}]}{(\gamma - 1)[1 - 1/r_v]} \tag{11.5}$$

External heat addition per unit mass consists of both constant-volume heat addition from 2–3 and constant-pressure heat addition from 3–4. The constant-volume heat addition per unit mass $(q_H)_V$ is given as

$$\left(q_H\right)_V = {}_2q_3 = C_v[T_3 - T_2]$$

which, in dimensionless form, is

$$\left(\frac{q_H}{C_vT_1}\right)_V = [\alpha r_v^{\gamma-1} - r_v^{\gamma-1}]$$

and the constant-pressure external heat addition from 3–4 is given as

$$\left(q_H\right)_P = {}_3q_4 = C_p[T_4 - T_3] = \gamma C_v[T_4 - T_3] \tag{11.6}$$

which, in dimensionless form, equals

$$\left(\frac{q_H}{C_vT_1}\right)_P = \gamma[\tau - \alpha r_v^{\gamma-1}]$$

The total external heat addition now is written as

$$q_H = (q_H)_V + (q_H)_P = C_vT_1[\alpha r_v^{\gamma-1} - r_v^{\gamma-1} + \gamma(\tau - \alpha r_v^{\gamma-1})]$$

and

$$\frac{q_H}{C_v T_1} = [(1-\gamma)\alpha r_v^{\gamma-1} + \gamma\tau - r_v^{\gamma-1}] \tag{11.7}$$

An indicated thermal efficiency η, for the Dual cycle using Equations (11.4) and (11.6), is defined as

$$\eta = \frac{\text{net cycle work}}{\text{net external heat addition}}$$

$$\eta = \frac{C_v T_1[1 - r_v^{\gamma-1} + \gamma(\tau - \alpha r_v^{\gamma-1}) + \alpha\{r_v^{\gamma-1} - (\tau/\alpha r_v^{\gamma-1})^{\gamma}\}]}{C_v T_1[\alpha r_v^{\gamma-1}(1-\gamma) + \gamma\tau - r_v^{\gamma-1}]}$$

$$\eta = \frac{[1 - r_v^{\gamma-1} + \gamma(\tau - \alpha r_v^{\gamma-1}) + \alpha\{r_v^{\gamma-1} - (\tau/\alpha r_v^{\gamma-1})^{\gamma}\}]}{[\alpha r_v^{\gamma-1}(1-\gamma) + \gamma\tau - r_v^{\gamma-1}]} \tag{11.8}$$

The classic constant-pressure Diesel cycle can be determined from these relationships if one assumes no constant volume heat addition (i.e., $\alpha = 1$, or $T_3 = T_2$). In this instance, the net work for a compression ignition $P = c$ heat addition cycle using Equation (11.4) equals

$$w_{net} = \frac{RT_1}{(\gamma - 1)}\left[1 - r_v^{\gamma-1} + \gamma(\tau - r_v^{\gamma-1}) + r_v^{\gamma-1} - (\tau/r_v^{\gamma-1})^{\gamma}\right]$$

$$w_{net} = \frac{RT_1}{(\gamma - 1)}\left[1 + \gamma(\tau - r_v^{\gamma-1}) - (\tau/r_v^{\gamma-1})^{\gamma}\right] \tag{11.9}$$

Ideal *IMEP* for the classic Diesel cycle then equals

$$IMEP = \frac{P_1}{(1 - 1/r_v)(\gamma - 1)}\left[1 + \gamma(\tau - r_v^{\gamma-1}) - (\tau/r_v^{\gamma-1})^{\gamma}\right] \tag{11.10}$$

and dimensionless specific work ξ is expressed as

$$\xi = \frac{1}{(1 - 1/r_v)(\gamma - 1)}\left[1 + \gamma(\tau - r_v^{\gamma-1}) - (\tau/r_v^{\gamma-1})^{\gamma}\right] \tag{11.11}$$

External heat addition for the $P = c$ Diesel cycle can be written as

$$\frac{q_H}{C_v T_1} = \gamma(\tau - r_v^{\gamma-1})$$

$$\frac{q_H}{C_p T_1} = (\tau - r_v^{\gamma-1}) \tag{11.12}$$

By combining Equations (11.9) and (11.12), an equation for the Diesel cycle indicated thermal efficiency can be stated as

$$\eta = \frac{[RT_1/(\gamma - 1)][1 + \gamma(\tau - r_v^{\gamma-1}) - (\tau/r_v^{\gamma-1})^{\gamma}]}{(C_pT_1)(\tau - r_v^{\gamma-1})}$$

or

$$\eta = \frac{[1 + \gamma(\tau - r_v^{\gamma-1}) - (\tau/r_v^{\gamma-1})^{\gamma}]}{\gamma(\tau - r_v^{\gamma-1})} \tag{11.13}$$

The dimensionless relationships developed for CI engines in this chapter and SI engines in Chapter 10 are summarized in Table 11.1, while Figure 11.2 illustrates ideal *T-s* and *P-v* diagrams for air standard CI thermodynamic cycles.

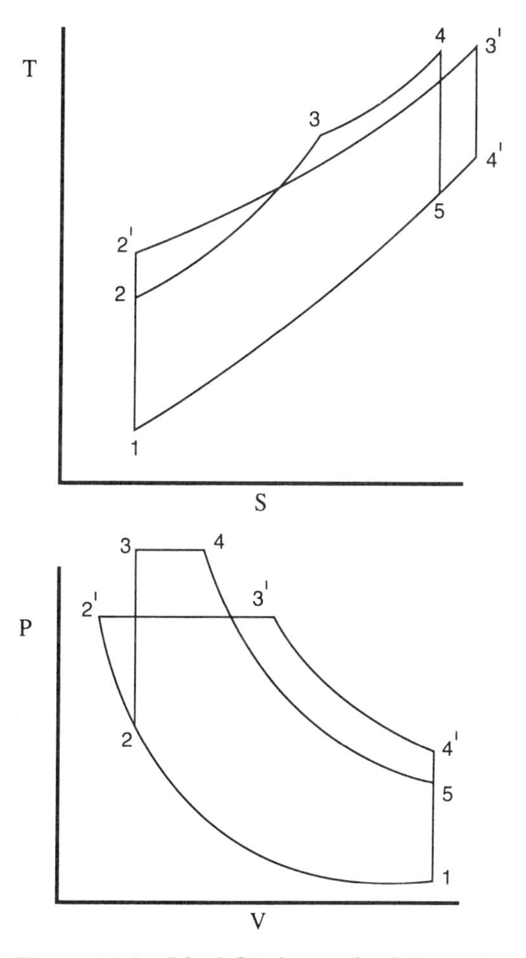

Figure 11.2 Ideal CI air standard *T-s* and *P-v* diagrams.

Table 11.1 Dimensionless Thermodynamic Relationships for Air Standard IC Engines

---

Net work: w_{net}

Otto cycle:

$$= \frac{RT_1}{(\gamma - 1)}\left[1 - r_v^{\gamma-1} + \tau - (\tau/r_v^{\gamma-1})\right] \qquad (10.1)$$

Dual cycle:

$$= \frac{RT_1}{(\gamma - 1)}\left[1 - r_v^{\gamma-1} + \gamma(\tau - \alpha r_v^{\gamma-1}) + \alpha\{r_v^{\gamma-1} - (\tau/\alpha r_v^{\gamma-1})^{\gamma}\}\right] \qquad (11.4)$$

Diesel cycle:

$$= \frac{RT_1}{(\gamma - 1)}\left[1 + \gamma(\tau - r_v^{\gamma-1}) - (\tau/r_v^{\gamma-1})^{\gamma}\right] \qquad (11.9)$$

External heat addition: $\dfrac{q_H}{C_v T_1}$

Otto cycle

$$= \tau - r_v^{\gamma-1} \qquad (10.4)$$

Dual cycle:

$$= [(1 - \gamma)\alpha r_v^{\gamma-1} + \gamma\tau - r_v^{\gamma-1}] \qquad (11.7)$$

Diesel cycle:

$$= \gamma\,(\tau - r_v^{\gamma-1}) \qquad (11.12)$$

Specific work: $\xi = \dfrac{IMEP}{P_1}$

Otto cycle:

$$= \frac{[(1 - r_v^{\gamma-1}) + \tau - \tau/r_v^{\gamma-1}]}{(\gamma - 1)[1 - 1/r_v]} \qquad (10.3)$$

Dual cycle:

$$= \frac{[1 - r_v^{\gamma-1} + \gamma(\tau - \alpha r_v^{\gamma-1}) + \alpha\{r_v^{\gamma-1} - (\tau/\alpha r_v^{\gamma-1})^{\gamma}\}]}{(\gamma - 1)[1 - 1/r_v]} \qquad (11.5)$$

Diesel cycle:

$$= \frac{1}{(1 - 1/r_v)(\gamma - 1)}\left[1 + \gamma(\tau - r_v^{\gamma-1}) - (\tau/r_v^{\gamma-1})^{\gamma}\right] \qquad (11.11)$$

Table 11.1 Continued

Indicated thermal efficiency: η

Otto cycle:

$$= 1 - \left(\frac{1}{r_v}\right)^{\gamma-1} \quad (10.5)$$

Dual cycle:

$$= \frac{[1 - r_v^{\gamma-1} + \gamma(\tau - \alpha r_v^{\gamma-1}) + \alpha\{r_v^{\gamma-1} - (\tau/\alpha r_v^{\gamma-1})^{\gamma}\}]}{[\alpha r_v^{\gamma-1}(1 - \gamma) + \gamma\tau - r_v^{\gamma-1}]} \quad (11.8)$$

Diesel cycle:

$$= \frac{[1 + \gamma(\tau - r_v^{\gamma-1}) - (\tau/r_v^{\gamma-1})^{\gamma}]}{\gamma(\tau - r_v^{\gamma-1})} \quad (11.13)$$

Variations in particular IC engine performance characteristics often need to be studied or compared for specific design and/or operational purposes. These types of investigations most often require certain parameters to be fixed, with all others then being dependent variables. The relationships in Table 11.1 may be of use in such an effort. Two thermodynamic cycles may be required to have equal speed, displacement volumes, compression ratios, heat inputs, work outputs, peak pressures, and/or peak temperatures. The relationships in Table 11.1, for example, can be used to compare ideal indicated thermal efficiencies for Diesel, Dual, and Otto cycles. An inspection of Equations (10.5), (11.8), and (11.13) reveals that all three cycles have thermal efficiencies that depend on compression ratio and also ν, α, and τ. Using the stated relationships and with the use of a *P-v* and/or *T-s* diagram, such as found in Figure 11.3, one can show that ideal indicated thermal efficiencies for classic air standard Diesel, Dual, and Otto cycles for the specific conditions below are ordered as follows:

Cycle constraint	Thermal efficiency
Fixed heat input and compression ratio	Otto > Dual > Diesel
Fixed heat input and maximum pressure	Diesel > Dual > Otto
Fixed work output and maximum pressure	Diesel > Dual > Otto

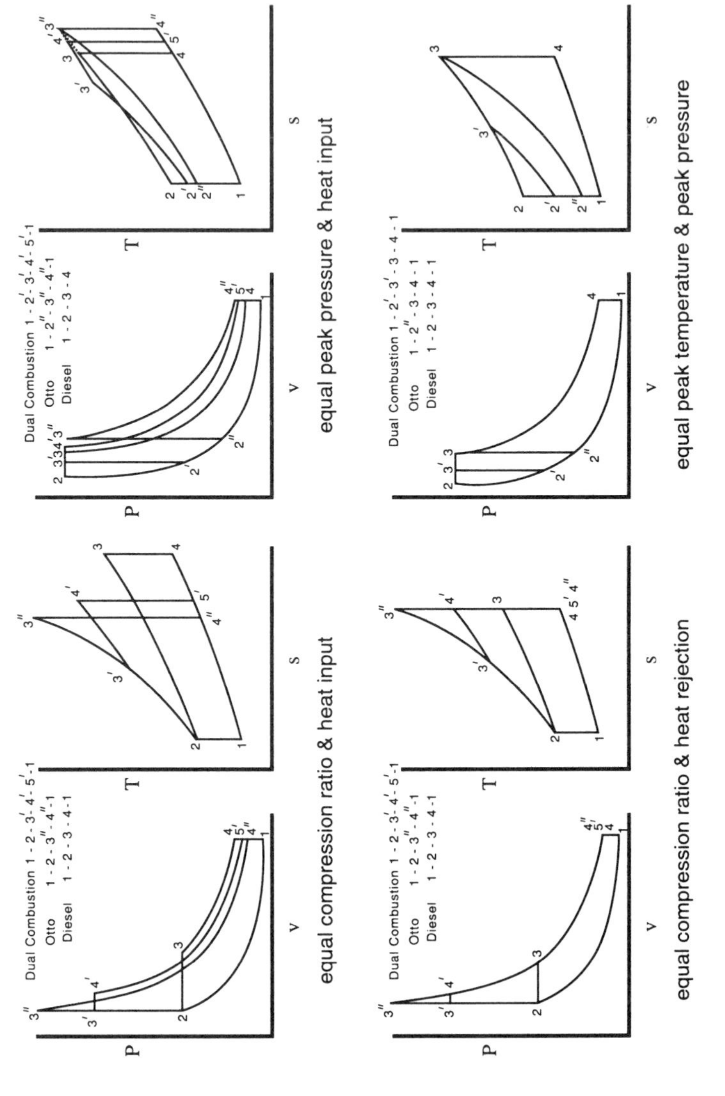

Figure 11.3 Comparison of various air standard cycles.

Frequently, one hears it stated that a diesel has a higher thermal efficiency than an SI engine. This is a somewhat misleading and truly incomplete statement. For the same compression ratio, the Diesel will have a lower efficiency than the Otto and Dual cycles. Diesel engines, because of their very different knock limitations, can be built to operate at higher compression ratios ($r_v > 12:1$) than knock-limited SI engine ($r_v < 11:1$). In this way, diesel engines may be produced having a higher thermal efficiency due chiefly to their higher attainable compression ratio. The relationships in Table 11.1 also can be used to study additional design combinations, such as mean effective and peak pressures as a function of heat input.

Zeroth-order air standard Diesel and Dual-cycle modeling can be expanded to include the intake and exhaust processes developed in Chapter 10. Fuel-air thermochemistry and residual gas influences also can be included using variable specific heats and equilibrium charts. Heat loss, finite blowdown rates, and physics of heat release in an actual diesel engine make dual-cycle thermal modeling less useful as a general predictive tool for CI engines than the constant-volume modeling was for describing general characteristics within actual SI engines. Real diesel combustion generates *P-v* indicator cards that may be fitted to a combined cycle diagram, but it is not yet feasible to develop a real engine combustion process from a proposed theoretical *P-v* indicator diagram.

EXAMPLE 11.1 Using a dimensionless Diesel, Dual, and Otto cycle analysis, determine the relative values for thermal efficiencies as a function of compression ratio under the following constraints: $\tau = 10$, fixed peak pressure ratio $P_3/P_1 = 30$, and fixed heat addition per unit mass $q_H/C_vT_1 = 10$. Assume that $\gamma = 1.4$

Solution:

1. Thermal efficiency, Diesel cycle:

$$\eta = \frac{[1 + \gamma(\tau - r_v^{\gamma-1}) - (\tau/r_v^{\gamma-1})^{\gamma}]}{\gamma(\tau - r_v^{\gamma-1})}$$

2. Thermal efficiency, Dual cycle:

$$\eta = \frac{[1 - r_v^{\gamma-1} + \gamma(\tau - \alpha r_v^{\gamma-1}) + \alpha\{r_v^{\gamma-1} - (\tau/\alpha r_v^{\gamma-1})^{\gamma}]}{[\alpha r_v^{\gamma-1}(1 - \gamma) + \gamma\tau - r_v^{\gamma-1}]}$$

3. Thermal efficiency, Otto cycle:

$$\eta = 1 - (1/r_v)^{\gamma-1}$$

4. The constant-volume heating ratio α for the Dual cycle is a function of total heat input, γ, τ, and compression ratio; see Equation (11.7). Thus,

$$q_H/C_vT_1 = 10 = [(1 - \gamma)\alpha r_v^{\gamma-1} + \gamma\tau - r_v^{\gamma-1}]$$

or

$$\alpha = \frac{10 - \gamma\tau + r_v^{\gamma-1}}{(1 - \gamma)r_v^{\gamma-1}}$$

5. Cycle thermal efficiency η:

r_c	α	Diesel	Dual	Otto
6	2.3836	0.2626	0.4494	0.5116
7	2.0916	0.3199	0.4711	0.5408
8[a]	1.8528[a]	0.3658[a]	0.4874[a]	0.5647[a]
9	1.6524	0.4036	0.4997	0.5848
10	1.4811	0.4355	0.5088	0.6019
11	1.3322	0.4628	0.5152	0.6168
12	1.2011	0.4865	0.5195	0.6299
13	1.0845	0.5073	0.5218	0.6416
14	0.9798	0.5259	0.5223	0.6520
15	0.8850	0.5425	0.5211	0.6615
16	0.7988	0.5575	0.5183	0.6701
17	0.7197	0.5711	0.5138	0.6780
18	0.6470	0.5835	0.5075	0.6853
19	0.5796	0.5949	0.4993	0.6920
20	0.5171	0.6055	0.4891	0.6983

[a] $\alpha = \frac{10 - 14 + 8^{0.4}}{(1 - 1.4)8^{0.4}} = 1.8528$

Diesel:

$$\eta = \frac{1 + 1.4(10 - 8^{0.4}) - (10/8^{0.4})^{1.4}}{1.4(10 - 8^{0.4})} = 0.3658$$

Dual:

$$\eta = \frac{1-8^{0.4}+1.4(10-1.8528 \times 8^{0.4})+1.8528(8^{0.4}-(10/(1.8528 \times 8^{0.4}))^{1.4}}{1.8528 \times 8^{0.4} \times (1-1.4)+1.4 \times 10-8^{0.4}}$$

$= 0.4874$

Otto:

$\eta = 1 - (1/8)^{0.4} = 0.5647$

Comments A review of the results for these *specific* constraints reveal the following facts;

For the same compression ratio, peak cycle temperature, and total heat input,

$$\eta_{\text{Otto}} > \eta_{\text{Diesel}}$$

$$\eta_{\text{Otto}} > \eta_{\text{Dual}}$$

Thermal efficiencies of the Diesel and Otto cycles increase with increasing compression ratio.

Thermal efficiency for the Dual cycle increases and then decreases with increasing compression ratio.

For the same total heat input, peak cycle temperature, and equal compression ratios,

$$\eta_{\text{Diesel}} > \eta_{\text{Dual}} \text{ for } r_v > 13$$

$$\eta_{\text{Diesel}} < \eta_{\text{Dual}} \text{ for } r_v < 13$$

Thermal efficiency of the Otto cycle at 8:1 compression for the same total heat input, peak cycle temperature can be surpassed by a Diesel cycle having a compression ratio greater than 17 to one.

11.3 FUEL THERMOCHEMISTRY AND COMPRESSION-IGNITION ENGINES

Diesel fuels undergo a series of events that enable CI engines to produce useful power output: storage, pumping and handling, filtering, preheating (required for the slow speed engine heavy residual fuels), ingestion, atomization, mixing with air, and finally combustion. These processes and their resulting impact on performance characteristics are governed by particular machinery, thermochemical properties of diesel fuels utilized, and specific CI fuel-engine interactions. Among the more dominant fuel properties are volatility, viscosity, pour point, distillation curve, self-ignition temperature, latent heat, and heat of combustion; see Table 11.2.

Diesel engines can burn a wider range of crude oil distillate liquid fuels than conventional SI engines and are therefore considered by many to be a broad-cut fuel IC engine. Low-speed diesels operate on a variety of hydrocarbon fuels, ranging from heavy residuals to power kerosines. However, high-speed diesels only are able to utilize a much narrower band of the light distillate fuel oils discussed in Chapter 5.

Fuel viscosity must be low enough to ensure proper fuel pump lubricity and also guarantee that atomization produces a suitable distribution of small droplets. Viscosity, however, should be high enough to prevent fuel

Table 11.2 Nominal Properties for Certain CI Fuels[a]

	Fuel type			
Fuel property	Kerosene	Premium diesel	Railroad diesel	Marine distillate diesel
Cetane number	50	47	40	38
Boiling range, °C (°F)	163–288 (325–550)	182–357 (360–675)	176–357 (350–675)	176–250 (350–500) (90%)
Viscosity, SSU at 38°C (100°F)	33	35	36	47
Gravity, °API	42	37	34	26
Sulfur, wt %	0.12	0.30	0.50	1.2
Uses	High-speed city buses	High speed: Buses, Trucks, Tractors, Light marine engines	Medium speed: RR engines, Marine engines, Stationary engines	Low speed: Heavy marine engines, Large stationary engines

[a]Based on ASTM D975 Classification of Diesel Fuel Oils.

oil injector leakage at the end of injection. Diesel fuel viscosity makes injection in large low-speed engines much easier and more efficient than in small high-speed machines, where the number, size, and placement of injection orifices are severely limited. Diesel volatility requirements are stated in terms of a 90% distillation temperature, i.e., a temperature at which 90% of the fuel will vaporize. High-speed engines with less time for fuel chemistry require a more volatile fuel than low-speed CI engines.

Fuel flash point temperatures should exceed 77°C (150°F) to ensure proper fuel compression ignition but, more importantly, to keep fuel safe during storage and handling. The pour point, or congealing temperature of a fuel, is a low-temperature performance factor ensuring fuel flow without the need for preheating, usually less than −17.4 °C (0°F). Paraffinic waxes that separate out from fuels at low temperatures are a chief cause of diesel fuel freezing problems. Heat of combustion is the major thermochemical fuel property that controls the power output of a CI engine. Fuel latent heat, i.e., the energy necessary to vaporize diesel liquid fuel droplets, must be provided by heat transfer from hot compressed air. Cold-starting difficulties may sometimes be circumvented by the use of a glow plug, an in-cylinder heater that can raise a cold air temperature to ensure proper charge preparation.

Diesel fuel ignition quality is stated in terms of a *cetane index,* an experimental measure of autoignition characteristics that is a complex function of fuel volatility, self-ignition temperature, and combustion chamber geometry. Cetane numbers for diesel fuels, like octane numbers for gasolines, are relative numbers that are obtained using a specific ASTM test and the CFR engine; see Chapter 9. Values for cetane numbers suitable for use in high-speed diesels range from 50 to 60, with a higher scale value being a more easily ignitable fuel. Detonation, or diesel knock, measured by the cetane number is different from SI knock, measured by an octane number. Recall that high octane number fuels resist autoignition, whereas high cetane number fuels have low autoignition characteristics. Note also that straight-chain paraffin compounds with single C–H bonding have high cetane values but low octane numbers, whereas aromatic ring compounds have low cetane but high octane numbers.

EXAMPLE 11.2 The temperature of the air in a diesel engine after compression must be greater than the self-ignition temperature of the injected fuel. Assume that ambient air conditions of 101 kPa and 27°C represent the initial state of the uncompressed air charge. For ideal compression, using both constant and variable specific heats, calculate (a) the compressed air temperature, K; and (b) the compressed air pressure, MPa, as a function of compression ratio. The minimum self-ignition temperature of diesel fuels is approximately 510°C. Determine values of (a) and (b) for this condition.

Solution

1. Isentropic compression of air, constant specific heats:

$$r_v = \frac{V_{BDC}}{V_{TDC}} = \left[\frac{T_2}{T_1}\right]^{1/(\gamma-1)}$$

$$T_2 = T_1 r_v^{\gamma-1} \qquad P_2 = P_1 r_v^{\gamma}$$

2. Isentropic compression of air, variable specific heats:

$$r_v = \frac{V_{BDC}}{V_{TDC}} = \frac{V_r<T_1>}{V_r<T_2>} \qquad V_r<T_2> = \frac{V_r<T_1>}{r_v}$$

$$T_2 = T_2<V_r> \qquad P_2 = \left[\frac{P_r<T_2>}{P_r<T_1>}\right] \times P_1$$

$$V_r = <300> \times 10^{10} = 18{,}028 \qquad P_r <300> \times 10^{-10} = 1.366$$

3. Compressed air properties vs. compression ratio:

r_v	T_2	P_2	$V_r \times 10^{10}$	$P_r \times 10^{-10}$	T_2	P_2
8	689	1.86	2253	25.627	678	1.90
9	723	2.19	2002	29.078	704	2.15
10	754[a]	2.54[a]	1802[b]	35.031[b]	736[b]	2.59[b]
11	783	2.90	1638	39.869	762	2.95
12	811	3.28	1502	43.961	784	3.25
13	837	3.66	1386	47.750	803	3.53
14	862	4.06	1287	54.520	828	4.03
15	886	4.48	1201	60.207	849	4.45
16	909	4.90	1126	65.352	868	4.83
17	932	5.33	1060	69.955	885	5.17
18	953	5.78	1001	73.746	899	5.45
19	974	6.23	948	81.245	919	6.01
20	994	6.70	901	87.712	936	6.49
21	1014	7.17	858	94.180	953	6.96
22	1033	7.65	819	99.506	967	7.36

[a]Constant specific heats

$T_2 = (300\text{K})(10)^{0.4} = 753.56 = 754\text{K}$

$P_2 = (0.101\ \text{MPa})(10)^{1.4} = 2.537 = 2.5\ \text{MPa}$

[b]Variable specific heats, using Example 3.2

$$V_r\langle T_2\rangle = \frac{18{,}020 \times 10^{-10}}{10} = 1802 \times 10^{-10}$$

$$T_2\langle V_r\rangle = T\langle 1802 \times 10^{-10}\rangle = 736\text{K}$$

$$P_r\langle 736\rangle = 35.031 \times 10^{10}$$

$$P_2 = \left[\frac{35.031}{1.366}\right] \times 0.101 \text{ MPa} = 2.59 = 2.6 \text{ MPa}$$

4. Diesel fuel ignition temperature, $T_{\text{ignit}} = 510 + 273 = 783\text{K}$:

$$r_v\langle 783\text{K}\rangle = 11{:}1 \text{ (const specific heats)}$$
$$= 12{:}1 \text{ (variable specific heats)}$$

Comment These calculations show that $\gamma = 1.4$ will overpredict the compressed air temperature at TDC. This overprediction increases with compression ratio. Pressure predictions for the two techniques are not that different and suggest that care should be taken when inferring gas temperature from cylinder pressure data.

Variations in CI fuel composition, as specified by an ultimate analysis, will not affect the heat of combustion and heat release as much as it influences the formation of certain undesirable emissions. These byproducts can severely impact the health and longevity of diesel engine components. For example, carbon residue is a prime source of carbon deposits on diesel engine parts and is usually limited to less than 0.01%. Sulfur content in diesel fuels most frequently must be kept to 0.5–1.5% to limit the formation of corrosive combustion products. Ash and sediment, abrasive constituents of fuel mixtures, should be kept to 0.01–0.05% to protect injection systems and piston-cylinder wear. Ash tends to be concentrated when refining heavy residual fuel oil.

Use of an *FA* ratio is somewhat nondescript when considering stratified charge CI combustion. Initially, a nonhomogeneous fuel-rich reactive mixture exists near the injector tip and specific points throughout the combustion chamber, with air throughout the remaining volume. Time-dependent injection spray patterns and physics produce a complex three-dimensional, unsteady reactive region in which locally an *FA* ratio is indefinite. One can speak of an overall *AF* ratio and, in this sense, diesel engines operate burning rich mixtures of 20:1 at full load to lean mixtures of 100:1 at light load.

11.4 COMPRESSION-IGNITION INTERNAL COMBUSTION ENGINE COMBUSTION

Proper charge preparation is necessary for efficient CI engine combustion, as was the case for SI engines as discussed in Chapter 10; see Figure 11.4.

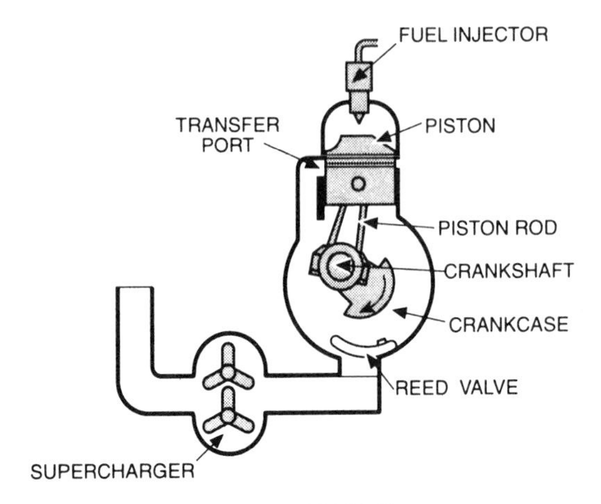

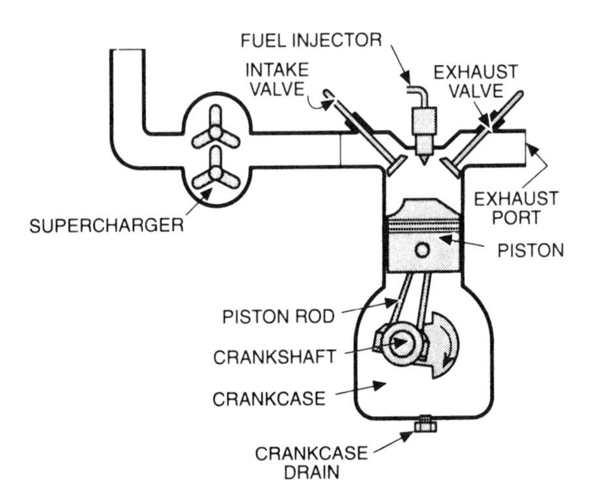

Figure 11.4 Charge preparation and the compression CI engine.

Classic SI engine combustion requires a premixed flame that propagates through a homogeneous mixture, whereas classic CI engine combustion is based on diffusion flame chemistry occurring within a heterogeneous mixture and initiates around small burning centers of vaporized liquid fuel. Very distinct stages of thermochemistry and heat release can be identified in classic CI combustion: (1) a period of ignition delay, (2) a period of ignition and rapid pressure rise, (3) a period of continued combustion and gradual pressure change; and (4) a period of postflame reactions. A pressure-crank angle or pressure-time diagram can be useful to help visualize these various stages of heat release; see Figure 11.5. Heat release rates for CI combustion in low- (under 300 rpm), medium- (300–1000 rpm), and high-speed (over 1000 rpm) engines are different. Greater mechanical and kinetic time is available during low-speed heat release (a $P=c$ heat addition thermodynamic model) than during medium- and high-speed heat release (a $V=c$ and $P=c$ heat addition thermodynamic model). The following comments will focus chiefly on high-speed diesel engines to illustrate broadly the principles at work during compression-ignition IC engine combustion chemistry.

The first stage of heat release, or *ignition delay,* is a time of charge preparation during which injected liquid fuel is atomized, mixed, and dispersed within hot compressed air. This delay period involves liquid fuel vaporization and a chemical kinetics stage during which fuel pyrolysis reactions begin. During this portion of the overall burn, there is a time of accumulation of unburned fuel and apparent inactivity. Several factors can influence the length of this induction and delay period, which occurs prior to ignition and rapid pressure rise, including:

Compression ratio
Inlet air temperature
Coolant temperature
Piston speed
Injection pressure
Fuel droplet size
Fuel injection rate
Fuel latent heat of vaporization
Fuel cetane number

Increasing the compression ratio will raise the initial charge reaction air temperature, i.e., molecular collisions, and chemical kinetic rates at TDC and shorten the delay. Increasing inlet air temperature for a specified compression ratio may also shorten this delay. Higher coolant temperatures may imply less heat transfer loss to cylinder walls and allow higher charge air temperature and greater wall initiation reactions, resulting in a reduced

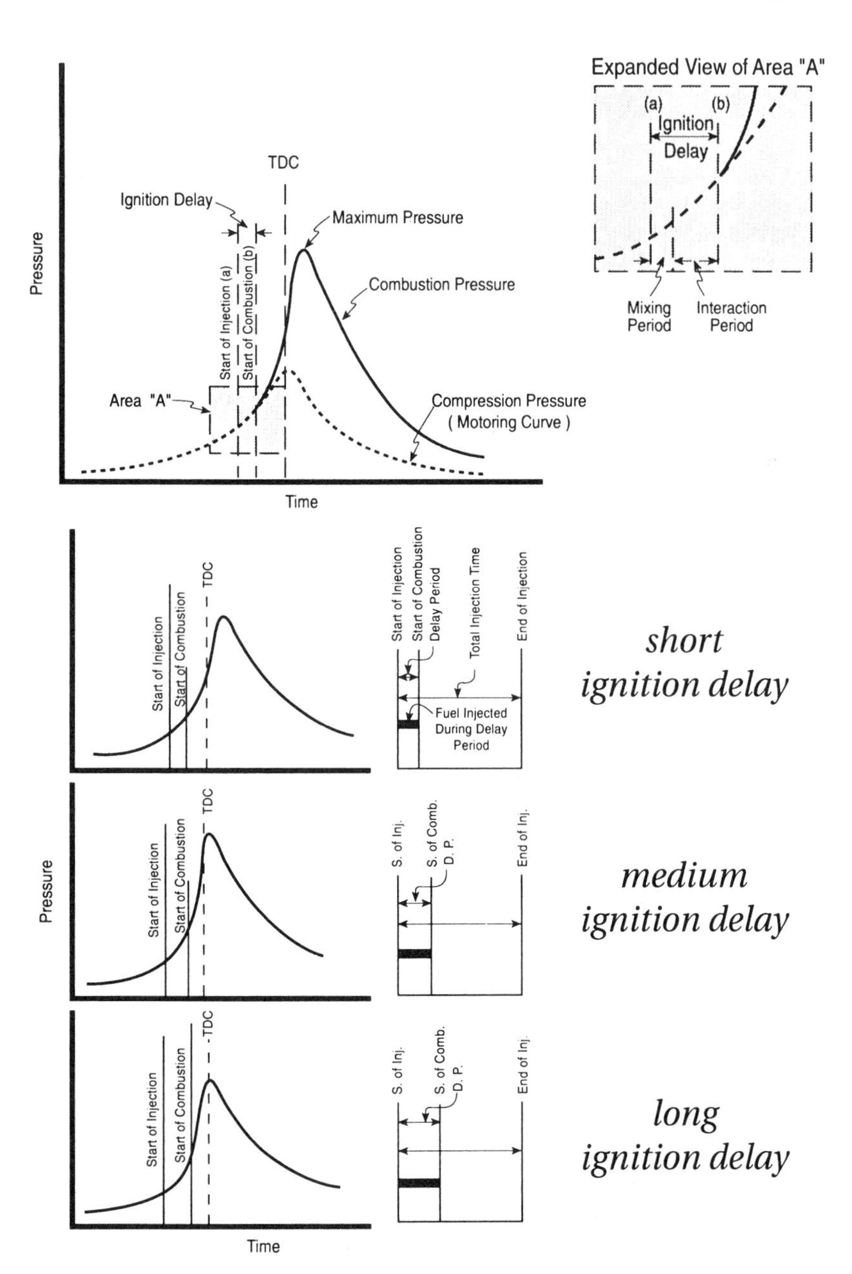

Figure 11.5 Combustion-ignition combustion–pressure–crank angle diagram.

ignition delay period. Engine speed affects combustion chamber turbulence and indirectly influences the delay period; i.e., a greater piston speed might produce greater turbulence and, hence, shorten delay.

Injection pressure influences the size, velocity, and dispersion of fuel droplets and, in general, higher injection pressures should shorten delay. Fuel droplet size will influence delay; i.e., larger droplets will result in longer delays, while small droplets may have insignificant momentum to provide proper fuel dispersion. Fuel injection rate is also a factor in delay since initial injection occurs in a high-oxygen environment with low reaction rates whereas, at the end of fuel injection, there is a lower-oxygen environment but higher temperatures and reaction rates.

Ignition delay can also be a function of such fuel properties as fuel molecular structure, volatility, and latent heat of vaporization. Cracked hydrocarbon and ring compounds tend to have long delay periods, while straight-run, single-bonded paraffin hydrocarbon chains have short delay periods. Fuel additives such as ethyl nitrate and thionitrate can shorten this delay period.

When sufficient droplets of fuel have vaporized, reached their self-ignition temperature, and undergone pyrolysis, the second stage of combustion begins. Oxidation chemical kinetics are dominated by diffusion flame physics. A sheath of vaporized fuel surrounds atomized droplets, which burn with oxygen from the compressed air diffusing into the vapor layer. This portion of rapid burn increases cylinder temperature and pressure and further accelerates evaporation, ignition, and heat release from the continuing fuel injection process. This second stage of burn is usually designed so that peak cylinder pressure occurs near TDC. The audible sound of CI diesel engine combustion also occurs during this second stage of combustion.

Peak pressure is a function of delay since, in general, the longer a delay, the greater the peak pressure. For an ideal short ignition delay, fuel vaporization is instantaneous, little fuel is accumulated, and major oxidation chemistry consists of degenerate chain-branching reactions. For a long ignition delay, fuel accumulation is severe, a fuel-rich mixture exists, and chemistry consists of chain-branching explosions. It is the long ignition delay and accumulation of unburned fuel that can result in detonation or knock in a diesel engine. Diesel knock can occur at the *beginning of the rapid combustion period* and therefore differs from SI knock, which occurs at the end of combustion.

Autoignition and detonation oxidation chemistry is common to both SI and CI engine combustion. Spark knock is undesirable, but it is the autoignition nature of a diesel fuel that is essential for its proper ignition. Compression-ignition knock can be minimized by early ignition and rapid burning of injected fuel, reducing the amount of accumulated fuel that would be involved in autocatalytic autoignition during the period of rapid pressure rise. A fuel with high volatility and low autoignition temperature will help minimize CI engine knock.

The third and final stage of CI combustion occurs toward the end of fuel injection. Reaction is not completed at TDC, continuing past the peak pressure point but at a lower rate during piston expansion. Cylinder pressure during expansion will be slightly greater than that for a nonreactive adiabatic expansion and will depend on fuel injection advance and cutoff. At the end of fuel injection, the last portion of fuel spray may not completely evaporate, and lower-temperature chemistry will produce a free carbon luminous flame that increases radiation heat transfer losses.

EXAMPLE 11.3 The indicated engine performance of a two-stroke, six-cylinder 4-in. × 4-in. diesel is to be modeled using an air standard cycle analysis. The engine operates at 1400 rpm, and a blower supplies 100% scavenging air at engine inlet conditions of 70°F and 14.7 psi. The heating value of the fuel is 18,500 Btu/lbm, and the overall mass AF ratio is 15:1. For a compression ratio of 14:1, calculate: (a) the cycle peak temperature and pressure; (b) the ideal indicated mean effective pressure, psi; (c) the indicated thermal efficiency, %; (d) the ideal indicated engine power, hp; and (e) the indicated specific fuel consumption, lbm/hp hr.

Solution:

1. State 1:

$$P_1 = 14.7 \text{ psia}$$

$$T_1 = 530°\text{R}$$

$$V_D = \frac{\pi}{4}\left(\frac{4}{12}\text{ ft}\right)^2\left(\frac{4}{12}\text{ ft}\right) = 0.02909 \text{ ft}^3/\text{stroke}$$

$$V_c = \frac{\pi}{4}\left(\frac{4}{12}\right)^2\left(\frac{4}{12}\right)\left(\frac{1}{13}\right) = 0.00224 \text{ ft}^3/\text{stroke}$$

$$V_1 = V_D + V_c = 0.02909 + 0.00224 = 0.03133$$

$$M_1 = \frac{P_1V_1}{RT_1} = \frac{(14.7 \text{ lbf/in.}^2)(144 \text{ in.}^2/\text{ft}^2)(0.03133 \text{ ft}^3)}{(53.34 \text{ ft lbf/lbm °R})(530°\text{R})}$$

$$= 0.002346 \text{ lbm}$$

$$v_1 = \frac{V_1}{M_1} = \frac{0.03133}{0.002346} = 13.3546 \text{ ft}^3/\text{lbm}$$

2. Process $1 \rightarrow 2$, $s = c$ compression:

   Compression ratio:

$$r_v = \frac{V_1}{V_2} = \frac{0.03133}{0.00224} = 14.0$$

   or

$$r_v = 14{:}1$$

$$V_2 = V_c = 0.00224 \text{ ft}^3$$

$$v_2 = \frac{0.00224}{0.002346} = 0.9548 \text{ ft}^3/\text{lbm}$$

$$P_2 = \left(\frac{V_1}{V_2}\right)^{\gamma} P_1 = \left(\frac{0.03133}{0.00224}\right)^{1.4} (14.7)$$

   a. $P_2 = 590.6$ psi

$$T_2 = \left(\frac{V_1}{V_2}\right)^{\gamma-1} T_1 = \left(\frac{0.03133}{0.00224}\right)^{0.4} (530°\text{R}) = 1522.5°\text{R}$$

   Work of compression:

$$\delta Q - \delta W = dU = MC_v\, dT$$

$$_1W_2 = MC_v(T_1 - T_2)$$

$$= (0.002346 \text{ lbm})(0.171 \text{ Btu/lbm °R})(530 - 1522.5°\text{R})$$

$$= -0.400 \text{ Btu}$$

3. Process 2–3, $P = c$ heat addition:

$$_2Q_3 = \frac{HV\text{<Btu/lbm fuel>}M\text{<lbm air>}}{AF\text{<lbm air/lbm fuel>}}$$

$$= \frac{(18{,}500)(0.002346)}{(15)} = 2.893 \text{ Btu}$$

$$\delta Q - \delta W = dU$$

$$\delta Q = dH$$

$$_2Q_3 = MC_p(T_3 - T_2)$$

   b. $T_3 = 1522.5 + \dfrac{2.893}{(0.240)(0.002346)} = 6661°\text{R}$

$$P_3 = P_2 = 590.6 \text{ psi}$$

$$V_3 = \frac{MRT_3}{P_3} = \frac{(0.002346)(53.34)(6661)}{(590.6)(144)} = 0.00980 \text{ ft}^3$$

$$v_3 = \frac{0.00980}{0.002346} = 4.1773 \text{ ft}^3/\text{lbm}$$

4. Process 3–4, $s = c$ expansion:

$$V_4 = V_1 = 0.03133 \text{ ft}^3$$

$$v_4 = v_1 = 13.3546 \text{ ft}^3/\text{lbm}$$

$$T_4 = \left(\frac{V_3}{V_4}\right)^{\gamma-1} T_3 = \left(\frac{0.009780}{0.03133}\right)^{0.4} (6661) = 4184.5$$

$$_3W_4 = MC_v(T_3 - T_4) = (0.002346)(0.171)(6661 - 4185)$$
$$= 0.993 \text{ Btu}$$

5. Process 4–1, assuming that $f = 1.0$ (100% scavenging air):

$$_4Q_1 = MC_v(T_1 - T_4)$$
$$= (0.002346)(0.171)(530 - 4185) = -1.466 \text{ Btu}$$

6. Cycle performance (per cylinder):

$$W_{net} = {_1W_2} + {_2W_3} + {_3W_4} + {_4W_1}$$
$$= {_1Q_2} + {_2Q_3} + {_3Q_4} + {_4Q_1}$$
$$= 2.893 - 1.466 = 1.427 \text{ Btu}$$

Mean effective pressure:

$$\overline{P} = \frac{W_{net}}{V_D} = \frac{(1.427 \text{ Btu})(778 \text{ ft lbf/Btu})}{(0.02909 \text{ ft}^3)(144 \text{ in.}^2/\text{ft}^2)}$$

c. $IMEP = 265$ psi

Thermal efficiency:

$$\eta = \frac{\text{desired energy output}}{\text{required energy output}} = \frac{W_{net}}{Q_{add}}$$

$$= 1 + \frac{_4Q_1}{_2Q_3} = 1 - \frac{1.466}{2.893} = 0.493$$

$$\eta = 49.3\%$$

7. Indicated engine performance:

Indicated power:

$$\dot{W}_I = \frac{\overline{P}<\text{lbf/ft}^2\text{ stroke}>\ L<\text{ft}>\ A<\text{ft}^2>\ N<\text{rev/min}>\ C<\text{no. cylinders}>}{(33{,}000\text{ ft lbf/hp min})\ n<\text{rev/stroke}>}$$

$$= \frac{(265)(144)(4/12)(\pi/4)(4/12)^2(1400)(6)}{(33{,}000)(1)} = 282.55\text{ ihp}$$

Indicated specific fuel consumption:

$$\dot{M}_{\text{fuel}} = \frac{M<\text{lbm air/stroke}>\ N<\text{rev/min}>\ C<\text{no. cylinders}>}{AF<\text{lbm air/lbm fuel}>\ n<\text{rev/stroke}>}$$

$$\dot{M}_{\text{fuel}} = \frac{(0.002346)(1400)(6)}{(14)(1)} = 1.407\text{ lbm fuel/min}$$

Indicated specific fuel consumption:

$$\text{ISFC} = \frac{\dot{M}_{fuel}\ (\text{lbm/min})}{\dot{W}_I\ (\text{hp})} = \frac{1.407}{282.55}$$

$$= 0.00498\text{ lbm/hp min} = 0.299\text{ lbm/hp hr}$$

Major diesel engine emissions are carbon monoxide, oxides of nitrogen, unburned hydrocarbons, odor, particulates, and smoke. Recall that carbon monoxide is due to insufficient oxygen for complete oxidation of carbon. Since diesel fuel combustion occurs in excess air environments, CO emissions are much lower for CI engines than SI engines. Further reduction in CO levels can be accomplished using oxidation catalysts similar to those used in SI engine technology.

Oxides of nitrogen (NOx) are a result of excess air and high–compression ratio operation of modern CI engines. Any reduction in peak temperature or excess air helps to reduce NOx emissions. Exhaust gas recirculation (EGR) is used in both SI and CI engines as a means of reducing NOx. Three-way catalysts, used to reduce SI NOx emissions, require stoichiometric or oxygen-deficient exhaust gas compositions to work and will not work with lean-burn diesel engines.

Unburned hydrocarbons, (UHCs) are due to specific CI fuel-engine interface characteristics, including (1) fuel injection, i.e., spray pattern, length of injection, and degree of atomization; (2) particular combustor geometry, i.e., shape of combustion chamber, fuel quenching from spray contacting the wall, and chamber burn pattern; and (3) combustion itself, i.e., fuel structure, locally rich mixture chemistry, and time available for burn. The UHC levels of diesel engines depend largely on load and speed but, with the high overall AF ratios of diesels, are generally lower than those of comparable SI engines. Diesel UHC emissions contain original, decomposed, and recombined diesel fuel constituents and, therefore, differ from those compounds generated by SI combustion of automotive fuels. Odor is a byproduct of the incomplete burning of diesel fuels. Compression-ignition engine chemistry produces partially oxidized, as well as decomposed and recombined, fuel fractions. The hydrocarbon composition of diesel fuels promotes incomplete combustion, odor-producing, aromatic ring hydrocarbons and oxygenated aldehyde compounds such as formaldehyde.

Particulates are exhaust emissions from a CI engine other than water that can be filtered. Certain of these diesel particulates are carcinogenic. Smoke, either a solid or liquid (aerosol) particulate suspended in an exhaust stream, is an easily observed and well-known diesel engine emission. Several types of smoke can be identified with specific CI engine operating conditions. Cold-starting, idle, and low-load engine operation produce white smoke, liquid particulates consisting of unburned, partially burned, or cracked fuel with a small fraction of lubricating oil. Black-gray smoke occurs at maximum loads and rich combustion and consists chiefly of solid carbon particulates resulting from incomplete combustion. Carbon is opaque and will therefore yield black smoke, whereas liquid hydrocarbon droplets have high optical transmissions and, hence, make smoke appear grayer. The intensity of smoke generated by CI engines is influenced by fuel cetane number, injection rate, cutoff, and atomization. Particulate emissions may also include fuel additives and blue smoke, a common characteristic of lubricating oil aerosols. Smoke can be reduced by burning supplementary fuel additives, derating (reducing the maximum fuel flow), avoiding engine overload, afterburning the exhaust gases, and proper maintenance.

Design requirements for diesel combustors are similar to many of the design objectives set for SI configurations, as discussed in Chapter 10.

Development of turbulence within the charge
Promoting a rapid but smooth rise in pressure versus time (crank angle) during burn
Achieving peak cylinder pressure as close to TDC as possible
Reducing the occurrence of detonation or knock
Promoting complete combustion
Minimizing heat loss

Several illustrative chamber geometries that have been developed in an effort to achieve proper CI combustion are shown in Figure 11.6. These CI combustion chambers may be classified as either *direct-injection* (*open-chamber*) or *divided-chamber* (*indirect-injection*) CI geometries. Particular power applications will dictate which of these CI combustion chambers is more suitable for use. In automotive applications, where a high power-to-

Figure 11.6 Various CI combustion chamber designs.

weight requirement exists, for example, a high-speed divided-chamber engine is usually needed whereas, in marine and industrial power applications, where weight and size are less critical than fuel economy, large slow-speed open-chamber engines are more often used.

The *open combustion chamber* uses space above the piston crown and head for combustion control. Fuel injected directly into this single-chamber combustor requires a high-pressure injection process to ensure proper atomization and penetration of fuel spray throughout the chamber volume. Air motion is relatively nonturbulent, so that intake valve location and piston crown geometries are varied in an attempt to introduce motion and air rotation within a chamber. Chamber geometry is frequently shaped to conform to the fuel spray pattern, with a fuel injector generally positioned close to the piston centerline. Open combustion chamber geometry may have less influence on combustion than air motion. A squish, or reduced area, configuration allows the volume between piston crown and clearance volume at the edge of a piston when at TDC to be quite small. This geometry is an attempt to cause an increased inward motion of a charge at the end of the compression stroke. An open-chamber CI engine requires considerable excess air but, in general, is easy to start. Smaller surface-to-volume ratios for an open chamber result in a lower heat loss than for other geometries. This design is utilized most often in large, low-speed diesel engines that burn diesel fuels having longer ignition delays, i.e., lower cetane numbers. Open-chamber CI engines generally require compression ratios in a range 12:1–16:1 and have better fuel consumption, higher pressure rise, and less throttling losses of all the CI chambers. High-speed open-chamber CI engines have a tendency to develop diesel knock.

The *divided combustion chamber* diesel uses two-chamber combustion in an effort to circumvent problems encountered with open-chamber fuel injection designs. A variety of configurations and combustion methods have been developed, including precombustion chambers, swirl or turbulent chambers, and air or energy cell chambers. Divided chambers are most frequently used with small high-speed engine applications in an effort to help shorten the time needed for combustion. The additional chamber volume can be located in the piston crown, cylinder wall, or head. Secondary chamber volumes can be as much as 50% of clearance volume.

Precombustion-chamber CI engines are indirect injection configurations consisting of a main chamber and a separate fuel ignition volume. Air enters the main and prechamber volumes during a compression stroke, but fuel is injected only into the smaller uncooled prechamber. Deep fuel spray penetration is not necessary, allowing a lower injection pressure and shorter ignition delay. Initial ignition occurs in the prechamber and produces a relatively rich homogeneous and turbulent preconditioned charge with jet spray expansion of reactive gases through orifice openings into the main chamber for complete reaction. Prechamber designs do not depend on air motion within their secondary chamber to burn diesel fuels.

A *swirl,* or *turbulent,* chamber is another indirect injection divided-chamber design. Air enters both chambers during compression, and fuel is injected only into the secondary volume, much like the prechamber. However, significant air motion and turbulence are purposely generated within the swirl chamber during compression. Turbulent-chamber diesels reduce stringent fine spray injection atomization requirements needed for open chambers and allow larger fuel droplets to be burned. The boundary distinguishing a prechamber from a turbulent-chamber design is somewhat arbitrary but, in general, prechamber volumes are relatively smaller and burn only a minor portion of the charge therein, whereas swirl-chamber volumes are larger, with a major portion of the burn occurring therein.

The *air,* or *energy, cell* is a divided combustion chamber configuration but, in this instance, fuel is injected into the main chamber. The air cell is located directly across from an injector, and fuel is sprayed across the main chamber and enters into an energy cell prior to ignition. This method offers benefits of both the open and turbulent geometries. Air-cell engines rely on their small antechamber and the ensuing air motion and charge interaction to reduce the injection pressure required for an open-chamber design. The piston motion during one mechanical cycle causes a turbulent flow to occur between the antechamber and main chamber. The rapid pressure rise and peak pressure in the figure-eight–shaped main is gas dynamically and thermally controlled by the energy cell. Burn in an air-cell engine occurs with slow initial combustion of injected fuel, followed by a rapid secondary burn resulting from flow jetting back into the main chamber during an expansion stroke. Since the energy cell is not cooled, the design regenerates a charge by returning some energy released by combustion to the charge at a later stage in the burn process. Energy-cell CI engines are generally characterized by larger heat losses to the wall, starting difficulties, and lower thermal efficiency but higher *IMEP*. Air- or energy-cell chambers are easier to start than the swirl or turbulent chamber since fuel is injected into the main chamber.

Divided-chamber combustors have larger surface-to-volume ratios, more fluid motion, greater heat transfer coefficients and, therefore, higher heat loss and lower charge temperatures relative to comparable open-chamber geometries. Indirect injection CI engines usually require compression ratios of 18:1 to 24:1 to ensure reliable ignition, but they are often smoother-running engines. Additional general characteristics of a divided-chamber indirect injection CI engine include higher volumetric efficiency, lower peak pressure, lower *IMEP,* lower thermal efficiency, and higher specific fuel consumption. Two-chamber CI configurations are not as compatible with two-stroke or large-displacement volume diesels.

Compression-ignition combustion is slower than SI combustion, which means that maximum engine speeds are lower for diesel engines. Power output, therefore, cannot be raised by increasing engine speed, so that CI engine output is usually increased by turbocharging, a technique that can also produce improvements in fuel economy. Turbocharging may mean that a reduction in compression ratio is required in order to maintain the peak pressure and temperature of the basic engine design. In particular, two stroke diesel engine designs are often turbocharged.

Diesel engines control load by simply regulating the amount of fuel injected into the combustion air. This approach eliminates throttling as required for SI engine power control. Because of the very localized fuel-rich nature of CI heat release, an engine operating at constant speed, and hence constant airflow rate, can meet a wide range of power demand by simply controlling the rate and amount of fuel injected. Overall rich or stoichiometric conditions are necessary for high power output. In this instance, the rich combustion and high carbon-to-hydrogen ratio of diesel fuels produce carbon atoms and generate black smoke, an objectionable pollutant. Idle engine operation means lower peak reaction temperatures, less air turbulence, incomplete combustion, and the formation of white smoke.

The higher thermal efficiency and greater fuel economy of diesel engines compared to SI engines are chiefly a result of the higher compression ratios required for diesels. To a lesser degree, diesel performance gains are due to the simplicity of the autoignition process, lower pumping losses as a result of removing the throttle valve, and overall lean *AF* mixtures required for combustion. Diesel engines also have the advantage of being able to maintain a higher sustained torque than a spark-ignition engine.

11.5 COMPRESSION-IGNITION ENGINE FUEL ALTERNATIVES

Diesel fuels for commercial usage are graded by the American Society for Testing Materials: Grade 1-D is a volatile fuel for use in engines that experience frequent speed and load changes; grade 2-D is a low-volatility fuel for use in large mobile and industrial engine applications; and grade 4-D is a fuel oil for low- and medium-speed engines. Special diesel fuels are also available for railroad, bus, marine, and military CI machinery; see Table 11.2. High-speed diesels can burn a gas fuel known as straight-run gas oil, a crude oil fraction that comes directly from the distillation column.

Declining quality of crude oil, increased demand for diesel fuel, and greater use of blended distillate fuels containing greater amounts of cracked components have contributed to a lowering of the average cetane number of diesel fuels. Cetane values are declining from their mid-50s values during the 1960s and are projected to reach values in the 40s by the turn of the century. One possible technique for burning low-cetane fuels is to use a *bifuel* engine, a CI engine having dual liquid fuel injection, i.e., a secondary or auxiliary fuel injected separately during the compression stroke and primary fuel injected later near TDC.

Diesel engines have been run on a variety of solid, liquid, and gaseous fuels. By the early 1900s, CI engines had been designed to burn coal dust and even gunpowder. Gasoline also has been used as a CI engine fuel; however, high-octane SI fuels are low-cetane CI fuels, and CI combustion of gasoline results in noisy and rough engine operation. In addition, gasoline, with its lower viscosity, causes wear and seizure of diesel injection pumps. Operation can be improved by using 10–15% diesel fuel/90–85% gasoline blends. Most CI fuel alternatives have been burned in large, slow-speed machines, while their use in intermediate- and high-speed units has not yet proved to be as successful.

EXAMPLE 11.4 Steady-state operation of a diesel engine test produces the following experimental data:

Engine inlet air temperature	25°C
Fuel inlet temperature	27°C
Exhaust gas temperature	800K
Engine indicated thermal efficiency	58%
Exhaust gas composition, ppm	

CO_2	62,279
CO	14,013
UHC	1,557
H_2	9,861
O_2	28,609
N_2	647,437
H_2O	71,361
NOx	164,883

Diesel fuel is *n*-dodecane, and the UHC can be assumed to be equivalent to CH_4 and the NOx equivalent to N_2O. For these conditions, find: (a) the equivalence ratio for the overall fuel-air chemistry; (b) the indicated work, kJ/kg fuel; (c) the overall heat loss as a percent of indicated work, %; and (d) the indicated specific fuel consumption, g/W hr.

Solution

1. Stoichiometric equation:

$$C_{12}H_{26} + 18.5[O_2 + 3.76N_2] \rightarrow 12CO_2 + 13H_2O + 69.56N_2$$

$$\overline{AF}_{stoic} = (18.5)(4.76)/1 = 88.06$$

2. Actual reaction:

$$aC_{12}H_{26} + b[O_2 + 3.76N_2] \rightarrow 62{,}279CO_2 + 14{,}013CO + 1557CH_4 + 9861H_2 + 28{,}609O_2 + 647{,}437N_2 + 71{,}361H_2O + 164{,}883N_2O$$

Carbon balance:

$$12a = 62{,}279 + 14{,}013 + 1557 \qquad a = 6487.4$$

Nitrogen balance:

$$3.76b = 647{,}437 + 164{,}883 \qquad b = 216{,}043$$

$$\overline{AF}_{act} = 216{,}043 \times \left[\frac{4.76}{6487.4}\right] = 158.52$$

3. Equivalence ratio Φ:

a. $$\Phi = \frac{\overline{AF}_{stoic}}{\overline{AF}_{act}} = \frac{88.06}{158.5} = 0.556$$

$$C_{12}H_{26} + (1.8)(18.5)[O_2 + 3.76N_2] \rightarrow 9.6CO_2 + 2.16CO + 0.24CH_4 + 1.52H_2 + 4.41O_2 + 99.8N_2 + 11H_2O + 25.4N_2O$$

4. Energy balance:

$$\bar{q} - \bar{w} = \Sigma N_i\bar{h}_i - \Sigma N_j\bar{h}_j$$

$$\bar{q} - \bar{w} = \{9.6[\bar{h}_f^0 + \Delta\bar{h}<800>]_{CO_2} + 2.16[\bar{h}_f^0 + \Delta\bar{h}<800>]_{CO} + 0.24[\bar{h}_f^0 + \Delta\bar{h}<800>]_{CH_4} + 1.52[\bar{h}_f^0 + \Delta\bar{h}<800>]_{H_2} + 4.41[\bar{h}_f^0 + \Delta\bar{h}<800>]_{O_2} + 99.8[\bar{h}_f^0 + \Delta\bar{h}<800>]_{N_2} + 11[\bar{h}_f^0 + \Delta\bar{h}<800>]_{H_2O_g} + 25.4[\bar{h}_f^0 + \Delta\bar{h}<800>]_{N_2O}\} - \{33.3[\bar{h}_f^0 + \Delta\bar{h}<298>]_{O_2} + (33.3)(3.76)[\bar{h}_f^0 + \Delta\bar{h}<298>]_{N_2} + [\bar{h}_f^0 + \Delta\bar{h}<300>]_{C_{12}H_{26}}\}$$

$$\bar{q} - \bar{w} = 9.6[-94{,}054 + 5453]_{CO_2} + 2.16[-26{,}417 + 3627]_{CO} + 0.24[-17{,}895 + 5897]_{CH_4} + 1.52[0 + 3514]_{H_2} + 4.41[0 + 3786]_{O_2} + 99.8[0 + 3596]_{N_2} + 11[-57{,}798 + 4300]_{H_2O_g} + 25.4[19{,}610 + 5589]_{N_2O} - [-71{,}014 + 126]_{C_{12}H_{26}} = -399{,}293$$

5. Thermal efficiency:

$$\eta = \frac{\bar{w}<\text{cal/g mole fuel}>}{LHV<\text{cal/g mole fuel}>} = 0.58$$

where

$$LHV<\text{cal/g mole}> = HHV - \bar{n} < \frac{\text{mole } H_2O}{\text{mole fuel}} > h_{fg}<\text{cal/g mole}>$$

$$LHV = 1{,}931{,}360 \text{ cal/g mole fuel}$$

$$- \left\{ \frac{\left(\frac{18 \text{ g } H_2O}{\text{g mole}}\right)\left(\frac{13 \text{ g mole } H_2O}{\text{g mole fuel}}\right)\left(\frac{1050 \text{ Btu}}{\text{lbm } H_2O}\right)\left(\frac{251.98 \text{ cal}}{\text{Btu}}\right)}{(453.6 \text{ g/cal})} \right\}$$

$$LHV = -\ 1{,}794{,}871 \text{ cal/g mole fuel}$$

and

$$\bar{w} = (0.58)(1{,}794{,}871) = 1{,}041{,}025 \text{ cal/g mole}$$

$$w = (1{,}041{,}025)(4.187 \text{ kJ/kg mole fuel})/(170 \text{ kg/kg mole fuel})$$

b. $|w| = 25{,}640 = 25{,}640$ kJ/kg fuel

6. Heat loss:

$$\bar{q} - \bar{w} = -399{,}293 \text{ cal/g mole fuel}$$

$$\bar{q} = -399{,}293 + 1{,}041{,}025 = 641{,}732 \text{ cal/g mole}$$

c. $\%\ \text{work} = (q/w) \times 100 = \left[\frac{641{,}732}{1{,}041{,}025} \right] \times 100 = 61.6\ \%$

7. Indicated specific fuel consumption:

$$ISFC = \frac{\dot{m}_{\text{fuel}}}{\dot{W}} = \frac{\dot{m}_{\text{fuel}}}{\dot{m}_{\text{fuel}}\eta LHV} = \frac{1}{\eta LHV}$$

$$ISFC = \frac{(170 \text{ g/g mole fuel})(860 \text{ cal/W hr})}{(0.58)(1{,}794{,}871 \text{ cal/g mole})}$$

d. $ISFC = 0.140$ g/W hr

Large diesel engines are available that burn both natural or synthetic gas as well as fuel oil. *Gas fuel engines* can operate as a standard CI machine in which compressed gas is injected into hot compressed air. Methane, a gas fuel with a critical compression ratio for autoignition of approximately 12.5:1 measured using a CFR engine, can be used as a spark-ignition engine fuel.

These types of gas-fueled SI engines are sometimes grouped with CI engines simply because of their "high" SI compression ratio. Dual-fuel engines compress gas-air mixtures but require a small fraction of injected fuel oil near TDC as a pilot ignition source. Gas diesel combustion is characterized by a lower maximum pressure and slower rate of pressure rise versus crank angle.

Nonpetroleum CI fuels derived from coal, shale, tar sands, and vegetable oils have had limited application, chiefly as a research and development fuel, and will require further fuel-engine interface engineering before they see commercial application as diesel fuels. Water injection in CI combustion has been utilized as a means of controlling peak temperature and NOx emissions, although some argue that spraying water droplets into the combustion volume can provide additional energetics for the ignition process. Other exotic CI fuel concepts have included pulverized coal-oil and oil-water mixtures.

EXAMPLE 11.5 A ten-cylinder, two-stroke low-speed marine diesel operates at a continuous speed of 130 rpm while burning a stoichiometric mixture of liquid dodecane and air. The 30-in. × 55-in. engine has a 13:1 compression ratio with conditions at BDC after air intake of 17.5 psi and 80°F. Fuel is injected at TDC at a temperature of 125°F. Based on the fuel-air Diesel cycle and JANAF data for the engine thermochemistry, determine (a) the ideal indicated net work, ft lbf; (b) the ideal indicated mean effective pressure, psi; (c) the ideal indicated thermal efficiency; and (d) the ideal indicated specific fuel consumption, lbm/hp hr.

Solution:

1. Engine geometry:

$$V_{\text{displacement}} = \frac{\pi}{4}\left[\frac{30}{12}\text{ ft}\right]^2\left[\frac{55}{12}\text{ ft}\right] = 22.498\text{ ft}^3$$

$$r_v = \frac{V_{\text{BDC}}}{V_{\text{TDC}}} = \frac{V_D + cV_D}{cV_D} = 13$$

$$c = 0.08333$$

$$V_{\text{TDC}} = (0.08333)(22.498) = 1.875\text{ ft}^3$$

$$V_{\text{BDC}} = (1.08333)(22.498) = 24.373\text{ ft}^3$$

2. Conditions at BDC state 1:

$$T_1 = 540°\text{R} = 300\text{K}$$

$$N_{\text{tot}} = \frac{P_1V_1}{\overline{R}T_1} = \frac{(17.5\text{ lbf/in.}^2)(144\text{ in.}^2/\text{ft}^2)(24.373\text{ ft}^3)}{(1545\text{ ft lbf/lb mole °R})(540°\text{R})}$$
$$= 0.0736\text{ lb mole air/cyl}$$

with

$$U\langle T_1\rangle = \sum_{i=1}^{N} \bar{x}_i N_{\text{tot}}[\bar{h}_f^0 + \Delta\bar{h}\langle T_1\rangle - \bar{R}T_1]_i$$

At BDC, assume air only and, using JANAF data from Appendix B for air,

$$U_1\langle 300\text{K}\rangle = \{(0.0736 \text{ lb mole air/cyl})(1.8001)\} \times [0 + 35 - (1.987)(300)] \text{ Btu/lb mole}$$

$$U_1\langle 300\text{K}\rangle = -74.34 \text{ Btu/cyl}$$

3. Process 1–2, isentropic compression of air only:

$$\frac{V_2}{V_1} = \frac{V_r\langle T_2\rangle}{V_r\langle T_1\rangle} = \frac{1}{13}$$

Using the material from Example 3.2,

$$V_r\langle 300\text{K}\rangle = 18{,}020 \times 10^{10}$$

$$V_r\langle T_2\rangle = \frac{18{,}020 \times 10^{10}}{13} = 1386 \times 10^{10}$$

and

$$T_2 = 803\text{K} = 1446°\text{R}$$

$$P_2 = \frac{N_{\text{tot}}\bar{R}T_2}{V_2} = \frac{(0.0736 \text{ lb mole/cyl})(1545 \text{ ft lbf/lb mole°R})(1446°\text{R})}{(1.875 \text{ ft}^3)(144 \text{ in.}^2/\text{ft}^2)}$$

$$= 609 \text{ psi/cyl}$$

$$U\langle T_2\rangle = \Sigma \bar{x}_i N_{\text{tot}}[\bar{h}_f^0 + \Delta\bar{h}\ \langle T_2\rangle - \bar{R}T_2]_i$$

Using JANAF data for air from Appendix B,

$$U\langle 803\text{K}\rangle = \{(0.0736 \text{ mole})(1.8001)\} \times [3667 - (1.987)(803)] \text{ Btu/lb mole}$$

$$= 274.44 \text{ Btu/cyl}$$

4. Process 2–3, P = constant adiabatic flame:

Stoichiometric equation, assume complete combustion:

$$C_{12}H_{26} + 18.5[O_2 + 3.76N_2] \rightarrow 12CO_2 + 13H_2O + 69.56N_2$$

$$\delta Q - P\,dV = dU$$

$$\delta Q = dH = 0 \qquad H\langle T_3\rangle = H\langle T_2\rangle$$

$$H\langle T_3\rangle = \sum_{\text{prod}} N_i[\bar{h}_f^0 + \Delta\bar{h}]_i$$

$$H\langle T_2\rangle = \sum_{\text{react}} N_j[\bar{h}_f^0 + \Delta\bar{h}]_j$$

Using JANAF data for the reactants,

$$H\langle T_2\rangle = \left\{ \frac{(0.0736 \text{ lb mole air/cyl})}{(18.5)(4.76) \text{ lb mole air/lb mole fuel}} [-71{,}014 + 2032.5] \right.$$

$$\left. + (0.0736)(3667) \right\}(1.8001) \text{ Btu/lb mole} = 382 \text{ Btu/cyl}$$

and for the products,

$$382 \frac{\text{Btu}}{\text{cyl}} = \left[\frac{(12 \text{ lb mole CO}_2\text{/lb mole fuel})(0.0736 \text{ lb mole air/cyl})}{(18.5)(4.76) \text{ lb mole air/lb mole fuel}} \right] \times$$

$$\left\{ [-94{,}054 + \Delta\bar{h}\langle T_3\rangle]_{CO_2} + \frac{(13)(0.0736)}{(18.5)(4.76)} [-57{,}798 + \Delta\bar{h}\langle T_3\rangle]_{H_2O} \right.$$

$$\left. + \frac{(69.56)(0.0736)}{(18.5)(4.76)} [\Delta\bar{h}\langle T_3\rangle]_{N_2} \right\}(1.8001)$$

$$2{,}133{,}925 = (12\,\Delta\bar{h}\langle T_3\rangle)_{CO_2} + (13\Delta\bar{h}\langle T_3\rangle)_{H_2O} + (69.56\Delta\bar{h}\langle T_3\rangle)_{N_2}$$

By trial and error,

$$T_3 = 2752\text{K} = 4954°\text{R}$$

5. Conditions at 3, $P_3 = P_2 = 609$ psi:

$$V_3 = \frac{N_3\bar{R}T_3}{P_3}$$

$$V_3 = \frac{\left(0.0736 \dfrac{\text{lb mole air}}{\text{cyl}}\right)\left(13 + 12 + 69.56 \dfrac{\text{lb mole prod}}{\text{lb mole fuel}}\right)(1545)(4954)}{(18.5)(4.76 \text{ lb mole air/lb mole fuel})(609)(144)}$$

$$= 6.898 \text{ ft}^3$$

$$U_3 = H_3 - P_3V_3$$

$$= 382 \text{ Btu} - \frac{(609 \text{ lbf/in.}^2)(144 \text{ in.}^2\text{/ft}^2)(6.898 \text{ ft}^3)}{778 \text{ ft lbf/Btu}}$$

$$= -395.5 \text{ Btu/cyl}$$

6. Process 3–4, $s = c$ expansion of products:

$$P_r\langle T_3\rangle = \exp\left\{ \frac{\Sigma \bar{x}_i \bar{s}^0\langle T_3\rangle}{\bar{R}} \right\}$$

$$= \exp\left\{ \frac{(12/94.56)(78.566) + (13/94.56)(67.280) + (69.56/94.56)(63.001)}{1.987} \right\}$$

$$= 2.1395 \times 10^{14}$$

and

$$V_r\langle T_3\rangle = \frac{\overline{R}T_3}{P_r\langle T_3\rangle} = \frac{(1.987)(2752)}{(2.1395 \times 10^{14})} = 2.556 \times 10^{-11}$$

but

$$\frac{V_r\langle T_4\rangle}{V_r\langle T_3\rangle} = \frac{V_1}{V_3} = \frac{24.373}{6.898} = 3.5333$$

$$V_r\langle T_4\rangle = (3.5333)(92.556 \times 10^{-11}) = 9.031 \times 10^{-11}$$

or

$$P_r\langle T_4\rangle = \frac{\overline{R}T_4}{V_r\langle T_4\rangle} = \frac{1.987\ T_4}{9.031 \times 10^{-11}}$$

where

$$P_r\langle T_4\rangle = \exp\left\{ \frac{(12/94.56)\bar{s}^0\langle T_4\rangle_{CO_2} + (13/94.56)\bar{s}^0\langle T_4\rangle_{H_2O} + (69.56/94.56)\bar{s}^0\langle T_4\rangle_{N_2}}{1.987} \right\}$$

By trial and error,

$$T_4 = 2000\text{K} = 3600°\text{R}$$

7. Conditions at 4:

$$V_4 = 24.3728\ \text{ft}^3$$

$$T_4 = 2000\text{K} = 3600°\text{R}$$

$$P_4 = \frac{(0.0736)(94.56)(1545)(3600)}{(18.5)(4.76)(144)(24.373)} = 125\ \text{psi}$$

$$U\langle T_4\rangle = \sum_{i=1} N_i[\bar{h}_f^0 + \Delta\bar{h}\langle T_4\rangle - \overline{R}T_4]_i$$

$$= \left[\frac{(0.0736)(1.8001)}{(18.5)(4.76)}\right] \times \{[(12)(-94{,}054 + 21{,}857 - (1.987)(2000))]_{CO_2}$$

$$+ [(13)(-57{,}798 + 17{,}373 - (1.987)(2000))]_{H_2O}$$

$$+ [(69.56)(+13{,}418 - (1.987)(2000))]_{N_2}\}$$

$$= -1255\ \text{Btu/cyl}$$

8. Ideal indicated work per cylinder:

$$W_{net} = {}_1W_2 + {}_2W_3 + {}_3W_4 + {}_4W_1$$

with process i–j,

$$\delta Q - \delta W = dU$$

or

$${}_1W_2 = U<T_1> - U<T_2> = -74.34 - 274.44 = -348.78 \text{ Btu/cyl}$$

and

$${}_2W_3 = P_3(V_3 - V_2) = (609)(144)(5.023)/778 = 566.19$$

$${}_3W_4 = U<T_3> - U<T_4> = -395.5 + 1255 = 859.50 \text{ Btu/cyl}$$

$$W_{net} = -348.78 + 566.19 + 859.50 = 1076.9 \text{ Btu/cyl}$$

a. $W_{net} = (1076.9)(778) = 837{,}828 \text{ ft lbf/cyl}$

9. Ideal indicated mean effective pressure:

b. $$IMEP = \frac{W_{net}}{V_D} = \frac{837{,}828 \text{ ft lbf/cyl}}{(22.498 \text{ ft}^3)(144 \text{ in.}^2/\text{ft}^2)} = 259 \text{ psi/cyl}$$

10. Ideal indicated thermal efficiency:

$$\eta = \frac{W_{net}}{\dot{M}_{fuel} HV}$$

$$= \frac{(1076.9 \text{ Btu/cyl})(18.5)(4.76 \text{ lb mole air/lb mole fuel})}{(1{,}931{,}360 \times 1.8001 \text{ Btu/lb mole fuel})(0.0736 \text{ lb mole air/cyl})}$$

c. $\eta = 0.370 = 37.0\%$

11. Ideal indicated engine power:

$$\dot{W}_I = \frac{\overline{P} LANc}{33{,}000\, n}$$

$$= \frac{(259 \text{ psi/cyl})(55/12 \text{ ft/stroke})(\pi/4)(30 \text{ in.})^2(130 \text{ rev/min})(10 \text{ cyl})}{(33{,}000 \text{ ft lbf/hp min})(1 \text{ rev/power stroke})}$$

$$= 33{,}055 \text{ ihp}$$

12. Ideal indicated specific fuel consumption:

$$\dot{M}_{fuel} = \left[\frac{(0.0736 \text{ lb mole air/cyl})(170 \text{ lbm/lb mole})}{(18.5)(4.76 \text{ lb mole air/lb mole fuel})}\right] \times$$

$$\left[\left(130 \frac{\text{rev}}{\text{min}}\right)\left(1 \frac{\text{intake}}{\text{rev}}\right)(10 \text{ cyl})\right] = 184.7 \frac{\text{lbm fuel}}{\text{min}}$$

$$ISFC = \frac{\dot{M}_{fuel}}{\dot{W}_I} = \frac{(184.7 \text{ lbm fuel/min})(60 \text{ min/hr})}{(33{,}055 \text{ hp})}$$

d. $= 0.335 \text{ lbm/hp hr}$

Alcohol fuels have seen little commercial use in CI engines chiefly because of their high self-ignition temperatures and low cetane numbers. In addition, alcohols would require extremely high compression ratios to achieve autoignition using conventional CI engines. Alcohols also present potential lubrication problems but show promise of emissions reduction when used with suitable CI combustion systems.

11.6 ADVANCED SPARK- AND COMPRESSION-IGNITION COMBUSTION CONCEPTS

Advances are being made today to improve the fuel-engine interface characteristics of modern internal combustion engines, which will make the differences between SI and CI combustion machinery less distinct. The constant-volume SI combustion originally envisioned by Otto and constant-temperature CI combustion of Diesel have matured into a number of overlapping combustion concepts, including stratified charge, two-chamber, or direct-chamber, fuel-injected spark-ignited machinery, as well as homogeneous charge and spark-assisted compression-ignited IC engines. Higher-efficiency engines, which have a wider tolerance to a variety of fuel alternatives and which minimize formation of critical pollutants, is a major goal of the mobility propulsion industry. These programs will face many specific and diverse issues, including engine applications, economics, manufacturing technologies, advanced materials, and the use of electronics.

A growing and detailed knowledge of combustion chemistry and physics has encouraged many to pursue the goal of developing a generic engine, i.e., one that would burn a wide range of portable fuels, thereby removing a demand on the oil refining industry to produce both SI and CI fuels from the same portion of a barrel of crude oil. Others are pursuing the feasibility of new engines running on nonpetroleum fuels. Improvements in SI, CI, or new hybrid IC engines require that engineers understand the rudimentary principles of thermochemistry and energy conversion that are at work in both SI and CI combustion. In addition, future engineering efforts to reduce engine heat loss, to better model engine heat transfer and gas dynamics, and to develop regenerative engine technologies will necessitate including more details of chemical energy conversion physics.

Chapters 10 and 11 have attempted to provide an overview of the essential fundamentals of applied combustion in SI and CI engines. There are additional engineering aspects of this issue that must be understood in the design, fabrication, operation, and maintenance of these machines, including charge preparation and ingestion systems, intake and exhaust manifold and valve mechanisms, turbochargers and superchargers, along with piston and engine block cooling and lubrication.

PROBLEMS

11.1 Consider a CI thermodynamic cycle engine model consisting of the the following idealized processes: 1–2 isentropic compression from BDC to TDC; 2–3 isothermal heat addition; 3–4 isentropic expansion to BDC; 4–1 isometric heat rejection at BDC. Using an air standard constant specific heat analysis, develop expressions for (a) net work per unit mass, (b) heat addition per unit mass, (c) *IMEP*, and (d) thermal efficiency as a function of compression ratio and cutoff ratio $r_L \equiv v_3/v_2$.

11.2 Repeat Example 11.1 for (a) a monatomic gas, $\gamma = 1.67$, (b) $\gamma = 1.3$, and (c) $\gamma = 1.2$.

11.3 Repeat Example 11.1 but, instead, solve for the specific work for the three values of γ.

11.4 Repeat Example 11.1 but, instead, solve for the heat rejection for the three values of γ.

11.5 Repeat Example 11.1, but solve for the thermal efficiency for the same constant-pressure heating ratio, $P_3/P_1 = 50$, and heat addition per unit mass, $q_H/C_vT_1 = 10$.

11.6 Diesel engine combustion is measured in terms of the cutoff ratio, $r_L = v_3/v_2$ (volume at the end of heat addition, or fuel injection, to that at the beginning of heat addition, or fuel injection). Show that the indicated thermal efficiency of the diesel cycle can be expressed as

$$\eta = 1 - \frac{1}{r_c^{\gamma-1}} \left[\frac{r_L^{\gamma} - 1}{k(r_L - 1)} \right]$$

11.7 Use Problem 11.6 to determine the effect of varying the cutoff ratio r_L on thermal efficiency for a given diesel cycle with a compression ratio of 18:1. Calculate the thermal efficiency for cutoff ratios of: (a) 1.5, (b) 2, (c) 4, and (d) 6.

11.8 Repeat Problem 11.7, but solve for compression ratios of: (a) 12:1, (b) 14:1, and (c) 16:1.

11.9 Using the results of Problems 11.6, determine the effect of varying the cutoff ratio r_L for the specific work of a given diesel cycle with a compression ratio of 18:1. Calculate the specific work for cutoff ratios of: (a) 1.5, (b) 2, (c) 4, and (d) 6.

11.10 Repeat Problem 11.9, but solve for compression ratios of: (a) 12:1, (b) 14:1, and (c) 16:1.

11.11 Consider the design of a CI engine that has a maximum design limit of 1700°C and 55 atm. If the compression ratio is to be maintained at 16:1 and if ambient *STP* conditions are taken to approximate charge conditions at BDC prior to compression, calculate the following for cutoff ratios of 1.5, 3, and 6: (a) specific work ξ; (b) heat input; (c) temperature at the end of the constant-volume burn, K; and (d) thermal efficiency, %.

11.12 Several fuel alternatives for compression-ignition engine utilization are frequently suggested. Using Table 8.6 and assuming that the charge prior to compression is that of air at *STP,* estimate the critical compression ratios for: (a) semibituminous coal dust, (b) hydrogen, (c) methane, and (d) gasoline.

11.13 A limited-pressure diesel engine with a 16:1 compression ratio has a maximum pressure of 900 psi. The equivalent air-fuel ratio is 25:1, and the fuel has a lower heating value of 19,500 Btu/lbm. Intake temperature and pressure are 60°F and 29.92 in.Hg, respectively. Using an open system variable specific heat air standard analysis, calculate (a) the residual fraction; (b) temperature at BDC at the beginning of compression; and (c) thermal efficiency, %.

11.14 A four-stroke, eight-cylinder diesel engine connects directly to an AC generator developing 150 kW full-load generator output (generator efficiency is 91%). The 14-cm × 18-cm engine, running at 1200 rev/min, uses a 35°API diesel fuel having an ultimate analysis of 84% carbon and 16% hydrogen. Dry exhaust gas analysis yields 9.593% CO_2 and 7.085% O_2. The full-load engine volumetric efficiency is 93%, with an air inlet temperature of 27°C, an exhaust gas temperature of 127°C, and a fuel temperature of 25°C. For these engine test data, find the following: (a) air mass flow rate, kg/hr; (b) brake *AF* ratio; (c) brake mean effective pressure, kPa; (d) brake specific fuel consumption, g/kW hr; and (e) brake thermal efficiency, %.

11.15 A four-stroke 4.75-in. × 6.5-in. diesel engine generates the following test data while running at 1160 rpm:

Dynamometer load	120 lbf
Brake arm	1.75 ft
Fuel consumption	21 lbm/hr
Friction horsepower	57.3 lbf
Mass *AF* ratio	13.5:1
Fuel oil specific gravity	0.82 at 60°F
Ambient conditions	14.7 psi and 68°F

Using these data, determine the following: (a) heating value of the fuel, Btu/lbm; (b) engine *BHP, FHP,* and *IHP*; (c) engine mechanical efficiency, %; (d) brake specific fuel consumption, lbm/hp hr; (e) brake mean effective pressure, psia; (f) engine brake torque, ft lbf; and (g) brake engine thermal efficiency, %

11.16 A low-speed, two-stroke research diesel engine operating at full-load conditions burns a stoichiometric mixture of liquid methanol injected into air. The blower supplies 100% scavengine air at 100°F and 15 psi at the beginning of the compression stroke. The compressed air temperature required for compression ignition is 1650°R. Using a variable specific heat air cycle analysis, calculate (a) ideal compression ratio; (b) peak pressure, psi; (c) peak temperature, °R; (d) net work, ft lbf/lbm; and (e) thermal efficiency, %

11.17 An eight-cylinder diesel engine burns #2 fuel oil and requires 200% theoretical air. The fuel, API<60°F>= 32, has an ultimate analysis of 86.4% carbon, 12.7% hydrogen, 0.4% sulfur, 0.3% oxygen, and 0.2% nitrogen. The engine is rated at a maximum brake power output of 5220 bhp and a brake specific fuel consumption of 0.284 lbm/hp hr. For these conditions, calculate: (a) mass *FA* ratio, lbm fuel/lbm air; (b) fuel density, lbm/ft^3; (c) fuel lower heating value, Btu/lbm; (d) fuel consumption, gal/hr; and (e) engine thermal efficiency, %

11.18 A prototype diesel is designed to burn methanol, CH_3OH, in 45% excess air. The conditions after compression of the air are given as 270 psi and 1250°R. Constant-pressure combustion occurs after compression ignition of the fuel in air, causing the gas mixture temperature to be tripled. For this engine, assuming ambient *STP* conditions, determine: (a) compression ratio, (b) cutoff ratio, and (c) exhaust gas dew point temperature, °F. Assume complete combustion.

11.19 A four-cylinder, 12-in. × 25-in., single-acting four-cycle gas diesel engine operates at 225 rpm and is designed to burn a variety of gaseous fuels, including: natural, biomass, town, and coal gas fuels. The engine is found to have a 78% mechanical efficiency, a volumetric efficiency of 90%, and an indicated thermal efficiency of 30% when burning methane in 30% excess air. Assuming ambient conditions of *STP,* calculate (a) brake power output, hp; (b) fuel gas consumption, ft^3/h; (c) *BMEP,* psi; and (d) volume of exhaust gas, assuming 840°F and 14.5 psi, ft^3/hr.

11.20 An expression for estimating the ignition delay in CI engines has been reported by Wolfer as

$$t<\text{msec}> = \frac{0.44 \exp\{4650/T<\text{K}>\}}{P<\text{atm}>^{1.19}}$$

Using this relationship, estimate the ignition delay for a diesel having the following conditions at the beginning of injection: a cylinder pressure of 48 atm and charge temperature of (a) 500K, (b) 1000K, (c) 1500K, and (d) 2000K.

11.21 Repeat Problem 11.20 for a charge temperature of 1500K but a charge pressure of (a) 15 atm, (b) 25 atm, (c) 35 atm, and (d) 45 atm.

11.22 A two-stroke diesel engine burns 3.0 lbm/min of fuel. Assuming an ideal 5° crank angle period of constant fuel injection rate, estimate the time available, sec, and injection rate, lbm/°, for an engine speed of (a) 500 rpm, (b) 1000 rpm, (c) 2000 rpm, and (d) 4000 rpm.

11.23 An open-chamber diesel engine of bore D, clearance $D/10$, and stroke $2D$ has a half-spherical volume of diameter $D/2$ centered in the face of the piston. For these conditions, determine: (a) the compression ratio, (b) the surface to volume ratio, and (c) the ratio of the bowl volume to the clearance volume. Calculate the required clearance in terms of D for an open chamber of bore D having the same compression ratio but without the bowl. Repeat parts (a), (b), and (c) for this geometry. (Assume that $D = 5$ in.)

11.24 A divided-chamber diesel engine of bore D, clearance $D/10$, and stroke $2D$ has a spherical prechamber of diameter $D/4$. For this geometry, determine the following: (a) compression ratio, (b) surface-to-volume ratio, and (c) ratio of the prechamber volume to the main chamber volume. Find the required clearance in terms D for a simple open chamber of bore D having the same compression ratio. but without the prechamber. Repeat parts (a), (b), and (c) for this configuration.

11.25 An eight-cylinder diesel engine burns #2 fuel oil and 200% theoretical air. The fuel, API<60> = 32, has an ultimate analysis of 86.4% carbon, 12.7% hydrogen, 0.4% sulfur, 0.3% oxygen, and 0.3% nitrogen. The engine is rated at a maximum brake power output of 5220 hp and a brake specific fuel consumption of 0.284 lbm/hp hr. For these conditions, calculate: (a) FA ratio; (b) fuel density, lbm/ft^3; (c) LHV of the fuel, Btu/lbm; (d) fuel consumption, gal/hr; and (e) brake thermal efficiency, %.

12

Gas Turbine Engine Combustion

12.1 INTRODUCTION

Chapters 10 and 11 discussed essential features of intermittent combustion that occur in spark- and compression-ignition IC engines. Some nineteenth-century inventors even put forth efforts to try to adapt the highly explosive nature of gunpowder ignition and knowledge of gun manufacturing to IC engine development, an early illustration of technology transfer. Another important class of gas-power engines includes continuous fluid flow machines such as the basic gas turbine unit consisting of separate compressor-combustor-turbine components; see Figure 12.1. Windmills are considered by many to be forerunners of today's power turbomachines. Historically, windmills were used by different cultures throughout the world, including that of ancient China. Hero of Alexandria, Greece, constructed a steam turbine device that was driven by hot flue gases rising from an open fire. The idea of an efficient power-generating gas turbine machine was envisioned during the Age of Steam, but the earliest recorded patent for a gas turbine was filed by the Englishman John Barber in 1791.

During the nineteenth century, a number of thermodynamic cycles and hot-air piston engines, such as those proposed by Ericsson and Joule, preceded gas turbine engines. Later, some early naval torpedoes were propelled by simple gas turbine devices. By the late 1930s, successful elements needed for an economical gas turbine began to become available. Useful

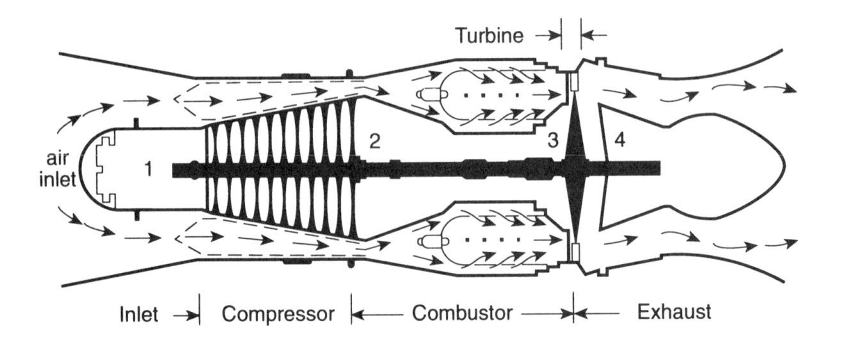

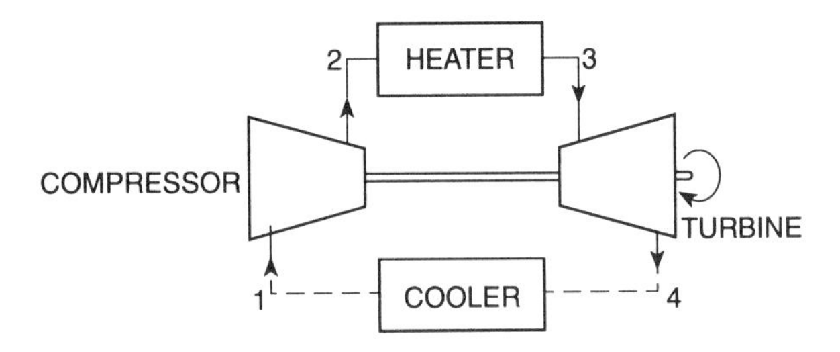

Figure 12.1 Basic gas turbine power system.

propulsion machinery has resulted from an availability and compatibility of both suitable fuels and engines, a thesis repeatedly stated throughout this text. In addition, an application or need for particular fuel-engine combinations must exist, manned flight provided an impetus for gas turbine engine development.

The first practical aircraft gas turbine engines appeared during the 1940s, although experimental machinery was operated a decade earlier. Sir Frank Wittle was a leader in developing aircraft turbojet propulsion engines. Work begun in the early 1930s resulted in his W1, a flying engine that operated in 1941 and produced 634 N (859 ft) of thrust. His original unit was built by Rover but, by 1943, Rover was taken over by Rolls-Royce. Several American and German companies also developed and built early jet engines;

however, only a few were operational by the end of World War II. Military use of jet engine aircraft was common by the end of the Korean War, and the first commercial use of a jet engine occurred in 1953. It was not until the 1960s, however, that jet engines entered into satisfactory commercial use. By the 1970s, aircraft-derivative gas turbines began to find a variety of industrial applications, including marinized aircraft designs for ship propulsion and Hovercraft applications. Limited use in railroad and automotive applications, most notably Rover and Chrysler, were attempted in the late 1950s.

Many factors helped to slow development of a reliable gas turbine, including the low compressor and turbine component efficiencies of early designs, lack of suitable materials able to withstand locally high temperatures in hot sections resulting from continuous combustion, and unacceptably large total stagnation pressure loss across the combustor. Worldwide gas turbine industries today manufacture civilian as well as military power plants for air, land, and sea applications. Engineering publications and standards are available that support all professional aspects relevant to the science and technology of gas turbine components, including compressors, combustors, turbines, intercoolers, reheaters, and/or regenerators. Reference material also deals with various power and thrust configurations of gas turbine engines, such as turboshaft, turboprop, turbofan, electric power generators, and turbochargers.

Chapter 12 will cover only essential and important fuel-engine energetic characteristics of gas turbine engines; see Figure 12.2. Material discussion, therefore, will focus primarily on the major combustion component, the *combustor.* This chapter will be limited to the following issues: the thermochemical nature of the reactive mixture, general gas turbine fuel characteristics, basic combustion processes, and related emissions problems generally associated with gas turbine operation, as well as basic combustor design. A brief description of two unique continuous-combustion power systems, the free piston and Stirling engines, will be reviewed as well.

12.2 THERMODYNAMICS AND GAS TURBINE ENGINE MODELING

The Brayton cycle is a zeroth-order thermodynamic model for an ideal basic gas turbine engine. An air standard Brayton cycle, similar to IC engine cycle developments found in Chapters 10 and 11, describes steady-state and steady-flow energy characteristics of turbomachinery components in terms of the following ideal processes:

1–2 Isentropic compression
2–3 Constant-pressure heat addition

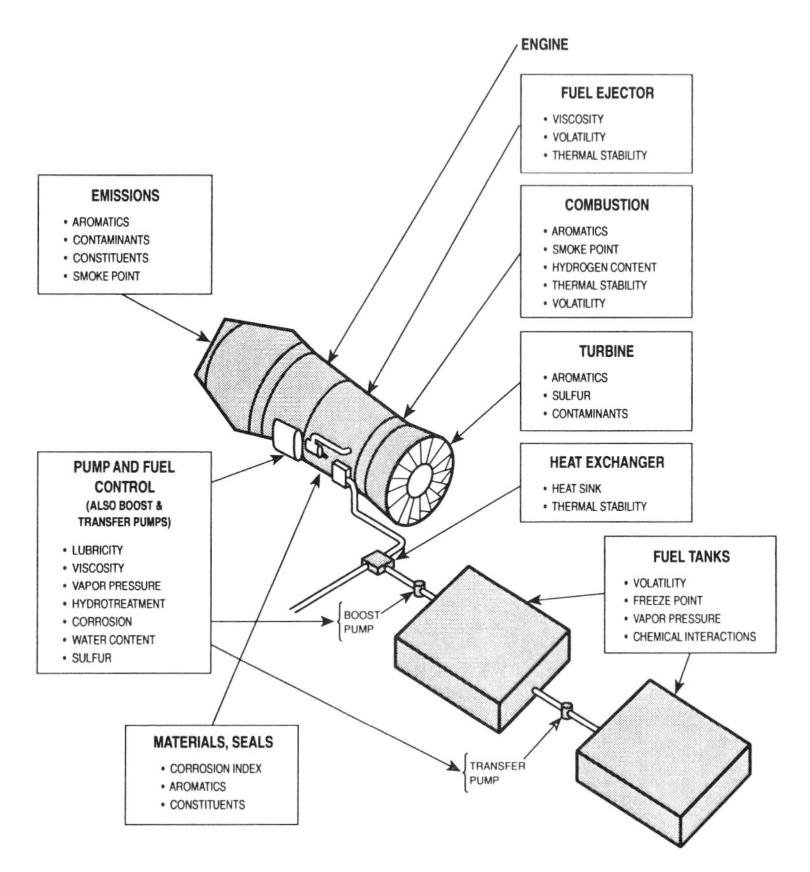

Figure 12.2 Basic gas turbine fuel-engine interface.

3–4 Isentropic expansion
4–1 Constant-pressure heat rejection

Now, the Brayton cycle has processes identical to those Joule proposed in the late nineteenth century for his two-cylinder, reciprocating piston-cylinder, positive-displacement hot-air engine. The historical success of a gas turbine and the failure of the Joule piston-cylinder engine illustrate a basic principle of successful power machinery design: Postulate achievable thermodynamic events that satisfy principles of energy conservation and conversion; then, develop actual hardware that can successfully achieve those postulated thermodynamic events.

A dimensionless Brayton cycle formulation can be expressed based on the following dimensionless parameters:

$\tau = T_3/T_1$ = maximum cycle temperature ratio
$r_p = P_2/P_1$ = pressure ratio (note $r_p \neq r_v$)
$\beta = T_3/T_2$ = constant-pressure heating temperature ratio
$\rho = T_4/T_1$ = constant-pressure cooling temperature ratio
η = indicated thermal efficiency

Using a constant specific heat analysis and the dimensionless relationships shown above, the following equations are valid. Note that, from 1 → 2, $Pv = RT$ and Pv^γ = const, or

$$T_2 = T_1 r_p^{(\gamma-1)/\gamma} \tag{12.1}$$

and

$$T_3 = \beta T_2 = \beta T_1 r_p^{(\gamma-1)/\gamma} = \tau T_1 \tag{12.2}$$

For an ideal Brayton cycle, $P_3 = P_2$ and $P_4 = P_1$, which means

$$P_4/P_3 = P_1/P_2 = 1/r_p \tag{12.3}$$

and

$$T_4 = (1/r_p)^{(\gamma-1)/\gamma} T_3 = \beta(1/r_p)^{(\gamma-1)/\gamma} T_2 = \rho T_1 \tag{12.4}$$

Net cycle work per unit mass of air flowing through the open compression and expansion components during steady-state and steady-flow operation is expressed as

$$\begin{aligned} w_{\text{net}} &= {}_1w_2 + {}_3w_4 = h_1 - h_2 + h_3 - h_4 \\ &= C_p(T_1 - T_2 + T_3 - T_4) \end{aligned} \tag{12.5}$$

or, in dimensionless form,

$$\xi = \frac{w_{\text{net}}}{C_p T_1} = \left[1 - r_p^{(\gamma-1)/\gamma} + \tau - \rho\right] \tag{12.5a}$$

External heat addition per unit mass flowing, q_H, is

$$q_H = {}_2q_3 = C_p[T_3 - T_2] = C_p[\tau - r_p^{(\gamma-1)/\gamma}]\, T_1 \tag{12.6}$$

and

$$q_H/C_p T_1 = \left[\tau - r_p^{(\gamma-1)/\gamma}\right] \tag{12.6a}$$

Indicated thermal efficiency η is defined as

$$\eta = \frac{w_{net}}{q_H} = \frac{1 - r_p^{(\gamma-1)/\gamma} + \tau - \rho}{\tau - r_p^{(\gamma-1)/\gamma}} \tag{12.7}$$

$$= \frac{\tau - r_p^{(\gamma-1)/\gamma} + 1 - \rho}{\tau - r_p^{(\gamma-1)/\gamma}}$$

$$= 1 + \frac{1 - \rho}{\tau - r_p^{(\gamma-1)/\gamma}}$$

Substituting Equation (12.2) for τ gives the expression

$$\eta = 1 + \frac{1 - \rho}{\beta r_p^{(\gamma-1)/\gamma} - r_p^{(\gamma-1)/\gamma}}$$

and

$$= 1 + \frac{1 - \rho}{r_p^{(\gamma-1)/\gamma}[\beta - 1]}$$

or

$$\eta = 1 - \frac{1}{r_p^{(\gamma-1)/\gamma}}\left[\frac{1 - \rho}{1 - \beta}\right]$$

Recall that $P_3 = P_2$ and $P_4 = P_1$ or $P_3/P_4 = P_2/P_1 = r_p$; then,

$$\frac{T_3}{T_4} = \frac{T_2}{T_1} = r_p^{(\gamma-1)/\gamma} \tag{12.8}$$

or

$$\frac{T_3}{T_2} = \frac{T_4}{T_1} \tag{12.9}$$

and, therefore,

$$\beta = \rho \tag{12.9a}$$

This allows a Brayton cycle thermal efficiency to be written as

$$\eta = 1 - \frac{1}{r_p^{(\gamma-1)/\gamma}} \tag{12.10}$$

EXAMPLE 12.1 Using the dimensionless Brayton cycle relations, determine the dimensionless performance parameters as a function of pressure ratio for: (a) fixed peak temperature ratio, $\tau = 10$; (b) fixed dimensionless net work per unit mass, $\xi = 4.0$; and (c) fixed dimensionless heat addition per unit mass, $q_H/C_pT_1 = 8.0$ For each of these conditions, determine the corresponding values of τ, ξ, and q_H/C_pT_1. Assume that $\gamma = 1.4$

Solution

1. Dimensionless Brayton cycle relationships:

$$\xi = [1 - r_p^{(\gamma-1)/\gamma} + \tau - \rho]$$
$$\rho = \beta = \tau / r_p^{(\gamma-1)/\gamma}$$
$$q_H/C_pT_1 = [\tau - r_p^{(\gamma-1)/\gamma}]$$
$$\eta = 1 - 1/r_p^{(\gamma-1)/\gamma}$$

2. Fixed peak temperature ratio, $\tau = 10$:

a.

r_p	ξ	q_H/C_pT_1	η
2	[a]1.5776	[a]8.7810	[a]0.1797
3	2.3253	8.6313	0.2694
4	2.7845	8.5140	0.3270
5	3.1023	8.4162	0.3686
6	3.3381	8.3315	0.4007
7	3.5212	8.2564	0.4265
8	3.6681	8.1886	0.4480
9	3.7888	8.1266	0.4662
10	3.8898	8.0693	0.4821

[a] $\xi = [1 - 2^{0.4/1.4} + 10(1 - ½^{0.4/1.4})]$
$= 1.5776$
$q_H/C_pT_1 = [10 - 2^{0.4/1.4}] = 8.7809$
$\eta = 1 - ½^{0.4/1.4} = 0.1719$

3. Fixed dimensionless net work per unit mass, $\xi = 4$:

b.

r_p	τ	q_H/C_pT_1	η
2	[b]23.4827	[b]22.2637	[b]0.1797
3	16.2165	14.8478	0.2694
4	13.7165	12.2305	0.3270
5	12.4353	10.8514	0.3686
6	11.6520	9.9835	0.4007
7	11.1226	9.3790	0.4265
8	10.7409	8.9295	0.4480
9	10.4530	8.5796	0.4662
10	10.2285	8.2979	0.4821

[b] $\tau = \dfrac{4 - 1 + 2^{0.4/1.4}}{(1 - ½^{0.4/1.4})} = 23.4827$
$q_H/C_pT_1 = [23.4827 - 2^{0.4/1.4}]$
$= 22.2037$
$\eta = 1 - ½^{0.4/1.4} = 0.1797$

4. Fixed dimensionless heat addition per unit mass, $q_H/C_pT_1 = 8$:

c. r_p	τ	ξ	η
2	ᶜ9.2190	ᶜ1.4373	ᶜ0.1797
3	9.3687	2.1552	0.2694
4	9.4860	2.6164	0.3270
5	9.5838	2.9489	0.3686
6	9.6685	3.2053	0.4007
7	9.7436	3.4119	0.4265
8	9.8114	3.5836	0.4480
9	9.8734	3.7298	0.4662
10	9.9307	3.8564	0.4821

ᶜ$\tau = 8.2^{0.4/1.4} = 9.2190$
$\xi = [1 - 2^{0.4/1.4} + 9.2190(1 - ½^{0.4/1.4})]$
$= 1.4373$
$\eta = 1 - ½^{0.4/1.4} = 0.1797$

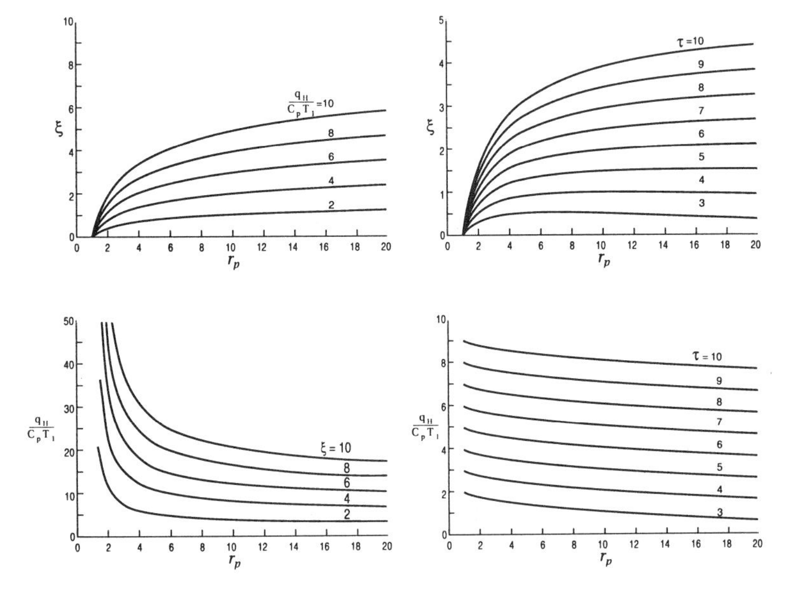

Figure 12.3 Ideal air standard Brayton cycle thermodynamic characteristics.

Comments This problem illustrates how the dimensionless relationships can be used to do a parametric analysis of a basic gas turbine engine. Modeling can be easily compared to Otto, diesel, and dual cycle predictions using material developed in Chapters 10 and 11.

The previous relationships can be readily utilized during any initial analysis of a basic gas turbine, as was done for both SI and CI engines in Chapters 10 and 11. Figure 12.3 shows dimensionless characteristics of an ideal constant specific heat air standard basic Brayton cycle. Thermodynamic modeling can be expanded to include both variable specific heats and a fuel-air cycle analysis. Adding compressor and turbine component efficiencies, as well as a combustion efficiency, will bring these simple thermodynamic predictions more in line with the actual energetics observed during turbomachinery operation. Modifying any basic engine configuration by adding more compressors, turbines, intercoolers, regenerators, and afterburners will impact the gas turbine fuel-engine interface but is not considered in this text.

12.3 GAS TURBINE FUEL THERMOCHEMISTRY

Theoretically, gas turbine engines can burn a wide variety of fuel alternatives. Commercial gas turbine fuels, much like current standard IC engine fuels, are based on specific hydrocarbon crude oil reserves, with turbine fuels coming from crude components that fall between gasoline and power kerosene portions of the distillation curve; see Chapter 5. Many of the fuel issues discussed in Chapters 10 and 11 are also relevant to the proper operation of gas turbine machinery; however, aviation gas turbine fuels, in addition, should specifically meet the following requirements:

Pump easily, and ensure good flow at all times
Maintain highly efficient combustion
Establish a high heat release rate
Minimize fire hazards associated with fuel handling
Permit easy burner ignition and good relight characteristics
Provide adequate lubrication to moving parts of the fuel system
Prohibit corrosive effects on fuel system components
Minimize harmful effects on the combustor and turbine

An understanding of specific thermochemical properties of gas turbine fuels can help engineers address most of the issues listed above.

Viscosity, for example, impacts both liquid fuel atomization and dispersion processes, essential to complete combustion within any given burner. In addition, pumping characteristics of jet fuel depend on viscosity, and sufficient fluid flow should be maintained down to −50°C (−111°F), the waxing point temperature restriction for most aircraft fuels. This requirement arises from the very low environmental temperatures as well as heat transfer losses from aircraft fuel tanks and fuel lines encountered during flight.

Fuel temperatures, such as pour point, flash point, and self-ignition, are also significant to proper and efficient combustion. Operational considerations and safety constraints will require a minimum flash point to preclude potential fires and/or explosions. Fuel temperature in aviation applications also depends on altitude, rate of climb, duration at altitude, and kinetic heating due to flight speed. Limiting fuel boil-off during rapid climbing, as

EXAMPLE 12.2 A marine split-shaft gas turbine has a rated brake thermal efficiency of 23.7% while developing an indicated power output of 15,000 ihp at full load. The unit burns a fuel oil having a specific gravity of 0.92 (at 60°/60°F). The compressor volumetric flow rate measured at the inlet of 64°F and 14.5 psi is 1.16×10^5 ft^3/min. If the combustor has a 95% efficiency and the power plant has a 90% mechanical efficiency, calculate (a) the fuel consumption rate, gal/hr; (b) the *FA* ratio, lbm fuel/lbm air; (c) the brake specific fuel consumption, lbm/bhp hr; and (d) the minimum tank capacity required for a 10,000-mi. range operating at a constant speed of 20 knots (1 knot = 1.151 mph), gal.

Solution:

1. Brake thermal efficiency:

$$\eta = \frac{\text{desired energy output}}{\text{required energy input}} = \left(\frac{\dot{W}_{\text{net}}}{\dot{Q}_{\text{add}}}\right)_{\text{actual}}$$

or

$$\dot{Q}_{\text{add}} = \frac{(15{,}000\text{ hp})\,(2544\text{ Btu/hp hr})}{(0.237)}$$

$$= 1.610 \times 10^8 \text{ Btu/hr}$$

$$= 2.68 \times 10^6 \text{ Btu/min}$$

2. Combustor efficiency:

$$\eta_{\text{comb}} = \frac{\dot{Q}_{\text{add}}}{\dot{m}_{\text{fuel}} LHV}$$

$$\dot{m}_{\text{fuel}} = \frac{2.68 \times 10^6 \text{ Btu/min}}{(0.95)(LHV\text{<Btu/lbm fuel>})}$$

From Table 5.3,

$$SG = 0.92 \qquad \rho = 7.64 \text{ lbm/gal} \qquad LHV = 18{,}030 \text{ Btu/lbm}$$

$$\dot{m}_{\text{fuel}} = \frac{(2.68 \times 10^6 \text{ Btu/min})}{(0.95)\,(18{,}030 \text{ Btu/lbm})} = 156.46 \text{ lbm/min}$$

$$\dot{V}_{\text{fuel}} = \frac{\dot{m}_{\text{fuel}}}{\rho_{\text{fuel}}} = \frac{156.46 \text{ lbm fuel/min}}{7.64 \text{ lbm fuel/gal}}$$

a. $\dot{V}_{\text{fuel}} = 20.5 \text{ gal/min} = 1230 \text{ gal/hr}$

3. Mass *FA* ratio:

$$FA = \frac{\dot{m}_{\text{fuel}}}{\dot{m}_{\text{air}}} = \left(\frac{RT}{P\dot{V}}\right)_{\text{air}} \dot{m}_{\text{fuel}}$$

$$FA = \frac{(53.34 \text{ ft lbf/lbm air °R})\,(524\text{°R})\,(156.46 \text{ lbm fuel/min})}{(14.5 \text{ lbf/in.}^2)\,(144 \text{ in.}^2/\text{ft}^2)\,(1.16 \times 10^5 \text{ ft}^3/\text{min})}$$

b. $FA = 0.0181 \text{ lbm fuel/lbm air}$

4. Brake specific fuel consumption:

$$BSFC = \frac{\dot{m}_{\text{fuel}}}{\left(\dot{W}_{\text{net}}\right)_{\text{brake}}} = \frac{\dot{m}_{\text{fuel}}}{\left(\dot{W}_{\text{net}}\right)_{\text{indicated}} \times \eta_{\text{mech}}}$$

$$= \frac{(156.46 \text{ lbm fuel/min})\,(60 \text{ min/hr})}{(15{,}000 \text{ hp})\,(0.90)}$$

c. $BSFC = 0.695 \text{ lbm/hp hr}$

5. Tank capacity:

$$V_{\text{tank}} = \frac{(1230 \text{ gal/hr})\,(10{,}000 \text{ mi.})}{(20 \text{ knots})\,(1.151 \text{ mph/knot})}$$

d. $V_{\text{tank}} = 534{,}000 \text{ gal}$

required for many military aircraft, can be a formidable challenge. When rapid boiling does occur, as it does with a wide-cut gasoline distillate fuel, vapor locking of the engine's fuel system may result. Gas turbine fuel temperature will not vary as greatly in land and/or marine applications and is therefore less of an issue in those operational instances.

Effective utilization of any gas turbine fuel is also influenced by both its latent heat and heat of combustion. For an aircraft with fixed fuel tank volume and specifying range as a limiting design factor, one should use a fuel with an energy density per unit volume as high as possible to obtain the greatest energy from a unit volume of fuel. However, for cases in which aircraft payload is the limiting factor, fuel energy density per unit mass should be as high as possible to achieve the greatest energy from a minimum weight of fuel.

The ability of turbine fuels to vaporize, i.e., their volatility, is critical, especially at the low temperatures often encountered in jet engine operation. Aircraft turbine engines can burn either a kerosene-base or wide-cut gasoline-base distilate hydrocarbon fuel with the distinctions between the two categories being basically in their volatility. Kerosene fuels have distillation-controlled volatility curves and fuel flash points that allow these fuels to have low vapor pressures and thus to boil only at extremely high altitudes. The wide-cut gasoline-type fuels have distillation-controlled volatility curves with Reid vapor pressures that ensure that fuel will boil off at much lower altitudes.

Gas turbine fuel suitability also requires there be low levels of water, sulfur, and sodium concentrations to minimize corrosion. In addition, minimal mineral materials, such as phosphorous, vanadium, and ash, must be maintained as well since they can clog or be abrasive to fuel components such as fuel-oil pumps and fuel nozzles, as well as erode power-generating machinery such as turbine blading. Vanadium pentoxide, V_2O_5, and sodium sulfate, Na_2SO_4, are produced in the highly oxidizing gas turbine exhaust and will cause fluxing of all protective oxide films covering hot steel components. These components, if cooled, produce sticky, insoluble deposits on turbine blading. Turbine fuels, when used in aircraft applications, also require precise concentrations of gum and rust inhibitors, as well as anti-icing, antistatic, and antismoke additives to make them more compatible.

Early turbojet fuel development was supported by the military, who helped to establish the JP specifications for those fuels. Table 12-1 lists properties of JP fuels. These fuels are similar in many characteristics but differ dramatically in their volatility, i.e., boiling point range. Commercial jet aircraft utilize fuel specifications set by ASTM. Three ASTM aircraft turbine fuel grades exist and include Jet-A Aviation Turbine Fuel, a kerosene fuel similar to JP-5, with a flash point of 66°C (150°F) and a freezing point of −40°C (−40°F). Jet A-1 fuel is used when turbine operation requires a lower freezing point temperature, −50°C (−58°F). Jet B is commercial JP-4.

Kerosenes, with their low viscosities and medium volatilities, do provide both efficient atomization and combustion over a wide range of inlet air temperatures, pressures, and velocities and therefore make good jet fuels. Land turbine units can burn a range of fuel phases as well as types, including natural gas, propane, blast-furnace gas, butane, petroleum distillates, residual fuel oil, and alcohols. Continuous combustion allows less expensive gas oils and diesel fuels to be burned in industrial and marinized aircraft-derivative engines. Burning heavy residual fuel oils in gas turbines does result in some negative effects:

Heating required prior to atomizing this highly viscous fuel
Tendency to polymerize and form tar or sludge when overheated
Incompatible with other hydrocarbon oils and potential for forming jellylike substances that can clog fuel systems

Table 12.1 Thermochemical Properties of Aviation Turbine Fuels[a]

	AV gas	(Jet B) JP-4	(Jet A) JP-8	JP-5	JP-7
Distillation					
°C	40–148	61–239	167–266	182–258	189–251
°F	103–280	141–462	332–510	359–495	372–483
Reid vapor					
At 38°C, kPa	46	18	1.4	0.6	18.6
At 100°F, psi	6.7	2.6	0.2	0.09	2.7
SG max, API	51 (0.7753)	57 (0.7507)	51 (0.7753)	48 (0.7880)	50 (0.7790)
SG min, API	39 (0.8299)	45 (0.8017)	39 (0.8299)	36 (0.8450)	44 (0.8060)
Min flash point					
°C	—	—	43	66	60
°F	—	—	110	150	140
Max flash point					
°C	—	—	66	66	70
°F	—	—	150	150	158
1 atm autoignition					
°C	443	246	238	241	241
°F	829	474	460	465	465
Heat of combustion					
kJ/kg	44,150	43,570	43,240	43,050	43,682
Btu/lbm	18,980	18,730	18,590	18,508	18,780
MJ/m^3	31,130	33,190	35,060	35,200	39,700
Btu/gal	111,750	119,102	125,785	126,320	142,494

[a]Data from the *Handbook of Aviation Fuel Properties*, CRC Rept #530, 1980

High carbon content fuel, producing excessive combustion chamber carbon deposits
Excessive vanadium, alkali metals, and ash, which impact on the reliability and availability of the hot section of the plant

EXAMPLE 12.3 Material constraints require that gas turbine combustion can exhaust temperatures be limited to 2000°F. Consider a gas turbine combustor burning a lean *n*-pentane–air mixture. Compressed air enters the can at 90.5 psia and 440°F while the fuel is injected at 77°F. For an exhaust gas temperature of 1880°F, estimate (a) the required mass fuel-air ratio; (b) excess air, %; (c) the exhaust gas molecular weight and specific heat ratio; and (d) exhaust gas specific volume, ft^3/lbm gas.

Solution:

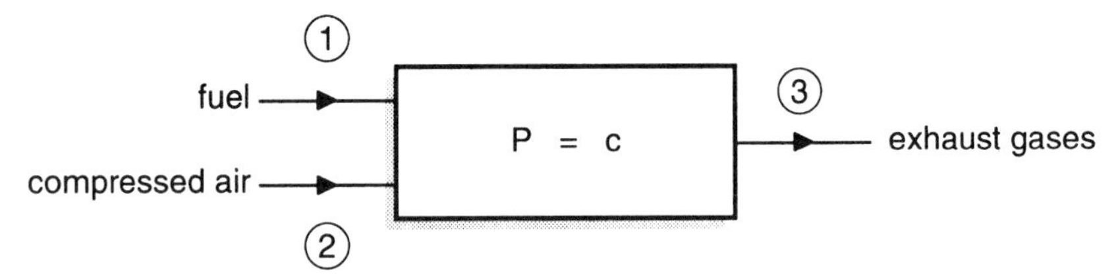

1. Stoichiometric equation:

$$C_5H_{12} + 8O_2 \rightarrow 5CO_2 + 6H_2O$$

2. Actual equation, assume excess air and complete combustion:

$$C_5H_{12} + 8a[O_2 + 3.76N_2] \rightarrow bCO_2 + cH_2O_g + dO_2 + 30.08aN_2$$

Carbon atom balance:

$$b = 5$$

Hydrogen atom balance:

$$12 = 2c \qquad c = 6$$

Oxygen atom balance:

$$16a = 2b + c + 2d \qquad d = 8[a - 1]$$

3. Energy equation, $P = c$ adiabatic combustion:

$$\sum_{jprod} N_j[\bar{h}_f^0 + \{\bar{h}<T_j> - \bar{h}<T_0>\}]_j = \sum_{ireact} N_i[\bar{h}_f^0 + \{h<T_i> - \bar{h}<T_0>\}]_i$$

where

$$\sum_{jprod} N_j\bar{h}_j = 5[\bar{h}_f^0 + \Delta\bar{h}<T_3>]_{CO_2} + 6[\bar{h}_f^0 + \Delta\bar{h}<T_3>]_{H_2O_g}$$
$$+ 8(a - 1)[\cancel{\bar{h}_f^0} + \Delta\bar{h}<T_3>]_{O_2} + 30.08a[\cancel{\bar{h}_f^0} + \Delta\bar{h}<T_3>]_{N_2}$$

and

$$\sum_{i\,react} N_i\bar{h}_i = 1[\bar{h}_f^0 + \Delta\bar{h}\langle \cancel{T_1}\rangle]_{C_5H_{12}} + 8a[\cancel{\bar{h}_f^0} + \Delta h\langle T_2\rangle]_{O_2}$$
$$+ 30.08a[\cancel{\bar{h}_f^0} + \Delta\bar{h}\langle T_2\rangle]_{N_2}$$

$$T_1 = T_0 = 537°R \qquad T_2 = 900°R \qquad T_3 = 2340°R$$

Using JANAF data found in Appendix B,

$$5[-94{,}054 + 11{,}988]_{CO_2} + 6[-57{,}798 + 9298]_{H_2O_g}$$
$$+ 8(a - 1)\,[7971]_{O_2} + 30.08a[7529]_{N_2}$$
$$= [-41{,}890]_{C_5H_{12}} + 8a[1455]_{O_2} + 30.08a[1413]_{N_2}$$
$$a = 3.06$$

4. Mass air-fuel ratio:

$$C_5H_{12} + (8)\,(3.06)\,[O_2 + 3.76N_2] \rightarrow \text{products}$$

or

$$FA = \frac{(1.0)\,(72.151)\text{ lbm fuel}}{(8)\,(3.06)\,(4.76)\,(28.97)\text{ lbm air}}$$

a. $FA = 0.0214$ lbm fuel/lbm air

5. Excess air, %;

stoichiometric air:

$$(8)\,(4.76) = 38.08$$

Actual air:

$$(8)\,(4.76)\,(3.06) = 116.52$$

Theoretical air = 306%

b. Excess air = 206%

6. Product molecular weight:

$$MW = \Sigma\bar{x}_i MW_i$$

where

$$\bar{x}_i = \frac{N_i}{N_{tot}}$$

$$N_{CO_2} = 5$$
$$N_{H_2O} = 6$$
$$N_{O_2} = 16.48$$
$$N_{N_2} = 92.04$$

$$N_{tot} = 119.52$$

$$MW = \frac{1}{119.52} [(5)(44) + (6)(18) + (16.48)(32) + (92.04)(28)]$$

c. $MW_{prod} = 28.72$ lbm/lb mole

7. Exhaust gas specific volume:

$$P V = m R T$$

$$v = \frac{(1545 \text{ ft lbf/lb mole °R})(2340\text{°R})}{(90.5 \text{ lbf/in.}^2)(144 \text{ in.}^2/\text{ft}^2)(28.72 \text{ lbm/lb mole})}$$

d. $v = 9.659$ ft^3/lbm gas

Gas turbine combustors also have been proposed to burn solid fuels, such as coal and wood. Techniques for burning solid fuels can include direct combustion, slurry combustion, fluidized bed combustion, and indirect combustion. These methods were discussed in Chapter 4 when solid fuels were reviwed and materials were presented that covered their use in systems such as steam generators. Gas turbine combustors designed to burn fuels such as coal face significant fuel preparation problems, such as pulverizing coal and introduction of coal powder into high-pressure and high gas flow rates. Note also that these solid fuels must burn completely during the short time they reside within a burner in order to ensure that no tar depositing occurs during gas expansion through the turbine. Solid fuel burners for gas turbines may require the development of a solid-liquid slurry combustion, or indirect gasification. Indirect gasification would use efficient heat exchangers to supply energy to the airstream instead of the combustors and thus permit solid fuels to burn separately, thereby protecting turbine sections.

Synthetic liquid fuels derived from coal, shale oil, and tar sands may serve as long-term gas turbine fuel replacements for the crude oil–derived fuels currently in use. Alcohol fuels, most probably methanol, may also find some future use in land and/or marine gas turbine engines. Aircraft use would be difficult because of the constant-temperature boiling characteristics of alcohols; see Chapter 5. Burning an alcohol fuel in a specific combustor designed for a hydrocarbon distillate fuel would result in a greater fuel consumption. Water solubility of alcohol will seriously impact its use as a marine fuel. Burning an alcohol fuel may partially derate an engine, but thermochemical as well as combustion properties of alcohols, i.e., viscosity, latent heat of vaporization, specific heats, flame speed, and flammability limits, do not preclude burner redesigns for optimum use of alcohol fuels.

Hydrogen was considered as a potential fuel resource in Chapter 6. Several properties make hydrogen an attractive prospect as a long-term aviation fuel replacement, including:

Considerably higher specific energy than kerosene fuels
Atomization and vaporization not required if burned as a gas
Good heat sink to help cool engine parts if stored as a liquid
Wide flammability range and high flame speeds, which are both compatible with small combustor designs
Absence of carbon lowers radiant heat loss, smoke, soot, and particulates
No formation of CO, CO_2, UHC
Absence of eroding or corroding containments
No carbon buildup on critical engine components

Note that any of the proposed alternative and/or synthetic turbine fuels are still required to meet the same performance characteristics described that present fuels must meet.

12.4 GAS TURBINE COMBUSTORS

Heat addition to a gas turbine working fluid may result from either an external or internal combustion process. Most conventional engines require continuous internal combustion heat release within compressed air where temperature is raised prior to expansion through a power turbine by adding a supply of fuel to the air stream and burning the mixture as it flows through the burner. Reliable operation, long burner life, and combustion consistency require burner designs that satisfy several conflicting but important issues:

Compact and lightweight geometry, especially in aircraft applications
Minimal stagnation pressure loss across the combustor, i.e., between burner inlet and outlet
Prevention of burner liner erosion during operation
Easy initial and relight ignition characteristics
Uniform combustor exhaust temperature distribution
Continuous flame stability over a wide AF range

Isolating and separating combustion from compression and expansion processes by the use of a separate burner give gas turbines their unique fuel-engine interface. For example, there are no octane or cetane detonation indices for the constant-pressure continuous-combustion process as was the case for intermittent burn associated with IC engines discussed in Chapters 9–11. High energy release rates, however, are necessary. For example, most turbojet propulsion machines typically require 373,000–2,980,000 kW/m^3 (10,000–80,000 Btu/ft^3/sec) of heat release, whereas stationary steam power plants have nominal values that range between 22,400 and 112,000 kW/m^3 (600–3000 Btu/ft^3/sec). Stable burn must occur within flows requiring bulk velocities of 30–60 m/sec (100–500 ft/sec). These are only two fuel-engine design requirements for a combustor configuration to meet in order to assure a high combustor efficiency. To achieve an efficiency of approximately 95%, a burner configuration should:

Sustain complete combustion in a high mass flux environment
Provide sufficient residence time for complete fuel combustion to occur
Prevent combustion processes from continuing into turbine components
Promote combustion reactions that generate no coking or deposition on combustor or turbine surfaces
Produce clean, smokeless exhaust gases having minimal eroding particulates
Limit heat losses from hot reacting gases through burner walls
Prevent hot spots, flameouts, and unreliable light-offs

Furnace combustion is at atmospheric conditions and surrounded by thick walls of firebrick and other materials; whereas high-pressure gas turbine thermochemistry is confined in a burner by a relatively small thickness of metal. A large portion of compressed air must thus bypass the reaction zone and be used to cool burner surfaces as well as to mix with, and to cool, burned gases entering the turbine. Approximately 25% of total airflow by weight is used for burning fuel. Combustor performance depends strongly on the aerothermochemistry of the reacting fluid flow passing through a given burner geometry. Burner configurations have evolved chiefly from experimentation and testing.

An electrically activated igniter, or spark plug, is used to initiate reaction as the fuel-air mixture flows through a burner. After initial ignition, reaction is self-sustaining unless any relighting is required to re-establish a reactive flow after a flameout. This process is different from spark-ignition engine ignition, in which each mechanical cycle requires a separate spark plug firing to continue engine operation.

Gas turbine combustion is basically a two-zone, high-pressure reaction process in which a primary zone is utilized to complete most of the combustion and a secondary zone is required for burner exhaust gas management. Primary combustion is usually a stoichiometric or rich regime and may require 20–30% of the compressed air. The secondary zone does provide for limited chemical kinetics but is predominantly used to produce an acceptable exhaust gas flow having a lower and more uniform temperature profile and is produced by mixing primary zone combustion products with excess air. Sufficient time must be provided within the combustor to ensure fuel droplet vaporization, primary reactant mixing, and kinetics. Combustion aerochemistry further requires proper fluid flow to ensure establishment and stabilization of a stationary primary flame zone that does not get swept downstream and out of the combustor or upstream into the compressor, causing a flameout. Water injection does not improve combustion efficiency; however, it may improve gas turbine performance by increasing total mass flow, while the latent heat of vaporization of water can help control peak temperature.

The principal gas turbine combustion pollutants are carbon monoxide, unburned hydrocarbons, aldehydes, particulates, smoke, and oxides of nitrogen. Carbon monoxide is produced in the stoichiometric or slightly fuel-rich primary combustion zone and results from both incomplete combustion and carbon dioxide dissociation. Unburned hydrocarbons result from unburned fuel and fuel pyrolysis products that exit the combustor without being oxidized. Soot, produced in the primary zone and partially consumed in the secondary zone, is predominantly carbon. Soot formation is a function of fuel structure, with aromatics more prone to soot than paraffin compounds. Smoke emissions contain exhaust soot, water vapor, and hydrocarbon vapors. Oxides of nitrogen such as nitric oxide (NO) result from three sources: oxidation of the dissociated atmospheric nitrogen, chemical kinetic nitric oxide, and oxidation of fuel nitrogen. Nitrogen dioxide (NO_2) is also produced within the combustor, and the total NOx emission from a gas turbine consists of both NO and NO_2.

Conditions that contribute to the production of CO and UHC are poor fuel atomization and fuel cracking. It should be noted that any action taken in the primary zone to alleviate CO, UHC, and soot, such as stoichiometry and reaction temperature shifts, oppose those that would reduce NOx. Simple can burner geometry provides a reasonable efficiency of combustion and fuel economy, but emission requirements for land-based operation cannot be easily achieved.

EXAMPLE 12.4 A gas turbine combustor can with a 6% pressure loss has inlet and discharge conditions as shown below:

$$T_2 = 444\text{K} \qquad P_2 = 40{,}500 \text{ kPa} \qquad V_2 = 61 \text{ m/sec}$$
$$T_3 = 1200\text{K}$$

Assuming that the working fluid is air having mean constant specific heats and an equivalent heat addition of 850 kJ/kg air, calculate (a) the inlet Mach number; (b) the inlet stagnation state; (c) the discharge Mach number; (d) the discharge stagnation state; and (e) the decrease in stagnation pressure across the combustor due solely to heating, the air, %

Solution

1. Properties for air:

$$T_{\text{mean}} = \frac{(444 + 1200)}{2} = 822\text{K}$$

From Appendix B,

$$\overline{C}_p\langle 822\text{K}\rangle = 7.633 \text{ cal/g mole K}$$

$$\gamma = \frac{\overline{C}_p}{\overline{C}_p - \overline{R}} = \frac{7.633}{(7.633-1.987)} = 1.352$$

and

$$C_p = \frac{(7.633 \text{ cal/g mole K})(4.187 \dfrac{\text{kJ/kg mole K}}{\text{cal/g mole K}})}{28.97 \text{ kg/kg mole}}$$

$$= 1.103 \text{ kJ/kg K} = 1103 \text{ N m/kg K}$$

2. Inlet Mach number (Example 7.1):

$$c_2 = (\gamma R T_2 g_0)^{1/2}$$
$$= [(1.352)\ (287\text{N m/kg K})\ (444\text{K})\ (1.0 \text{ kg m/N sec}^2)]^{1/2} = 415 \text{ m/sec}$$

a. $N_{m_2} = \dfrac{V_2}{c_2} = \dfrac{61}{415} = 0.147$

3. Inlet stagnation conditions (Example 7.2):

b. $T_{0_2} = T_2\left[1 + N_{m_2}^2(\gamma-1)/2\right]$

$$= 444\left[1 + (0.147)^2 \frac{0.352}{2}\right] = 445.7\text{K}$$

$$P_{0_2} = P_2\left[1 + N_{m_2}^2\,(\gamma-1)/2\right]^{\gamma/(\gamma-1)}$$

$$= 40{,}500\left[1 + (0.147)^2 \frac{0.352}{2}\right]^{1.352/0.352}$$

$$= 41{,}095 \text{ kPa}$$

4. Energy:

$${}_2q_3 + \frac{V_2^2}{2g_0} + h_2 = \frac{V_3^2}{2g_0} + h_3$$

$$\frac{V_3^2}{2g_0} = 850{,}000 \frac{\text{N m}}{\text{kg}} + \left(1103 \frac{\text{N m}}{\text{kg K}}\right)(444 - 1200\text{K}) + \frac{(61 \frac{\text{m}}{\text{sec}})^2}{(2)(1.0 \frac{\text{kg m}}{\text{N sec}^2})}$$

c. $V_3 = 189.7 = 190$ m/sec

5. Discharge Mach number:

$$c_3 = (\gamma R T_3 g_0)^{1/2}$$
$$= [(1.352)(287 \text{ N m/kg K})(1200\text{K})(1.0 \text{ kg m/N sec}^2)]^{1/2}$$
$$= 682 \text{ m/sec}$$

d. $N_{m_2} = \dfrac{V_2}{c_2} = \dfrac{190}{682} = 0.279$

6. Discharge stagnation conditions:

e. $T_{0_3} = 1200\left[1 + (0.279)^2 \dfrac{(0.352)}{2}\right] = 1216\text{K}$

$$P_{0_3} = 38{,}070\left[1 + (0.279)^2 \frac{(0.352)}{2}\right]^{1.352/0.352}$$
$$= 40{,}113 \text{ kPa}$$

7. Percent decrease in stagnation pressure:

f. $\% = \left[\dfrac{41{,}095 - 40{,}113}{41{,}095}\right](100) = 2.4\%$

Comments A pressure loss in a gas turbine will result in a loss in net work and hence in thermal efficiency. From the analysis, approximately 2.4% of the 6% drop in pressure is due solely to heating. A lower velocity would make this stagnation loss less. Additional losses result from mixing and turbulence.

Consider the simple combustor geometry shown in Figure 12.4a. Several limitations are found to exist with this simple arrangement:

Large stagnation pressure loss resulting from the entropy of reaction
Potential flame extinction or movement from the burner exit in high mass flow rate regimes
Possible flame extinction or movement toward the fuel nozzles in low mass flow rate regimes
High heat loss from the combustor because of the steady and highly localized nature of the burn
Stoichiometric AF necessary for complete combustion but excess air re-

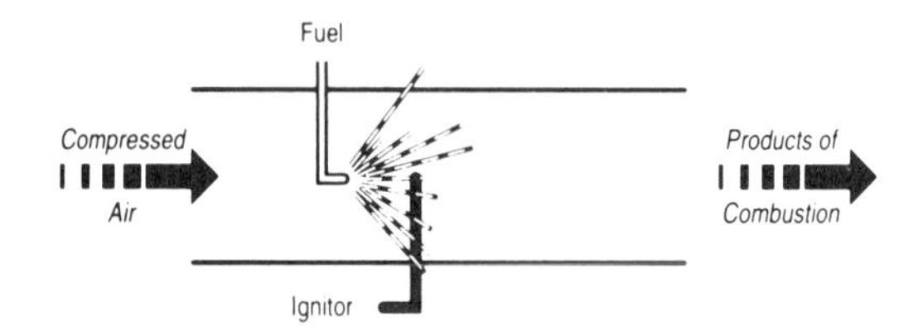

(a) Simple Combustor Geometry

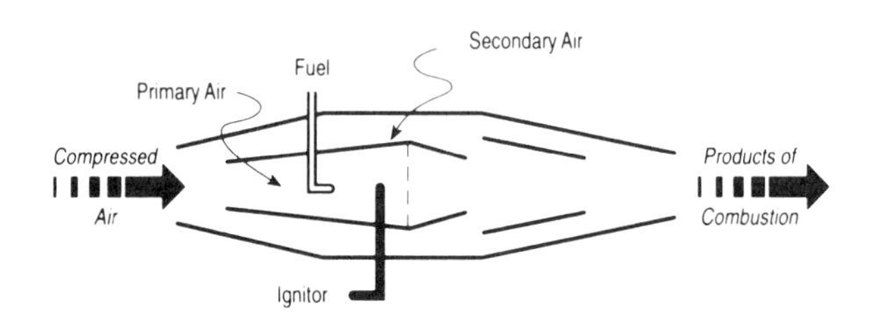

(b) Modified Combustor Geometry

Figure 12.4 Basic gas turbine combustor geometry.

quired to reduce the exhaust gas temperature to acceptable levels at the burner exit
Gas temperatures that can melt even high-alloy steels

Figure 12.4b illustrates modifications to this simple geometry that may improve the original burner performance. A diffuser is added to reduce supply air inlet velocity. A flameholder is inserted to allow a reaction zone, or flame, to attach and to be stabilized. Geometry provides a stoichiometric reactive gas flow reactant mixture at the nozzles for complete combustion as well as excess air at the combustor exit to provide a lower and more uniform discharge gas temperature. Additional features of this modified burner include:

Outer shell to insulate high-temperature reactive flow
Ports and slots to allow air for stoichiometric burn near the fuel nozzle and excess air after the initial burn
Fuel nozzle designed to provide a conical spray of fuel droplets
Torch or spark-igniter ignition source

Many specific combustors have been developed in an attempt to produce units having superior performance. These various designs are usually classified as *can* or *tubular, annular,* or *cannular. Can-type* burners are basically cylindrical-geometry combustors and are compatible with constant-discharge pressure centrifugal compressors that have speed-dependent variable-discharge capacities. Can-type burners are usually found on small gas turbine units. A single fuel nozzle and igniter plug are provided for each can. Individual cans are often placed around a compressor-turbine power shaft. A gas path through a tubular burner may utilize a once-through or a single reverse pass flow. These combustors can be easily separated from an operating gas turbine unit for maintenance, testing, or modification. Can combustors are mechanically rugged but are usually heavy and bulky configurations. High can strength results from the cylindrical shape although high stagnation pressure losses also occur with this burner geometry. Nonuniform light-off temperature distributions, as well as variations in can-to-can burn characteristics, occur in multican usage. Tubular geometries have smaller diameters than annular chambers.

EXAMPLE 12.5 Natural gas (CH_4) is pumped using the gas turbine configuration shown below. Methane enters the fuel compressor at 22 psia and 77°F, with an inlet volumetric flow rate of 500 ft^3/sec, and discharges from the unit at 88 psia. Air is supplied by a separate air compressor having suction conditions of 14.7 psia and 77°F and a discharge pressure of 88 psia. The combustor burns only a percentage of the total methane supply in 350% theoretical air, being supplied by the air compressor. Exhaust gases expanding through the turbine produce only enough power to run the two compressors. The turbine exhausts to a pressure of 14.7 psia. For ideal conditions, determine (a) the required CH_4 bleed, %; (b) the air compressor inlet volumetric flow rate, ft^3/min; (c) the exhaust gas mass flow rate, lbm/hr; and (d) the required turbine and compressor power, hp.

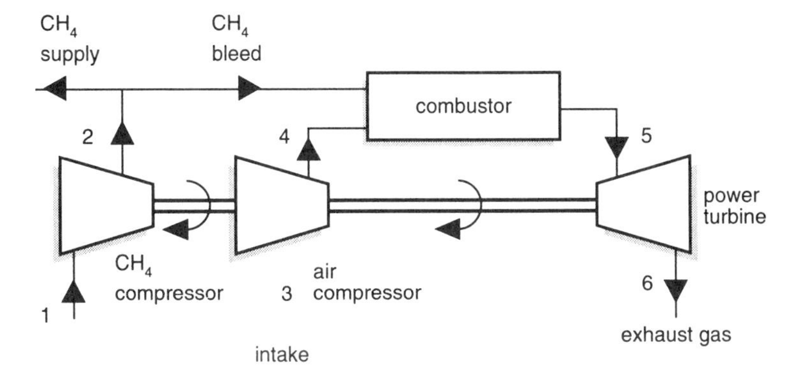

Solution:

1. Methane compressor:

$$P_1 = 22 \text{ psia} \qquad T_1 = 537°\text{R} \qquad \dot{V}_1 = 500 \text{ ft}^3/\text{sec}$$

Mass flow rate:

$$\dot{m} = \frac{P_1\dot{V}_1}{RT_1} = \frac{(22 \text{ lbf/in.}^2)\,(144 \text{ in.}^2/\text{ft}^2)\,(500 \text{ ft}^3/\text{sec})\,(16 \text{ lbm/lb mole})}{(1545 \text{ ft lbf/lb mole °R})\,(537°\text{R})}$$

$$= 30.5 \text{ lbm } CH_4/\text{sec}$$

Isentropic compression process, $1 \rightarrow 2$:

$$\frac{P_r\langle T_2\rangle}{P_r\langle T_1\rangle} = \frac{P_2}{P_1} = \frac{88}{22} = 4$$

Using CH_4 data found in Appendix B, $\Delta\bar{h}\langle T_1\rangle = 0$

$$P_r\langle T_1\rangle = \exp\left\{\frac{\bar{s}^0\langle 537\rangle}{\bar{R}}\right\} = \exp\left\{\frac{44.490}{1.987}\right\} = 5.2977 \times 10^9$$

or

$$P_r\langle T_2\rangle = (4)\,(5.2977 \times 10^9) = 2.1191 \times 10^{10}$$

$$\frac{\bar{s}^0\langle T_2\rangle}{\bar{R}} = \ln(2.1191 \times 10^{10}) = 23.7768$$

$$\bar{s}^0\langle T_2\rangle = (1.987)\,(23.7768) = 47.245$$

Interpolating,

$$\text{At } T_2 = 720°\text{R} \qquad \bar{s}^0 = 47.144 \qquad \Delta\bar{h}\langle T_2\rangle = 923 \text{ cal/g mole}$$

Compressor power:

$$\dot{W}_{CH_4} = \dot{m}[\Delta\bar{h}\langle T_1\rangle - \Delta\bar{h}\langle T_2\rangle]$$

$$= (30.5 \text{ lbm/sec})\left[\frac{(0 - 923) \text{ cal/g mole}}{16 \text{ lbm/lb mole}}\right]\left(1.8001\,\frac{\text{Btu/lb mole}}{\text{cal/g mole}}\right)$$

$$\dot{W}_{CH_4} = -3167 \text{ Btu/sec}$$

2. Air compressor:

$$P_3 = 14.7 \text{ psia} \qquad T_3 = 537°\text{R} \qquad \Delta\bar{h}\langle T_3\rangle = 0$$

Isentropic compression process:

$$\frac{P_r\langle T_4\rangle}{P_r\langle T_3\rangle} = \frac{P_4}{P_3} = \frac{88}{14.7} = 5.986$$

Using air data found in Example 3.2,

$$P_r\langle T_4\rangle = (5.986)\,(1.288 \times 10^{10}) = 7.710 \times 10^{10}$$

$P_r \times 10^{-10}$	T
3.749	720
8.286	900

Interpolating, $T_4 = 877°\text{R}$

and for air from Appendix B,

$$\Delta\bar{h}\langle T_4\rangle = 1349 \text{ cal/g mole}$$

Compressor work:

$$\begin{aligned} w_{air} &= [\Delta\bar{h}\langle T_3\rangle - \Delta\bar{h}\langle T_4\rangle] \\ &= \left[\frac{(0 - 1349)\text{ cal/g mole}}{28.96\text{ lbm/lb mole}}\right]\left(1.8001\frac{\text{Btu/lb mole}}{\text{cal/g mole}}\right) \\ &= -83.85\text{ Btu/lbm} \end{aligned}$$

3. Combustor $P = c$ adiabatic combustion:

Reaction equation, assume complete reaction:

$$CH_4 + (2)\,(3.5)\,[O_2 + 3.76N_2] \rightarrow CO_2 + (2)\,(2.50)O_2 + 2H_2O + 26.32\ N_2$$
$$1 \text{ mole } CH_4 \rightarrow 1 + 5 + 2 + (7)\,(3.76) = 34.32 \text{ moles product}$$

Product mole fractions:

$$\bar{x}_{CO_2} = \frac{1}{34.32} = 0.0291$$

$$\bar{x}_{O_2} = \frac{5}{34.32} = 0.1457$$

$$\bar{x}_{H_2O} = \frac{2}{34.32} = 0.0583$$

$$\bar{x}_{N_2} = \frac{(7)\,(3.76)}{34.32} = 0.7669$$

$$\begin{aligned} MW_{prod} = \Sigma\bar{x}_i MW_i &= (0.0291)\,(44) + (0.1457)\,(32) + (0.0583)\,(18) \\ &\quad + (0.7669)\,(28) = 28.465 \text{ lbm/lb mole} \end{aligned}$$

Combustor energy balance:

$$T_2 = 720°\text{R} \qquad T_4 = 877°\text{R}$$

$$\{1[\bar{h}_f^0 + \Delta\bar{h}<T_2>]_{CH_4} + (2)\,(3.5)\,[\bar{h}_f^0 + \Delta\bar{h}<T_4>]_{O_2}$$
$$+ (2)\,(3.5)\,(3.76)\,[\bar{h}_f^0 + \Delta\bar{h}<T_4>]_{N_2}\} = \{1[\bar{h}_f^0 + \Delta\bar{h}<T_5>]_{CO_2}$$
$$+ (2)\,(2.5)\,[\bar{h}_f^0 + \Delta\bar{h}<T_5>]_{O_2} + 2[\bar{h}_f^0 + \Delta\bar{h}<T_5>]_{H_2O}$$
$$+ (2)\,(3.5)\,(3.76)\,[\bar{h}_f^0 + \Delta\bar{h}<T_5>_{N_2}\}$$

Using JANAF data found in Appendix B for the reactants and products,

$$\{[-17{,}895 + 923]_{CH_4} + (2)\,(3.5)\,[1362]_{O_2} + (26.32)\,[1323]_{N_2}\}$$
$$= \{[-94{,}054 + \Delta\bar{h}<T_5>]_{CO_2} + (2)\,(2.5)\,[\Delta\bar{h}<T_5>]_{O_2}$$
$$+ 2[-57{,}798 + \Delta\bar{h}<T_5>]_{H_2O} + (2)\,(3.5)\,(3.76)\,[\Delta\bar{h}<T_5>]_{O_2}\}$$
$$237{,}033 = \Delta\bar{h}<T_5>)_{CO_2} + 5\,\Delta\bar{h}<T_5>)_{O_2} + 2\Delta\bar{h}<T_5>)_{H_2O}$$
$$+ (2)\,(3.5)\,(3.76)\,\Delta\bar{h}<T_5>)_{N_2}$$

where, by trial and error,

$$T_5 = 2145°\text{R}$$

4. Turbine:

$$P_5 = 88 \text{ psia} \qquad T_5 = 2145°\text{R}$$

Product isentropic expansion:

$$\frac{P_r<T_6>}{P_r<T_5>} = \frac{P_6}{P_5} = \frac{14.7}{88} = 0.167$$

where

$$\bar{s}^0<T_5> = \bar{x}_i\bar{s}^0{}_i<T_5>$$

Using JANAF data for the products,

$$\bar{s}^0<T> = [(0.0291)\,(66.659) + (0.1457)\,(59.668) + (0.0583)\,(57.366)$$
$$+ (0.7669)\,(55.897)] = 56.845 \text{ cal/g mole K}$$

and

$$P_r<T_5> = \exp\left\{\frac{\bar{s}^0<T_5>}{\bar{R}}\right\} = \exp\left\{\frac{56.845}{1.987}\right\} = 2.6576 \times 10^{12}$$

or

$$P_r\langle T_6 \rangle = (0.167)\,(2.6576 \times 10^{12}) = 4.4382 \times 10^{11}$$

$$\frac{\bar{s}^0\langle T_6 \rangle}{\bar{R}} = \ln(4.4382 \times 10^{11}) = 26.8187$$

$$\bar{s}^0\langle T_6 \rangle = (1.987)\,(26.8187) = 53.2888$$

$$53.2888 = \sum_{\text{prod}} \bar{x}_i \bar{s}^0{}_i \langle T_6 \rangle$$

By trial and error,

$$T_6 = 1384°\text{R}$$

Turbine work:

$$\bar{w}_{\text{prod}} = [\Sigma \bar{x}_i \Delta\bar{h}\langle T_5 \rangle - \Sigma \bar{x}_i \Delta\bar{h}\langle T_6 \rangle]$$

$$= (0.0291)\,[10{,}521 - 5077] + (0.1457)\,[7043 - 3538]$$

$$+ (0.0583)\,[8154 - 4017] + (0.7669)\,[6651 - 3365]$$

$$= 3430 \text{ cal/g mole products}$$

$$= \left[\frac{3430 \text{ cal/g mole}}{28.465 \text{ lb/lb mole}}\right]\left(1.8001\,\frac{\text{Btu/lb mole}}{\text{cal/g mole}}\right) = 216.91 \text{ Btu/lbm products}$$

5. Net power:

$$|\dot{W}_{\text{turb}}| = |\dot{W}_{CH_4}| + |\dot{W}_{\text{air}}|$$

or

$$\dot{m}_{\text{prod}} w_{\text{turb}} = \dot{m}_{CH_4} w_{CH_4} + \dot{m}_{\text{air}} w_{\text{air}}$$

with

$$\dot{m}_{\text{prod}} = \left[\frac{34.32 \text{ moles prod}}{1 \text{ mole } CH_4 \text{ burned}}\right]\left[\frac{28.465 \text{ lbm/lb mole prod}}{16 \text{ lbm/lb mole fuel}}\right] \times \dot{m}_{CH_4} \text{ burned}$$

$$\dot{m}_{\text{prod}} = 61.057 \dot{m}_{CH_4} \text{ burned}$$

and

$$\dot{m}_{\text{air}} = \left[\frac{(7)\,(4.76 \text{ moles air})}{1 \text{ mole } CH_4 \text{ burned}}\right]\left[\frac{28.97 \text{ lbm/lb mole air}}{16 \text{ lbm/lb mole fuel}}\right] \times \dot{m}_{CH_4} \text{ burned}$$

$$\dot{m}_{\text{air}} = 60.33 \dot{m}_{CH_4} \text{burned}$$

substituting for the various terms,

$$(216.91)\,(61.057)\dot{m}_{CH_4} = 3167 \text{ Btu/sec} + (83.85)\,(60.33)\dot{m}_{CH_4}$$

$$\dot{m}_{CH_4} \text{ burned} = 0.387 \text{ lbm } CH_4\text{/sec}$$

6. Methane bleed fraction:

$$\% \text{ bleed} = \frac{\dot{m} \text{ burned}}{\dot{m} \text{ supply}} \times 100 = \left[\frac{0.387 \text{ lbm CH}_4 \text{ burned/sec}}{30.5 \text{ lbm CH}_4 \text{ supply/sec}} \right] \times 100$$

a. = 1.3% of supply bleed

7. Air compressor volumetric flow rate:

$$\dot{m}_{air} = (60.33)\ (0.387 \text{ lbm/sec})$$

$$\dot{V}_3 = \frac{\dot{m}RT_3}{P_3} = \frac{(60.33)\ (0.387 \text{ lbm/sec})\ (1545 \text{ ft lbf/lb mole °R})\ (537\text{°R})}{(22 \text{ lbf/in.}^2)\ (144 \text{ in.}^2/\text{ft}^2)\ (28.97 \text{ lbm/lb mole})}$$

$$= 211 \text{ ft}^3/\text{sec}$$

$$= (211 \text{ ft}^3/\text{sec})\ (60 \text{ sec/min})$$

b. $\dot{V}_{air} = 12{,}660 \text{ ft}^3/\text{min}$

8. Exhaust gas mass flow rate:

$$\dot{m}_{prod} = (61.093)\ (0.387 \text{ lbm/sec})\ (3600 \text{ sec/hr}) = 85{,}115 \text{ lbm/hr}$$

9. Compressor and turbine power:

$$\dot{W}_{CH_4} = \frac{(3167 \text{ Btu/sec})\ (3600 \text{ sec/hr})}{(2545 \text{ Btu/hp hr})}$$

c. $\dot{W}_{CH_4} = 4480 \text{ hp}$

$$\dot{W}_{air} = \frac{(60.33)\ (0.387 \text{ lbm/sec})\ (3600 \text{ sec/hr})\ (83.85 \text{ Btu/lbm})}{(2545 \text{ Btu/hp hr})}$$

d. $\dot{W}_{air} = 277 \text{ hp}$

$$\dot{W}_{turb} = \frac{(85{,}115 \text{ lbm/hr})\ (216.91 \text{ Btu/lbm})}{(2545 \text{ Btu/hp hr})} = 7250 \text{ hp}$$

The *annular burner* is a gas turbine combustor design that is most compatible with *axial compressors,* i.e., constant-capacity compressors with a speed-dependent variable-discharge state. These designs are most frequently found on larger gas turbine configurations in which a single annulus is wrapped around the compressor-turbine power shaft. This type of burner arrangement requires multiple nozzles and igniter plugs. Fluid flow can be similar to that for a can, that is, once-through or a single pass reverse gas flow arrangement. Annular configurations are more compact geometries than cans and have low surface-to-volume ratios. Annular burners require less length, are lighter, and have a lower stagnation pressure loss across the combustor than comparable tubular arrangements. These combustor configurations are easier to light off, but it is more difficult to establish a uniform exhaust gas temperature profile. Burner liners are short-lived components, with outer flame tubes subject to buckling. Unlike can elements, these components are difficult to remove for inspection, maintenance, or modifications. *Can-annular* combustor designs are a compromise between can and annular geometries. Can-annular burners are used with large–pressure ratio gas turbine units. Again, a single annulus is placed around a centerline shafting, but a series of tubular flame tubes are placed inside the annulus. A can-annular configuration is shorter and lighter than a can-shape burner but less compact than an annular configuration. These burners may have a lower stagnation pressure loss than a simple can design. Can-annular systems provide good structural stability for flame tube geometry. With separate tubular flame tubes, these units are less difficult to service, test, or modify than the annular type of burner.

12.5 THE FREE PISTON AND STIRLING ENGINES

Future combustion science and engine technology will arise from the various machines and mechanisms discussed in the last three chapters. Two unique engine concepts that illustrate how combustion engine characteristics can be categorized in such a manner are the *free piston engine* and the *Stirling engine.* A free piston engine, shown in Figure 12.5, incorporates features of both IC engine combustion and gas turbine power output. A Stirling engine, illustrated in Figure 12.6, combines gas turbine continuous combustion and reciprocating IC engine power output.

The free piston engine is actually a diesel combustion engine gasifier that incorporates a gas turbine expander for power output. A special piston design, consisting of a large compression piston at one end and a small but integral gas expansion piston at the other end, is basic to successful operation of this unique opposed-piston diesel configuration. Two of these special pistons are needed, and they are arranged so that the smaller pistons face inward toward each other. Both large pistons serve a twofold purpose; the inner face is utilized to compress air, and the outer piston face to work with a special closed air-bounce chamber.

One mechanical cycle of a free piston engine begins with air compression. With transfer ports uncovered, air is compressed and moved into a clearance volume defined by the opposed gas expansion pistons. During compression, engine intake and exhaust gas valves are all closed. By the end of compres-

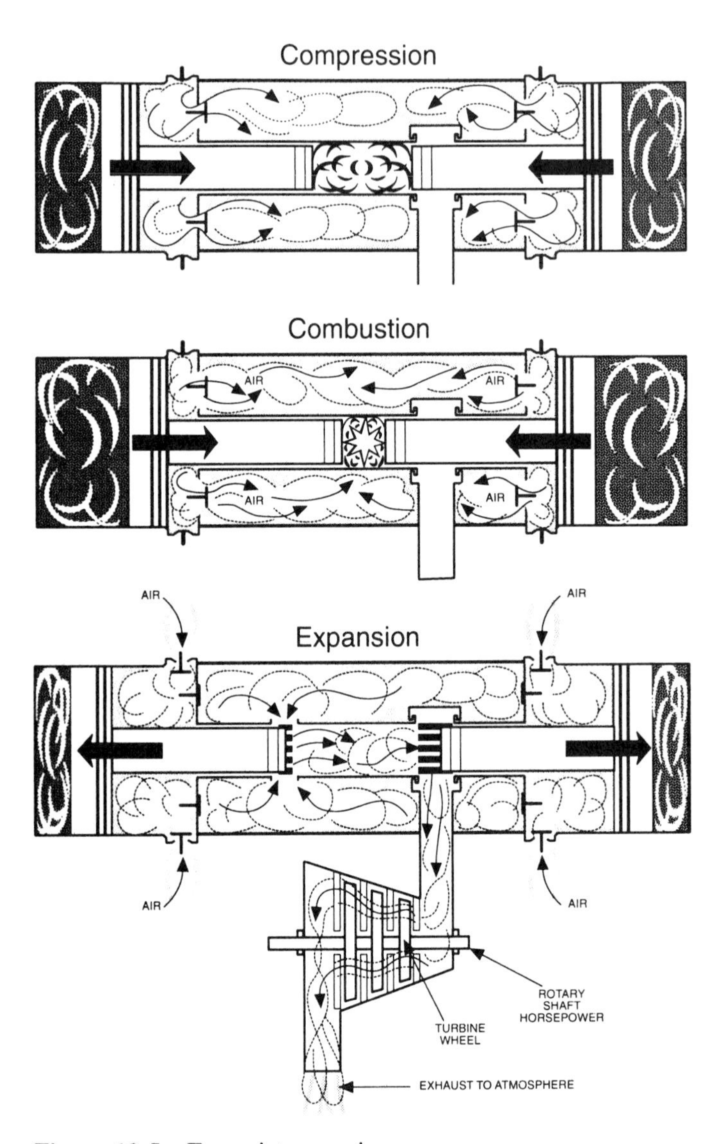

Figure 12.5 Free piston engine.

sion, air transfer ports are covered by the pistons, and combustion air is centered within the engine. This process helps to scavenge and supercharge this two-stroke type of diesel combustion air. Fuel injection into the high-pressure and high-temperature air results in compression-ignition combustion, with ensuing gas expansion against both small gas expansion pistons. During gas expansion, exhaust ports are exposed, allowing high-pressure gases to expand through a power turbine wheel to produce useful power output. Expansion of the small gas cylinders drives the large compression pistons apart, compressing air trapped in the bounce chambers which, in turn, force the piston assemblies back toward center and thus begin the mechanical cycle again. Exhaust and air intake valves are open while the exhaust gases expand through the turbine, and a new supply of air is drawn into the engine. Since there are no connecting rods or crankshafts, the pistons are kept in place by means of connecting linkages.

Successful engines must find a niche in which they can provide a viable answer to a real need. In this regard, the free piston engine has not found a suitable application, in part because of the following:

Fixed costs too high to pay for efficiency gains
Mechanical problems associated with starting and control during operation
Synchronization of the pistons in single units and multiunits

The Stirling engine, a positive-displacement engine with external heat addition, was developed by Dr. Robert Stirling, a Scottish clergyman, in 1816, approximately 80 years before the diesel. A Stirling engine is considerably more complex than SI and/or CI IC engines. In its original geometry, two separate cylinders having relatively small communicating passageways were required to run. This path was filled with a wire coil or mesh to allow the region to act as a regenerator when the working fluid was displaced from one cylinder to the other by the pistons. Stirling engines operate as closed loops using a gas, such as helium and/or hydrogen, and therefore require indirect external heating to produce useful power. It is this separation of fuel-air chemistry from events occurring within the engine that allows continuous combustion to be utilized.

A variety of fuels are potentially compatible with Stirling engine combustion, including diesel fuel, unleaded gasoline, alcohol, kerosene, butane, propane, and natural gas. During the 1970s, the U.S. Department of Energy supported pilot projects to stimulate the development of new engine technologies that would be 30% more efficient than the current conventional mobility power plants, meet the 1985 federal emissions standards, be capable of burning a wide variety of fuels, and be cost-competitive with standard production engines. Stirling engines were considered by many as a strong candidate to meet these criteria.

The ideal thermodynamic events required for a Stirling engine to produce useful power are as follows:

1–2 $T = c$ Heat rejection during isothermal compression in cylinder 1
2–3 $v = c$ Internal regenerative heat transfer (gain) during flow between cylinders 1 and 2
3–4 $T = c$ External heat addition during isothermal expansion in cylinder 2
4–1 $v = c$ internal regenerative heat transfer (loss) during flow between cylinders 2 and 1

Cycle performance is strongly influenced by the regenerator heat exchanger design and high effectiveness necessary to return energy from gases during the process 4–1 to the gases during the process 2–3. Mechanical and thermal losses penalize this machine and prevent it from making a breakthrough and finding a general application. The actual P-v diagram from current Stirling engines departs considerably from the ideal processes, yet the maximum operating efficiencies can exceed those for comparable SI and CI engines. This regenerative machine for the ideal events, i.e., $T = c$ heat transfer, would approach the efficiency of the Carnot cycle, the most efficient thermodynamic cycle described in Chapter 1.

The advantages of a Stirling engine include:

Decoupling of fuel-air combustion from engine operation
Low emissions (continuous combustion)
Multifuel
Improved fuel economy over Otto cycle
Quiet, low-vibration engine
High thermal efficiency
Can operate with a pressurized burner

The disadvantages of a Stirling engine include:

Complex heat transfer engine
Bulky (50% volume is heat transfer equipment)
Expensive
Seals and gas leakage
Hydrogen leaks and embrittlement
Questionable durability

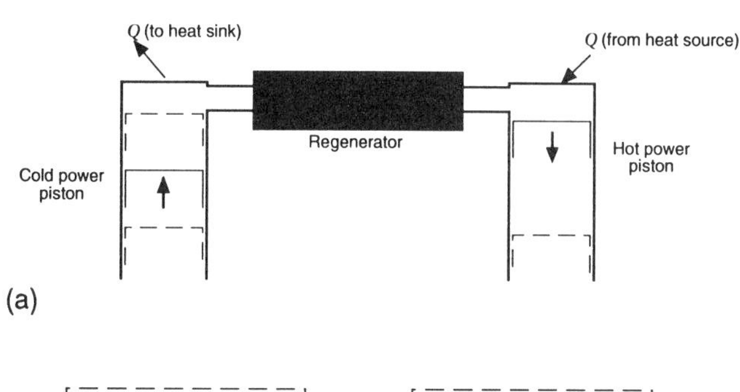

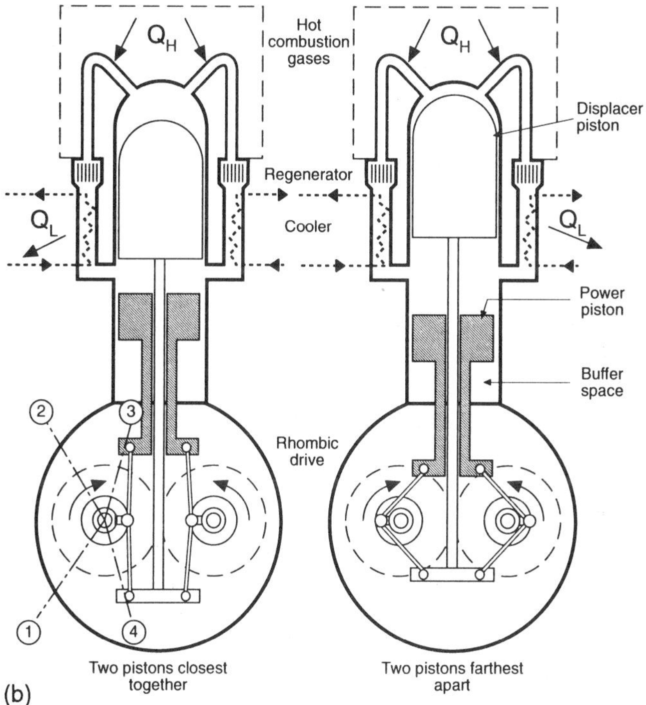

Figure 12.6 The Stirling engine: (a) ideal Stirling cycle engine; (b) practical Stirling cycle engine.

PROBLEMS

12.1 Repeat Example 12.1 for (a) a monatomic gas, $\gamma = 1.67$; (b) $\gamma = 1.3$; and (c) $\gamma = 1.2$

12.2 A gas turbine power plant burns a hydrocarbon fuel having an unknown composition. To determine the fuel composition, a sample is burned in air, and the Orsat analysis of the products of combustion yields 10.5% CO_2, 5.3% O_2, and 84.2% N_2. For this fuel, estimate (a) the mass composition of the fuel, %C and %H; (b) the percent of excess air for this reaction, %; and (c) the overall chemical formula, CxHy.

12.3 A blend of liquid ethanol and methanol is being proposed as a potential stationary gas turbine fuel alternative. If the fuel mixture consists of 25% ethanol–75% methanol by weight, determine (a) the fuel mixture density at 60°F, lbm/ft^3; (b) the fuel mixture specific gravity at 60°F; (c) the stoichiometric *FA* ratio, lbm fuel/lbm air; and (d) the fuel lower heating value, Btu/gal.

12.4 A propulsion gas power turbine has inlet conditions of 1440°R and 30 psi, with a discharge temperature of 1260°R. The burner can exhaust gas composition on a molar basis is 3.6% CO_2, 3.6% H_2O, 10.0% O_2, and 76.8% N_2. For a turbine inlet volumetric flow rate of 119,450 ft^3/min, find (a) the exhaust gas molecular weight, lbm/lbmole; (b) the turbine inlet mass flow rate, lbm exhaust gases/min; (c) the total enthalpy change across the turbine, Btu/lb mole; and (d) the turbine power, hp.

12.5 An adiabatic propulsion turbine has the following operational data: inlet conditions P_3 = 1175 kPa, T_3 = 1100K, $\dot{m}$ = 66 kg/sec; discharge conditions T_4 = 700K, gas composition (by volume) 6.57% CO_2, 7.12% H_2O, 10.13% O_2, 76.18% N_2. For these conditions, determine (a) the gas mixture molecular weight, kg/kg mole; (b) the gas inlet volumetric flow rate, m^3/sec; (c) the turbine specific work, kJ/kg gas; and (d) the indicated turbine power output, kW.

12.6 Marks' *Handbook* gives an empirical equation for the *LHV* of fuel oils that may be expressed as

$$LHV = 19{,}960 + 1360\ \mathrm{SG} <60/60> - 3780\ \mathrm{SG}^2 <60/60>\ \frac{\mathrm{Btu}}{\mathrm{lbm}}$$

Estimate the *LHV* of kerosene having a specific gravity of 50°API gravity using (a) the empirical equation; (b) the relationship in Chapter 5 given by Equation 5.5b; and (c) the tabulated values found in Appendix D.

12.7 The following data are obtained from the test of a gas turbine burner:

Fuel	$(C_4H_{10})g$ at 77°F and 1 atm
Oxidant	300% theoretical air at 77°F and 1 atm
Products	1160°F and 1 atm
Inlet air velocity	200 ft/sec
Discharge gas velocity	2200 ft/sec

Assuming complete combustion, determine (a) the net heat transfer per mole of fuel, Btu/lb mole; and (b) the net increase in entropy per mole of fuel, Btu/lb mole °R.

12.8 A fuel with an H/C ratio of 0.16 is burned in a combustor with an *FA* ratio of 0.02 lbm fuel/lbm air. The fuel temperature enters the can at 100°F, and the air inlet temperature is 290°F. The *LHV* of the fuel is 18,500 Btu/lbm, and the specific heat of the fuel is 0.5 Btu/lbm °R. Calculate (a) the adiabatic or ideal temperature rise across the burner, °F; (b) the ideal temperature entering the turbine, °F; and (c) the combustion efficiency if the actual turbine inlet temperature is 1995°R, %.

12.9 Liquid C_8H_{18} has a lower heating value at constant pressure of 19,100 Btu/lbm at 68°F. Estimate the product temperature if 60.4 lbm of air at 440°F are supplied to a gas turbine combustion chamber for each pound of liquid C_8H_{18} supplied at 120°F. The specific heat of the fuel is 0.5 Btu/lbm °R, and the combustion losses are estimated to be 5% of the *LHV* of the fuel.

12.10 JP5, a power kerosene used in military jet aircraft, is burned in a gas turbine combustor at 1000 kPa. The fuel has an API specific gravity at 15°C of 42 and is supplied to the burner with 400% theoretical air. For ideal complete combustion, find (a) the *FA* ratio, kg fuel/kg air; (b) the exhaust gas molecular weight, kg/kgmole; (c) the exhaust dew point temperature, K; (d) the fuel density at 15°C, kg/m^3; and (e) the fuel *LHV* at 15°C, kJ/kg fuel.

12.11 A gas turbine engine having a 34% brake thermal efficiency develops 450-hp net brake power output. The unit burns a distillate fuel having a specific gravity at 60°F of 0.81. Find (a) the fuel specific gravity at 60°F, API; (b) the fuel density at 60°F, lbm/ft^3; (c) the fuel consumption, gal/hr; and (d) the brake specific fuel consumption, lbm/hp hr.

12.12 Repeat Example 12.3 using hydrogen as the fuel.

12.13 Repeat Example 12.3 using methanol as the fuel.

12.14 Supercharging is a means of boosting power output from an IC engine by admitting air at a density greater than ambient. One version uses hot exhaust gases from the engine to drive a turbine which, in turn, powers a supercharging compressor. This simple gas turbine with no net power output is termed a constant-pressure turbocharger. Consider such a unit having compressor and turbine conditions as given below:

Turbine:	Compressor:
T_{in} = 990K	T_{in} = 300K
T_{out} = 955K	P_{in} = 101 kPa
P_{out} = 101 kPa	$\dot{m}_{turbine}/\dot{m}_{compressor}$ = 1.05
r_p = 1.18	

Neglecting turbomachinery and mechanical efficiencies and assuming air to represent the gas mixture passing through both components, calculate for this operating condition (a) the turbine work output per unit mass flowing, kJ/kg; (b) the compressor boost discharge temperature, K; (c) the compressor boost discharge pressure, kPa; and (d) the turbocharger boost, i.e., density ratio across the compressor.

12.15 A natural gas supply is reported as follows: CO_2 = 0.5%, O_2 = 0.3%, N_2 = 4.6%, CH_4 = 74.0%, C_2H_4 = 20.6%. The gas is used as a fuel for a gas turbine in a pipeline pumping station. For this fuel, determine (a) the combustion equation of this gas for complete reaction with 400% theoretical air; (b) the *FA* ratio, lbm fuel/lbm air; (c) the product molecular weight, lbm/lb mole; (c) the adiabatic temperature if the gas enters at 536°R and the air enters at 900°R, °R; and (d) the actual discharge temperature if the combustion chamber has a combustion efficiency of 95%, °R.

12.16 A combustion chamber is supplied with octene, C_8H_{16}, at 77°F and air at 720°R. Exhaust gas analysis of the stack gases generates the following reaction equal for the unit:

$$C_8H_{16} + 48.0O_2 + 180.5N_2 \rightarrow 7.6CO_2 + 7.4H_2O + 180.5N_2 + 36.6O_2 + 0.2CO + 0.2H_2 + 0.2CH_4$$

If heat losses from the combustion can are negligible, estimate the combustion efficiency.

12.17 An open cycle gas turbine plant is designed to deliver 4000 bhp under the following conditions:

Atmospheric temperature	80.3°F	Peak temperature	1400°F
Atmospheric pressure	14 psia	Compressor efficiency	80 %
Pressure ratio	5:1	Turbine efficiency	84 %
Pressure drop in combustion can	5 psi	Mechanical efficiency	90 %
		Fuel *LHV*	18,000 Btu/lbm

Use a variable specific heat analysis for the compressor and turbine and the isentropic tables for air found in Example 3.2. Calculate (a) the internal state points through the unit; (b) the compressor power, ihp; (c) the turbine power, ihp; and (d) the fuel consumption, lbm/hr.

12.18 A turbojet engine is designed for a speed of 400 mph at 30,000 ft. Intake air is diffused isentropically from a relative velocity of 400 mph at the diffuser inlet to zero velocity at the diffuser exit. Air is then compressed isentropically in a compressor having a pressure ratio of 4. Octane, C_8H_{18}, is introduced into the combustor at a temperature of 540°R and 25% theoretical air. The constant-pressure adiabatic combustion process in the burner goes to completion with negligible dissociation. The products then expand isentropically through a turbine such that power output is equal to the required compressor power input. Exhaust gases then expand through a nozzle reversibly and adiabatically to atmospheric pressure. Use a variable specific heat analysis for the compressor and turbine and the isentropic tables for air found in Example 3.2. Calculate (a) the isentropic compressor discharge temperature, °R; (b) the compressor work, Btu/lbm; (c) the adiabatic flame peak combustor temperature, °R; (d) the turbine discharge temperature, °R; and (e) the nozzle discharge velocity, ft/sec.

Appendixes

APPENDIX A: DIMENSIONS AND UNITS

Combustion calculations require a proper use of all physical laws and, therefore, any particular problem must be correctly formulated both qualitatively, i.e., algebraically, and quantitatively, i.e., numerically.

All fundamental laws and relationships can be expressed algebraically using a consistent set of terms, or *dimensions*. A list of principal dimensions includes:

Force	$[F]$
Mass	$[M]$
Length	$[L]$
Time	$[t]$
Temperature	$[T]$

As an example, velocity has the dimensions of length divided by time, $[L/t]$ or $[Lt^{-1}]$.

Standards for mass, length, time, and temperature have been established and maintained under the supervision of an international body, the General Conference on Weights and Measures. The magnitude of any dimension is expressed numerically in terms of *units*. Some familiar unit choices for the basic dimension of length include:

$[L]$
inches
miles
angstroms
kilometers

Table A.I summarizes several systems of primary dimensions and corresponding units. Two dimension and unit systems are currently used in most engineering and scientific work in combustion: the new International System of Units (SI) and the English Engineers. Length $[L]$ and time $[t]$ are primary dimensions in both these measurement systems; see Table A.I. In the SI, mass $[M]$ is taken as a primary dimension and force $[F]$ is a derived, or secondary, dimension. The Engineers system differs from SI in that force $[F]$ as well as mass $[M]$ are *both* taken as primary dimensions.

Newton's second law defines the interrelationship between force and mass as

$$F \sim Ma \tag{A.1}$$

or, written dimensionally, as

$$[F] \sim [M][L][t^{-2}]$$

Since force, mass, length, and time are distinct dimensions, Newton's second law is more properly written as

$$F = \frac{Ma}{g_0} \tag{A.2}$$

where g_0 is an arbitrary constant necessary to provide both dimensional and unit homogeneity. Thus,

$$[g_0] = \frac{[M][a]}{[F]}$$

or

$$[g_0] = [M][L][t^{-2}][F^{-1}] \tag{A.3}$$

Substituting dimensionally for g_0, Newton's second law then yields proper dimensions on both the left and right sides of the equation or

$$[F] = \frac{[M][L][t^{-2}]}{[M][L][t^{-2}][F^{-1}]} = [F]$$

Table A.1 Some Systems of Primary Dimensions with Corresponding Units

Dimensions		Basic Units: Système International (SI)	Basic Units: Metric (CGS)	Basic Units: English Engineers	Basic Units: British Gravitational
PRIMARY					
Force	$[F]$	—	—	Pound force (lbf)	Pound force (lbf)
Mass	$[M]$	Kilogram (kg)	Gram (g)	Pound mass (lbm)	—
Length	$[L]$	Meter (m)	Centimeter (cm)	Feet (ft)	Feet (ft)
Time	$[t]$	Second (sec)	Second (sec)	Second (sec)	Second (sec)
Temperature	$[T]$	Kelvin (K)	Kelvin (K)	Rankine (R)	Rankine (R)
SECONDARY					
Force	$[M][L]/[t^2]$	Newton (kg m/sec^2)	Dyne (g cm/sec^2)	—	—
Mass	$[M]$	—	—	—	Slug
Pressure	$[F]/[L^2]$	Pascal (N/m^2)	Bar (dyne/cm^2)	psf (lbf/ft^2)	psf (lbf/ft^2)
Energy	$[F][L]$	Joule (N m)	Erg (dyne cm)	ft lbf	ft lbf
Power	$[F][L][t]$	Watt (J/sec)	—	Horsepower (ft lbf/sec)	Horsepower (ft lbf/sec)
g_0		$1.0 \frac{\text{kg m}}{\text{N sec}^2} = 1.0$	$1.0 \frac{\text{g cm}}{\text{dyne sec}^2} = 1.0$	$32.2 \frac{\text{lbm ft}}{\text{lbf sec}^2}$	$1.0 \frac{\text{slug ft}}{\text{lbf sec}^2} = 1.0$

In the SI, the secondary dimension of force, specified as a newton [N], is defined as the force that would accelerate 1 kg at a rate of 1 m/sec^2. For the SI, then,

$$g_0 = \frac{Ma}{F} = \frac{(1\ \text{kg})\ (1\ \text{m/sec}^2)}{1\ \text{N}} = 1\ \frac{\text{kg m}}{\text{N sec}^2}$$

Note, however, that the newton is defined in terms of $[M]$, $[L]$, and $[t]$ as

$$1\ \text{N} \equiv 1\ \text{kg m/sec}^2$$

and

$$g_0 = \frac{\text{kg m}}{[\text{kg m/sec}^2]\ \text{sec}^2} = 1.0$$

For the SI, g_0 has a magnitude of unity and dimensionless character.

For the English Engineers system, force, mass, length, and time are all primary quantities. From physics, the pound force [lbf] is that force that will accelerate a pound mass [lbm] at 32.1740 ft/sec^2. Using the above values, the value of g_0 in the English Engineers system becomes

$$g_0 = \frac{(1\ \text{lbm})\ (32.1740\ \text{ft/sec}^2)}{(1\ \text{lbf})} = 32.1740\ \frac{\text{lbm ft}}{\text{lbf sec}^2}$$

It is important to emphasize that g_0 in Engineers units is a dimensional constant having *both* magnitude and dimensions.

Engineering analysis in this text is expressed in either SI or Engineers units of measurement. Often, calculations are compounded by the fact that these two systems of dimensions and units both may appear. Calculations that involve two sets of dimensions and units require conversion factors that allow an easy interchange between units. An equation can be multiplied by a conversion factor without affecting the dimensional characteristics because conversion factors are unities (dimensionless). Two familiar examples are:

$$\frac{2.54\ \text{cm}}{\text{in.}} \equiv 1 \qquad \frac{60\ \text{sec}}{\text{min}} \equiv 1$$

Table A.2 gives some general conversion factors, while Table A.4 lists some useful conversion factors between the English Engineers and SI units. Table A.3 is a set of 16 prefixes given below for use with the SI units for form multiples and submultiples of basic SI units.

Table A.2 Useful Conversion Factors

<u>Length</u>				
$10^{10}\ \frac{\mathring{A}}{m}$	$2.54\ \frac{cm}{in.}$	$5280\ \frac{ft}{mi}$	$3\ \frac{ft}{yd}$	$1.609\ \frac{km}{mi}$
$30.48\ \frac{cm}{ft}$	$3.28\ \frac{ft}{m}$	$6080.2\ \frac{ft}{n.mi}$	$12\ \frac{in.}{ft}$	$10^4\ \frac{\mu}{cm}$
<u>Area</u>				
$640\ \frac{acres}{mi^2}$	$929\ \frac{cm^2}{ft^2}$	$6.452\ \frac{cm^2}{in.^2}$	$10.76\ \frac{ft^2}{m^2}$	$144\ \frac{in.^2}{ft^2}$
<u>Volume</u>				
$28.317\ \frac{cm^3}{ft^3}$	$35.31\ \frac{ft^3}{m^3}$	$1728\ \frac{in.^3}{ft^3}$	$61.025\ \frac{in.^3}{liter}$	$3.7854\ \frac{liter}{gal}$
$10^3\ \frac{cm^3}{liter}$	$7.481\ \frac{gal}{ft^3}$	$231\ \frac{in.^3}{gal}$	$28.317\ \frac{liter}{ft^3}$	$8\ \frac{pt}{gal}$
<u>Time</u>				
$24\ \frac{hr}{day}$	$1440\ \frac{min}{day}$	$60\ \frac{min}{hr}$	$3600\ \frac{sec}{hr}$	$60\ \frac{sec}{min}$
<u>Speed, Rotational and Angular Velocity</u>				
$152.4\ \frac{cm/min}{in./sec}$	$1.689\ \frac{ft/sec}{knot}$	$0.3048\ \frac{m/sec}{ft/sec}$	$0.44704\ \frac{m/sec}{mph}$	$0.6215\ \frac{mph}{km/hr}$
$88\ \frac{ft/min}{mph}$	$1.467\ \frac{ft/sec}{mph}$	$0.5144\ \frac{m/sec}{knot}$	$0.6818\ \frac{mph}{ft/sec}$	$1.152\ \frac{mph}{knot}$
$57.3\ \frac{deg}{rad}$	$\frac{6.2832}{2\pi}\ \frac{rad}{rev}$	$\frac{1}{2\pi}\ \frac{rev/min}{rad/min}$	$9.549\ \frac{rev/min}{rad/sec}$	
<u>Mass and Force</u>				
$980.665\ \frac{dynes}{gmf}$	$28.35\ \frac{gm}{oz}$	$14.594\ \frac{kg}{slug}$	$2.205\ \frac{lbm}{kg}$	$4.4482\ \frac{N}{lbf}$
$444{,}820\ \frac{dynes}{lbf}$	$453.6\ \frac{gm\ mole}{lb\ mole}$	$907.18\ \frac{kg}{ton}$	$32.174\ \frac{lbm}{slug}$	$16\ \frac{oz}{lbm}$
$32.174\ \frac{poundals}{lbf}$	$7000\ \frac{grains}{lbm}$	$1.0\ \frac{kilopond}{kg}$	$2000\ \frac{lbm}{ton}$	$10^5\ \frac{dynes}{N}$
$453.6\ \frac{g}{lbm}$	$1000\ \frac{kg}{metric\ ton}$	$1000\ \frac{lbf}{kip}$	$9.80665\ \frac{N}{kgf}$	
<u>Density</u>				
$0.51538\ \frac{g/cm^3}{slug/ft^3}$	$32.174\ \frac{lbm/ft^3}{slug/ft^3}$	$16.018\ \frac{kg/m^3}{lbm/ft^3}$	$1728\ \frac{lbm/ft^3}{lbm/in.^3}$	$1000\ \frac{kg/m^3}{g/cm^3}$

Table A.2 Continued

Pressure

$0.0361\ \frac{\text{psi}}{\text{in.}H_2O(60^\circ F)}$	$\frac{9.869}{10^7}\ \frac{\text{atm}}{\text{dyne/cm}^2}$	$29.921\ \frac{\text{in. Hg}(^\circ C)}{\text{atm}}$	$0.731\ \frac{\text{kg/m}^2}{\text{psi}}$	1.01325
$406.79\ \frac{\text{in. }H_2O(39.2^\circ F)}{\text{atm}}$	$101{,}325\ \frac{\text{N/m}^2}{\text{atm}}$	$14.504\ \frac{\text{psi}}{\text{bar}}$	$703.07\ \frac{\text{kg/m}^2}{\text{psi}}$	$6894.8\ \frac{\text{N/m}^2}{\text{psi}}$
$0.4898\ \frac{\text{psi}}{\text{in. Hg}(60^\circ F)}$	$10^6\ \frac{\text{dyne/cm}^2}{\text{bar}}$	$133.3\ \frac{\text{N/m}^2}{\text{torr}}$	$760\ \frac{\text{mm Hg}(0^\circ C)}{\text{atm}}$	$47.88\ \frac{\text{N/m}^2}{\text{psf}}$
$33.934\ \frac{\text{ft }H_2O(60^\circ F)}{\text{atm}}$	$10^5\ \frac{\text{N/m}^2}{\text{bar}}$	$1.0332\ \frac{\text{kg/cm}^2}{\text{atm}}$	$14.696\ \frac{\text{psi}}{\text{sym}}$	$760\ \frac{\text{torr}}{\text{atm}}$
$13.57\ \frac{\text{in. }H_2O(60^\circ F)}{\text{in. Hg}(60^\circ F)}$	$0.1\ \frac{\text{dyne/cm}^2}{\text{N/m}^2}$	$13.6\ \frac{\text{kg}}{\text{mm Hg}(0^\circ C)}$	$51.75\ \frac{\text{mm Hg}(0^\circ C)}{\text{psf}}$	

Energy and Power

$\frac{11.817}{10^{12}}\ \frac{\text{ft lbf}}{\text{MeV}}$	$2.7194\ \frac{\text{Btu}}{\text{atm ft}^3}$	$1.8\ \frac{\text{Btu/lbm}}{\text{cal/g}}$	$\frac{1.6021}{10^{12}}\ \frac{\text{ergs}}{\text{eV}}$	$1.055\ \frac{\text{kJ}}{\text{Btu}}$
$0.430\ \frac{\text{Btu/lb mole}}{\text{J/g mole}}$	$2544.4\ \frac{\text{Btu}}{\text{hp hr}}$	$10^7\ \frac{\text{ergs}}{\text{J}}$	$5050\ \frac{\text{hp h}}{\text{ft lbf}}$	$4.1868\ \frac{\text{kJ}}{\text{kcal}}$
$1800\ \frac{\text{Btu/lb mole}}{\text{kcal/g mole}}$	$42.4\ \frac{\text{Btu}}{\text{hp min}}$	$778.16\ \frac{\text{ft lbf}}{\text{Btu}}$	$1.3558\ \frac{\text{J}}{\text{ft lbf}}$	$3600\ \frac{\text{J}}{\text{kW hr}}$
$3412.2\ \frac{\text{Btu}}{\text{kW hr}}$	$251.98\ \frac{\text{cal}}{\text{Btu}}$	$550\ \frac{\text{ft lbf}}{\text{hp sec}}$	$\frac{16.021}{10^{12}}\ \frac{\text{J}}{\text{Mev}}$	$0.746\ \frac{\text{kW}}{\text{hp}}$
$737.562\ \frac{\text{ft lbf}}{\text{kW sec}}$	$860\ \frac{\text{cal}}{\text{W hr}}$	$56.87\ \frac{\text{Btu}}{\text{kW min}}$	$101.92\ \frac{\text{kg m}}{\text{kJ}}$	$1.0\ \frac{\text{kW sec}}{\text{kJ}}$

Specific Heat and Entropy

$1.0\ \frac{\text{Btu/lb}^\circ R}{\text{cal/g K}}$	$1.0\ \frac{\text{Btu/lb }^\circ R}{\text{kcal/kg K}}$	$1.0\ \frac{\text{Btu/lbm mole}^\circ R}{\text{cal/g mole K}}$	$0.2389\ \frac{\text{Btu/lbm mole}^\circ R}{\text{J/g mole}}$	$4.187\ \frac{\text{kJ/kg K}}{\text{Btu/lbm }^\circ R}$

Table A.3 SI Prefixes

Factor	Prefix	Symbol
10^{18}	exa	E
10^{15}	peta	P
10^{12}	tera	T
10^{9}	giga	G
10^{6}	mega	M
10^{3}	kilo	k
10^{2}	hecto	h
10^{1}	deka	da
10^{-1}	deci	d
10^{-2}	centi	c
10^{-3}	milli	m
10^{-6}	micro	μ
10^{-9}	nano	n
10^{-12}	pico	p
10^{-15}	femto	f
10^{-18}	atto	a

Table A.4 Engineers–SI Conversion Factors[a]

To convert from English Engineers units	To SI units	Multiply by
Acceleration (ft/sec^2)	meter/second^2 (m/sec^2)	3.048 000 E − 01
Btu	joules (J)	1.055 056 E + 03
Btu/gal (U.S. liquid)	kJ/m^3	2.787 200 E + 05
Btu/lbm	joules/kilogram (J/kg)	2.326 000 E + 03
Btu/lbm ^{0}R	joules/kilogram Kelvin (J/kg K)	4.184 000 E + 03
Btu/ft^3	joules/meter^3 (J/m^3)	3.725 895 E + 04
Btu/hr	watt (W)	2.930 711 E − 01
degree Celsius	kelvin (K)	t K = t°C + 273.15
degree Fahrenheit	degree Celsius	t°C = (t°F − 32)/1.8
degree Fahrenheit	kelvin (K)	t K = (t°F + 459.67)/1.8
degree Rankine	kelvin (K)	t K = t°R/1.8
foot	meter (m)	3.048 000 E − 01
foot of H_2O	pascal (Pa)	2.988 980 E + 03
ft^2	meter^2 (m^2)	9.290 304 E − 02
ft^3 (volume)	meter^3 (m^3)	2.831 685 E − 02
ft^3/lbm	meter^3/kg (m^3/kg)	6.242 797 E − 02
ft lbf	joule (J)	1.355 818 E + 00
ft lbf/lbm	joules/kilogram (J/kg)	2.989 067 E + 00
ft lbf/hr	watt (W)	3.766 161 E − 04
ft/sec^2	meter/sec (m/s^2)	3.048 000 E − 01
gallon (U.S. liquid)	meter^3 (m^3)	3.785 412 E − 03
horsepower (550 ft lbf/sec)	watt (W)	7.456 999 E + 02
inch	meter (m)	2.540 000 E − 02
inch of Hg (32°F)	pascal (Pa)	3.386 380 E + 03
inch^2	meter^2 (m^2)	6.451 600 E − 04
kelvin	degree Celsius	t°C = t K − 273.15
kilowatt hour	joule (J)	3.600 000 E + 06
liter	meter^3 (m^3)	1.000 000 E − 03
mile (statute)	meter (m)	1.609 300 E + 03
mile (U.S. nautical)	meter (m)	1.852 000 E + 03
pound mass (lbm avoirdupois)	kilogram (kg)	4.535 924 E − 01
lbm/ft^3	kilogram/meter^3 (kg/m^3)	1.601 846 E + 01
lbm/hp hr	kilogram/kilowatt hour (kg/kW hr)	1.689 659 E − 07
lbm/gal (U.S. liquid)	kg/m^3	1.198 264 E + 02
pound force (lbf)	newton (N)	4.448 222 E + 00
lbf/in.^{2} (psi)	pascal (Pa)	6.894 757 E + 03
ton (long, 2240 lbm)	kilogram (kg)	1.016 047 E + 03
ton (short, 2000 lbm)	kilogram (kg)	9.071 847 E + 02
velocity, ft/sec	meter/sec (m/sec)	3.048 000 E − 01
volume flow rate, ft^3/sec	meter^3/sec (m^3/sec)	2.831 685 E − 02

[a]Reference ASTM Standard for Metric Practice E380-76, 1976.

APPENDIX B. SOME THERMOCHEMICAL PROPERTIES

Table B.1 Standard State Heats of Formation and Combustion (Higher Heating Value) for Various Compounds

Substance	Formula	Molecular Weight	State	$\overline{h}^0_f$	$-\Delta\overline{H}^0_i$ [a]
Carbon	C_s	12.0111	solid	0.000	94,054
Carbon	C_g	12.0111	g	170,890	—
Methyl radical	CH_3	15.0350	g	34,820	—
Carbon monoxide	CO	28,0105	g	–26,420	67,636
Carbon dioxide	CO_2	44,0099	g	–94,054	—
Monatomic hydrogen	H	1.00797	g	52,100	—
Diatomic hydrogen	H_2	2.0160	g	0.000	68,317
Water	H_2O	18.016	ℓ	–68,317	—
Water	H_2O	18.016	g	–57,798	—
Hydrogen peroxide	H_2O_2	34.015	g	–32,576	—
Hydrogen sulfide	H_2S	34.080	g	–4,815	—
Monatomic nitrogen	N	14.0067	g	113,000	—
Nitric oxide	NO	30.008	g	21,580	—
Nitrogen dioxide	NO_2	46.008	g	7,910	—
Diatomic nitrogen	N_2	28.0134	g	0.000	—
Ammonia	NH_3	17.036	g	–10,970	—
Hydrazine	N_2H_4	32.045	g	22,790	—
Dinitrogen monoxide	N_2O	44.016	g	19,610	—
Dinitrogen tetroxide	N_2O_4	92.016	g	2,170	—
Monatomic oxygen	O	15.9994	g	59,559	—
Hydroxyl radical	OH	17.0074	g	9,432	—
Diatomic oxygen	O_2	31.9988	g	0.000	—
Ozone	O_3	47.9982	g	34,200	—
Monatomic sulfur	S	32.064	g	66,680	—
Sulfur	S	32.064	solid	0.000	—
Sulfur dioxide	SO_2	64.066	g	–70,947	—
		Paraffins			
Methane	CH_4	16.041	g	–17,889	212,800
Ethane	C_2H_6	30.067	g	–20,236	372,820
Propane	C_3H_8	44.092	g	–24,820	530,600
n–Butane	C_4H_{10}	58.118	ℓ	–35,390	682,410
n–Butane	C_4H_{10}	58.118	g	–30,150	687,640
iso–Butane	C_4H_{10}	58.118	ℓ	–37,470	680,330
iso–Butane	C_4H_{10}	58.118	g	–32,400	685,400
n–Pentane	C_5H_{12}	72.144	ℓ	–41,890	851,340
n–Pentane	C_5H_{12}	72.144	g	–35,718	845,160
iso–Pentane	C_5H_{12}	72.144	ℓ	–43,590	836,580
iso–Pentane	C_5H_{12}	72.144	g	–37,700	842,470
n–Hexane	C_6H_{14}	86.169	ℓ	–46,850	995,690
n–Hexane	C_6H_{14}	86.169	g	–39,960	1,002,570
		g/g mole		cal/g mole	

Table B.1 Continued

Substance	Formula	Molecular Weight	State	$\overline{h}_f^0$	$-\Delta\overline{H}_i^0$ [a]
n–Heptane	C_6H_{14}	100.198	ℓ	–54,380	1,150,430
n–Heptane	C_6H_{14}	100.198	g	–45,645	1,159,170
n–Octane	C_8H_{18}	114.224	ℓ	–59,740	1,306,590
n–Octane	C_8H_{18}	114.224	g	–49,820	1,316,490
iso–Octane	C_8H_{18}	114.224	ℓ	–62,885	1,304,400
iso–Octane	C_8H_{18}	114.224	g	–55,495	1,311,790
n–Decane	$C_{10}H_{22}$	142.276	ℓ	–72,875	1,618,930
n–Decane	$C_{10}H_{22}$	142.276	g	–60,697	1,631,180
n–Dodecane	$C_{12}H_{26}$	170.328	ℓ	–85,370	1,931,220
n–Dodecane	$C_{12}H_{26}$	170.328	g	–70,800	1,945,790
(increment per above C_6)	C atom	—	g	(~4,925)	(157,440)
		Olefins			
Ethylene	C_2H_4	28.054	g	–12,496	337,150
Propylene	C_3H_6	42.081	g	–4,879	491,990
n–Butene	C_4H_8	56.108	g	–30	649,380
n–Pentene	C_5H_{10}	70.135	g	–5,000	806,700
n–Hexene	C_6H_{12}	84.162	g	–9,970	964,240
n–Heptene	C_7H_{14}	98.189	g	–89,176	1,047,400
n–Octene	C_8H_{16}	112.208	g	–20,740	1,278,230
n–Nonene	C_9H_{18}	126.243	g	–120,262	1,341,050
n–Decene	$C_{10}H_{20}$	140.270	g	–135,810	1,487,870
(increment per above C_6)	C atom	—	g	(~4,925)	(157,400)
	Additional Hydrocarbon Fuels				
Acetylene	C_2H_2	26.038	g	–54,194	310,620
1–3–Butadiene	C_4H_6	54.092	g	–26,330	607,490
Benzene	C_6H_6	78.114	ℓ	–11,720	780,980
Benzene	C_6H_6	78.114	g	–19,820	789,080
Cyclohexane	C_6H_{12}	84.162	ℓ	–37,340	936,860
Cyclohexane	C_6H_{12}	84.162	g	–29,430	944,770
Toluene	C_7H_8	92.142	ℓ	–2,870	934,500
Toluene	C_7H_8	92.142	g	–11,950	943,580
Methylcyclohexane	C_7H_{14}	98.189	ℓ	–45,450	1,091,130
Methylcyclohexane	C_7H_{14}	98.189	g	–36,990	1,099,580
Styrene	C_8H_8	104.153	g	–35,220	1,060,900
		g/g mole		cal/g mole	

Table B.1 Continued

Substance	Formula	Molecular Weight	State	$\overline{h}_f^0$	$-\Delta \overline{H}_c^0$ [a]
		Alcohol fuels			
Methanol	CH_3OH	32.042	ℓ	–57,110	173,600
Methanol	CH_3OH	32.042	g	–48,050	182,640
Ethanol	C_2H_5OH	46.069	ℓ	–66,200	326,860
Ethanol	C_3H_5OH	46.069	g	–56,030	337,020
Propanol	C_3H_7OH	60.096	ℓ	–71,340	484,080
Butanol	C_4H_9OH	74.124	ℓ	–79,540	638,250
		g/g mole		cal/g mole	

Sources:
- ASTM Special Technical Publications No. 109A, 1963, and No. 225, 1958
- Selected Values of Physical and Thermodynamic Properties of Hydrocarbons and Related Compounds, American Petroleum Institute Research Project 44, Carnegie Press, 1953
- Selected Values of Chemical Thermodynamic Properties, National Bureau of Standards, U.S. Circ. 500, 1952
- JANAF Thermochemical Tables, 2nd Edition, National Bureau of Standards, Publication NSRDS–NBS37, 1971

[a]The heat of combustion assumes the produces are H_2O_ℓ and CO_{2_g}.
NOTE: To convert the molar values above to Btu/lb mole and kJ/kg mole, multiply by 1.8001 and 4.187, respectively.
To convert the above molar values to a mass basis, divide by the molecular weight, i.e.,

[MW] = lbm/lb mole (kg/kg mole)

Examples:

$$(h_f^0)_{H_2O_g} = (-57{,}798 \text{ cal/g mole})\left(1.8001 \frac{\text{Btu/lb mole}}{\text{cal/g mole}}\right)$$

$$= -104{,}040 \text{ Btu/lb mole}$$

$$(\Delta H_c^0)_{CH_4} = \frac{(212{,}800)\,(4.187) \text{ kJ/kg mole}}{16.043 \text{ kg/kg mole}}$$

$$= 55{,}540 \text{ kJ/kg}$$

Table B.2 Thermochemical Properties of Air

Air[a]
$MW = 28.964$
$\bar{h}^0_f = 0.000$ kcal/g mole

T		$\bar{C}^0_p$	$\bar{h}<T>-\bar{h}<T_0>$	$\bar{s}^0<T>$	$\Delta G^0<T>$	$\log K_p$
0	0	0.000	−2.050	0.000	0.000	0.000
100	180	7.050	−1.359	38.704	0.000	0.000
200	360	6.956	−0.666	43.556	0.000	0.000
298	536	6.947	0.000	46.255	0.000	0.000
300	540	6.949	0.035	46.372	0.000	0.000
400	720	7.010	0.732	48.378	0.000	0.000
500	900	7.122	1.439	49.954	0.000	0.000
600	1080	7.268	2.158	51.265	0.000	0.000
700	1260	7.432	2.893	52.397	0.000	0.000
800	1440	7.598	3.644	53.400	0.000	0.000
900	1620	7.756	4.412	54.305	0.000	0.000
1000	1800	7.893	5.195	55.129	0.000	0.000
1100	1980	8.006	5.990	55.887	0.000	0.000
1200	2160	8.109	6.796	56.588	0.000	0.000
1300	2340	8.203	7.611	57.241	0.000	0.000
1400	2520	8.289	8.436	57.852	0.000	0.000
1500	2700	8.367	9.269	58.427	0.000	0.000
1600	2880	8.437	10.109	58.969	0.000	0.000
1700	3060	8.502	10.956	59.482	0.000	0.000
1800	3240	8.560	11.809	59.970	0.000	0.000
1900	3420	8.612	12.668	60.434	0.000	0.000
2000	3600	8.661	13.532	60.877	0.000	0.000
2100	3780	8.704	14.400	61.301	0.000	0.000
2200	3960	8.743	15.272	61.707	0.000	0.000
2300	4140	8.779	16.148	62.097	0.000	0.000
2400	4320	8.812	17.028	62.471	0.000	0.000
2500	4500	8.842	17.911	62.831	0.000	0.000
2600	4680	8.869	18.796	63.178	0.000	0.000
2700	4860	8.895	19.684	63.513	0.000	0.000
2800	5040	8.918	20.575	63.837	0.000	0.000
2900	5220	8.940	21.468	64.150	0.000	0.000
3000	5400	8.962	22.363	64.454	0.000	0.000
3100	5580	8.987	23.249	64.545	0.000	0.000
3200	5760	9.008	24.149	64.740	0.000	0.000
3300	5940	9.027	25.051	65.018	0.000	0.000
3400	6120	9.047	25.955	65.288	0.000	0.000
3500	6300	9.065	26.860	65.550	0.000	0.000
K	°R	cal/(g mole K)	kcal/g mole	cal/(g mole K)	kcal/g mole	—

[a]78.11% N_2; 20.96% O_2; 0.93% argon by volume

Sources: JANAF Thermochemical Tables, 2nd Edition, National Bureau of Standards, Publication NSRDS–NBS37, 1971

Table B.3 Thermochemical Properties of Carbon

Carbon (C_S)
$MW = 12.011$
$\bar{h}_f^0 = 0.000$ kcal/g mole

T		$\bar{C}_p^0$	$\bar{h}<T>-\bar{h}<T_0>$	$\bar{s}^0<T>$	$\Delta G^0<T>$	$\log K_p$
0	0	0.000	−0.252	0.000	0.000	0.000
100	180	0.395	−0.238	0.210	0.000	0.000
200	360	1.202	−0.160	0.720	0.000	0.000
298	536	2.038	0.000	1.359	0.000	0.000
300	540	2.054	0.004	1.372	0.000	0.000
400	720	2.851	0.250	2.075	0.000	0.000
500	900	3.496	0.569	2.784	0.000	0.000
600	1080	4.038	0.947	3.471	0.000	0.000
700	1260	4.440	1.372	4.126	0.000	0.000
800	1440	4.740	1.831	4.739	0.000	0.000
900	1620	4.970	2.318	5.311	0.000	0.000
1000	1800	5.149	2.824	5.844	0.000	0.000
1100	1980	5.304	3.347	6.342	0.000	0.000
1200	2160	5.430	3.883	6.809	0.000	0.000
1300	2340	5.527	4.432	7.248	0.000	0.000
1400	2520	5.605	4.988	7.661	0.000	0.000
1500	2700	5.669	5.552	8.050	0.000	0.000
1600	2880	5.721	6.122	8.417	0.000	0.000
1700	3060	5.765	6.696	8.765	0.000	0.000
1800	3240	5.803	7.275	9.096	0.000	0.000
1900	3420	5.836	7.857	9.411	0.000	0.000
2000	3600	5.865	8.442	9.711	0.000	0.000
2100	3780	5.891	9.029	9.998	0.000	0.000
2200	3960	5.914	9.620	10.272	0.000	0.000
2300	4140	5.936	10.212	10.536	0.000	0.000
2400	4320	5.956	10.807	10.789	0.000	0.000
2500	4500	5.974	11.403	11.032	0.000	0.000
2600	4680	5.992	12.002	11.267	0.000	0.000
2700	4860	6.009	12.602	11.493	0.000	0.000
2800	5040	6.026	13.203	11.712	0.000	0.000

K	°R	$\frac{cal}{g\ mole\ K}$	$\frac{kcal}{g\ mole}$	$\frac{cal}{g\ mole\ K}$	$\frac{kcal}{g\ mole}$	—
2900	5220	6.042	13.807	11.924	0.000	0.000
3000	5400	6.057	14.412	12.129	0.000	0.000
3100	5580	6.073	15.018	12.328	0.000	0.000
3200	5760	6.088	15.626	12.521	0.000	0.000
3300	5940	6.103	16.236	12.708	0.000	0.000
3400	6120	6.119	16.847	12.891	0.000	0.000
3500	6300	6.134	17.460	13.068	0.000	0.000
3600	6480	6.150	18.074	13.241	0.000	0.000
3700	6660	6.165	18.690	13.410	0.000	0.000
3800	6840	6.181	19.307	13.575	0.000	0.000
3900	7020	6.197	19.926	13.736	0.000	0.000
4000	7200	6.213	20.546	13.893	0.000	0.000
4100	7380	6.230	21.168	14.046	0.000	0.000
4200	7560	6.247	21.792	14.197	0.000	0.000
4300	7740	6.264	22.418	14.344	0.000	0.000
4400	7920	6.281	23.045	14.488	0.000	0.000
4500	8100	6.299	23.674	14.629	0.000	0.000
4600	8280	6.317	24.305	14.768	0.000	0.000
4700	8460	6.335	24.937	14.904	0.000	0.000
4800	8640	6.354	25.572	15.038	0.000	0.000
4900	8820	6.373	26.208	15.169	0.000	0.000
5000	9000	6.392	26.846	15.298	0.000	0.000
5100	9180	6.412	27.487	15.424	0.000	0.000
5200	9360	6.432	28.129	15.549	0.000	0.000
5300	9540	6.452	28.773	15.672	0.000	0.000
5400	9720	6.473	29.419	15.793	0.000	0.000
5500	9900	6.494	30.068	15.912	0.000	0.000
5600	10080	6.516	30.718	16.029	0.000	0.000
5700	10260	6.538	31.371	16.144	0.000	0.000
5800	10440	6.560	32.026	16.258	0.000	0.000
5900	10620	6.583	32.683	16.371	0.000	0.000
6000	10800	6.606	33.342	16.481	0.000	0.000

Sources: JANAF Thermochemical Tables, 2nd Edition, National Bureau of Standards, Publication NSRDS–NBS37, 1971

Table B.4 Thermochemical Properties of Methane

Methane (CH_4)
$MW = 16.043$
$\bar{h}^0_f = -17.895$ kcal/g mole

T		$\bar{C}^0_p$	$\bar{h}<T>-\bar{h}<T_0>$	$\bar{s}^0<T>$	$\Delta G^0<T>$	$\log K_p$
0	0	0.000	−2.396	0.000	−15.991	infinite
100	180	7.949	−1.601	35.706	−15.400	33.656
200	360	8.001	−0.805	41.222	−13.909	15.198
298	536	8.518	0.000	44.490	−12.145	8.902
300	540	8.535	0.016	44.543	−12.110	8.822
400	720	9.680	0.923	47.144	−10.066	5.500
500	900	11.076	1.960	49.453	−7.845	3.429
600	1080	12.483	3.138	51.597	−5.493	2.001
700	1260	13.813	4.454	53.622	−3.046	0.951
800	1440	15.041	5.897	55.548	−0.533	0.146
900	1620	16.157	7.458	57.385	2.029	−0.493
1000	1800	17.160	9.125	59.141	4.625	−1.011
1100	1980	18.052	10.887	60.819	7.247	−1.440
1200	2160	18.842	12.732	62.424	9.887	−1.801
1300	2340	19.538	14.652	63.960	12.535	−2.107
1400	2520	20.150	16.637	65.431	15.195	−2.372
1500	2700	20.688	18.679	66.840	17.859	−2.602
1600	2880	21.161	20.772	68.191	20.520	−2.803
1700	3060	21.579	22.910	69.486	23.189	−2.981
1800	3240	21.947	25.086	70.730	25.854	−3.139
1900	3420	22.273	27.298	71.926	28.522	−3.281
2000	3600	22.562	29.540	73.076	31.187	−3.408
2100	3780	22.820	31.809	74.183	33.851	−3.523
2200	3960	23.050	34.103	75.250	36.511	−3.627
2300	4140	23.256	36.418	76.279	39.173	−3.722
2400	4320	23.441	38.753	77.273	41.833	−3.809
2500	4500	23.608	41.106	78.233	44.483	−3.889
2600	4680	23.758	43.474	79.162	47.141	−3.962
2700	4860	23.894	45.857	80.062	49.791	−4.030
2800	5040	24.018	48.253	80.933	52.440	−4.093

K	°R	$\frac{cal}{g\ mole\ K}$	$\frac{kcal}{g\ mole}$	$\frac{cal}{g\ mole\ K}$	$\frac{kcal}{g\ mole}$	—
2900	5220	24.131	50.660	81.778	55.093	−4.152
3000	5400	24.233	53.079	82.597	57.736	−4.206
3100	5580	24.327	55.507	83.394	60.381	−4.257
3200	5760	24.413	57.944	84.167	63.026	−4.304
3300	5940	24.493	60.389	84.920	65.669	−4.349
3400	6120	24.565	62.842	85.652	68.309	−4.391
3500	6300	24.633	65.302	86.365	70.951	−4.430
3600	6480	24.695	67.768	87.060	73.589	−4.467
3700	6660	24.752	70.241	87.737	76.231	−4.503
3800	6840	24.806	72.719	88.398	78.872	−4.536
3900	7020	24.855	75.202	89.043	81.511	−4.568
4000	7200	24.901	77.690	89.673	84.150	−4.598
4100	7380	24.944	80.182	90.288	86.785	−4.626
4200	7560	24.984	82.678	90.890	89.429	−4.653
4300	7740	25.022	85.179	91.478	92.063	−4.679
4400	7920	25.057	87.683	92.054	94.700	−4.704
4500	8100	25.090	90.190	92.617	97.335	−4.727
4600	8280	25.121	92.701	93.169	99.983	−4.750
4700	8460	25.150	95.214	93.710	102.625	−4.772
4800	8640	25.177	97.730	94.240	105.268	−4.793
4900	8820	25.203	100.249	94.759	107.912	−4.813
5000	9000	25.227	102.771	95.268	110.552	−4.832
5100	9180	25.250	105.295	95.768	113.198	−4.851
5200	9360	25.272	107.821	96.259	115.844	−4.869
5300	9540	25.292	110.349	96.740	118.501	−4.886
5400	9720	25.311	112.879	97.213	121.145	−4.903
5500	9900	25.330	115.411	97.678	123.799	−4.919
5600	10080	25.347	117.945	98.134	126.449	−4.935
5700	10260	25.364	120.481	98.583	129.106	−4.950
5800	10440	25.379	123.018	99.024	131.762	−4.965
5900	10620	25.394	125.557	99.458	134.428	−4.979
6000	10800	25.409	128.097	99.885	137.081	−4.993

Sources: JANAF Thermochemical Tables, 2nd Edition, National Bureau of Standards, Publication NSRDS–NBS37, 1971

Table B.5 Thermochemical Properties of Carbon Monoxide

Carbon Monoxide (CO)
$MW = 28.01055$

$\bar{h}_f^0 = -26.417$ kcal/g mole

T		$\bar{C}_p^0$	$\bar{h}<T>-\bar{h}<T_0>$	$\bar{s}^0<T>$	$\Delta G^0<T>$	$\log K_p$
0	0	0.000	-2.072	0.000	-27.200	infinite
100	180	6.956	-1.379	39.613	-28.741	62.809
200	360	6.957	-0.683	44.435	-30.718	33.566
298	536	6.965	0.000	47.214	-32.783	24.029
300	540	6.965	0.013	47.257	-32.823	23.910
400	720	7.013	0.711	49.265	-34.975	19.109
500	900	7.121	1.417	50.841	-37.144	16.235
600	1080	7.276	2.137	52.152	-39.311	14.318
700	1260	7.450	2.873	53.287	-41.468	12.946
800	1440	7.624	3.627	54.293	-43.612	11.914
900	1620	7.786	4.397	55.200	-45.744	11.108
1000	1800	7.931	5.183	56.028	-47.859	10.459
1100	1980	8.057	5.983	56.790	-49.962	9.926
1200	2160	8.168	6.794	57.496	-52.049	9.479
1300	2340	8.263	7.616	58.154	-54.126	9.099
1400	2520	8.346	8.446	58.769	-56.189	8.771
1500	2700	8.417	9.285	59.348	-58.241	8.485
1600	2880	8.480	10.130	59.893	-60.284	8.234
1700	3060	8.535	10.980	60.409	-62.315	8.011
1800	3240	8.583	11.836	60.898	-64.337	7.811
1900	3420	8.626	12.697	61.363	-66.349	7.631
2000	3600	8.664	13.561	61.807	-68.353	7.469
2100	3780	8.698	14.430	62.230	-70.346	7.321
2200	3960	8.728	15.301	62.635	-72.335	7.185
2300	4140	8.756	16.175	63.024	-74.311	7.061
2400	4320	8.781	17.052	63.397	-76.282	6.946
2500	4500	8.804	17.931	63.756	-78.247	6.840
2600	4680	8.825	18.813	64.102	-80.202	6.741
2700	4860	8.844	19.696	64.435	-82.153	6.649
2800	5040	8.863	20.582	64.757	-84.093	6.563
2900	5220	8.879	21.469	65.069	-86.028	6.483

K	°R					
3000	5400	8.895	22.357	65.370	−87.957	6.407
3100	5580	8.910	23.248	65.662	−89.878	6.336
3200	5760	8.924	24.139	65.945	−91.795	6.269
3300	5940	8.937	25.032	66.220	−93.707	6.206
3400	6120	8.949	25.927	66.487	−95.609	6.145
3500	6300	8.961	26.822	66.746	−97.509	6.088
3600	6480	8.973	27.719	66.999	−99.400	6.034
3700	6660	8.984	28.617	67.245	−101.286	5.982
3800	6840	8.994	29.516	67.485	−103.164	5.933
3900	7020	9.004	30.416	67.718	−105.039	5.886
4000	7200	9.014	31.316	67.946	−106.908	5.841
4100	7380	9.024	32.218	68.169	−108.774	5.798
4200	7560	9.033	33.121	68.387	−110.630	5.756
4300	7740	9.042	34.025	68.599	−112.483	5.717
4400	7920	9.051	34.930	68.807	−114.333	5.679
4500	8100	9.059	35.835	69.011	−116.177	5.642
4600	8280	9.068	36.741	69.210	−118.012	5.607
4700	8460	9.076	37.649	69.405	−119.845	5.573
4800	8640	9.084	38.557	69.596	−121.672	5.540
4900	8820	9.092	39.465	69.784	−123.497	5.508
5000	9000	9.100	40.375	69.967	−125.315	5.477
5100	9180	9.107	41.285	70.148	−127.132	5.448
5200	9360	9.115	42.196	70.325	−128.941	5.419
5300	9540	9.123	43.108	70.498	−130.741	5.391
5400	9720	9.130	44.021	70.669	−132.542	5.364
5500	9900	9.138	44.934	70.836	−134.336	5.338
5600	10080	9.145	45.849	71.001	−136.129	5.312
5700	10260	9.153	46.763	71.163	−137.919	5.288
5800	10440	9.160	47.679	71.322	−139.698	5.264
5900	10620	9.167	48.595	71.479	−141.473	5.240
6000	10800	9.175	49.513	71.633	−143.249	5.218
K	°R	$\frac{cal}{g\ mole\ K}$	$\frac{kcal}{g\ mole}$	$\frac{cal}{g\ mole\ K}$	$\frac{kcal}{g\ mole}$	—

Sources: JANAF Thermochemical Tables, 2nd Edition, National Bureau of Standards, Publication NSRDS–NBS37, 1971

Table B.6 Thermochemical Properties of Carbon Dioxide

Carbon Dioxide (CO_2)

$MW = 44.00995$

$\bar{h}^0_f = -94.054$ kcal/g mole

T		$\bar{C}^0_p$	$\bar{h}\langle T\rangle - \bar{h}\langle T_0\rangle$	$\bar{s}^0\langle T\rangle$	$\Delta G^0\langle T\rangle$	$\log K_p$
0	0	0.000	−2.238	0.000	−93.965	infinite
100	180	6.981	−1.543	42.758	−94.100	205.645
200	360	7.734	−0.816	47.769	−94.191	102.922
298	536	8.874	0.000	51.072	−94.265	69.095
300	540	8.896	0.016	51.127	−94.267	68.670
400	720	9.877	0.958	53.830	−94.335	51.540
500	900	10.666	1.987	56.122	−94.399	41.260
600	1080	11.310	3.087	58.126	−94.458	34.405
700	1260	11.846	4.245	59.910	−94.510	29.506
800	1440	12.293	5.453	61.522	−94.556	25.830
900	1620	12.667	6.702	62.992	−94.596	22.970
1000	1800	12.980	7.984	64.344	−94.628	20.680
1100	1980	13.243	9.296	65.594	−94.658	18.806
1200	2160	13.466	10.632	66.756	−94.681	17.243
1300	2340	13.656	11.988	67.841	−94.701	15.920
1400	2520	13.815	13.362	68.859	−94.716	14.785
1500	2700	13.953	14.750	69.817	−94.728	13.801
1600	2880	14.074	16.152	70.722	−94.739	12.940
1700	3060	14.177	17.565	71.578	−94.746	12.180
1800	3240	14.269	18.987	72.391	−94.750	11.504
1900	3420	14.352	20.418	73.165	−94.751	10.898
2000	3600	14.424	21.857	73.903	−94.752	10.353
2100	3780	14.489	23.303	74.608	−94.746	9.860
2200	3960	14.547	24.755	75.284	−94.744	9.411
2300	4140	14.600	26.212	75.931	−94.735	9.001
2400	4320	14.648	27.674	76.554	−94.724	8.625
2500	4500	14.692	29.141	77.153	−94.714	8.280
2600	4680	14.734	30.613	77.730	−94.698	7.960
2700	4860	14.771	32.088	78.286	−94.683	7.664
2800	5040	14.807	33.567	78.824	−94.662	7.388

2900	5220	14.841	35.049	79.344	−94.639	7.132
3000	5400	14.873	36.535	79.848	−94.615	6.892
3100	5580	14.902	38.024	80.336	−94.587	6.668
3200	5760	14.930	39.515	80.810	−94.560	6.458
3300	5940	14.956	41.010	81.270	−94.531	6.260
3400	6120	14.982	42.507	81.717	−94.495	6.074
3500	6300	15.006	44.006	82.151	−94.462	5.898
3600	6480	15.030	45.508	82.574	−94.421	5.732
3700	6660	15.053	47.012	82.986	−94.379	5.574
3800	6840	15.075	48.518	83.388	−94.331	5.425
3900	7020	15.097	50.027	83.780	−94.286	5.283
4000	7200	15.119	51.538	84.162	−94.237	5.149
4100	7380	15.139	53.051	84.536	−94.186	5.020
4200	7560	15.159	54.566	84.901	−94.130	4.898
4300	7740	15.179	56.082	85.258	−94.072	4.781
4400	7920	15.197	57.601	85.607	−94.015	4.670
4500	8100	15.216	59.122	85.949	−93.954	4.563
4600	8280	15.234	60.644	86.284	−93.885	4.460
4700	8460	15.254	62.169	86.611	−93.818	4.362
4800	8640	15.272	63.695	86.933	−93.746	4.268
4900	8820	15.290	65.223	87.248	−93.678	4.178
5000	9000	15.306	66.753	87.557	−93.603	4.091
5100	9180	15.327	68.285	87.860	−93.528	4.008
5200	9360	15.349	69.819	88.158	−93.450	3.927
5300	9540	15.371	71.355	88.451	−93.361	3.850
5400	9720	15.393	72.893	88.738	−93.280	3.775
5500	9900	15.415	74.433	89.021	−93.190	3.703
5600	10080	15.437	75.976	89.299	−93.104	3.633
5700	10260	15.459	77.521	89.572	−93.017	3.566
5800	10440	15.481	79.068	89.841	−92.918	3.501
5900	10620	15.503	80.617	90.106	−92.820	3.438
6000	10800	15.525	82.168	90.367	−92.724	3.377
K	°R	cal/(g mole K)	kcal/(g mole)	cal/(g mole K)	kcal/(g mole)	—

Sources: JANAF Thermochemical Tables, 2nd Edition, National Bureau of Standards, Publication NSRDS–NBS37, 1971

Table B.7 Thermochemical Properties of Acetylene

Acetylene
$MW = 26.038$

$\bar{h}_f^0 = 54.190$ kcal/g mole

T		$\bar{C}_p^0$	$\bar{h}<T>-\bar{h}<T_0>$	$\bar{s}^0<T>$	$\Delta G^0<T>$	log K_p
0	0	0.000	−2.393	0.000	54.325	infinite
100	180	7.014	−1.698	39.002	52.814	−115.418
200	360	8.505	−0.938	44.213	51.383	−56.146
298	536	10.539	0.000	48.004	49.993	−36.644
300	540	10.571	0.020	48.069	49.966	−36.399
400	720	12.065	1.155	51.326	48.567	−26.534
500	900	13.114	2.418	54.139	47.181	−20.622
600	1080	13.931	3.771	56.604	45.813	−16.687
700	1260	14.615	5.199	58.805	44.466	−13.882
800	1440	15.239	6.693	60.798	43.137	−11.784
900	1620	15.801	8.245	62.625	41.821	−10.155
1000	1800	16.318	9.852	64.317	40.522	−8.856
1100	1980	16.789	11.507	65.895	39.234	−7.795
1200	2160	17.221	13.208	67.375	37.960	−6.913
1300	2340	17.613	14.950	68.769	36.690	−6.168
1400	2520	17.968	16.729	70.087	35.432	−5.531
1500	2700	18.291	18.543	71.338	34.177	−4.979
1600	2880	18.582	20.387	72.528	32.923	−4.497
1700	3060	18.845	22.258	73.663	31.679	−4.072
1800	3240	19.085	24.155	74.747	30.436	−3.695
1900	3420	19.302	26.074	75.785	29.199	−3.358
2000	3600	19.504	28.015	76.780	27.962	−3.055
2100	3780	19.684	29.974	77.736	26.730	−2.782
2200	3960	19.853	31.951	78.656	25.493	−2.532
2300	4140	20.004	33.944	79.541	24.266	−2.306
2400	4320	20.151	35.952	80.396	23.037	−2.098
2500	4500	20.282	37.974	81.221	21.804	−1.906
2600	4680	20.404	40.008	82.019	20.579	−1.730
2700	4860	20.519	42.055	82.791	19.349	−1.566
2800	5040	20.625	44.112	83.540	18.124	−1.415

K	°R					
2900	5220	20.726	46.179	84.265	16.901	-1.274
3000	5400	20.820	48.257	84.969	15.674	-1.142
3100	5580	20.910	50.343	85.654	14.451	-1.019
3200	5760	20.996	52.439	86.319	13.227	-0.903
3300	5940	21.078	54.542	86.966	12.000	-0.795
3400	6120	21.154	56.664	87.596	10.779	-0.693
3500	6300	21.225	58.773	88.211	9.554	-0.597
3600	6480	21.297	60.899	88.810	8.331	-0.506
3700	6660	21.367	63.032	89.394	7.111	-0.420
3800	6840	21.431	65.172	89.965	5.894	-0.339
3900	7020	21.494	67.319	90.522	4.675	-0.262
4000	7200	21.557	69.471	91.067	3.455	-0.189
4100	7380	21.615	71.630	91.600	2.230	-0.119
4200	7560	21.670	73.794	92.122	1.017	-0.053
4300	7740	21.728	75.964	92.632	-0.205	0.010
4400	7920	21.782	78.139	93.133	-1.425	0.071
4500	8100	21.835	80.320	93.623	-2.648	0.129
4600	8280	21.883	82.506	94.103	-3.858	0.183
4700	8460	21.935	84.697	94.574	-5.073	0.236
4800	8640	21.985	86.893	95.037	-6.286	0.286
4900	8820	22.036	89.094	95.490	-7.500	0.335
5000	9000	22.077	91.300	95.936	-8.715	0.381
5100	9180	22.129	93.510	96.374	-9.935	0.426
5200	9360	22.174	95.725	96.804	-11.144	0.468
5300	9540	22.219	97.945	97.227	-12.348	0.509
5400	9720	22.263	100.169	97.642	-13.559	0.549
5500	9900	22.309	102.397	98.051	-14.767	0.587
5600	10080	22.349	104.630	98.454	-15.977	0.624
5700	10260	22.393	106.867	98.850	-17.188	0.659
5800	10440	22.433	109.108	99.239	-18.394	0.693
5900	10620	22.474	111.354	99.623	-19.588	0.726
6000	10800	22.521	113.603	100.001	-20.802	0.758
K	°R	$\frac{cal}{g\ mole\ K}$	$\frac{kcal}{g\ mole}$	$\frac{cal}{g\ mole\ K}$	$\frac{kcal}{g\ mole}$	—

Sources: JANAF Thermochemical Tables, 2nd Edition, National Bureau of Standards, Publication NSRDS–NBS37, 1971

Table B.8 Thermochemical Properties of Ethylene

Ethylene (C_2H_4)
MW = **28.05418**
$\bar{h}^0_f$ = **12.540 kcal/g mole**

T		$\bar{C}^0_p$	$\bar{h}<T>-\bar{h}<T_0>$	$\bar{s}^0<T>$	$\Delta G^0<T>$	log K_p
0	0	0.000	−2.514	0.000	14.578	infinite
100	180	7.952	−1.719	43.125	14.434	−31.544
200	360	8.451	−0.909	48.721	15.227	−16.638
298	536	10.250	0.000	52.396	16.338	−11.975
300	540	10.292	0.019	52.459	16.361	−11.918
400	720	12.679	1.167	55.745	17.752	−9.699
500	900	14.933	2.550	58.821	19.319	−8.444
600	1080	16.889	4.143	61.721	21.008	−7.652
700	1260	18.574	5.918	64.454	22.788	−7.114
800	1440	20.039	7.851	67.033	24.628	−6.728
900	1620	21.320	9.920	69.468	26.514	−6.438
1000	1800	22.443	12.109	71.774	28.431	−6.213
1100	1980	23.427	14.404	73.960	30.373	−6.034
1200	2160	24.290	16.791	76.036	32.334	−5.889
1300	2340	25.044	19.258	78.011	34.302	−5.766
1400	2520	25.706	21.797	79.892	36.282	−5.664
1500	2700	26.285	24.397	81.686	38.266	−5.575
1600	2880	26.794	27.051	83.399	40.246	−5.497
1700	3060	27.242	29.753	85.037	42.237	−5.430
1800	3240	27.636	32.498	86.605	44.224	−5.369
1900	3420	27.986	35.279	88.109	46.215	−5.316
2000	3600	28.296	38.094	89.552	48.204	−5.267
2100	3780	28.571	40.937	90.940	50.192	−5.223
2200	3960	28.818	43.807	92.275	52.174	−5.183
2300	4140	29.038	46.700	93.561	54.163	−5.146
2400	4320	29.236	49.614	94.801	56.148	−5.113
2500	4500	29.414	52.546	95.998	58.124	−5.081
2600	4680	29.575	55.496	97.155	60.109	−5.052
2700	4860	29.721	58.461	98.274	62.085	−5.025
2800	5040	29.853	61.440	99.357	64.064	−5.000

K	°R	$\frac{cal}{g\ mole\ K}$	$\frac{kcal}{g\ mole}$	$\frac{cal}{g\ mole\ K}$	$\frac{kcal}{g\ mole}$	—
2900	5220	29.973	64.431	100.407	66.047	−4.977
3000	5400	30.083	67.434	101.425	68.019	−4.955
3100	5580	30.184	70.447	102.413	69.996	−4.934
3200	5760	30.276	73.470	103.373	71.971	−4.915
3300	5940	30.360	76.502	104.305	73.947	−4.897
3400	6120	30.438	79.542	105.213	75.920	−4.880
3500	6300	30.510	82.590	106.096	77.894	−4.864
3600	6480	30.577	85.644	106.957	79.864	−4.848
3700	6660	30.638	88.705	107.795	81.844	−4.834
3800	6840	30.695	91.772	108.613	83.822	−4.821
3900	7020	30.748	94.844	109.411	85.798	−4.808
4000	7200	30.797	97.921	110.190	87.775	−4.796
4100	7380	30.843	101.003	110.951	89.746	−4.784
4200	7560	30.886	104.090	111.695	91.729	−4.773
4300	7740	30.926	107.180	112.422	93.703	−4.762
4400	7920	30.964	110.275	113.134	95.678	−4.752
4500	8100	30.999	113.373	113.830	97.653	−4.742
4600	8280	31.032	116.475	114.512	99.643	−4.734
4700	8460	31.063	119.579	115.180	101.628	−4.725
4800	8640	31.093	122.687	115.834	103.617	−4.718
4900	8820	31.120	125.798	116.475	105.608	−4.710
5000	9000	31.146	128.911	117.104	107.593	−4.703
5100	9180	31.171	132.027	117.721	109.582	−4.696
5200	9360	31.194	135.145	118.327	111.574	−4.689
5300	9540	31.216	138.266	118.921	113.583	−4.683
5400	9720	31.236	141.388	119.505	115.577	−4.677
5500	9900	31.256	144.513	120.078	117.583	−4.672
5600	10080	31.275	147.639	120.641	119.586	−4.667
5700	10260	31.292	150.768	121.195	121.591	−4.662
5800	10440	31.309	153.898	121.740	123.597	−4.657
5900	10620	31.325	157.030	122.275	125.625	−4.653
6000	10800	31.340	160.163	122.802	127.627	−4.649

Sources: JANAF Thermochemical Tables, 2nd Edition, National Bureau of Standards, Publication NSRDS–NBS37, 1971

Table B.9 Thermochemical Properties of *n*-Octane

n-Octane (C_8H_{18})
MW = 114.224
$\bar{h}_f^0$ = −49.820 kcal/g mole

T		$\bar{c}_p^0$	$\bar{h}<T>-\bar{h}<T_0>$	$\bar{s}^0<T>$	$\Delta G^0<T>$	log K_p
0	0	0.000	−8.610	0.000	−38.198	infinite
100	180	19.640	−6.633	76.855	−28.733	62.801
200	360	33.837	−3.948	94.896	−24.937	27.252
298	536	45.149	0.000	111.807	3.801	−2.788
300	540	45.368	0.082	111.853	4.228	−3.080
400	720	57.371	5.219	126.555	22.743	−12.427
500	900	68.338	11.523	140.574	41.973	−18.348
600	1080	77.668	18.813	153.816	61.677	−22.468
700	1260	85.661	26.975	166.387	81.706	−25.512
800	1440	92.509	35.907	178.320	101.857	−27.828
900	1620	98.429	45.477	189.540	122.195	−29.675
1000	1800	103.612	55.576	200.202	142.536	−31.154
K	°R	cal/g mole K	kcal/g mole	cal/g mole K	kcal/g mole	—

Sources: Tables based on coefficients provided with

- NASA Computer Program for Calculation of Complex Chemical Equil Compositions, Rocket Performance, Incident and Reflected Shocks, in Chapman–Jouquet Detonations, S. Gordon and B. McBride, NASA L Center, 1967
- Revised NASA Burn Program and Thermodynamic Library, W. Shul APRAPCOM Chemical Systems Laboratory, 1981.

Table B.10 Thermochemical Properties of *n*-Dodecane

n-Dodecane ($C_{12}H_{26}$)
$MW = 170.328$
$\bar{h}^0_f = -71.014$ kcal/g mole

T (K)	T (°R)	$\bar{C}^0_p$	$\bar{h}\langle T\rangle - \bar{h}\langle T_0\rangle$	$\bar{s}^0\langle T\rangle$	$\Delta G^0\langle T\rangle$	$\log K_p$
0	0	0.000	−12.578	0.000	−54.256	infinite
100	180	29.911	−9.967	96.413	−39.366	86.041
200	360	50.841	−5.914	123.673	−32.469	35.483
298	536	67.000	0.000	148.767	10.413	−7.637
300	540	67.326	0.126	149.207	10.951	−7.978
400	720	85.103	7.752	171.028	38.980	−21.299
500	900	101.294	17.100	191.803	68.022	−29.735
600	1080	115.031	27.895	211.423	97.716	−35.596
700	1260	126.751	39.973	230.022	127.872	−39.927
800	1440	136.751	53.194	247.669	158.199	−43.222
900	1620	145.399	67.326	264.279	188.734	−45.835
1000	1800	152.876	82.256	280.006	219.330	−47.939
K	°R	cal/g mole K	kcal/g mole	cal/g mole K	kcal/g mole	—

Sources: Tables based on coefficients provided with

- NASA Computer Program for Calculation of Complex Chemical Equil Compositions, Rocket Performance, Incident and Reflected Shocks, in Chapman–Jouquet Detonations, S. Gordon and B. McBride, NASA L Center, 1967
- Revised NASA Burn Program and Thermodynamic Library, W. Shul APRAPCOM Chemical Systems Laboratory, 1981.

Table B.11 Thermochemical Properties of Methanol

Methanol $(CH_3OH)_g$
$MW = 32.042$
$\bar{h}^0_f = -48.050$ kcal/g mole

T		$\bar{c}^0_p$	$\bar{h}<T>-\bar{h}<T_0>$	$\bar{s}^0<T>$	$\Delta G^0<T>$	$\log K_p$
0	0	0.000	−2.315	0.000	−45.028	infinite
100	180	7.748	−1.771	47.773	−44.171	96.544
200	360	8.883	−0.945	53.458	−44.334	48.450
298	536	10.447	0.000	57.280	−38.813	28.467
300	540	10.482	0.021	57.350	−38.751	28.232
400	720	12.315	1.160	60.614	−35.516	19.407
500	900	14.197	2.485	63.564	−32.077	14.022
600	1080	15.996	3.996	66.314	−28.496	10.381
700	1260	17.625	5.679	68.904	−24.807	7.746
800	1440	19.049	7.514	71.353	−21.064	5.755
900	1620	20.280	9.482	73.669	−17.263	4.192
1000	1800	21.378	11.566	75.864	−13.430	2.935
K	°R	$\frac{cal}{g\ mole\ K}$	$\frac{kcal}{g\ mole}$	$\frac{cal}{g\ mole\ K}$	$\frac{kcal}{g\ mole}$	—

Sources: Tables based on coefficients provided with

- NASA Computer Program for Calculation of Complex Chemical Equil Compositions, Rocket Performance, Incident and Reflected Shocks, in Chapman–Jouquet Detonations, S. Gordon and B. McBride, NASA L Center, 1967
- Revised NASA Burn Program and Thermodynamic Library, W. Shul APRAPCOM Chemical Systems Laboratory, 1981.

Table B.12 Thermochemical Properties of Ethanol

Ethanol $(C_2H_5OH)_g$
$MW = 46.069$
$\bar{h}_f^0 = -056.030$ kcal/g mole

T		$\bar{C}_p^0$	$\bar{h}<T>-\bar{h}<T_0>$	$\bar{s}^0<T>$	$\Delta G^0<T>$	$\log K_p$
0	0	0.000	-2.916	0.000	-51.332	infinite
100	180	7.838	-2.325	55.467	-49.512	infinite
200	360	11.791	-1.344	62.109	-49.098	53.656
298	536	15.622	0.000	67.529	-41.427	30.384
300	540	15.699	0.031	67.634	-40.037	29.169
400	720	19.390	1.788	72.662	-34.515	18.860
500	900	22.743	3.898	77.356	-28.720	12.555
600	1080	25.689	6.323	81.770	-22.744	8.285
700	1260	28.210	9.021	85.925	-16.633	5.194
800	1440	30.341	11.952	89.835	-10.459	2.858
900	1620	32.168	15.080	93.516	-4.220	1.025
1000	1800	33.827	18.380	96.992	2.054	-0.449
K	°R	$\frac{\text{cal}}{\text{g mole K}}$	$\frac{\text{kcal}}{\text{g mole}}$	$\frac{\text{cal}}{\text{g mole K}}$	$\frac{\text{kcal}}{\text{g mole}}$	—

Sources: Tables based on coefficients provided with

- NASA Computer Program for Calculation of Complex Chemical Equil Compositions, Rocket Performance, Incident and Reflected Shocks, in Chapman–Jouquet Detonations, S. Gordon and B. McBride, NASA L Center, 1967
- Revised NASA Burn Program and Thermodynamic Library, W. Shul APRAPCOM Chemical Systems Laboratory, 1981.

Table B.13 Thermochemical Properties of Hydrogen

Hydrogen (H_2)
$MW = 2.016$
$\bar{h}^0_f = 0.000$ kcal/g mole

T		$\bar{C}^0_p$	$\bar{h}\langle T\rangle - \bar{h}\langle T_0\rangle$	$\bar{s}^0\langle T\rangle$	$\Delta G^0\langle T\rangle$	log K_p
0	0	0.000	–2.024	0.000	0.000	0.000
100	180	5.393	–1.265	24.387	0.000	0.000
200	360	6.518	–0.662	28.520	0.000	0.000
298	536	6.892	0.000	31.208	0.000	0.000
300	540	6.894	0.013	31.251	0.000	0.000
400	720	6.975	0.707	33.247	0.000	0.000
500	900	6.993	1.406	34.806	0.000	0.000
600	1080	7.009	2.106	36.082	0.000	0.000
700	1260	7.036	2.808	37.165	0.000	0.000
800	1440	7.087	3.514	38.107	0.000	0.000
900	1620	7.148	4.226	38.946	0.000	0.000
1000	1800	7.219	4.944	39.702	0.000	0.000
1100	1980	7.300	5.670	40.394	0.000	0.000
1200	2160	7.390	6.404	41.033	0.000	0.000
1300	2340	7.490	7.148	41.628	0.000	0.000
1400	2520	7.600	7.902	42.187	0.000	0.000
1500	2700	7.720	8.668	42.716	0.000	0.000
1600	2880	7.823	9.446	43.217	0.000	0.000
1700	3060	7.921	10.233	43.695	0.000	0.000
1800	3240	8.016	11.030	44.150	0.000	0.000
1900	3420	8.108	11.836	44.586	0.000	0.000
2000	3600	8.195	12.651	45.004	0.000	0.000
2100	3780	8.279	13.475	45.406	0.000	0.000
2200	3960	8.358	14.307	45.793	0.000	0.000
2300	4140	8.434	15.146	46.166	0.000	0.000
2400	4320	8.506	15.993	46.527	0.000	0.000
2500	4500	8.575	16.848	46.875	0.000	0.000
2600	4680	8.639	17.708	47.213	0.000	0.000
2700	4860	8.700	18.575	47.540	0.000	0.000
2800	5040	8.757	19.448	47.857	0.000	0.000
2900	5220	8.810	20.326	48.166	0.000	0.000

K	°R	$\frac{cal}{g\ mole\ K}$	$\frac{kcal}{g\ mole}$	$\frac{cal}{g\ mole\ K}$	$\frac{kcal}{g\ mole}$	—
3000	5400	8.859	21.210	48.465	0.000	0.000
3100	5580	8.911	22.098	48.756	0.000	0.000
3200	5760	8.962	22.992	49.040	0.000	0.000
3300	5940	9.012	23.891	49.317	0.000	0.000
3400	6120	9.061	24.794	49.586	0.000	0.000
3500	6300	9.110	25.703	49.850	0.000	0.000
3600	6480	9.158	26.616	50.107	0.000	0.000
3700	6660	9.205	27.535	50.359	0.000	0.000
3800	6840	9.252	28.457	50.605	0.000	0.000
3900	7020	9.297	29.385	50.846	0.000	0.000
4000	7200	9.342	30.317	51.082	0.000	0.000
4100	7380	9.386	31.253	51.313	0.000	0.000
4200	7560	9.429	32.194	51.540	0.000	0.000
4300	7740	9.472	33.139	51.762	0.000	0.000
4400	7920	9.514	34.088	51.980	0.000	0.000
4500	8100	9.555	35.042	52.194	0.000	0.000
4600	8280	9.595	35.999	52.405	0.000	0.000
4700	8460	9.634	36.961	52.612	0.000	0.000
4800	8640	9.673	37.926	52.815	0.000	0.000
4900	8820	9.711	38.895	53.015	0.000	0.000
5000	9000	9.748	39.868	53.211	0.000	0.000
5100	9180	9.785	40.845	53.405	0.000	0.000
5200	9360	9.822	41.825	53.595	0.000	0.000
5300	9540	9.859	42.809	53.783	0.000	0.000
5400	9720	9.895	43.797	53.967	0.000	0.000
5500	9900	9.930	44.788	54.149	0.000	0.000
5600	10080	9.965	45.783	54.328	0.000	0.000
5700	10260	10.000	46.781	54.505	0.000	0.000
5800	10440	10.034	47.783	54.679	0.000	0.000
5900	10620	10.067	48.788	54.851	0.000	0.000
6000	10800	10.100	49.796	55.020	0.000	0.000

Sources: JANAF Thermochemical Tables, 2nd Edition, National Bureau of Standards, Publication NSRDS–NBS37, 1971

Table B.14 Thermochemical Properties of Water Vapor

Water Vapor (H_2O)

$MW = 18.016$

$\bar{h}^0_f = -57.798$ kcal/g mole

T		$\bar{C}^0_p$	$\bar{h}<T>-\bar{h}<T_0>$	$\bar{s}^0<T>$	$\Delta G^0<T>$	$\log K_p$
0	0	0.000	−2.367	0.000	−57.103	infinite
100	180	7.961	−1.581	36.396	−56.557	123.600
200	360	7.969	−0.784	41.916	−55.635	60.792
298	536	8.025	0.000	45.106	−54.636	40.048
300	540	8.027	0.015	45.155	−54.617	39.786
400	720	8.186	0.825	47.484	−53.519	29.240
500	900	8.415	1.654	49.334	−52.361	22.886
600	1080	8.676	2.509	50.891	−51.156	18.633
700	1260	8.954	3.390	52.249	−49.915	15.583
800	1440	9.246	4.300	53.464	−48.646	13.289
900	1620	9.547	5.240	54.570	−47.352	11.498
1000	1800	9.851	6.209	55.592	−46.040	10.062
1100	1980	10.152	7.210	56.545	−44.712	8.883
1200	2160	10.444	8.240	57.441	−43.371	7.899
1300	2340	10.723	9.298	58.288	−42.022	7.064
1400	2520	10.987	10.384	59.092	−40.663	6.347
1500	2700	11.233	11.495	59.859	−39.297	5.725
1600	2880	11.462	12.630	60.591	−37.927	5.180
1700	3060	11.674	13.787	61.293	−36.549	4.699
1800	3240	11.869	14.964	61.965	−35.170	4.270
1900	3420	12.048	16.160	62.612	−33.786	3.886
2000	3600	12.214	17.373	63.234	−32.401	3.540
2100	3780	12.366	18.602	63.834	−31.012	3.227
2200	3960	12.505	19.846	64.412	−29.621	2.942
2300	4140	12.634	21.103	64.971	−28.229	2.682
2400	4320	12.753	22.372	65.511	−26.832	2.443
2500	4500	12.863	23.653	66.034	−25.439	2.224
2600	4680	12.965	24.945	66.541	−24.040	2.021
2700	4860	13.059	26.246	67.032	−22.641	1.833
2800	5040	13.146	27.556	67.508	−21.242	1.658
2900	5220	13.228	28.875	67.971	−19.838	1.495

3000	5400	13.304	30.201	68.421	−18.438	1.343
3100	5580	13.374	31.535	68.858	−17.034	1.201
3200	5760	13.441	32.876	69.284	−15.630	1.067
3300	5940	13.503	34.223	69.698	−14.223	0.942
3400	6120	13.562	35.577	70.102	−12.818	0.824
3500	6300	13.617	36.936	70.496	−11.409	0.712
3600	6480	13.669	38.300	70.881	−10.000	0.607
3700	6660	13.718	39.669	71.256	−8.589	0.507
3800	6840	13.764	41.043	71.622	−7.177	0.413
3900	7020	13.808	42.422	71.980	−5.766	0.323
4000	7200	13.850	43.805	72.331	−4.353	0.238
4100	7380	13.890	45.192	72.673	−2.938	0.157
4200	7560	13.927	46.583	73.008	−1.522	0.079
4300	7740	13.963	47.977	73.336	−0.105	0.005
4400	7920	13.997	49.375	73.658	1.311	−0.065
4500	8100	14.030	50.777	73.973	2.729	−0.133
4600	8280	14.061	52.181	74.281	4.154	−0.197
4700	8460	14.091	53.589	74.584	5.576	−0.259
4800	8640	14.120	55.000	74.881	6.998	−0.319
4900	8820	14.148	56.413	75.172	8.422	−0.376
5000	9000	14.174	57.829	75.459	9.844	−0.430
5100	9180	14.201	59.248	75.740	11.275	−0.483
5200	9360	14.228	60.669	76.016	12.700	−0.534
5300	9540	14.254	62.093	76.287	14.135	−0.583
5400	9720	14.279	63.520	76.553	15.560	−0.630
5500	9900	14.303	64.949	76.816	16.995	−0.675
5600	10080	14.328	66.381	77.074	18.426	−0.719
5700	10260	14.351	67.815	77.327	19.862	−0.762
5800	10440	14.375	69.251	77.577	21.299	−0.803
5900	10620	14.398	70.690	77.823	22.736	−0.842
6000	10800	14.422	72.131	78.065	24.174	−0.880
K	°R	$\frac{cal}{g\ mole\ K}$	$\frac{kcal}{g\ mole}$	$\frac{cal}{g\ mole\ K}$	$\frac{kcal}{g\ mole}$	—

Sources: JANAF Thermochemical Tables, 2nd Edition, National Bureau of Standards, Publication NSRDS–NBS37, 1971

Table B.15 Thermochemical Properties of Nitric Oxide

Nitric Oxide (NO)
$MW = 30.008$

$\bar{h}_f^0 = 21.580$ kcal/g mole

T		$\bar{C}_p^0$	$\bar{h}<T>-\bar{h}<T_0>$	$\bar{s}^0<T>$	$\Delta G^0<T>$	$\log K_p$
0	0	0.000	−2.197	0.000	21.456	infinte
100	180	7.721	−1.451	42.286	21.256	−46.453
200	360	7.271	−0.705	47.477	20.984	−22.929
298	536	7.133	0.000	50.347	20.697	−15.171
300	540	7.132	0.013	50.392	20.692	−15.073
400	720	7.157	0.727	52.444	20.394	−11.142
500	900	7.287	1.448	54.053	20.095	−8.783
600	1080	7.466	2.186	55.397	19.795	−7.210
700	1260	7.655	2.942	56.562	19.494	−6.086
800	1440	7.832	3.716	57.596	19.192	−5.243
900	1620	7.988	4.507	58.528	18.890	−4.587
1000	1800	8.123	5.313	59.377	18.588	−4.062
1100	1980	8.238	6.131	60.157	18.285	−3.633
1200	2160	8.336	6.960	60.878	17.981	−3.275
1300	2340	8.419	7.798	61.548	17.678	−2.972
1400	2520	8.491	8.644	62.175	17.373	−2.712
1500	2700	8.552	9.496	62.763	17.069	−2.487
1600	2880	8.605	10.354	63.317	16.765	−2.290
1700	3060	8.651	11.217	63.840	16.461	−2.116
1800	3240	8.692	12.084	64.335	16.156	−1.962
1900	3420	8.727	12.955	64.806	15.853	−1.823
2000	3600	8.759	13.829	65.255	15.548	−1.699
2100	3780	8.788	14.706	65.683	15.244	−1.586
2200	3960	8.813	15.587	66.092	14.941	−1.484
2300	4140	8.837	16.469	66.484	14.637	−1.391
2400	4320	8.858	17.354	66.861	14.336	−1.305
2500	4500	8.877	18.241	67.223	14.033	−1.227
2600	4680	8.895	19.129	67.571	13.732	−1.154
2700	4860	8.912	20.020	67.908	13.432	−1.087
2800	5040	8.927	20.911	68.232	13.132	−1.025
2900	5220	8.941	21.805	68.545	12.834	−0.967

K	°R	cal/(g mole K)	kcal/(g mole)	cal/(g mole K)	kcal/(g mole)	—
3000	5400	8.955	22.700	68.849	12.535	−0.913
3100	5580	8.968	23.596	69.143	12.237	−0.863
3200	5760	8.980	24.493	69.427	11.940	−0.815
3300	5940	8.991	25.392	69.704	11.644	−0.771
3400	6120	9.002	26.291	69.973	11.349	−0.729
3500	6300	9.012	27.192	70.234	11.054	−0.690
3600	6480	9.022	28.094	70.488	10.762	−0.653
3700	6660	9.032	28.997	70.735	10.470	−0.618
3800	6840	9.041	29.900	70.976	10.179	−0.585
3900	7020	9.050	30.805	71.211	9.889	−0.554
4000	7200	9.058	31.710	71.440	9.598	−0.524
4100	7380	9.066	32.616	71.664	9.311	−0.496
4200	7560	9.074	33.523	71.882	9.024	−0.470
4300	7740	9.082	34.431	72.096	8.739	−0.444
4400	7920	9.090	35.340	72.305	8.452	−0.420
4500	8100	9.097	36.249	72.509	8.169	−0.397
4600	8280	9.105	37.159	72.709	7.888	−0.375
4700	8460	9.112	38.070	72.905	7.605	−0.354
4800	8640	9.119	38.982	73.097	7.324	−0.333
4900	8820	9.125	39.894	73.285	7.040	−0.314
5000	9000	9.132	40.807	73.470	6.763	−0.296
5100	9180	9.139	41.720	73.651	6.484	−0.278
5200	9360	9.145	42.634	73.828	6.207	−0.261
5300	9540	9.152	43.549	74.002	5.932	−0.245
5400	9720	9.158	44.465	74.173	5.654	−0.229
5500	9900	9.164	45.381	74.342	5.383	−0.214
5600	10080	9.170	46.298	74.507	5.107	−0.199
5700	10260	9.176	47.215	74.669	4.835	−0.185
5800	10440	9.182	48.133	74.829	4.566	−0.172
5900	10620	9.188	49.051	74.986	4.292	−0.159
6000	10800	9.194	49.970	75.140	4.024	−0.147

Sources: JANAF Thermochemical Tables, 2nd Edition, National Bureau of Standards, Publication NSRDS–NBS37, 1971

Table B.16 Thermochemical Properties of Nitrogen Dioxide

Nitrogen Dioxide (NO_2)

$MW = 46.008$

$\bar{h}_f^0 = 7.910$ kcal/g mole

T		$\bar{C}_p^0$	$\bar{h}<T>-\bar{h}<T_0>$	$\bar{s}^0<T>$	$\Delta G^0<T>$	$\log K_p$
0	0	0.000	−2.435	0.000	8.586	infinite
100	180	7.953	−1.640	48.387	9.545	−20.859
200	360	8.218	−0.835	53.954	10.853	−11.859
298	536	8.837	0.000	57.343	12.247	−8.977
300	540	8.850	0.016	57.398	12.274	−8.941
400	720	9.601	0.939	60.046	13.751	−7.513
500	900	10.327	1.936	62.268	15.258	−6.669
600	1080	10.955	3.001	64.208	16.778	−6.111
700	1260	11.469	4.123	65.937	18.302	−5.714
800	1440	11.881	5.291	67.496	19.828	−5.417
900	1620	12.208	6.496	68.915	21.355	−5.185
1000	1800	12.468	7.730	70.215	22.879	−5.000
1100	1980	12.677	8.988	71.414	24.400	−4.848
1200	2160	12.847	10.265	72.524	25.921	−4.721
1300	2340	12.985	11.556	73.558	27.438	−4.612
1400	2520	13.099	12.861	74.525	28.951	−4.519
1500	2700	13.193	14.176	75.432	30.464	−4.438
1600	2880	13.273	15.499	76.286	31.975	−4.367
1700	3060	13.340	16.830	77.093	33.484	−4.304
1800	3240	13.398	18.167	77.857	34.992	−4.248
1900	3420	13.447	19.509	78.853	36.498	−4.198
2000	3600	13.490	20.856	79.274	38.002	−4.152
2100	3780	13.527	22.207	79.933	39.507	−4.111
2200	3960	13.560	23.561	80.563	41.010	−4.074
2300	4140	13.588	24.919	81.166	42.514	−4.040
2400	4320	13.614	26.279	81.745	44.019	−4.008
2500	4500	13.636	27.641	82.301	45.520	−3.979
2600	4680	13.656	29.006	82.836	47.025	−3.953
2700	4860	13.674	30.373	83.352	48.529	−3.928
2800	5040	13.690	31.741	83.850	50.031	−3.905
2900	5220	13.705	33.111	84.330	51.540	−3.884

3000	5400	13.718	34.482	84.795	53.045	–3.864
3100	5580	13.730	35.854	85.245	54.551	–3.846
3200	5760	13.741	37.228	85.681	56.058	–3.828
3300	5940	13.751	38.602	86.104	57.565	–3.812
3400	6120	13.760	39.978	86.515	59.075	–3.797
3500	6300	13.768	41.354	86.914	60.583	–3.783
3600	6480	13.776	42.731	87.302	62.097	–3.770
3700	6660	13.783	44.109	87.679	63.613	–3.757
3800	6840	13.790	45.488	88.047	65.128	–3.746
3900	7020	13.796	46.867	88.405	66.643	–3.734
4000	7200	13.801	48.247	88.755	68.158	–3.724
4100	7380	13.806	49.627	89.096	69.678	–3.714
4200	7560	13.811	51.008	89.428	71.201	–3.705
4300	7740	13.816	52.390	89.753	72.727	–3.696
4400	7920	13.820	53.771	90.071	74.247	–3.688
4500	8100	13.824	55.154	90.382	75.772	–3.680
4600	8280	13.828	56.536	90.686	77.304	–3.673
4700	8460	13.831	57.919	90.983	78.833	–3.666
4800	8640	13.834	59.302	91.274	80.366	–3.659
4900	8820	13.837	60.686	91.559	81.894	–3.652
5000	9000	13.840	62.070	91.839	83.428	–3.646
5100	9180	13.843	63.454	92.113	84.966	–3.641
5200	9360	13.846	64.838	92.382	86.501	–3.635
5300	9540	13.848	66.223	92.646	88.044	–3.630
5400	9720	13.850	67.608	92.905	89.579	–3.625
5500	9900	13.852	68.993	93.159	91.128	–3.621
5600	10080	13.854	70.379	93.408	92.673	–3.617
5700	10260	13.856	71.764	93.654	94.215	–3.612
5800	10440	13.858	73.150	93.895	95.767	–3.608
5900	10620	13.860	74.536	94.132	97.313	–3.605
6000	10800	13.862	75.922	94.365	98.866	–3.601
K	°R	$\frac{cal}{g\ mole\ K}$	$\frac{kcal}{g\ mole}$	$\frac{cal}{g\ mole\ K}$	$\frac{kcal}{g\ mole}$	—

Sources: JANAF Thermochemical Tables, 2nd Edition, National Bureau of Standards, Publication NSRDS–NBS37, 1971

Table B.17 Thermochemical Properties of Nitrogen

Nitrogen (N_2)
$MW = 28.0134$
$\bar{h}_f^0 = 0.000$ kcal/g mole

T		$\bar{C}_p^0$	$\bar{h}<T>-\bar{h}<T_0>$	$\bar{s}^0<T>$	$\Delta G^0<T>$	$\log K_p$
0	0	0.000	−2.072	0.000	0.000	0.000
100	180	6.956	−1.379	38.170	0.000	0.000
200	360	6.957	−0.683	42.992	0.000	0.000
298	536	6.961	0.000	45.770	0.000	0.000
300	540	6.961	0.013	45.813	0.000	0.000
400	720	6.990	0.710	47.818	0.000	0.000
500	900	7.069	1.413	49.386	0.000	0.000
600	1080	7.196	2.125	50.685	0.000	0.000
700	1260	7.350	2.853	51.806	0.000	0.000
800	1440	7.512	3.596	52.798	0.000	0.000
900	1620	7.670	4.355	53.692	0.000	0.000
1000	1800	7.815	5.129	54.507	0.000	0.000
1100	1980	7.945	5.917	55.258	0.000	0.000
1200	2160	8.061	6.718	55.955	0.000	0.000
1300	2340	8.162	7.529	56.604	0.000	0.000
1400	2520	8.252	8.350	57.212	0.000	0.000
1500	2700	8.330	9.179	57.784	0.000	0.000
1600	2880	8.398	10.015	58.324	0.000	0.000
1700	3060	8.458	10.858	58.835	0.000	0.000
1800	3240	8.512	11.707	59.320	0.000	0.000
1900	3420	8.559	12.560	59.782	0.000	0.000
2000	3600	8.601	13.418	60.222	0.000	0.000
2100	3780	8.638	14.280	60.642	0.000	0.000
2200	3960	8.672	15.146	61.045	0.000	0.000
2300	4140	8.703	16.015	61.431	0.000	0.000
2400	4320	8.731	16.886	61.802	0.000	0.000
2500	4500	8.756	17.761	62.159	0.000	0.000
2600	4680	8.779	18.638	62.503	0.000	0.000
2700	4860	8.800	19.517	62.835	0.000	0.000
2800	5040	8.820	20.398	63.155	0.000	0.000
2900	5220	8.838	21.280	63.465	0.000	0.000

3000	5400	8.855	22.165	63.765	0.000	0.000
3100	5580	8.871	23.051	64.055	0.000	0.000
3200	5760	8.886	23.939	64.337	0.000	0.000
3300	5940	8.900	24.829	64.611	0.000	0.000
3400	6120	8.914	25.719	64.877	0.000	0.000
3500	6300	8.927	26.611	65.135	0.000	0.000
3600	6480	8.939	27.505	65.387	0.000	0.000
3700	6660	8.950	28.399	65.632	0.000	0.000
3800	6840	8.962	29.295	65.871	0.000	0.000
3900	7020	8.972	30.191	66.104	0.000	0.000
4000	7200	8.983	31.089	66.331	0.000	0.000
4100	7380	8.993	31.988	66.553	0.000	0.000
4200	7560	9.002	32.888	66.770	0.000	0.000
4300	7740	9.012	33.788	66.982	0.000	0.000
4400	7920	9.021	34.690	67.189	0.000	0.000
4500	8100	9.030	35.593	67.392	0.000	0.000
4600	8280	9.039	36.496	67.591	0.000	0.000
4700	8460	9.048	37.400	67.785	0.000	0.000
4800	8640	9.057	38.306	67.976	0.000	0.000
4900	8820	9.066	39.212	68.162	0.000	0.000
5000	9000	9.074	40.119	68.346	0.000	0.000
5100	9180	9.083	41.027	68.525	0.000	0.000
5200	9360	9.091	41.935	68.702	0.000	0.000
5300	9540	9.100	42.845	68.875	0.000	0.000
5400	9720	9.109	43.755	69.045	0.000	0.000
5500	9900	9.118	44.667	69.213	0.000	0.000
5600	10080	9.127	45.579	69.377	0.000	0.000
5700	10260	9.136	46.492	69.539	0.000	0.000
5800	10440	9.145	47.406	69.698	0.000	0.000
5900	10620	9.155	48.321	69.854	0.000	0.000
6000	10800	9.165	49.237	70.008	0.000	0.000
K	°R	$\frac{cal}{g\ mole\ K}$	$\frac{kcal}{g\ mole}$	$\frac{cal}{g\ mole\ K}$	$\frac{kcal}{g\ mole}$	—

Sources: JANAF Thermochemical Tables, 2nd Edition, National Bureau of Standards, Publication NSRDS–NBS37, 1971

Table B.18 Thermochemical Properties of Dinitrogen Monoxide

Dinitrogen Monoxide (N_2O)

$MW = 44.016$

$\bar{h}^0_f = 19.610$ kcal/g mole

T		$\bar{C}^0_p$	$\bar{h}<T>-\bar{h}<T_0>$	$\bar{s}^0<T>$	$\Delta G^0<T>$	log K_p
0	0	0.000	−2.290	0.000	20.430	infinite
100	180	7.015	−1.594	43.988	21.573	−47.145
200	360	8.033	−0.849	49.108	23.185	−25.334
298	536	9.230	0.000	52.546	24.896	−18.248
300	540	9.250	0.017	52.603	24.928	−18.159
400	720	10.201	0.992	55.400	26.716	−14.596
500	900	10.953	2.051	57.761	28.514	−12.463
600	1080	11.565	3.178	59.813	30.311	−11.040
700	1260	12.070	4.360	61.635	32.097	−10.021
800	1440	12.486	5.589	63.275	33.873	−9.253
900	1620	12.830	6.855	64.766	35.638	−8.654
1000	1800	13.113	8.153	66.133	37.391	−8.171
1100	1980	13.348	9.476	67.394	39.132	−7.774
1200	2160	13.544	10.821	68.565	40.863	−7.442
1300	2340	13.707	12.184	69.655	42.583	−7.158
1400	2520	13.845	13.562	70.676	44.293	−6.914
1500	2700	13.961	14.952	71.635	45.996	−6.701
1600	2880	14.060	16.353	72.540	47.691	−6.514
1700	3060	14.145	17.764	73.395	49.377	−6.347
1800	3240	14.218	19.182	74.205	51.054	−6.198
1900	3420	14.282	20.607	74.976	52.728	−6.065
2000	3600	14.337	22.038	75.710	54.392	−5.943
2100	3780	14.385	23.474	76.411	56.049	−5.833
2200	3960	14.428	24.915	77.081	57.703	−5.732
2300	4140	14.466	26.359	77.723	59.350	−5.639
2400	4320	14.499	27.808	78.339	60.993	−5.554
2500	4500	14.529	29.259	78.932	62.630	−5.475
2600	4680	14.556	30.713	79.502	64.262	−5.401
2700	4860	14.580	32.170	80.052	65.894	−5.333
2800	5040	14.602	33.629	80.583	67.517	−5.270
2900	5220	14.621	35.090	81.095	69.140	−5.210

K	°R	$\frac{cal}{g\ mole\ K}$	$\frac{kcal}{g\ mole}$	$\frac{cal}{g\ mole\ K}$	$\frac{kcal}{g\ mole}$	—
3000	5400	14.639	36.533	81.591	70.757	−5.154
3100	5580	14.655	38.018	82.072	72.371	−5.102
3200	5760	14.670	39.484	82.537	73.981	−5.052
3300	5940	14.784	40.952	82.989	75.591	−5.006
3400	6120	14.696	42.421	83.427	77.196	−4.962
3500	6300	14.707	43.891	83.853	78.797	−4.920
3600	6480	14.718	45.363	84.268	80.400	−4.881
3700	6660	14.727	46.835	84.671	81.995	−4.843
3800	6840	14.736	48.308	85.064	83.591	−4.807
3900	7020	14.745	49.782	85.447	85.183	−4.773
4000	7200	14.752	51.257	85.820	86.774	−4.741
4100	7380	14.760	52.733	86.185	88.365	−4.710
4200	7560	14.766	54.209	86.541	89.951	−4.680
4300	7740	14.772	55.686	86.888	91.537	−4.652
4400	7920	14.778	57.163	87.228	93.120	−4.625
4500	8100	14.784	58.641	87.560	94.703	−4.599
4600	8280	14.789	60.120	87.885	96.288	−4.574
4700	8460	14.794	61.599	88.203	97.860	−4.550
4800	8640	14.798	63.079	88.515	99.441	−4.527
4900	8820	14.802	64.559	88.820	101.011	−4.505
5000	9000	14.806	66.039	89.119	102.590	−4.484
5100	9180	14.810	67.520	89.412	104.156	−4.463
5200	9360	14.814	69.001	89.700	105.733	−4.444
5300	9540	14.817	70.483	89.982	107.305	−4.425
5400	9720	14.820	71.965	90.259	108.872	−4.406
5500	9900	14.823	73.447	90.531	110.446	−4.388
5600	10080	14.826	74.929	90.798	112.010	−4.371
5700	10260	14.829	76.412	91.060	113.575	−4.354
5800	10440	14.831	77.895	91.318	115.142	−4.338
5900	10620	14.834	79.378	91.572	116.706	−4.323
6000	10800	14.836	80.862	91.821	118.267	−4.308

Sources: JANAF Thermochemical Tables, 2nd Edition, National Bureau of Standards, Publication NSRDS–NBS37, 1971

Table B.19 Thermochemical Properties of Monatomic Oxygen

Oxygen, Monatomic (O)
$MW = 16.000$

$\bar{h}_f^0 = 59.559$ **kcal/g mole**

T		$\bar{C}_p^0$	$\bar{h}\langle T \rangle - \bar{h}\langle T_0 \rangle$	$\bar{s}^0\langle T \rangle$	$\Delta G^0\langle T \rangle$	$\log K_p$
0	0	0.000	−1.608	0.000	58.989	infinite
100	180	5.666	−1.080	32.466	57.989	−126.730
200	360	5.434	−0.523	36.340	56.733	−61.992
298	536	5.237	0.000	38.468	55.395	−40.604
300	540	5.235	0.010	38.501	55.369	−40.334
400	720	5.135	0.528	39.991	53.946	−29.473
500	900	5.081	1.038	41.131	52.485	−22.940
600	1080	5.049	1.544	42.054	50.995	−18.574
700	1260	5.029	2.048	42.831	49.486	−15.449
800	1440	5.015	2.550	43.501	47.960	−13.101
900	1620	5.006	3.052	44.092	46.422	−11.272
1000	1800	4.999	3.552	44.619	44.875	−9.807
1100	1980	4.994	4.051	45.095	43.318	−8.606
1200	2160	4.990	4.551	45.529	41.755	−7.604
1300	2340	4.987	5.049	45.928	40.186	−6.755
1400	2520	4.984	5.548	46.298	38.611	−6.027
1500	2700	4.982	6.046	46.642	37.032	−5.395
1600	2880	4.981	6.544	46.963	35.448	−4.842
1700	3060	4.979	7.042	47.265	33.862	−4.353
1800	3240	4.979	7.540	47.550	32.271	−3.918
1900	3420	4.978	8.038	47.819	30.678	−3.529
2000	3600	4.978	8.536	48.074	29.082	−3.178
2100	3780	4.978	9.034	48.317	27.484	−2.860
2200	3960	4.979	9.532	48.549	25.884	−2.571
2300	4140	4.980	10.029	48.770	24.282	−2.307
2400	4320	4.981	10.527	48.982	22.679	−2.065
2500	4500	4.984	11.026	49.185	21.073	−1.842
2600	4680	4.986	11.524	49.381	19.467	−1.636
2700	4860	4.990	12.023	49.569	17.859	−1.446
2800	5040	4.994	12.522	49.751	16.251	−1.268
2900	5220	4.999	13.022	49.926	14.642	−1.103

K	°R	$\frac{cal}{g\ mole\ K}$	$\frac{kcal}{g\ mole}$	$\frac{cal}{g\ mole\ K}$	$\frac{kcal}{g\ mole}$	—
3000	5400	5.004	13.522	50.096	13.031	–0.949
3100	5580	5.010	14.023	50.260	11.420	–0.805
3200	5760	5.017	14.524	50.419	9.807	–0.670
3300	5940	5.025	15.026	50.573	8.194	–0.543
3400	6120	5.033	15.529	50.724	6.581	–0.423
3500	6300	5.041	16.033	50.870	4.967	–0.310
3600	6480	5.050	16.537	51.012	3.354	–0.204
3700	6660	5.060	17.043	51.150	1.739	–0.103
3800	6840	5.070	17.549	51.285	0.125	–0.007
3900	7020	5.081	18.057	51.417	–1.492	0.084
4000	7200	5.091	18.565	51.546	–3.107	0.170
4100	7380	5.103	19.075	51.672	–4.723	0.252
4200	7560	5.114	19.586	51.795	–6.339	0.330
4300	7740	5.126	20.098	51.915	–7.955	0.404
4400	7920	5.138	20.611	52.033	–9.573	0.475
4500	8100	5.150	21.126	52.149	–11.189	0.543
4600	8280	5.162	21.641	52.262	–12.805	0.608
4700	8460	5.174	22.158	52.373	–14.423	0.671
4800	8640	5.186	22.676	52.482	–16.041	0.730
4900	8820	5.198	23.195	52.589	–17.661	0.788
5000	9000	5.210	23.715	52.695	–19.279	0.843
5100	9180	5.222	24.237	52.798	–20.896	0.895
5200	9360	5.234	24.760	52.899	–22.517	0.946
5300	9540	5.246	25.284	52.999	–24.134	0.995
5400	9720	5.258	25.809	53.097	–25.755	1.042
5500	9900	5.269	26.335	53.194	–27.373	1.088
5600	10080	5.280	26.863	53.289	–28.995	1.132
5700	10260	5.292	27.392	53.383	–30.616	1.174
5800	10440	5.302	27.921	53.475	–32.234	1.215
5900	10620	5.313	28.452	53.565	–33.856	1.254
6000	10800	5.323	28.984	53.655	–35.476	1.292

Sources: JANAF Thermochemical Tables, 2nd Edition, National Bureau of Standards, Publication NSRDS–NBS37, 1971

Table B.20 Thermochemical Properties of Diatomic Oxygen

Oxygen, Diatomic (O_2)
MW = **31.9988**
$\bar{h}_f^0$ = **0.000 kcal/g mole**

T		$\bar{C}_p^0$	$\bar{h}<T>-\bar{h}<T_0>$	$\bar{s}^0<T>$	$\Delta G^0<T>$	$\log K_p$
0	0	0.000	−2.075	0.000	0.000	0.000
100	180	6.958	−1.381	41.395	0.000	0.000
200	360	6.961	−0.685	46.218	0.000	0.000
298	536	7.020	0.000	49.004	0.000	0.000
300	540	7.023	0.013	49.047	0.000	0.000
400	720	7.196	0.724	51.091	0.000	0.000
500	900	7.431	1.455	52.722	0.000	0.000
600	1080	7.670	2.210	54.098	0.000	0.000
700	1260	7.883	2.988	55.297	0.000	0.000
800	1440	8.063	3.786	56.361	0.000	0.000
900	1620	8.212	4.600	57.320	0.000	0.000
1000	1800	8.336	5.427	58.192	0.000	0.000
1100	1980	8.439	6.266	58.991	0.000	0.000
1200	2160	8.527	7.114	59.729	0.000	0.000
1300	2340	8.604	7.971	60.415	0.000	0.000
1400	2520	8.674	8.835	61.055	0.000	0.000
1500	2700	8.738	9.706	61.656	0.000	0.000
1600	2880	8.800	10.583	62.222	0.000	0.000
1700	3060	8.858	11.465	62.757	0.000	0.000
1800	3240	8.916	12.354	63.265	0.000	0.000
1900	3420	8.973	13.249	63.749	0.000	0.000
2000	3600	9.029	14.149	64.210	0.000	0.000
2100	3780	9.084	15.054	64.652	0.000	0.000
2200	3960	9.139	15.966	65.076	0.000	0.000
2300	4140	9.194	16.882	65.483	0.000	0.000
2400	4320	9.248	17.804	65.876	0.000	0.000
2500	4500	9.301	18.732	66.254	0.000	0.000
2600	4680	9.354	19.664	66.620	0.000	0.000
2700	4860	9.405	20.602	66.974	0.000	0.000
2800	5040	9.455	21.545	67.317	0.000	0.000
2900	5220	9.503	22.493	67.650	0.000	0.000

K	°R	$\frac{cal}{g\ mole\ K}$	$\frac{kcal}{g\ mole}$	$\frac{cal}{g\ mole\ K}$	$\frac{kcal}{g\ mole}$	—
3000	5400	9.551	23.446	67.973	0.000	0.000
3100	5580	9.596	24.403	68.287	0.000	0.000
3200	5760	9.640	25.365	68.592	0.000	0.000
3300	5940	9.682	26.331	68.889	0.000	0.000
3400	6120	9.723	27.302	69.179	0.000	0.000
3500	6300	9.762	28.276	69.461	0.000	0.000
3600	6480	9.799	29.254	69.737	0.000	0.000
3700	6660	9.835	30.236	70.006	0.000	0.000
3800	6840	9.869	31.221	70.269	0.000	0.000
3900	7020	9.901	32.209	70.525	0.000	0.000
4000	7200	9.932	33.201	70.776	0.000	0.000
4100	7380	9.961	34.196	71.022	0.000	0.000
4200	7560	9.988	35.193	71.262	0.000	0.000
4300	7740	10.015	36.193	71.498	0.000	0.000
4400	7920	10.039	37.196	71.728	0.000	0.000
4500	8100	10.062	38.201	71.954	0.000	0.000
4600	8280	10.084	39.208	72.176	0.000	0.000
4700	8460	10.104	40.218	72.393	0.000	0.000
4800	8640	10.123	41.229	72.606	0.000	0.000
4900	8820	10.140	42.242	72.814	0.000	0.000
5000	9000	10.156	43.257	73.019	0.000	0.000
5100	9180	10.172	44.274	73.221	0.000	0.000
5200	9360	10.187	45.292	73.418	0.000	0.000
5300	9540	10.200	46.311	73.613	0.000	0.000
5400	9720	10.213	47.332	73.803	0.000	0.000
5500	9900	10.225	48.353	73.991	0.000	0.000
5600	10080	10.237	49.377	74.175	0.000	0.000
5700	10260	10.247	50.401	74.356	0.000	0.000
5800	10440	10.258	51.426	74.535	0.000	0.000
5900	10620	10.267	52.452	74.710	0.000	0.000
6000	10800	10.276	53.479	74.883	0.000	0.000

Sources: JANAF Thermochemical Tables, 2nd Edition, National Bureau of Standards, Publication NSRDS–NBS37, 1971

Table B.21 Thermochemical Properties of Sulfur

Sulfur (S)
$MW = 32.064$

$\bar{h}_f^0 = 0.000$ **kcal/g mole**

T		$\bar{C}_p^0$	$\bar{h}<T>-\bar{h}<T_0>$	$\bar{s}^0<T>$	$\Delta G^0<T>$	$\log K_p$
0	0	0.000	−1.053	0.000	0.000	0.000
100	180	3.060	−0.889	2.965	0.000	0.000
200	360	4.639	−0.496	5.622	0.000	0.000
298	536	5.401	0.000	7.631	0.000	0.000
300	540	5.412	0.010	7.665	0.000	0.000
400	720	7.734	1.109	10.674	0.000	0.000
500	900	9.081	2.047	12.768	0.000	0.000
600	1080	8.200	2.904	14.333	0.000	0.000
700	1260	7.799	3.704	15.601	0.000	0.000
800	1440	4.368	17.529	31.363	0.000	0.000
900	1620	4.396	17.967	31.879	0.000	0.000
1000	1800	4.418	18.408	32.344	0.000	0.000
1100	1980	4.435	18.851	32.765	0.000	0.000
1200	2160	4.450	19.295	33.152	0.000	0.000
1300	2340	4.461	19.740	33.509	0.000	0.000
1400	2520	4.471	20.187	33.840	0.000	0.000
1500	2700	4.480	20.635	34.148	0.000	0.000
1600	2880	4.488	21.083	34.438	0.000	0.000
1700	3060	4.495	21.532	34.710	0.000	0.000
1800	3240	4.501	21.982	34.967	0.000	0.000
1900	3420	4.507	22.432	35.211	0.000	0.000
2000	3600	4.513	22.883	35.442	0.000	0.000
2100	3780	4.518	23.335	35.662	0.000	0.000
2200	3960	4.523	23.787	35.873	0.000	0.000
2300	4140	4.528	24.240	36.074	0.000	0.000
2400	4320	4.532	24.693	36.267	0.000	0.000
2500	4500	4.537	25.146	36.452	0.000	0.000
2600	4680	4.541	25.600	36.630	0.000	0.000
2700	4860	4.545	26.054	36.801	0.000	0.000
2800	5040	4.549	26.509	36.966	0.000	0.000
2900	5220	4.553	26.964	37.126	0.000	0.000

K	°R	$\frac{cal}{g\ mole\ K}$	$\frac{kcal}{g\ mole}$	$\frac{cal}{g\ mole\ K}$	$\frac{kcal}{g\ mole}$	—
3000	5400	4.557	27.420	37.281	0.000	0.000
3100	5580	4.561	27.875	37.430	0.000	0.000
3200	5760	4.565	28.332	37.575	0.000	0.000
3300	5940	4.568	28.788	37.715	0.000	0.000
3400	6120	4.572	29.245	37.852	0.000	0.000
3500	6300	4.575	29.703	37.984	0.000	0.000
3600	6480	4.579	30.160	38.113	0.000	0.000
3700	6660	4.583	30.619	38.239	0.000	0.000
3800	6840	4.586	31.077	38.361	0.000	0.000
3900	7020	4.590	31.536	38.480	0.000	0.000
4000	7200	4.593	31.995	38.597	0.000	0.000
4100	7380	4.596	32.454	38.710	0.000	0.000
4200	7560	4.600	32.914	38.821	0.000	0.000
4300	7740	4.603	33.374	38.929	0.000	0.000
4400	7920	4.607	33.835	39.035	0.000	0.000
4500	8100	4.610	34.296	39.139	0.000	0.000
4600	8280	4.613	34.757	39.240	0.000	0.000
4700	8460	4.617	35.218	39.339	0.000	0.000
4800	8640	4.620	35.680	39.436	0.000	0.000
4900	8820	4.624	36.142	39.532	0.000	0.000
5000	9000	4.627	36.605	39.625	0.000	0.000
5100	9180	4.630	37.068	39.717	0.000	0.000
5200	9360	4.633	37.531	39.807	0.000	0.000
5300	9540	4.637	37.994	39.895	0.000	0.000
5400	9720	4.640	38.458	39.982	0.000	0.000
5500	9900	4.643	38.922	40.067	0.000	0.000
5600	10080	4.647	39.387	40.151	0.000	0.000
5700	10260	4.650	39.852	40.233	0.000	0.000
5800	10440	4.653	40.317	40.314	0.000	0.000
5900	10620	4.656	40.782	40.393	0.000	0.000
6000	10800	4.660	41.248	40.472	0.000	0.000

Sources: JANAF Thermochemical Tables, 2nd Edition, National Bureau of Standards, Publication NSRDS–NBS37, 1971

Table B.22 Thermochemical Properties of Sulfur Dioxide

Sulfur Dioxide (SO_2)

MW = 64.066

$\bar{h}_f^0 = -70.947$ **kcal/g mole**

T		$\bar{C}_p^0$	$\bar{h}<T>-\bar{h}<T_0>$	$\bar{s}^0<T>$	$\Delta G^0<T>$	$\log K_p$
0	0	0.000	−2.522	0.000	−70.341	infinite
100	180	8.013	−1.725	49.932	−70.966	155.088
200	360	8.693	−0.893	55.670	−71.425	78.046
298	536	9.530	0.000	59.298	−71.741	52.585
300	540	9.547	0.018	59.357	−71.746	52.264
400	720	10.395	1.016	62.222	−71.947	39.308
500	900	11.132	2.093	64.623	−71.923	31.436
600	1080	11.723	3.237	66.707	−71.790	26.148
700	1260	12.180	4.433	68.550	−71.562	22.342
800	1440	12.532	5.669	70.200	−72.574	19.825
900	1620	12.806	6.937	71.693	−70.822	17.197
1000	1800	13.022	8.229	73.054	−69.071	15.095
1100	1980	13.194	9.540	74.303	−67.326	13.376
1200	2160	13.335	10.866	75.458	−65.582	11.943
1300	2340	13.451	12.206	76.530	−63.840	10.732
1400	2520	13.549	13.556	77.530	−62.102	9.694
1500	2700	13.632	14.915	78.468	−60.369	8.795
1600	2880	13.704	16.282	79.350	−58.635	8.009
1700	3060	13.767	17.656	80.183	−56.905	7.315
1800	3240	13.822	19.035	80.971	−55.178	6.699
1900	3420	13.872	20.420	81.720	−53.452	6.148
2000	3600	13.917	21.809	82.433	−51.731	5.653
2100	3780	13.958	23.203	83.113	−50.010	5.204
2200	3960	13.995	24.601	83.763	−48.290	4.797
2300	4140	14.030	26.002	84.386	−46.573	4.425
2400	4320	14.063	27.407	84.984	−44.855	4.084
2500	4500	14.093	28.815	85.558	−43.141	3.771
2600	4680	14.122	30.225	86.112	−41.426	3.482
2700	4860	14.149	31.639	86.645	−39.713	3.214
2800	5040	14.175	33.055	87.160	−38.002	2.966
2900	5220	14.200	34.474	87.658	−36.288	2.735

K	°R	cal/(g mole K)	kcal/(g mole)	cal/(g mole K)	kcal/(g mole)	—
3000	5400	14.224	35.895	88.140	−34.575	2.519
3100	5580	14.247	37.319	88.607	−32.864	2.317
3200	5760	14.270	38.745	89.059	−31.154	2.128
3300	5940	14.291	40.173	89.499	−29.446	1.950
3400	6120	14.312	41.603	89.926	−27.733	1.783
3500	6300	14.333	43.035	90.341	−26.026	1.625
3600	6480	14.353	44.469	90.745	−24.313	1.476
3700	6660	14.373	45.906	91.138	−22.602	1.335
3800	6840	14.392	47.344	91.522	−20.890	1.201
3900	7020	14.411	48.784	91.896	−19.183	1.075
4000	7200	14.430	50.226	92.261	−17.469	0.954
4100	7380	14.448	51.670	92.618	−15.758	0.840
4200	7560	14.467	53.116	92.966	−14.047	0.731
4300	7740	14.485	54.563	93.307	−12.333	0.627
4400	7920	14.502	56.013	93.640	−10.623	0.528
4500	8100	14.520	57.464	93.966	−8.908	0.433
4600	8280	14.537	58.917	94.285	−7.194	0.342
4700	8460	14.554	60.371	94.598	−5.482	0.255
4800	8640	14.572	61.828	94.905	−3.769	0.172
4900	8820	14.588	63.286	95.205	−2.056	0.092
5000	9000	14.605	64.745	95.500	−0.345	0.015
5100	9180	14.622	66.207	95.790	1.375	−0.059
5200	9360	14.639	67.670	96.074	3.087	−0.130
5300	9540	14.655	69.134	96.353	4.806	−0.198
5400	9720	14.672	70.601	96.627	6.518	−0.264
5500	9900	14.688	72.069	96.896	8.237	−0.327
5600	10080	14.704	73.538	97.161	9.952	−0.388
5700	10260	14.720	75.010	97.421	11.665	−0.447
5800	10440	14.736	76.482	97.677	13.387	−0.504
5900	10620	14.753	77.957	97.930	15.099	−0.559
6000	10800	14.769	79.433	98.178	16.823	−0.613

Sources: JANAF Thermochemical Tables, 2nd Edition, National Bureau of Standards, Publication NSRDS–NBS37, 1971

APPENDIX C: PROPERTIES OF SATURATED WATER

SI				
T_{sat}	P_{sat}[a]	C_p[b]	$\rho \times 10^{3a}$	h_{fg}[a]
32	0.00061	4.217	1.0000	2501.4
10	0.001227	4.193	0.9996	2477.7
20	0.002339	4.182	0.9982	2454.1
30	0.004246	4.179	0.9957	2430.5
40	0.007384	4.179	0.9923	2406.7
50	0.012349	4.181	0.9880	2382.7
60	0.019940	4.185	0.9831	2358.5
70	0.03119	4.190	0.9777	2333.8
80	0.04739	4.197	0.9717	2308.8
90	0.07014	4.205	0.9653	2283.2
100	0.10135	4.216	0.9583	2257.0
110	0.14327	4.229	0.9509	2230.2
120	0.19853	4.245	0.9431	2202.6
130	0.2701	4.263	0.9348	2174.2
140	0.3613	4.285	0.9262	2144.7
150	0.4758	4.310	0.9170	2114.3
160	0.6178	4.339	0.9074	2082.6
170	0.7917	4.371	0.8974	2049.5
180	1.0021	4.408	0.8870	2015.0
190	1.2544	4.449	0.8761	1978.8
200	1.5538	4.497	0.8647	1940.7
210	1.9062	4.551	0.8528	1900.7
220	2.318	4.614	0.8403	1858.5
230	2.795	4.686	0.8273	1813.8
240	3.344	4.770	0.8136	1766.5
250	3.973	4.869	0.7992	1716.2
260	4.688	4.985	0.7840	1662.5
270	5.499	5.13	0.7679	1605.2
280	6.412	5.30	0.7507	1543.6
290	7.436	5.51	0.7323	1477.1
300	8.581	5.77	0.7125	1404.9
°C	MgPa	kJ/kg K	kg/m³	kJ/kg

Sources: [a]*Steam Tables* (S.I. Units), Keenan, Keyes, Hill, and Moore, 1969, John Wiley & Sons, Inc.
[b]*Tables on the Thermophysical Properties of Liquids and Gases*, 2nd Edition, N.B. Vargaftik, Hemisphere Publishing Co., Washington, 1975

English Engineers				
T_{sat}	P_{sat} [a]	C_p [b]	ρ [a]	h_{fg} [a]
32	0.8859	1.007	62.414	1075.4
40	0.12166	1.005	62.422	1070.9
60	0.2563	1.000	62.364	1059.6
80	0.5073	0.998	62.216	1048.3
100	0.9503	0.997	61.996	1037.0
120	1.6945	0.997	61.709	1025.5
140	2.892	0.999	61.376	1014.0
160	4.745	1.001	60.994	1002.2
180	7.515	1.002	60.573	990.2
200	11.529	1.003	60.118	977.9
220	17.188	1.008	59.623	965.3
240	24.97	1.012	59.095	952.3
260	35.42	1.016	58.534	938.8
280	49.18	1.022	57.941	924.9
300	66.98	1.029	57.313	910.4
320	89.60	1.036	56.651	895.3
340	117.93	1.045	55.953	879.5
360	152.92	1.055	55.224	862.9
380	195.60	1.066	54.457	845.4
400	247.1	1.080	53.654	826.8
420	308.5	1.095	52.809	807.2
440	381.2	1.114	51.921	786.3
460	466.3	1.135	50.984	764.1
480	565.5	1.160	49.995	740.3
500	680.0	1.191	48.948	714.8
520	811.4	1.230	47.824	687.3
540	961.5	1.277	46.620	657.5
560	1131.8	1.337	45.310	625.0
580	1324.3	1.415	43.898	589.3
600	1541.0	1.524	42.319	549.7
°F	lbf/in^2	Btu/lbm°R	lbm/ft^3	Btu/lbm

Sources: [a]*Steam Tables* (English Units), Keenan, Keyes, Hill, and Moore, 1969, 1978, John Wiley & Sons, Inc.

[b]*Tables on the Thermophysical Properties of Liquids and Gases*, N.B. Vargaftik, Hemisphere Publishing Co., Washington, 1975

APPENDIX D: PROPERTIES OF FUEL OIL

Fuel Oil Specific Gravity, Density, and Heats of Combustion

Specific Gravity at 60/60°F		Density at 60°F		Higher Heating Value (Constant Volume)				Lower Heating Value (Constant Pressure)			
5	1.0366	1,036	8.643	4.395	42,450	157,700	18,250	4.164	40,220	149,400	17,290
6	1.0291	1,028	8.580	4.384	42,640	157,300	18,330	4.147	40,335	148,800	17,340
7	1.0217	1,021	8.518	4.365	42,775	156,600	18,390	4.128	40,450	148,100	17,390
8	1.0143	1,013	8.457	4.345	42,890	155,900	18,440	4.111	40,565	147,500	17,440
9	1.0071	1,006	8.397	4.329	43,000	155,300	18,490	4.094	40,680	146,900	17,490
10	1.0000	999.0	8.337	4.309	43,125	154,600	18,540	4.075	40,800	146,200	17,540
11	0.9930	992.1	8.279	4.290	43,240	153,900	18,590	4.058	40,890	145,600	17,580
12	0.9861	985.1	8.221	4.273	43,360	153,300	18,640	4.039	40,985	144,900	17,620
13	0.9792	978.3	8.164	4.253	43,470	152,600	18,690	4.019	41,100	144,200	17,670
14	0.9725	971.6	8.108	4.237	43,590	152,000	18,740	4.002	41,190	143,600	17,710
15	0.9659	965.0	8.053	4.217	43,705	151,300	18,790	3.983	41,285	142,900	17,750
16	0.9593	958.4	7.998	4.200	43,820	150,700	18,840	3.966	41,380	142,300	17,790
17	0.9529	951.9	7.944	4.181	43,940	150,000	18,890	3.947	41,450	141,600	17,820
18	0.9465	945.6	7.891	4.164	44,030	149,400	18,930	3.927	41,540	140,900	17,860
19	0.9402	939.4	7.839	4.147	44,150	148,800	18,980	3.910	41,635	140,300	17,900
20	0.9340	933.1	7.787	4.128	44,240	148,100	19,020	3.891	41,705	139,600	17,930
21	0.9279	927.0	7.736	4.111	44,330	147,500	19,060	3.874	41,775	139,000	17,960
22	0.9218	921.0	7.686	4.092	44,450	146,800	19,110	3.855	41,870	138,300	18,000
23	0.9159	915.0	7.636	4.075	44,540	146,200	19,150	3.838	41,940	137,700	18,030
24	0.9100	909.2	7.587	4.058	44,635	145,600	19,190	3.821	42,030	137,100	18,070
25	0.9042	903.3	7.538	4.041	44,730	145,000	19,230	3.802	42,100	136,400	18,100
26	0.8984	897.5	7.490	4.022	44,820	144,300	19,270	3.785	42,170	135,800	18,130
27	0.8927	891.9	7.443	4.005	44,915	143,700	19,310	3.768	42,240	135,200	18,160
28	0.8871	886.3	7.396	3.988	45,010	143,100	19,350	3.752	42,310	134,600	18,190
29	0.8816	880.8	7.350	3.972	45,080	142,500	19,380	3.732	42,380	133,900	18,220
30	0.8762	875.4	7.305	3.952	45,170	141,800	19,420	3.715	42,450	133,300	18,250
31	0.8708	870.0	7.260	3.936	45,240	141,200	19,450	3.699	42,520	132,700	18,280
32	0.8654	864.6	7.215	3.919	45,330	140,600	19,490	3.682	42,590	132,100	18,310
33	0.8602	859.3	7.171	3.902	45,405	140,000	19,520	3.665	42.635	131,500	18,330
34	0.8550	854.2	7.128	3.885	45,495	139,400	19,560	3.648	42,705	130,900	18,360
35	0.8498	849.0	7.085	3.869	45,565	138,800	19,590	3.632	42,775	130,300	18,390
36	0.8448	844.0	7.043	3.852	45,635	138,200	19,620	3.615	42,820	129,700	18,410
37	0.8398	838.9	7.001	3.835	45,705	137,600	19,650	3.598	42,870	129,100	18,430
38	0.8348	834.0	6.960	3.818	45,775	137,000	19,680	3.582	42,940	128,500	18,460
39	0.8299	829.2	6.920	3.802	45,870	136,400	19,720	3.565	42,985	127,900	18,480
40	0.8251	824.3	6.879	3.785	45,940	135,800	19,750	3.548	43,050	127,300	18,510
41	0.8203	819.5	6.839	3.768	46,010	135,200	19,780	3.531	43,100	126,700	18,530
42	0.8155	814.7	6.799	3.754	46,080	134,700	19,810	3.517	43,170	126,200	18,560
43	0.8109	810.1	6.760	3.738	46,125	134,100	19,830	3.501	43,220	125,600	18,580
44	0.8063	805.5	6.722	3.721	46,195	133,500	19,860	3.484	43,265	125,000	18,600
45	0.8017	800.9	6.684	3.704	46,265	132,900	19,890	3.467	43,310	124,400	18,620
46	0.7972	796.4	6.646	3.690	46,335	132,400	19,920	3.453	43,360	123,900	18,640
47	0.7927	792.0	6.609	3.676	46,380	131,900	19,940	3.437	43,405	123,300	18,660
48	0.7883	787.5	6.572	3.657	46,450	131,200	19,970	3.423	43,450	122,800	18,680
49	0.7839	783.2	6.536	3.643	46,520	130,700	20,000	3.406	43,495	122,200	18,700
API		kg/m^3	lbm/gal	$kJ/m^3 \times 10^{-10}$	kJ/kg	Btu/gal	Btu/lbm	$kJ/m^3 \times 10^{-10}$	kJ/kg	Btu/gal	Btu/lbm

Sources: Adapted from National Bureau of Standards Miscellaneous Publication No. 97, Thermal Properties of Petroleum Products, April 28, 1933, and Technical Bulletins from major oil companies, e.g., Exxon, Gulf, Mobil, Shell.

Bibliography

THERMODYNAMICS REFERENCES

Burghardt, M.D., *Engineering Thermodynamics with Applications, Second Edition,* Harper and Row, New York, NY, 1982, ISBN 0-06-041042-6.

Faires, V.M. and Simmang, C.M., *Thermodynamics, Sixth Edition,* MacMillan Publishing Co., New York, NY, 1978, ISBN 0-02-335530-1.

Johnston, R.M., Brockett, W.A., Bock, A.E. and Keating, E.L., *Elements of Applied Thermodynamics,* Naval Institute Press, Annapolis, MD, 1978, ISBN 0-87021-169-2.

Keenan, J.H., Chao, J. and Kaye, J., *Gas Tables,* John Wiley & Sons, New York, NY, 1980, ISBN 0-471-02207-1.

Keenan, J.H., Keyes, F.G., Hill, P.G. and Moore, J.G., *Steam Tables,* John Wiley & Sons, New York, NY, 1969, SBN 471 46500 3.

Smith, J.M. and Van Ness, H.C., *Introduction to Chemical Engineering Thermodynamics, Third Edition,* McGraw-Hill, New York, NY, 1975, ISBN 0-07-058701-9.

Stull, D.R. and Prophet, H., project directors, *JANAF Thermochemical Tables, Second Edition,* U.S. Dept. of Commerce, National Bureau of Standards, Washington, D.C., 1971.

Wark, K., *Thermodynamics, Fourth Edition,* McGraw Hill, New York, NY, 1983, ISBN 0-07-068284-4.

POWERPLANT ENERGY REFERENCES

Burkhardt, C.H., *Domestic and Commercial Oil Burners, Third Edition,* McGraw Hill, New York, NY, 1969.

Culp, Jr., A.W., *Principles of Energy Conversion,* McGraw-Hill, New York, NY, 1979, ISBN 0-07-014892-9.

Department of the Navy Energy Fact Book, Navy Publications and Forms Center, Philadelphia, PA, 1979.

El-Wakil, M.M., *Powerplant Technology,* McGraw-Hill, New York, NY, 1984, ISBN 0-07-019288-X.

Faulkner, Jr., E.A., *Guide to Efficient Burner Operation: Gas, Oil, and Dual Fuel,* Fairmont Press, Atlanta, GA, 1981, ISBN 0-915586-35-5.

Flack, J., Bennett, A.J.S., Strong, R. and Culver, L.J., *Marine Combustion Practice,* Pergamon Press, London, UK, 1969.

Krenz, J.H., *Energy Conversion and Utilization, Second Edition,* Allyn and Bacon, Boston, MA, 1984, ISBN 0-205-08021-9.

Sionger, J.G., editor, *Combustion: Fossil Power Systems,* Combustion Engineering, Inc., Windsor, CT, 1981.

Sorensen, H.A., *Energy Conversion Systems,* John Wiley & Sons, New York, NY, 1983, ISBN 0-471-08872-2.

Staff of Research and Education Association, *Modern Energy Technology, Vol. I and II,* Research and Education Association, New York, NY, 1975.

GENERAL FUELS REFERENCES

Allinson, J.P., editor, *Criteria for Quality of Petroleum Products,* Halsted Press, New York, NY, 1973, ISBN 0-470-02500-X.

Anderson, L.L. and Tillman, D.A., *Synthetic Fuels from Coal,* John Wiley & Sons, New York, NY, 1979, ISBN 0-471-01784-1.

Bastress, E.K., editor, *Gas Turbine Combustion and Fuels Technology,* ASME, New York, NY, 1977.

Benn, F.R., Edewor, J.O. and McAuliffe, C.A., *Production and Utilization of Synthetic Fuels—An Energy Economics Study,* Halsted Press, New York, NY, 1981, ISBN 0-470-27171-X.

Blackmore, D.R. and Thomas, A., *Fuel Economy of the Gasoline Engine, Fuel, Lubricant and Other Effects,* Halsted Press, New York, NY, 1977, ISBN 0-470-99132-1.

Bungay, H.R., *Energy, The Biomass Option,* John Wiley & Sons, New York, NY, 1981, ISBN 0-471-04386-9.

Colucci, J. and Gallopoulos, editors, *Future Automotive Fuels,* Plenum Press, New York, 1977, ISBN 0-306-31017-1.

Coordinating Research Council, *Handbook of Aviation Fuel Properties,* SAE, Warrendale, PA, 1983.

Ezra, D., *Coal and Energy,* Ernest Benn, London, UK, 1978, ISBN 0-510-00004-5.

Francis, W. and Peters, M.C., *Fuels and Fuel Technology,* Pergamon Press, New York, NY, 1980, ISBN 0-08-025250-8.

Goodger, E.M., *Alternative Fuels: Chemical Energy Resources,* Halsted Press, New York, NY, 1980, ISBN 0-470-26952-9.

Goodger, E.M., *Hydrocarbon Fuels Production, Properties and Performance of Liquids and Gases,* Halsted Press, New York, NY, 1975, ISBN 0-470-31365-X.

Haslam, R.T. and Russell, R.B., *Fuels and Their Combustion,* McGraw-Hill, New York, NY, 1926.

Hirao, O. and Pefley, R.K., *Present and Future Automotive Fuels,* Wiley Interscience, New York, 1988, ISBN 0-471-80259-X.

Odgers, J. and Kretschmer, D., *Gas Turbine Fuels and Their Influence on Combustion,* Abacus Press, Cambridge, MA, 1986, ISBN 0-85626-342-7.

Owen, K. and Coley, T., *Automotive Fuels Handbook,* SAE, Warrendale PA, 1990, ISBN 1-56091-064-X.

Paul, J.K., editor, *Ethyl Alcohol Production and Use as a Motor Fuel,* Noyes Data Corporation, Park Ridge, NJ, 1979, ISBN 0-08155-0780-1.

Paul, J.K., editor, *Methanol Technology and Application in Motor Fuels,* Noyes Data Corporation, Park Ridge, NJ, 1979, ISBN 0-08155-0719-4.

Porteous, A., *Refuse Derived Fuels,* Applied Science Publishers Ltd, London, UK, 1981, ISBN 0-85334-937-1.

Probstein, R.F. and Hicks, R.E., *Synthetic Fuels,* McGraw Hill, New York, NY, 1982, ISBN 0-07-050908-5.

Rider, D.K., *Energy: Hydrocarbon Fuels and Chemical Resources,* John Wiley & Sons, New York, NY, 1981, ISBN 0-471-05915-3.

Smith, M.L. and Stinson, K.W., *Fuels and Combustion,* McGraw-Hill, New York, NY, 1952.

Smoot, L.D. and Pratt, D.T., *Pulverized-Coal Combustion and Gasification,* Plenum Press, New York, NY, 1979, ISBN 0-306-40084-7.

Vegetable Oil Fuels, Proceeding of the International Conference on Plant and Vegetable Oils as Fuels, American Society of Agricultural Engineers, St. Joseph, MI, 1982, ISBN 0-916150-46-1.

CHEMICAL KINETICS AND GAS DYNAMICS REFERENCES

Barnard, J.A. and Bradley, J.N., *Flame and Combustion, Second Edition,* Chapman and Hall, London, UK, 1985, ISBN 0-412-23040-2.

Benson, S.W., *The Foundations of Chemical Kinetics,* McGraw-Hill, New York, NY, 1960.

Bowen, J.R., Manson, N., Oppenheim, A.K. and Soloukhin, R.I., editors *Combustion in Reactive Systems,* AIAA, New York, NY, 1981, ISBN 0-915928-47-7.

Bowman, C.T. and Birkeland, J., *Alternative Hydrocarbon Fuels: Combustion and Chemical Kinetics,* AIAA, 1978, ISBN 0-915928-25-6.

Chigier, N., *Energy, Combustion and Environment,* McGraw Hill, New York, NY, 1981, ISBN 0-07-010766-1.

Chomiak, J., *Combustion: A Study in Theory, Fact and Application,* Abacus Press, New York, NY, 1990, ISBN 0-85626-453-9.

Fristrom, R.M. and Westenberg, A.A., *Flame Structure,* McGraw-Hill, New York, NY, 1965.

Glassman, I., *Combustion,* Academic Press, New York, NY, 1977, ISBN 0-12-285850-6.

Hucknell, D.J. *Chemistry of Hydrocarbon Combustion,* Chapman and Hall, London, UK, 1985, ISBN 0-412-26110-3.

Kondrat'ev, V.N., *Chemical Kinetics of Gas Reactions,* Pergamon Press, London, UK, 1964.

Levenspiel, O., *Chemical Reaction Engineering,* John Wiley & Sons, New York, NY, 1967.

Lewis, B., Pease, R.N., and Taylor, H.S., *Combustion Processes, Vol. II,* Princeton University Press, Princeton, NJ, 1956.

Penner, S.S., *Chemistry Problems in Jet Propulsion,* Pergamon Press, New York, NY, 1957.

Smith, J.M., *Chemical Engineering Kinetics, Second Edition,* McGraw-Hill, New York, NY, 1970.

Zucker, R.D., *Fundamentals of Gas Dynamics,* Matrix Publishers, Portland, OR, 1977, ISBN 0-916460-12-6.

Zucrow, M.J. and Hoffman, J.D., *Volume I: Gas Dynamics,* John Wiley & Sons, New York, NY, 1976, ISBN 0-471-98440-X.

ENVIRONMENTAL REFERENCES

Chanlett, E.T., *Environmental Protection,* McGraw Hill, New York, NY, 1979, ISBN 0-07-010531-6.

Knoll, K.E. and Davis, W.T., *Power Generation: Air Pollution Monitoring and Control,* Ann Arbor Science Publishers Inc., Ann Arbor, MI, 1976, ISBN 0-250-40118-5.

Patterson, D.J. and Henein, N.A., *Emissions From Combustion Engines and Their Control,* Ann Arbor Science Publishers Inc., Ann Arbor, MI, 1981, ISBN 0-250-97514-9.

Perkins, H.C., *Air Pollution,* McGraw Hill, New York, NY, 1974, ISBN 0-07-049302-2.

Wark, K. and Warner, C.F., *AIR POLLUTION Its Origin and Control,* IEP Dun-Donnelley Publisher, New York, NY, 1976, ISBN 0-7002-2488-2.

I.C. ENGINE REFERENCES

Benson, R.S. and Whitehouse, N.D., *Internal Combustion Engines, Vol. I & II,* Pergamon Press, New York, NY, 1979, ISBN 0-08-022718 and ISBN 0-08-022720-1.

Blair, G.P., *The Basic Design of Two-Stroke Engines,* SAE, Warrendale, PA, 1990, ISBN 1-56091-008-9.

Burghardt, M.D., and Kingsley, G.D., *Marine Diesels,* Prentice Hall, Inc., Englewood Cliffs, NJ, 1981, ISBN 0-13-556985-0.

Calder, N., *Marine Diesel Engines: Maintenance, Troubleshooting, and Repair,* International Marine Publishing Co., Camden, MA, 1987, ISBN 0-87742-237-0.

Campbell, A.S., *Thermodynamic Analysis of Combustion Engines,* John Wiley & Sons, New York, NY, 1979, ISBN 0-471-03751-6.

Collie, M.J., editor, *Stirling Engine: Design and Feasibility for Automotive Use,* Noyes Data Corporation, Park Ridge, NJ, 1979, ISBN 0-8155-0763-1.

Cummins Jr., C.L., *Internal Fire,* Carnot Press, Lake Oswego, OR, 1976.

Dicksee, C.B., *The High-Speed Compression-Ignition Engine,* Blackie & Sons Ltd., London, 1946.

Fenton, J., editor, *Gasoline Engine Analysis,* Mechanical Engineering Publications Ltd, London, UK, 1986, ISBN 0-85298-6343.

Ferguson, C.R., *Internal Combustion Engines,* John Wiley & Sons, New York, NY, 1986, ISBN 0-471-88129-5.

Gill, P.W., Smith Jr., J.H., and Ziurys, E.J., *Fundamentals of Internal Combustion Engines,* Naval Institute Press, Annapolis, MD, 1959.

Heywood, J.B., *Internal Combustion Engine Fundamentals,* McGraw Hill, New York, NY, 1988, ISBN 0-07-028637-X.

Howarth, M.H., *The Design of High Speed Diesel Engines,* Constable & Company, London, UK, 1966.

Knak, C., *Diesel Motor Ships Engines and Machinery, Text and Drawings,* G-E-C GAD Publishers, Copenhagen, 1979, ISBN 87-12-46775-8 and ISBN 87-12-46777-4.

Lichty, L.C., *Combustion Engine Processes,* McGraw-Hill, New York, NY, 1967.

Lilly, L.C.R., editor, *Diesel Engine Reference Book,* Butterworths, London, UK, 1984, ISBN 0-408-00443-6.

Newton, K., Steeds, W. and Garrett, T.K., *The Motor Vehicle, 10th edition,* Butterworths, London, UK, 1983, ISBN 0-408-01157-2.

Norbye, J.P., *The Wankel Engine,* Chilton Book Co., Philadelphia, PA, 1972, ISBN 0-8019-5591-2.

Obert, E.F., *Internal Combustion Engines and Air Pollution,* Harper and Row, New York, NY, 1973, ISBN 0-352-04560-0.

Reader, G.T. and Hooper, C., *Stirling Engines,* E. & F. N. Spon, London, UK, 1983, ISBN 0-419-12400-4.

Ricardo, H.R. and Hempson, J.G.G., *The High-Speed Internal-Combustion Engine,* Blackie & Sons Ltd., London, UK, 1968.

Stephenson, R.R., *Should We Have a New Engine?, Vol. I and II,* Jet Propulsion Laboratory, California Institute of Technology, Pasadena, CA, 1979.

Stinson, K.W., *Diesel Engineering Handbook, 12th Edit.,* Business Journals, Inc., Stamford, CT, 1972.

Stone, R., *Introduction to Internal Combustion Engines,* Macmillan Education LTD, London, UK, 1985, ISBN 0-333-37594-7.

Taylor, C.F., *The Internal-Combustion Engine in Theory and Practice, Vol. I and II,* M.I.T. Press, Cambridge, MA, 1977, ISBN 0-262-70015-8 and ISBN 0-262-70016-6.

Walker, G., *Stirling Engines,* Clarendon Press, Oxford, UK, 1980, ISBN 0-19-856209-8.

Watson, N. and Janota, M.S., *Turbocharging the Internal Combustion Engine,* John Wiley & Sons, New York, NY, 1982, ISBN 0 471-87072-2.

GAS TURBINE TEXTS

Cohen, H., Rogers, G.F.C., and Saravanamuttoo, H.I.H., *Gas Turbine Theory,* Longman Group Ltd, London, UK, 1974, ISBN 0-582-44927-8.

Dusinberre, G.M. and Lester, J.C., *Gas Turbine Power,* International Textbook Company, 1962.

Gas Turbines For Autos and Trucks, (selected papers through 1980), SAE, Warrendale, PA, 1981, ISBN 0-89883-108-3.

Harman, R.T., *Gas Turbine Engineering,* Halsted Press Book, John Wiley & Sons, New York, NY, 1981, ISBN 0-470-27065-9.

Jennings, B.H. and Rogers, W.L., *Gas Turbine Analysis and Practice,* McGraw Hill, New York, NY, 1953.

Lefebvre, A.W., *Gas Turbine Combustion,* McGraw Hill, New York, NY, 1983, ISBN 0-07-037029-X.

O'Brien, J.P., editor, *Gas Turbines for Automotive Use,* Noyes Data Corporation, Park Ridge, New Jersey, 1980, ISBN 0-8155-0786-0.

Saarlas, M., *Steam and Gas Turbines for Marine Propulsion,* Naval Institute Press, Annapolis, MD, 1978, ISBN 0-87021-680-5.

The Aircraft Gas Turbine Engine and its Operation, United Technologies-Pratt & Whitney, 1982.

Whittle, F., *Gas Turbine Aero-Thermodynamics,* Pergamon Press, New York, NY, 1981, ISBN 0-8-026718-1.

REFERENCES FOR TEXT CHAPTERS

Chapter 1

Amann, C., Trends in engine design, Search, G.M. Research Laboratories, Vol. 19, #1, Mar–Apr 1984.

Huebner, Jr., G. J., Future automotive power plants, SAE Paper 760607.

Oppenheim, A. K., A rationale for advances in the technology of the I.C. engines, SAE Paper 820047.

Rain, C., Future auto engines: Competition heats up, High Technology, Vol. 2, #3, May–June 1982.

Chapter 3

Gordon, S., and McBride, B. J., Computer program for calculation of complex chemical equilibrium compositions, rock performance, incident and reflected shocks, and Chapman–Jouguet detonations, NASA SP-273, 1971.

Hottel, H., and Williams, G., Charts of thermodynamic properties of fluids encountered in calculations of internal combustion engine cycles, NACA TN 1026, May 1946.

Hottel, H., Williams, G., and Satterfield, C. N., *Thermodynamic Charts for Combustion Processes,* John Wiley, New York, 1949.

McCann, W., Thermodynamic charts for internal combustion engine fluids, NACA RB 3G28, 1943 and NACA TN 1883, July 1949.

Newhall, H., and Starkman, Thermodynamic properties of octane and air for engine performance calculations, SAE Progress in Technology, Vol. 7, 1964.

Powell, H., Applications of An Enthalpy-Fuel Air Ratio Program, ASME Trans., Vol. 79, July 1957.

Chapter 4

ASME Steam Generating Units Power Test Codes, ASME PTC4.1-1964 and ANSI PTC4.1-1974.

Steam, Its Generation and Use, 37th Edition, The Babcock & Wilcox Company, 1963.

Chapter 5

DiBella, C. A. W., U.S. Synthetic Fuel Corporation, private communications, Conference and Workshop on Implementing Transportation Fuel Alternatives for North America Into the 21st Century, November 13–15, 1985, Washington, D.C.

Standard Handbook for Mechanical Engineers, 7th Edition, Baumeister, T. and Marks, L., McGraw-Hill Book Co., New York, NY, 1967, ISBN 07-004122-9.

Chapter 8

Boccio, J. L., Weilerstein, G., and Edelman, R. B., A mathematical model for jet engine combustor pollutant emissions, NASA CR 1212083, 1973.

Cornelius, W., and Agnew, W. G. (eds.), *Emissions from Continuous Combustion Systems,* Plenum Press, New York, 1972.

Crowe, C. T., Conservation equations for vapor-droplet flows, *Proc. 25th HTFM Inst.* (McKillop, A.A., Baughn, J.W., and Dwyer, H.A., Eds.), Stanford University Press, 1976, p. 214.

Crowe, C. T., Vapour-droplet flow equations, Lawrence Livermore Laboratory, Report No. UCRL-51877, 1977.

Gibson, M. M., and Morgan, B. B., J. Inst. Fuel, Vol. 43, 1970, p. 517.

Gouldin, F. C., Comb. Sci. & Tech., Vol. 9, 1974, p. 17

Lilley, D. G., and Wendt, J. O. L., Modeling pollutant formation in coal combustion, Proc. 25 HTFM Inst. (McKillop, A.A., Baughn, J.W., and Dwyer, H.A., Eds.), Stanford University Press, 1976, p. 196.

Pratte, B. D., and Keffer, J. R., ASME J. of Basic Eng., Vol. 94, Dec. 1972, p. 739.

Rai, C., and Siegel, R.D. (Eds.), Air: II. Control of NO_X and SO_X Emissions, AIChE Symposium Series No. 148, Vol. 71, 1975.

Vincent, M. W., Ph.D. Thesis, University of Sheffield, 1973.

Westbrook, C. K., 16th Symp. (Intl.) on Comb., The Combustion Institute, 1977.

Williams, F. A., *Combustion Theory,* Addison-Wesley, 1965.

Chapter 9

1979 Annual Book of ASTM Standards, Part 47, Test Methods for Rating Motor, Diesel, Aviation Fuels, 01-47079-12, Philadelphia, PA 1979.

Federal Register 33(108), Part II, 1968; 35(219), Part II, 1970; 36(128), Part II, 1971.

Halstad, L. R., Automobiles and air pollution, Conference on Universities, National Laboratories and Man's Environment, in Argonne, IL, July 1969.

Chapter 10

A technical history of the automobile, Part 1, Vol. 98, #6, Automotive Engineering, June 1990.

A technical history of the automobile, Part 2, Vol. 98, #7, Automotive Engineering, July 1990.

A technical history of the automobile, Part 3, Vol. 98, #8, Automotive Engineering, Aug 1990.

Flink, J. J., Innovation in automotive technology, American Scientist, Vol. 73, Mar–Apr 1985.

Foster, D. E., and Myers, P. S., Can paper engines stand the heat? SAE Paper 840911, 1984.

Hottel, H., and Williams, G., Charts of thermodynamic properties of fluids encountered in calculations of internal combustion engine cycles, NACA TN 1026, May 1946.

Hottel, H. C., Williams, G. C., and Satterfield, C. N., *Thermodynamic Charts for Combustion Processes,* J. Wiley, New York, 1949.

McCann, W., Thermodynamic charts for internal combustion engine fluids, NACA RB3G28, 1943, and NACA TN 1883, July 1949.

Millar, Gordon, Conference and Workshop on Implementing Transportation Fuel Alternatives for North America into the 21st Century, November 13–15, 1985, Washington, D.C.

Newhall, H., and Starkman, E., Thermodynamic properties of octane and air for engine performance calculations, In *Digital Calculations of Engine Cycles,* SAE Progress in Technology, Vol. 7, 1964, p. 38.

Pouring, A. A., Sonex Corp., Chemical acoustic charge condition for low emission I.C. engine, Vol. I, 1st International Conference on Combustion Technologies for a Clean Environment, Vilamoura (Algarve), Portugal, September 1991.

Sawyer, R. F., The future of the automobile, A Series of Lectures on Energy and the Future of Automotive Transportation, Meakin Interdisciplinary Studies Center, College of Engineering, U.C. Berkley, CA, 1981.

Thring, R. H., Gasoline engines and their future, Mechanical Engineering, October 1983.

Zeleznik, F. J. and McBride, B. J., Modeling the internal combustion engine, NASA Reference Publication 1094, March 1985.

Chapter 11

50 years of Diesel progress, Diesel Progress, Vol. L1, #7, ISSN 0744-0073, July 1985.

Millar, G. H., Commercial engine development to year 2000 and beyond, SAE/IMechE Exchange Lecture, London, England, UK, May 1984.

Chapter 12

Standard Handbook for Mechanical Engineers, 7th Edition, Baumeister, T. and Marks, L., McGraw-Hill Book Co., New York, NY, 1967, ISBN 07-004122-9.

Wolfer, H. H., Ignition lag in the Diesel engine, VDI Forschung 392, 1938.

Index